T0210931

Gröbner Bases

Takayuki Hibi

Editor

Gröbner Bases

Statistics and Software Systems

Springer

Editor
Takayuki Hibi
Department of Pure and Applied
  Mathematics
Osaka University
Toyonaka, Osaka, Japan

ISBN 978-4-431-56215-3   ISBN 978-4-431-54574-3 (eBook)
DOI 10.1007/978-4-431-54574-3
Springer Tokyo Heidelberg New York Dordrecht London

Preface

The present volume is an English translation of the Japanese mathematics book "Gröbner Dojo" (Kyoritsu Shuppan Co., Ltd., September 2011). The *dojo* is a Japanese traditional term, which represents, in general, the place for the training of the *judo*, an Olympic sport. Our book "Gröbner Dojo" invites the reader to the Gröbner world, a fascinating research area of mathematics, where three aspects of Gröbner bases, viz., theory, application and computation, are linked effectively and systematically. A beginner including a first year graduate student can learn the ABC's of Gröbner bases from "Gröbner Dojo." In addition, "Gröbner Dojo" can be a how-to book for users of Gröbner bases such as scientists engaging in statistical problems as well as engineers being active in industrial society. This is the reason why we select the term *dojo* for the title of our book.

An idea of Gröbner bases was apparently studied by Francis Sowerby Macaulay in 1927; he succeeded in finding a combinatorial characterization of the Hilbert functions of homogeneous ideals of the polynomial ring. Later, current definition of Gröbner bases was independently introduced by Heisuke Hironaka in 1964 and Bruno Buchberger in 1965. However, after the discovery of the notion of Gröbner bases by Hironaka and Buchberger, no activity had been done for about twenty years. A first breakthrough was done by David Bayer and Michael Stillman in the middle of 1980s, who created the computer algebra system Macaulay with the help of Gröbner bases. In 1995 the second breakthrough was achieved by Bernd Sturmfels, who discovered the fascinating relation between regular triangulations of convex polytopes and Stanley–Reisner ideals of initial ideals of toric ideals. Furthermore, the third breakthrough arose in 1998 when Persi Diaconis and Bernd Sturmfels demonstrated an exciting application of Gröbner bases to algebraic statistics.

With these backgrounds, in October 2008, the JST[1] CREST[2] Hibi project started toward the progress of theory and application of Gröbner bases together with the

---

[1] Japan Science and Technology Agency.
[2] Core Research for Evolutional Science & Technology.

development of their algorithms. The publication of "Gröbner Dojo" was already announced in the original research plan of the project.

"Gröbner Dojo" is a comprehensive textbook to learn algebraic statistics based on Gröbner bases. First, in Chap. 1, starting from Dickson's Lemma, a classical result in combinatorics, we explain the division algorithm, Buchberger criterion and Buchberger algorithm. Then the theory of elimination follows and toric ideals are introduced. In addition, the basic theory of Hilbert functions is discussed. Moreover, the historical background of Gröbner bases is surveyed.

Chapter 2 is a warming-up drill for learning the basic ideas of using mathematical software. We choose the mathematical software environment named "MathLibre." It is a collection of mathematical software and free documents. MathLibre is a kind of Live Linux system. Linux is a system compatible with UNIX, a traditional OS for specialists. Many mathematical research systems are developed on UNIX. The basic usages and fundamental ideas of UNIX are introduced.

Chapter 3 discusses how to compute various objects related to Gröbner bases explained in Chap. 1. After introducing fundamental tools for efficient Gröbner basis computation, we illustrate fundamental computations related to Gröbner bases by using Macaulay2, SINGULAR, CoCoA and Risa/Asir.

In writing Chaps. 1–3, we do not assume that the reader is familiar with theory and computation of Gröbner bases. If the reader has an experience of handling Gröbner bases, then these chapters may be skipped partly. On the other hand, since the latter Chaps. 4–6 are written independently, after reading the former Chaps. 1–3, the reader can read Chaps. 4–6 in any order.

Chapter 4 is devoted to algebraic statistics. This field was initiated by the work of Diaconis and Sturmfels in 1998 and the work of Pistone and Wynn in 1996, both applying Gröbner basis theory to statistics. Since then the field has been developing rapidly with providing challenging problems to both statisticians and algebraists.

Chapter 5 plays the introduction to two fascinating rainbow bridges between the world of Gröbner bases and that of convex polytopes. One is the big theory of Gröbner fans and state polytopes. The other is the reciprocal relation between initial ideals of toric ideals and triangulations of convex polytopes.

Recently, Gröbner bases of rings of differential operators turn out to be useful to numerical evaluations of a broad class of normalizing constants in statistics. The method is called the holonomic gradient method, which is a rapidly growing area in algebraic statistics. Chapter 6 is a self-contained exposition to invite readers to these topics. Nobuki Takayama, the author of Chap. 6, thanks Professor Francisco Castro-Jiménez for providing useful comments to a draft of Chap. 6.

Finally, Chap. 7 provides a collection of rich problems and their answers by utilizing various software systems, such as Risa/Asir, 4ti2, polymake, R, and so on. Chapter 7 complements Chaps. 4–6, and is helpful for readers to understand how to use software systems to study or apply Gröbner bases.

On behalf of the JST CREST Hibi project, I would express our thanks to JST for providing financial support, which made it possible to organize international conferences and to employ promising young researchers. Finally, I am grateful to Ms. Kaoru Yamano for her administrative job for Hibi project.[3]

Toyonaka, Osaka, Japan   Takayuki Hibi

[3] All computer programs appearing in this volume and a list of corrections are available at http://www.math.kobe-u.ac.jp/OpenXM/Math/dojo-en

Contents

List of Contributors

Satoshi Aoki Department of Mathematics and Computer Science, Kagoshima University, Korimoto, Kagoshima, Japan

Tatsuyoshi Hamada Department of Applied Mathematics, Fukuoka University, Nanakuma, Fukuoka, Japan

Takayuki Hibi Department of Pure and Applied Mathematics, Graduate School of Information Science and Technology, Osaka University, Toyonaka, Osaka, Japan

Hiromasa Nakayama Department of Mathematics, Graduate School of Science, Kobe University, Rokko, Nadaku, Kobe, Japan

Kenta Nishiyama School of Management and Information, University of Shizuoka, Shizuoka, Japan

Masayuki Noro Department of Mathematics, Graduate school of Science, Kobe University, Kobe, Japan

Hidefumi Ohsugi Department of Mathematics, College of Science, Rikkyo University, Toshima-ku, Tokyo, Japan

Nobuki Takayama Department of Mathematics, Graduate School of Science, Kobe University, Rokko, Nadaku, Kobe, Japan

Akimichi Takemura Department of Mathematical Informatics, Graduate School of Information Science and Technology, University of Tokyo, Bunkyo, Tokyo, Japan

Chapter 1
A Quick Introduction to Gröbner Bases

Takayuki Hibi

Abstract Neither specialist knowledge nor extensive investment of time is required in order for a nonspecialist to learn fundamentals on Gröbner bases. The purpose of this chapter is to provide the reader with sufficient understanding of the theory of Gröbner bases as quickly as possible with the assumption of only a minimum of background knowledge. In Sect. 1.1, the story starts with Dickson's Lemma, which is a classical result in combinatorics. The Gröbner basis is then introduced and Hilbert Basis Theorem follows. With considering the reader who is unfamiliar with the polynomial ring, an elementary theory of ideals of the polynomial ring is also reviewed. In Sect. 1.2, the division algorithm, which is the framework of Gröbner bases, is discussed with a focus on the importance of the remainder when performing division. The highlights of the fundamental theory of Gröbner bases are, without doubt, Buchberger criterion and Buchberger algorithm. In Sect. 1.3 the groundwork of these two items are studied. Now, to read Sects. 1.1–1.3 is indispensable for being a user of Gröbner bases. Furthermore, in Sect. 1.4, the elimination theory, which is effective technique for solving simultaneous equations, is discussed. The toric ideal introduced in Sect. 1.5 is a powerful weapon for the application of Gröbner bases to combinatorics on convex polytopes. Clearly, without toric ideals, the results of Chaps. 4 and 5 could not exist. The Hilbert function studied in Sect. 1.6 is the most fundamental tool for developing computational commutative algebra and computational algebraic geometry. Section 1.6 supplies the reader with sufficient preliminary knowledge to read Chaps. 5 and 6. However, since the basic knowledge of linear algebra is required for reading Sect. 1.6, the reader who is unfamiliar with linear algebra may wish to skip Sect. 1.6 in his/her first reading. Finally, in Sect. 1.7, the historical background of Gröbner bases is surveyed with providing references for further study.

T. Hibi (✉)
Department of Pure and Applied Mathematics, Graduate School of Information Science and Technology, Osaka University, Toyonaka, Osaka 560-0043, Japan
e-mail: hibi@math.sci.osaka-u.ac.jp

T. Hibi (ed.), *Gröbner Bases: Statistics and Software Systems*,
DOI 10.1007/978-4-431-54574-3_1, © Springer Japan 2013

1.1 Polynomial Rings

The polynomial ring and the ring of integers are the origin of commutative algebra. If we imagine the theory of Gröbner bases to form a theatrical play, then the stage is the polynomial ring and the first act starts with Dickson's Lemma. After Dickson's Lemma, the Gröbner basis is introduced and then Hilbert Basis Theorem follows.

1.1.1 Monomials and Polynomials

Let $\mathbb{Q}$, $\mathbb{R}$ and $\mathbb{C}$ denote the set of rational, real and complex numbers, respectively. Throughout Chap. 1, the notation K stands for one of $\mathbb{Q}$, $\mathbb{R}$ and $\mathbb{C}$.

A *monomial* in the variables $x_1, x_2, \ldots, x_n$ is a product of the form

$$\prod_{i=1}^{n} x_i^{a_i} = x_1^{a_1} x_2^{a_2} \cdots x_n^{a_n},$$

where each a_i is a nonnegative integer. Its *degree* is $\sum_{i=1}^{n} a_i$. For example, the degree of $x_2^3 x_5 x_6^2$ is 6. In particular 1 $(= x_1^0 x_2^0 \cdots x_n^0)$ is a monomial of degree 0. A *term* is a monomial together with a nonzero *coefficient*. For example, $-5x_3^3 x_5 x_8 x_9^2$ is a term of degree 7 with -5 its coefficient. A *constant term* is the monomial 1 together with a nonzero coefficient.

A *polynomial* is a finite sum of terms. For example,

$$f = -5x_1^2 x_2 x_3^2 + \frac{2}{3} x_2 x_4^3 x_5^2 - x_3^3 - 7$$

is a polynomial with 4 terms. The monomials appearing in f are

$$x_1^2 x_2 x_3^2, \ x_2 x_4^3 x_5^2, \ x_3^3, \ 1$$

and the coefficients of f are

$$-5, \ \frac{2}{3}, \ -1, \ -7.$$

The degree of a polynomial is defined to be the maximal degree of monomials which appears in the polynomial. For example, the degree of the above polynomial f is 6. With an exception 0 is regarded as a polynomial, but the degree of 0 is undefined. If the degree of all monomials appearing in a polynomial is equal to q, then the polynomial is called a *homogeneous* polynomial of degree q. For example,

$$-7x_1^2 x_3 + \frac{3}{5} x_2 x_4 x_5 - x_4^3 + x_1 x_3 x_5$$

is a homogeneous polynomial of degree 3.

Let $K[x_1, x_2, \ldots, x_n]$ denote the set of all polynomials in the variables $x_1, x_2, \ldots, x_n$ with coefficients in K. If f and g are polynomials belonging to $K[x_1, x_2, \ldots, x_n]$, then the sum $f + g$ and the product fg can be defined in the obvious way. With emphasizing that $K[x_1, x_2, \ldots, x_n]$ possesses the structure of the sum and the product, we say that $K[x_1, x_2, \ldots, x_n]$ is the *polynomial ring* in n variables over K.

1.1.2 Dickson's Lemma

Let $\mathcal{M}_n$ denote the set of monomials in the variables $x_1, x_2, \ldots, x_n$. When we deal with monomials, we often use u, v and w instead of $\prod_{i=1}^n x_i^{a_i}$ unless confusion arises.

We say that a monomial $u = \prod_{i=1}^n x_i^{a_i}$ *divides* $v = \prod_{i=1}^n x_i^{b_i}$ if one has $a_i \leq b_i$ for all $1 \leq i \leq n$. We write $u \mid v$ if u divides v.

Let M be a nonempty subset of $\mathcal{M}_n$. A monomial $u \in M$ is called a *minimal element* of M if the following condition is satisfied: If $v \in M$ and $v \mid u$, then $v = u$.

Example 1.1.1. (a) Let $n = 1$. Then a minimal element of a nonempty subset M of $\mathcal{M}_1$ is unique. In fact, if q is the minimal degree of monomials belonging to M, then the monomial x_1^q is a unique minimal element of M.

(b) Let $n = 2$ and M a nonempty subset of $\mathcal{M}_2$. Then the number of minimal elements of M is at most finite. To see why this is true, suppose that $u_1 = x_1^{a_1} x_2^{b_1}, u_2 = x_1^{a_2} x_2^{b_2}, \ldots$ are the minimal elements of M with $a_1 \leq a_2 \leq \ldots$. If $a_i = a_{i+1}$, then either u_i or u_{i+1} cannot be minimal. Hence $a_1 < a_2 < \ldots$. Since u_i cannot divide u_{i+1}, one has $b_i > b_{i+1}$. Thus $b_1 > b_2 > \ldots$. Hence the number of minimal elements of M is at most finite, as desired.

Problem 1.1.2. Given an integer $s > 0$, show the existence of a nonempty subset $M \subset \mathcal{M}_2$ with exactly s minimal elements.

Example 1.1.1(b) will turn out to be true for every $n \geq 1$. This fact is called *Dickson's Lemma*, which is a classical result in combinatorics and which can be proved easily by using induction. On the other hand, however, Dickson's Lemma plays an essential role in the foundation of the theory of Gröbner bases. It guarantees that several important algorithms terminate after a finite number of steps.

Theorem 1.1.3 (Dickson's Lemma). *The set of minimal elements of a nonempty subset M of $\mathcal{M}_n$ is at most finite.*

Proof. We work with induction on the number of variables. First of all, it follows from Example 1.1.1 that Dickson's Lemma is true for $n = 1$ and $n = 2$. Let $n > 2$ and suppose that Dickson's Lemma is true for $n - 1$. Let $y = x_n$. Let N denote the set of monomials u in the variables $x_1, x_2, \ldots, x_{n-1}$ satisfying the condition that there exists $b \geq 0$ with $uy^b \in M$. Clearly $N \neq \emptyset$. The induction hypothesis says that the number of minimal elements of N is at most finite. Let $u_1, u_2, \ldots, u_s$ denote the minimal elements of N. Then by the definition of N, it follows that, for each u_i, there is $b_i \geq 0$ with $u_i y^{b_i} \in M$. Let b be the largest integer among $b_1, b_2, \ldots, b_s$. Moreover, given $0 \leq c < b$, we define a subset N_c of N by setting

$$N_c = \{u \in N \; : \; uy^c \in M\}.$$

Again, the induction hypothesis says that the number of minimal elements of N_c is at most finite. Let $u_1^{(c)}, u_2^{(c)}, \ldots, u_{s_c}^{(c)}$ denote the minimal elements of N_c. Then we claim that a monomial belonging to M can be divided by one of the monomials listed below:

$$u_1 y^{b_1}, \ldots, u_s y^{b_s}$$

$$u_1^{(0)}, \ldots, u_{s_0}^{(0)}$$

$$u_1^{(1)} y, \ldots, u_{s_1}^{(1)} y$$

$$\cdots$$

$$u_1^{(b-1)} y^{b-1}, \ldots, u_{s_{b-1}}^{(b-1)} y^{b-1}$$

In fact, for a monomial $w = uy^e \in M$, where u is a monomial in $x_1, x_2, \ldots, x_{n-1}$, one has $u \in N$. Hence if $e \geq b$, then w is divided by one of $u_1 y^{b_1}, \ldots, u_s y^{b_s}$. On the other hand, if $0 \leq e < b$, then, since $u \in N_e$, it follows that w can be divided by one of $u_1^{(e)} y^e, \ldots, u_{s_e}^{(e)} y^e$. Hence each minimal element of M must appear in the above list of monomials. In particular, the number of minimal elements of M is at most finite, as required. $\square$

1.1.3 Ideals

For the sake of the reader who is unfamiliar with the polynomial ring, we briefly review an elementary theory of ideals of the polynomial ring. In order to simplify the notation, we abbreviate the polynomial ring $K[x_1, x_2, \ldots, x_n]$ as $K[\mathbf{x}]$.

The *zero point* of a polynomial $f = f(x_1, x_2, \ldots, x_n)$ is a point $(a_1, a_2, \ldots, a_n)$ belonging to the space

$$K^n = \{(a_1, a_2, \ldots, a_n) \; : \; a_1, a_2, \ldots, a_n \in K\}$$

such that

$$f(a_1, a_2, \ldots, a_n) = 0.$$

Given a subset $V \subset K^n$, we write $I(V)$ for the set of those polynomials $f \in K[\mathbf{x}]$ such that $f(a_1, a_2, \ldots, a_n) = 0$ for all points $(a_1, a_2, \ldots, a_n)$ belonging to V. Clearly

- If $f \in I(V)$, $g \in I(V)$, then $f + g \in I(V)$;
- If $f \in I(V)$, $g \in K[\mathbf{x}]$, then $gf \in I(V)$.

With considering the above properties on $I(V)$, we now introduce the notion of ideals of the polynomial ring.

A nonempty subset I of $K[\mathbf{x}]$ is called an *ideal* of $K[\mathbf{x}]$ if the following conditions are satisfied:

- If $f \in I$, $g \in I$, then $f + g \in I$;
- If $f \in I$, $g \in K[\mathbf{x}]$, then $gf \in I$.

Example 1.1.4. The ideals of the polynomial ring $K[x]$ ($= K[x_1]$) in one variable can be easily determined. Let $I \subset K[x]$ be an ideal with at least one nonzero polynomial and d the smallest degree of nonzero polynomials belonging to I. Let $g \in I$ be a polynomial of degree d. Given an arbitrary polynomial $f \in I$, the division algorithm of $K[x]$, which is learned in the elementary algebra, guarantees the existence of unique polynomials q and r such that $f = qg + r$, where either $r = 0$ or the degree of r is less than d. Since f and g belong to the ideal I, it follows that $r = f - qg$ also belongs to I. If $r \neq 0$, then r is a nonzero polynomial belonging to I whose degree is less than d. This contradict the choice of d. Hence $r = 0$. Thus

$$I = \{qg : q \in K[x]\}.$$

Problem 1.1.5. Let $\{f_\lambda : \lambda \in \Lambda\}$ be a nonempty subset of $K[\mathbf{x}] = K[x_1, x_2, \ldots, x_n]$. Then show that the set of polynomials of the form

$$\sum_{\lambda \in \Lambda} g_\lambda f_\lambda,$$

where $g_\lambda \in K[\mathbf{x}]$ is 0 except for a finite number of λ's, is an ideal of $K[\mathbf{x}]$.

The ideal of Problem 1.1.5 is called the ideal *generated by* $\{f_\lambda : \lambda \in \Lambda\}$ and is written as

$$\langle\{f_\lambda : \lambda \in \Lambda\}\rangle.$$

Conversely, given an arbitrary ideal $I \subset K[\mathbf{x}]$, there exists a subset $\{f_\lambda : \lambda \in \Lambda\}$ of $K[\mathbf{x}]$ with $I = \langle\{f_\lambda : \lambda \in \Lambda\}\rangle$. The subset $\{f_\lambda : \lambda \in \Lambda\}$ is called a *system of generators* of the ideal I. In particular, if $\{f_\lambda : \lambda \in \Lambda\}$ is a finite set $\{f_1, f_2, \ldots, f_s\}$, then $\langle\{f_1, f_2, \ldots, f_s\}\rangle$ is abbreviated as

$$\langle f_1, f_2, \ldots, f_s \rangle.$$

A *finitely generated* ideal is an ideal with a system of generators consisting of a finite number of polynomials. In particular, an ideal with a system of generators consisting of only one polynomial is called a *principal ideal*. Example 1.1.4 says that every ideal of the polynomial ring in one variable is principal. However,

Problem 1.1.6. Show that the ideal $\langle x_1, x_2, \ldots, x_n \rangle$ of $K[x_1, x_2, \ldots, x_n]$ with $n \geq 2$ cannot be a principal ideal.

Now, a *monomial ideal* is an ideal with a system of generators consisting of monomials.

Lemma 1.1.7. *Every monomial ideal is finitely generated. More precisely if I is a monomial ideal and if $\{u_\lambda : \lambda \in \Lambda\}$ is its system of generators consisting of monomials, then there exists a finite subset $\{u_{\lambda_1}, u_{\lambda_2}, \ldots, u_{\lambda_s}\}$ of $\{u_\lambda : \lambda \in \Lambda\}$ such that $I = \langle u_{\lambda_1}, u_{\lambda_2}, \ldots, u_{\lambda_s} \rangle$.*

Proof. It follows from Theorem 1.1.3 that the number of minimal elements of the set of monomials $\{u_\lambda : \lambda \in \Lambda\}$ is at most finite. Let $\{u_{\lambda_1}, u_{\lambda_2}, \ldots, u_{\lambda_s}\}$ be the set of its minimal elements. We claim $I = \langle u_{\lambda_1}, u_{\lambda_2}, \ldots, u_{\lambda_s} \rangle$. In fact, each $f \in I$ can be expressed as $f = \sum_{\lambda \in \Lambda} g_\lambda u_\lambda$, where $g_\lambda \in K[\mathbf{x}]$ is 0 except for a finite number of λ's. Then, for each λ with $g_\lambda \neq 0$, we choose u_{λ_i} which divides u_λ and set $h_\lambda = g_\lambda(u_\lambda/u_{\lambda_i})$. Thus $g_\lambda u_\lambda = h_\lambda u_{\lambda_i}$. Hence f can be expressed as $f = \sum_{i=1}^s f_i u_{\lambda_i}$ with each $f_i \in K[\mathbf{x}]$. $\qquad \square$

Let I be a monomial ideal. A system of generators of I consisting of a finite number of monomials is called a *system of monomial generators* of I.

Lemma 1.1.8. *Let $I = \langle u_1, u_2, \ldots, u_s \rangle$ be a monomial ideal, where $u_1, u_2, \ldots, u_s$ are monomials. Then a monomial u belongs to I if and only if one of u_i's divides u.*

Proof. The sufficiency is clear. We prove the necessity. A monomial u belonging to I can be expressed as $u = \sum_{i=1}^s f_i u_i$ with each $f_i \in k[\mathbf{x}]$. Let $f_i = \sum_{j=1}^{s_i} a_j^{(i)} v_j^{(i)}$, where $0 \neq a_j^{(i)} \in K$ and where each $v_j^{(i)}$ is a monomial. Since $u = \sum_{i=1}^s f_i u_i = \sum_{i=1}^s (\sum_{j=1}^{s_i} a_j^{(i)} v_j^{(i)}) u_i$, there exist i and j with $u = v_j^{(i)} u_i$. In other words, there is u_i which divides u, as desired. $\qquad \square$

A system of generators of a monomial ideals does not necessarily consist of monomials. For example, $\{x_1^2 + x_2^3, x_2^2\}$ is a system of generators of the monomial ideal $\langle x_1^2, x_2^2 \rangle$.

Corollary 1.1.9. *Among all systems of monomial generators of a monomial ideal, there exists a unique system of monomial generators which is minimal with respect to inclusion.*

Proof. Lemma 1.1.7 guarantees the existence of a system of monomial generators of a monomial ideal I. If it is not minimal, then removing redundant monomials yields a minimal system of monomials generators.

Now, suppose that $\{u_1, u_2, \ldots, u_s\}$ and $\{v_1, v_2, \ldots, v_t\}$ are minimal systems of monomial generators of I. It follows from Lemma 1.1.8 that that, for each $1 \leq i \leq$

s, there is v_j which divides u_i. Similarly, there is u_k which divides v_j. Consequently, u_k divides u_i. Since $\{u_1, u_2, \ldots, u_s\}$ is minimal, one has $i = k$. Thus $u_i = v_j$. Hence $\{u_1, u_2, \ldots, u_s\} \subset \{v_1, v_2, \ldots, v_t\}$. Since $\{v_1, v_2, \ldots, v_t\}$ is minimal, it follows that $\{u_1, u_2, \ldots, u_s\}$ coincides with $\{v_1, v_2, \ldots, v_t\}$, as required. □

1.1.4 Monomial Orders

Recall that a *partial order* on a set Σ is a relation $\leq$ on Σ such that, for all $x, y, z \in \Sigma$, one has

(i) $x \leq x$ (reflexivity);
(ii) $x \leq y$ and $y \leq x \Rightarrow x = y$ (antisymmetry);
(iii) $x \leq y$ and $y \leq z \Rightarrow x \leq z$ (transitivity).

A *partially ordered set* is a set Σ with a partial order $\leq$ on Σ. It is custom to write $a < b$ if $a \leq b$ and $a \neq b$. A *total order* on Σ is a partial order $\leq$ on Σ such that, for any two elements x and y belonging to Σ, one has either $x \leq y$ or $y \leq x$. A *totally ordered set* is a set Σ with a total order $\leq$ on Σ.

Example 1.1.10. (a) Let T be a nonempty set and $\mathscr{B}_T$ the set of subsets of T. If A and B belong to $\mathscr{B}_T$, then we define $A \leq B$ if $A \subset B$. It turns out that $\leq$ is a partial order on $\mathscr{B}_T$. This partial order is called a partial order ordered by inclusion. We say that the partially ordered set $\mathscr{B}_T$ is a *boolean lattice*. In particular if T is a finite set $[d] = \{1, 2, \ldots, d\}$, then $\mathscr{B}_{[d]}$ is called the boolean lattice of *rank d*.

(b) Let $N > 0$ be an integer and $\mathscr{D}_N$ the set of divisors of N. If a and b are divisors of N, then we define $a \leq b$ if a divides b. Then $\leq$ is a partial order on $\mathscr{D}_N$, which is called a partial order by divisibility. The partially ordered set $\mathscr{D}_N$ is called a *divisor lattice*. If $p_1, p_2, \ldots, p_d$ are prime numbers with $p_1 < p_2 < \cdots < p_d$ and if $N = p_1 p_2 \cdots p_d$, then the divisor lattice $\mathscr{D}_N$ coincides with the boolean lattice $\mathscr{B}_{[d]}$.

Recall that $K[\mathbf{x}] = K[x_1, x_2, \ldots, x_n]$ is the polynomial ring in n variables over K and $\mathscr{M}_n$ is the set of monomials in the variables $x_1, x_2, \ldots, x_n$. A *monomial order*[1] on $K[\mathbf{x}]$ is a total order $<$[2] on $\mathscr{M}_n$ such that

(i) $1 < u$ for all $1 \neq u \in \mathscr{M}_n$;
(ii) if $u, v \in \mathscr{M}_n$ and $u < v$, then $uw < vw$ for all $w \in \mathscr{M}_n$.

Example 1.1.11. (a) Let $u = x_1^{a_1} x_2^{a_2} \cdots x_n^{a_n}$ and $v = x_1^{b_1} x_2^{b_2} \cdots x_n^{b_n}$ be monomials. We define the total order $<_{\text{lex}}$ on $\mathscr{M}_n$ by setting $u <_{\text{lex}} v$ if either (i) $\sum_{i=1}^{n} a_i < \sum_{i=1}^{n} b_i$, or (ii) $\sum_{i=1}^{n} a_i = \sum_{i=1}^{n} b_i$ and the leftmost nonzero component of

---

[1] A monomial order is also called a *term order(ing)*.
[2] Some authors prefer $\prec$ to $<$.

the vector $(b_1 - a_1, b_2 - a_2, \ldots, b_n - a_n)$ is positive. It follows that $<_{\text{lex}}$ is a monomial order on $K[\mathbf{x}]$, which is called the *lexicographic order* on $K[\mathbf{x}]$ induced by the ordering $x_1 > x_2 > \cdots > x_n$.

(b) Let $u = x_1^{a_1} x_2^{a_2} \cdots x_n^{a_n}$ and $v = x_1^{b_1} x_2^{b_2} \cdots x_n^{b_n}$ be monomials. We define the total order $<_{\text{rev}}$ on $\mathcal{M}_n$ by setting $u <_{\text{rev}} v$ if either (i) $\sum_{i=1}^{n} a_i < \sum_{i=1}^{n} b_i$, or (ii) $\sum_{i=1}^{n} a_i = \sum_{i=1}^{n} b_i$ and the rightmost nonzero component of the vector $(b_1 - a_1, b_2 - a_2, \ldots, b_n - a_n)$ is negative. It follows that $<_{\text{rev}}$ is a monomial order on $K[\mathbf{x}]$, which is called the *reverse lexicographic order*[3] on $K[\mathbf{x}]$ induced by the ordering $x_1 > x_2 > \cdots > x_n$.

(c) Let $u = x_1^{a_1} x_2^{a_2} \cdots x_n^{a_n}$ and $v = x_1^{b_1} x_2^{b_2} \cdots x_n^{b_n}$ be monomials. We define the total order $<_{\text{purelex}}$ on $\mathcal{M}_n$ by setting $u <_{\text{purelex}} v$ if the leftmost nonzero component of the vector $(b_1 - a_1, b_2 - a_2, \ldots, b_n - a_n)$ is positive. It follows that $<_{\text{purelex}}$ is a monomial order on $K[\mathbf{x}]$, which is called the *pure lexicographic order* on $K[\mathbf{x}]$ induced by the ordering $x_1 > x_2 > \cdots > x_n$.

Let $\pi = i_1 i_2 \cdots i_n$ be a permutation of $[n] = \{1, 2, \ldots, n\}$. How can we define the lexicographic order (or the reverse lexicographic order) induced by the ordering $x_{i_1} > x_{i_2} > \cdots > x_{i_n}$? First, given a monomial $u = x_1^{a_1} x_2^{a_2} \cdots x_n^{a_n} \in \mathcal{M}_n$, we set

$$u^{\pi} = x_1^{b_1} x_2^{b_2} \cdots x_n^{b_n}, \quad \text{where} \quad b_j = a_{i_j}.$$

Second, we introduce the total order $<_{\text{lex}}^{\pi}$ (resp. $<_{\text{rev}}^{\pi}$) on $\mathcal{M}_n$ by setting $u <_{\text{lex}}^{\pi} v$ (resp. $u <_{\text{rev}}^{\pi} v$) if $u^{\pi} <_{\text{lex}} v^{\pi}$ (resp. $u^{\pi} <_{\text{rev}} v^{\pi}$), where $u, v \in \mathcal{M}_n$. It then follows that $<_{\text{lex}}^{\pi}$ (reps. $<_{\text{rev}}^{\pi}$) is a monomial order on $K[\mathbf{x}]$. The monomial order $<_{\text{lex}}^{\pi}$ (reps. $<_{\text{rev}}^{\pi}$) is called the lexicographic order (resp. reverse lexicographic order) on $K[\mathbf{x}]$ induced by the ordering $x_{i_1} > x_{i_2} > \cdots > x_{i_n}$.

Unless otherwise stated, we usually consider monomial orders satisfying

$$x_1 > x_2 > \cdots > x_n.$$

Problem 1.1.12. Let $n = 3$ and $x_1 = x$, $x_2 = y$, $x_3 = z$. List the 21 monomials of degree 5 with respect to $<_{\text{lex}}$ and $<_{\text{rev}}$, respectively.

Example 1.1.13. Fix a nonzero vector $\mathbf{w} = (w_1, w_2, \ldots, w_n)$ with each $w_i \geq 0$. Let $<$ be a monomial order on $K[\mathbf{x}]$. We then define the total order $<_{\mathbf{w}}$ on $\mathcal{M}_n$ as follows: If $u = x_1^{a_1} x_2^{a_2} \cdots x_n^{a_n}$ and $v = x_1^{b_1} x_2^{b_2} \cdots x_n^{b_n}$ are monomials, then we define $u <_{\mathbf{w}} v$ if either (i) $\sum_{i=1}^{n} a_i w_i < \sum_{i=1}^{n} b_i w_i$, or (ii) $\sum_{i=1}^{n} a_i w_i = \sum_{i=1}^{n} b_i w_i$ and $u < v$. It follows that $<_{\mathbf{w}}$ is a monomial order on $K[\mathbf{x}]$.

Problem 1.1.14. Show that the total order $<_w$ of Example 1.1.13 is a monomial order on $K[\mathbf{x}]$.

We conclude this subsection with discussing (simple, but) indispensable lemmata on monomial orders on the polynomial ring $K[\mathbf{x}]$.

---

[3] A reverse lexicographic order is also called a *graded reverse lexicographic order(ing)*.

Lemma 1.1.15. *Let $<$ be a monomial order on $K[\mathbf{x}]$. Let u and v be monomials with $u \neq v$ and suppose that u divides v. Then $u < v$.*

Proof. Let w be a monomial with $v = wu$. Since $u \neq v$, one has $w \neq 1$. The definition of monomial orders says that $1 < w$. Hence, again, the definition of monomial orders says that $1 \cdot u < w \cdot u$. Thus $u < v$, as desired. □

Lemma 1.1.16. *Let $<$ be a monomial order on $K[\mathbf{x}]$. Then there exists no infinite descending sequence of the form*

$$u_0 > u_1 > u_2 > \cdots ,$$

where $u_0, u_1, u_2, \ldots$ are monomials.

Proof. Suppose on the contrary that such an infinite descending sequence exists. Let $M = \{u_0, u_1, u_2, \ldots\}$. Theorem 1.1.3 then guarantees that the number of minimal elements of M is at most finite. Let $u_{i_1}, u_{i_2}, \ldots, u_{i_s}$ be the minimal elements of M, where $i_1 < i_2 < \cdots < i_s$. Now, if $j > i_s$, then u_j must be divided by one of the minimal elements. Let, say, u_{i_k} divide u_j. Then Lemma 1.1.15 says that $u_{i_k} < u_j$. However, since $j > i_s \geq i_k$, it follows $u_{i_k} > u_j$ and a contradiction arises. □

1.1.5 Gröbner Bases

We are now in the position to introduce the notion of Gröbner bases. However, before studying Gröbner bases, it is required to discuss initial ideals of ideals of the polynomial ring.

We fix a monomial order $<$ on the polynomial ring $K[\mathbf{x}] = K[x_1, x_2, \ldots, x_n]$. Given a nonzero polynomial

$$f = a_1 u_1 + a_2 u_2 + \cdots + a_t u_t$$

of $K[\mathbf{x}]$, where $0 \neq a_i \in K$ and where $u_1, u_2, \ldots, u_t$ are monomials with

$$u_1 > u_2 > \cdots > u_t,$$

the *support* of f is the set of monomials appearing in f. It is written as $\mathrm{supp}(f)$. The *initial monomial* of f with respect to $<$ is the largest monomial belonging to $\mathrm{supp}(f)$ with respect to $<$. It is written as $\mathrm{in}_<(f)$. Thus

$$\mathrm{supp}(f) = \{u_1, u_2, \ldots, u_t\}$$

and

$$\mathrm{in}_<(f) = u_1.$$

Example 1.1.17. Let $n = 4$ and $f = x_1x_4 - x_2x_3$. Then $\mathrm{supp}(f) = \{x_1x_4, x_2x_3\}$. One has $\mathrm{in}_{<_\mathrm{lex}}(f) = x_1x_4$ and $\mathrm{in}_{<_\mathrm{rev}}(f) = x_2x_3$.

Problem 1.1.18. Let f and g be nonzero polynomials of $K[\mathbf{x}]$. Show that $\mathrm{in}_<(fg) = \mathrm{in}_<(f) \cdot \mathrm{in}_<(g)$. In particular if w is a monomial, then $\mathrm{in}_<(wg) = w \cdot \mathrm{in}_<(g)$.

Let I be an ideal of the polynomial ring $K[\mathbf{x}]$ with $I \neq \langle 0 \rangle$. The monomial ideal generated by $\{\mathrm{in}_<(f) : 0 \neq f \in I\}$ is called the *initial ideal* of I with respect to $<$ and is written as $\mathrm{in}_<(I)$. In other words,

$$\mathrm{in}_<(I) = \langle \{\mathrm{in}_<(f) : 0 \neq f \in I\} \rangle.$$

In general, however, even if $I = \langle \{f_\lambda\}_{\lambda \in \Lambda} \rangle$, it is not necessarily true that $\mathrm{in}_<(I)$ coincides with $\langle \{\mathrm{in}_<(f_\lambda)\}_{\lambda \in \Lambda} \rangle$.

Example 1.1.19. Let $n = 7$. Let $f = x_1x_4 - x_2x_3$, $g = x_4x_7 - x_5x_6$ and $I = \langle f, g \rangle$. Then $\mathrm{in}_{<_\mathrm{lex}}(f) = x_1x_4$, $\mathrm{in}_{<_\mathrm{lex}}(g) = x_4x_7$. Let $h = x_7 f - x_1 g = x_1x_5x_6 - x_2x_3x_7$. Since $h \in I$, it follows that $\mathrm{in}_{<_\mathrm{lex}}(h) = x_1x_5x_6 \in \mathrm{in}_{<_\mathrm{lex}}(I)$. However, $x_1x_5x_6 \notin \langle x_1x_4, x_4x_7 \rangle$. Hence $\langle x_1x_4, x_4x_7 \rangle \neq \mathrm{in}_{<_\mathrm{lex}}(I)$.

Now, Lemma 1.1.7 says that the monomial ideal $\mathrm{in}_<(I)$ is finitely generated. Thus there exists a finite subset

$$\{\mathrm{in}_<(f_1), \mathrm{in}_<(f_2), \ldots, \mathrm{in}_<(f_s)\}$$

of $\{\mathrm{in}_<(f) : 0 \neq f \in I\}$ which is a system of monomial generators of $\mathrm{in}_<(I)$.

Definition 1.1.20. We fix a monomial order $<$ on the polynomial ring $K[\mathbf{x}] = K[x_1, x_2, \ldots, x_n]$. Let I be an ideal of the polynomial ring $K[\mathbf{x}]$ with $I \neq \langle 0 \rangle$. Then a *Gröbner basis* of I with respect to $<$ is a finite set $\{g_1, g_2, \ldots, g_s\}$ of nonzero polynomials belonging to I such that $\{\mathrm{in}_<(g_1), \mathrm{in}_<(g_2), \ldots, \mathrm{in}_<(g_s)\}$ is a system of monomial generators of the initial ideal $\mathrm{in}_<(I)$.

A Gröbner basis exists. This follows from the argument just before Definition 1.1.20. However, a Gröbner basis cannot be unique. In fact, if $\{g_1, g_2, \ldots, g_s\}$ is a Gröbner basis of I, then any finite subset of $I \setminus \{0\}$ which contains $\{g_1, g_2, \ldots, g_s\}$ is again a Gröbner basis of I.

Corollary 1.1.9 says that the monomial ideal $\mathrm{in}_<(I)$ possesses a unique minimal system of monomial generators. We say that a Gröbner basis $\{g_1, g_2, \ldots, g_s\}$ of I is a *minimal Gröbner basis* of I if $\{\mathrm{in}_<(g_1), \mathrm{in}_<(g_2), \ldots, \mathrm{in}_<(g_s)\}$ is a minimal system of monomial generators of $\mathrm{in}_<(I)$ and if the coefficient of $\mathrm{in}_<(g_i)$ coincides with 1 for all $1 \leq i \leq s$. A minimal Gröbner basis exists. However, a minimal Gröbner basis may not be unique. For example, if $\{g_1, g_2, g_3, \ldots, g_s\}$, where $s > 1$, is a minimal Gröbner basis of I with $\mathrm{in}_<(g_1) < \mathrm{in}_<(g_2)$, then $\{g_1, g_2 + g_1, g_3, \ldots, g_s\}$ is again a minimal Gröbner basis of I.

1.1.6   Hilbert Basis Theorem

A fundamental theorem to be necessary for the development of the ideal theory of the polynomial ring is the Hilbert Basis Theorem, which guarantees that every ideal of the polynomial ring is finitely generated.

Theorem 1.1.21. *A Gröbner basis of an ideal of the polynomial ring is a system of generators of the ideal.*

Proof. Let I be an ideal of $K[\mathbf{x}]$ and $\{g_1, g_2, \ldots, g_s\}$ a Gröbner basis of I with respect to a monomial order $<$. Then

$$\mathrm{in}_<(I) = \langle \mathrm{in}_<(g_1), \mathrm{in}_<(g_2), \ldots, \mathrm{in}_<(g_s) \rangle$$

We claim $I = \langle g_1, g_2, \ldots, g_s \rangle$.

Let $0 \neq f \in I$. Since $\mathrm{in}_<(f) \in \mathrm{in}_<(I)$, there exist a monomial w together with $1 \leq i \leq s$ such that $\mathrm{in}_<(f) = w \cdot \mathrm{in}_<(g_i)$. Problem 1.1.18 says that $\mathrm{in}_<(f) = \mathrm{in}_<(wg_i)$. Let c_i be the coefficient of $\mathrm{in}_<(g_i)$ in g_i and c the coefficient of $\mathrm{in}_<(f)$ in f. Let $f^{(1)} = c_i f - c w g_i \in I$. If $f^{(1)} = 0$, then $f = (c/c_i) w g_i \in \langle g_1, g_2, \ldots, g_s \rangle$.

Let $f^{(1)} \neq 0$. Then $\mathrm{in}_<(f^{(1)}) < \mathrm{in}_<(f)$. In the case that $f^{(1)} \neq 0$, the same technique as we used for f can be applied to $f^{(1)}$ and we obtain $f^{(2)} \in I$. If $f^{(2)} = 0$, then $f^{(1)}$ belongs to $\langle g_1, g_2, \ldots, g_s \rangle$ and $f \in \langle g_1, g_2, \ldots, g_s \rangle$. If $f^{(2)} \neq 0$, then $\mathrm{in}_<(f^{(2)}) < \mathrm{in}_<(f^{(1)})$. In general, if $f^{(k-1)} \neq 0$, then the same technique as we used for f can be applied to $f^{(k-1)}$ and we obtain $f^{(k)} \in I$. If $f^{(k)} = 0$, then $f^{(k-1)}, f^{(k-2)}, \ldots, f^{(1)}$ belong to $\langle g_1, g_2, \ldots, g_s \rangle$ and $f \in \langle g_1, g_2, \ldots, g_s \rangle$. If $f^{(k)} \neq 0$, then $\mathrm{in}_<(f^{(k)}) < \mathrm{in}_<(f^{(k-1)})$.

Now, suppose that $f^{(k)} \neq 0$ for all $k \geq 1$. Then the infinite sequence

$$\mathrm{in}_<(f) > \mathrm{in}_<(f^{(1)}) > \cdots > \mathrm{in}_<(f^{(k-1)}) > \mathrm{in}_<(f^{(k)}) > \cdots$$

arises. However, Lemma 1.1.16 rejects the existence of such a sequence. In other words, there is $q > 0$ with $f^{(q)} = 0$, as desired.   □

Since a Gröbner basis is a finite set, Theorem 1.1.21 yields the so-called *Hilbert Basis Theorem*.

Corollary 1.1.22 (Hilbert Basis Theorem). *Every ideal of the polynomial ring is finitely generated. More precisely, given a system of generators $\{f_\lambda : \lambda \in \Lambda\}$ of an ideal I of $K[\mathbf{x}]$, there exists a finite subset of $\{f_\lambda : \lambda \in \Lambda\}$ which is a system of generators of I.*

Proof. Theorem 1.1.21 guarantees that every ideal of the polynomial ring is finitely generated. Let $I = \langle \{f_\lambda : \lambda \in \Lambda\} \rangle$ be an ideal of $K[\mathbf{x}]$ and $\{f_1, f_2, \ldots, f_s\}$ a system of generators of I consisting of a finite number of polynomials. Then, for each $1 \leq i \leq s$, there exists an expression of the form $f_i = \sum_{\lambda \in \Lambda} h_\lambda^{(i)} f_\lambda$, where $h_\lambda^{(i)} \in K[\mathbf{x}]$ is 0 except for a finite number of λ's. Let

$$\Lambda_i = \{\lambda \in \Lambda : h_\lambda^{(i)} \neq 0\}.$$

Then the finite set

$$\{f_\lambda \: : \: \lambda \in \cup_{i=1}^{s} \Lambda_i\}$$

is a system of generators of I.   □

Example 1.1.23. Let $n = 10$ and I the ideal of $K[x_1, x_2, \ldots, x_{10}]$ generated by

$$f_1 = x_1 x_8 - x_2 x_6, \quad f_2 = x_2 x_9 - x_3 x_7, \quad f_3 = x_3 x_{10} - x_4 x_8,$$

$$f_4 = x_4 x_6 - x_5 x_9, \quad f_5 = x_5 x_7 - x_1 x_{10}.$$

We claim that there exists *no* monomial order $<$ on $K[x_1, x_2, \ldots, x_{10}]$ such that $\{f_1, f_2, \ldots, f_5\}$ is a Gröbner basis of I with respect to $<$.

Suppose on the contrary that there exists a monomial order $<$ on $K[x_1, x_2, \ldots, x_{10}]$ such that $\mathscr{G} = \{f_1, f_2, \ldots, f_5\}$ is a Gröbner basis of I with respect to $<$. First, routine computation says that each of the five polynomials

$$x_1 x_8 x_9 - x_3 x_6 x_7, \quad x_2 x_9 x_{10} - x_4 x_7 x_8, \quad x_2 x_6 x_{10} - x_5 x_7 x_8,$$

$$x_3 x_6 x_{10} - x_5 x_8 x_9, \quad x_1 x_9 x_{10} - x_4 x_6 x_7$$

belongs to I. Let, say, $x_1 x_8 x_9 > x_3 x_6 x_7$. Since $x_1 x_8 x_9 \in \mathrm{in}_<(I)$, there is $g \in \mathscr{G}$ such that $\mathrm{in}_<(g)$ divides $x_1 x_8 x_9$. Such $g \in \mathscr{G}$ must be f_1. Hence $x_1 x_8 > x_2 x_6$. Thus $x_2 x_6 \notin \mathrm{in}_<(I)$. Hence there exists no $g \in \mathscr{G}$ such that $\mathrm{in}_<(g)$ divides $x_2 x_6 x_{10}$. Hence $x_2 x_6 x_{10} < x_5 x_7 x_8$. Thus $x_5 x_7 > x_1 x_{10}$. Continuing these arguments yields

$$x_1 x_8 x_9 > x_3 x_6 x_7, \quad x_2 x_9 x_{10} > x_4 x_7 x_8, \quad x_2 x_6 x_{10} < x_5 x_7 x_8,$$

$$x_3 x_6 x_{10} > x_5 x_8 x_9, \quad x_1 x_9 x_{10} < x_4 x_6 x_7$$

and

$$x_1 x_8 > x_2 x_6, \quad x_2 x_9 > x_3 x_7, \quad x_3 x_{10} > x_4 x_8,$$

$$x_4 x_6 > x_5 x_9, \quad x_5 x_7 > x_1 x_{10}.$$

Hence

$$(x_1 x_8)(x_2 x_9)(x_3 x_{10})(x_4 x_6)(x_5 x_7) > (x_2 x_6)(x_3 x_7)(x_4 x_8)(x_5 x_9)(x_1 x_{10}).$$

However, both sides of the above inequality coincide with $x_1 x_2 \cdots x_{10}$. This is a contradiction.

Problem 1.1.24. Let $n = 8$ and $I = \langle f_1, f_2, f_3 \rangle$ an ideal of $K[x_1, x_2, \ldots, x_8]$, where

$$f_1 = x_2 x_8 - x_4 x_7, \quad f_2 = x_1 x_6 - x_3 x_5, \quad f_3 = x_1 x_3 - x_2 x_4.$$

Show that there exists no monomial order $<$ on $K[x_1, x_2, \ldots, x_8]$ such that $\{f_1, f_2, f_3\}$ is a Gröbner basis of I with respect to $<$.

1.2   Division Algorithm

In Example 1.1.4, the division algorithm in one variable plays an important role. We now provide a more general algorithm for the case of more than one variable.

1.2.1   Division Algorithm

The *division algorithm* plays a fundamental role in the theory of Gröbner bases. In order to aid understanding of the proof of Theorem 1.2.1, the reader may wish to read Example 1.2.3.

Theorem 1.2.1 (The Division Algorithm). *We work with a fixed monomial order* $<$ *on the polynomial ring* $K[\mathbf{x}] = K[x_1, x_2, \ldots, x_n]$ *and with nonzero polynomials* $g_1, g_2, \ldots, g_s$ *belonging to* $K[\mathbf{x}]$. *Then, given a polynomial* $0 \neq f \in K[\mathbf{x}]$, *there exist* $f_1, f_2, \ldots, f_s$ *and* f' *belonging to* $K[\mathbf{x}]$ *with*

$$f = f_1 g_1 + f_2 g_2 + \cdots + f_s g_s + f' \tag{1.1}$$

such that the following conditions are satisfied:

- *If* $f' \neq 0$ *and* $u \in \mathrm{supp}(f')$, *then none of the initial monomials* $\mathrm{in}_<(g_i)$, $1 \leq i \leq s$, *divides* u. *In other words, if* $f' \neq 0$, *then no monomial* $u \in \mathrm{supp}(f')$ *belongs to the monomial ideal* $\langle \mathrm{in}_<(g_1), \mathrm{in}_<(g_2), \ldots, \mathrm{in}_<(g_s) \rangle$.
- *If* $f_i \neq 0$, *then*

$$\mathrm{in}_<(f) \geq \mathrm{in}_<(f_i g_i).$$

Definition 1.2.2. The right-hand side of (1.1) is said to be a *standard expression* of f with respect to $g_1, g_2, \ldots, g_s$ and f' a *remainder* of f with respect to $g_1, g_2, \ldots, g_s$.

Proof (of Theorem 1.2.1). Let $I = \langle \mathrm{in}_<(g_1), \mathrm{in}_<(g_2), \ldots, \mathrm{in}_<(g_s) \rangle$. If no monomial $u \in \mathrm{supp}(f)$ belongs to I, then the desired expression can be obtained by setting $f' = f$ and $f_1 = f_2 = \cdots = f_s = 0$.

Now, suppose that a monomial $u \in \mathrm{supp}(f)$ belongs to I and write u_0 for the monomial which is biggest with respect to $<$ among the monomials belonging to $\mathrm{supp}(f) \cap I$. Let, say, $\mathrm{in}_<(g_{i_0})$ divide u_0 and $w_0 = u_0/\mathrm{in}_<(g_{i_0})$. We rewrite

$$f = c'_0 c_{i_0}^{-1} w_0 g_{i_0} + h_1,$$

where c'_0 is the coefficient of u_0 in f and c_{i_0} is that of $\mathrm{in}_<(g_{i_0})$ in g_{i_0}. Then

$$\mathrm{in}_<(w_0 g_{i_0}) = w_0 \cdot \mathrm{in}_<(g_{i_0}) = u_0 \leq \mathrm{in}_<(f).$$

If either $h_1 = 0$ or if $h_1 \neq 0$ and no monomial $u \in \mathrm{supp}(h_1)$ belongs to I, then $f = c_0' c_{i_0}^{-1} w_0 g_{i_0} + h_1$ is a standard expression of f with respect to $g_1, g_2, \ldots, g_s$ and h_1 is a remainder of f.

If a monomial $u \in \mathrm{supp}(h_1)$ belongs to I and if u_1 is the monomial which is biggest with respect to $<$ among the monomials belonging to $\mathrm{supp}(h_1) \cap I$, then

$$u_1 < u_0$$

In fact, if a monomial u with $u > u_0 \; (= \mathrm{in}_<(w_0 g_{i_0}))$ belongs to $\mathrm{supp}(h_1)$, then u must belong to $\mathrm{supp}(f)$. This is impossible. Moreover, since the coefficient of u_0 in f coincides with that in $c_0' c_{i_0}^{-1} w_0 g_{i_0}$, it follows that u_0 cannot belong to $\mathrm{supp}(h_1)$.

Let, say, $\mathrm{in}_<(g_{i_1})$ divide u_1 and $w_1 = u_1 / \mathrm{in}_<(g_{i_1})$. Again, we rewrite

$$f = c_0' c_{i_0}^{-1} w_0 g_{i_0} + c_1' c_{i_1}^{-1} w_1 g_{i_1} + h_2,$$

where c_1' is the coefficient of u_1 in h_1 and c_{i_1} is that of $\mathrm{in}_<(g_{i_1})$ in g_{i_1}. Then

$$\mathrm{in}_<(w_1 g_{i_1}) < \mathrm{in}_<(w_0 g_{i_0}) \leq \mathrm{in}_<(f).$$

Continuing these procedures yields the descending sequence

$$u_0 > u_1 > u_2 > \cdots$$

Lemma 1.1.16 thus guarantees that these procedures will stop after a finite number of steps, say N steps, and we obtain an expression

$$f = \sum_{q=0}^{N-1} c_q' c_{i_q}^{-1} w_q g_{i_q} + h_N,$$

where either $h_N = 0$ or, in case of $h_N \neq 0$, no monomial $u \in \mathrm{supp}(h_N)$ belongs to I. Moreover, for each $1 \leq q \leq N - 1$, one has

$$\mathrm{in}_<(w_q g_{i_q}) < \cdots < \mathrm{in}_<(w_0 g_{i_0}) \leq \mathrm{in}_<(f).$$

Thus, by letting $\sum_{i=1}^s f_i g_i = \sum_{q=0}^{N-1} c_q' c_{i_q}^{-1} w_q g_{i_q}$ and $f' = h_N$, we obtain a standard expression $f = \sum_{i=1}^s f_i g_i + f'$ of f, as desired.   □

Example 1.2.3. Let $<_{\mathrm{lex}}$ denote the lexicographic order on $K[x, y, z]$ induced by $x > y > z$. Let $g_1 = x^2 - z$, $g_2 = xy - 1$ and $f = x^3 - x^2 y - x^2 - 1$. Each of

$$f = x^3 - x^2 y - x^2 - 1 = x(g_1 + z) - x^2 y - x^2 - 1$$

$$= xg_1 - x^2 y - x^2 + xz - 1 = xg_1 - (g_1 + z)y - x^2 + xz - 1$$

$$= xg_1 - yg_1 - x^2 + xz - yz - 1 = xg_1 - yg_1 - (g_1 + z) + xz - yz - 1$$

$$= (x - y - 1)g_1 + (xz - yz - z - 1)$$

and

$$
\begin{aligned}
f &= x^3 - x^2 y - x^2 - 1 = x(g_1 + z) - x^2 y - x^2 - 1 \\
  &= xg_1 - x^2 y - x^2 + xz - 1 = xg_1 - x(g_2 + 1) - x^2 + xz - 1 \\
  &= xg_1 - xg_2 - x^2 + xz - x - 1 = xg_1 - xg_2 - (g_1 + z) + xz - x - 1 \\
  &= (x - 1)g_1 - xg_2 + (xz - x - z - 1)
\end{aligned}
$$

is a standard expression of f with respect to g_1 and g_2, and each of $xz - yz - z - 1$ and $xz - x - z - 1$ is a remainder of f.

Example 1.2.3 shows that in the division algorithm a remainder of f is, in general, not unique. However,

Lemma 1.2.4. *If a finite set $\{g_1, g_2, \ldots, g_s\}$ consisting of polynomials belonging to $K[\mathbf{x}]$ is a Gröbner basis of the ideal $I = \langle g_1, g_2, \ldots, g_s \rangle$, then, for any polynomial $0 \neq f \in K[\mathbf{x}]$, a remainder of f with respect to $g_1, g_2, \ldots, g_s$ is unique.*

Proof. Suppose that each of the polynomials f' and f'' is a remainder of f with respect to $g_1, \ldots, g_s$. Let $f' \neq f''$. Since $0 \neq f' - f'' \in I$, the initial monomial $w = \mathrm{in}_<(f' - f'')$ belongs to $\mathrm{in}_<(I)$. On the other hand, since w belongs to either $\mathrm{supp}(f')$ or $\mathrm{supp}(f'')$, it follows that w cannot belong to $\langle \mathrm{in}_<(g_1), \mathrm{in}_<(g_2), \ldots, \mathrm{in}_<(g_s) \rangle$. However, since $\{g_1, \ldots, g_s\}$ is a Gröbner basis, the initial ideal $\mathrm{in}_<(I)$ coincides with $\langle \mathrm{in}_<(g_1), \mathrm{in}_<(g_2), \ldots, \mathrm{in}_<(g_s) \rangle$. This is a contradiction.   □

Corollary 1.2.5. *Suppose that a finite set $\{g_1, g_2, \ldots, g_s\}$ consisting of polynomials belonging to $K[\mathbf{x}]$ is a Gröbner basis of the ideal $I = \langle g_1, g_2, \ldots, g_s \rangle$ of $K[\mathbf{x}]$. Then a polynomial $0 \neq f \in K[\mathbf{x}]$ belongs to I if and only if a unique remainder of f with respect to $g_1, g_2, \ldots, g_s$ is 0.*

Proof. In general, if a remainder of a polynomial $0 \neq f \in K[\mathbf{x}]$ with respect to $g_1, g_2, \ldots, g_s$ is 0, then f belongs to the ideal $I = \langle g_1, g_2, \ldots, g_s \rangle$.

Now, suppose that $0 \neq f \in K[\mathbf{x}]$ belongs to I and that a standard expression of f with respect to $g_1, g_2, \ldots, g_s$ is $f = f_1 g_1 + f_2 g_2 + \cdots + f_s g_s + f'$. Since $f \in I$, one has $f' \in I$. If $f' \neq 0$, then $\mathrm{in}_<(f') \in \mathrm{in}_<(I)$. Since $\{g_1, g_2, \ldots, g_s\}$ is a Gröbner basis of I, one has $\mathrm{in}_<(I) = \langle \mathrm{in}_<(g_1), \mathrm{in}_<(g_2), \ldots, \mathrm{in}_<(g_s) \rangle$. However, since f' is a remainder, $\mathrm{in}_<(f') \in \mathrm{supp}(f')$ cannot belong to $\langle \mathrm{in}_<(g_1), \mathrm{in}_<(g_2), \ldots, \mathrm{in}_<(g_s) \rangle$. This is a contradiction.   □

1.2.2 Reduced Gröbner Bases

We work with a fixes monomial order $<$ on the polynomial ring $K[\mathbf{x}] = K[x_1, \ldots, x_n]$. A Gröbner basis $\{g_1, g_2, \ldots, g_s\}$ of an ideal of $K[\mathbf{x}]$ is called *reduced* if the following conditions are satisfied:

- The coefficient of $\mathrm{in}_<(g_i)$ in g_i is 1 for all $1 \leq i \leq s$;

- If $i \neq j$, then none of the monomials belonging to $\mathrm{supp}(g_j)$ is divided by $\mathrm{in}_<(g_i)$.

A reduced Gröbner basis is a minimal Gröbner basis. However, the converse is, of course, false.

Theorem 1.2.6. *A reduced Gröbner basis exists and is uniquely determined.*

Proof. (Existence) Let $\{g_1, g_2, \ldots, g_s\}$ be a minimal Gröbner basis of an ideal I of $K[\mathbf{x}]$. Then $\{\mathrm{in}_<(g_1), \mathrm{in}_<(g_2), \ldots, \mathrm{in}_<(g_s)\}$ is a unique minimal system of monomial generators of the initial ideal $\mathrm{in}_<(I)$. Thus, if $i \neq j$, then $\mathrm{in}_<(g_i)$ cannot be divided by $\mathrm{in}_<(g_j)$.

First, let h_1 be a remainder of g_1 with respect to $g_2, g_3, \ldots, g_s$. Since $\mathrm{in}_<(g_1)$ can be divided by none of $\mathrm{in}_<(g_j)$, $2 \leq j \leq s$, it follows that $\mathrm{in}_<(h_1)$ coincides with $\mathrm{in}_<(g_1)$. Thus $\{h_1, g_2, \ldots, g_s\}$ is a minimal Gröbner basis of I and each monomial belonging to $\mathrm{supp}(h_1)$ can be divided by none of $\mathrm{in}_<(g_j)$, $2 \leq j \leq s$.

Second, let h_2 be a remainder of g_2 with respect to $h_1, g_3, \ldots, g_s$. Since $\mathrm{in}_<(g_2)$ can be divided by none of $\mathrm{in}_<(h_1)(= \mathrm{in}_<(g_1)), \mathrm{in}_<(g_3), \ldots, \mathrm{in}_<(g_s)$, it follows that $\mathrm{in}_<(h_2)$ coincides with $\mathrm{in}_<(g_2)$ and $\{h_1, h_2, g_3, \ldots, g_s\}$ is a minimal Gröbner basis of I with the property that each monomial belonging to $\mathrm{supp}(h_1)$ can be divided by none of $\mathrm{in}_<(h_2), \mathrm{in}_<(g_3), \ldots, \mathrm{in}_<(g_s)$ and each monomial belonging to $\mathrm{supp}(h_2)$ can be divided by none of $\mathrm{in}_<(h_1), \mathrm{in}_<(g_3), \ldots, \mathrm{in}_<(g_s)$.

Continuing these procedures yields polynomials $h_3, h_4, \ldots, h_s$ and we obtain a reduced Gröbner basis $\{h_1, h_2, \ldots, h_s\}$ of I.

(Uniqueness) If $\{g_1, g_2, \ldots, g_s\}$ and $\{g'_1, g'_2, \ldots, g'_t\}$ are reduced Gröbner bases of I, then $\{\mathrm{in}_<(g_1), \mathrm{in}_<(g_2), \ldots, \mathrm{in}_<(g_s)\}$ and $\{\mathrm{in}_<(g'_1), \mathrm{in}_<(g'_2), \ldots, \mathrm{in}_<(g'_t)\}$ are minimal system of monomial generators of $\mathrm{in}_<(I)$. Lemma 1.1.9 then says that $s = t$ and, after rearranging the indices, we may assume that $\mathrm{in}_<(g_i) = \mathrm{in}_<(g'_i)$ for all $1 \leq i \leq s (= t)$. Let, say $g_i - g'_i \neq 0$. Then $\mathrm{in}_<(g_i - g'_i) < \mathrm{in}_<(g_i)$. Since $\mathrm{in}_<(g_i - g'_i)$ belongs to either $\mathrm{supp}(g_i)$ or $\mathrm{supp}(g'_i)$, it follows that none of $\mathrm{in}_<(g_j)$, $j \neq i$, can divide $\mathrm{in}_<(g_i - g'_i)$. Hence $\mathrm{in}_<(g_i - g'_i) \notin \mathrm{in}_<(I)$. This contradict the fact that $g_i - g'_i$ belongs to I. Hence $g_i = g'_i$ for all $1 \leq i \leq s$. □

We write $\mathscr{G}_{\mathrm{red}}(I; <)$ for *the* reduced Gröbner basis of an ideal I of $K[\mathbf{x}]$ with respect to a monomial order $<$.

Corollary 1.2.7. *Let I and J be ideals of $K[\mathbf{x}]$. Then $I = J$ if and only if $\mathscr{G}_{\mathrm{red}}(I; <) = \mathscr{G}_{\mathrm{red}}(J; <)$.*

1.3 Buchberger Criterion and Buchberger Algorithm

The highlights of the theory of Gröbner bases must be Buchberger criterion and Buchberger algorithm. A Gröbner basis of an ideal is its system of generators. It is then natural to ask: Given a system of generators of an ideal, how can we decide whether they form its Gröbner basis or not? The answer is Buchberger criterion, which also yields an algorithm called Buchberger algorithm. Starting from a system of generators of an ideal, the algorithm supplies the effective procedure to compute a Gröbner basis of the ideal. The discovery of the algorithm is the most important achievement of Buchberger.

1.3.1 S-Polynomials

Let, as before, $K[\mathbf{x}] = K[x_1, \ldots, x_n]$ denote the polynomial ring over K. We work with a fixed monomial order $<$ on $K[\mathbf{x}]$ and, for simplicity, omit the phrase "with respect to $<$", if there is no danger of confusion.

The least common multiple $\mathrm{lcm}(u, v)$ of two monomials $u = x_1^{a_1} x_2^{a_2} \cdots x_n^{a_n}$ and $v = x_1^{b_1} x_2^{b_2} \cdots x_n^{b_n}$ is the monomial $x_1^{c_1} x_2^{c_2} \cdots x_n^{c_n}$ with each $c_i = \max\{a_i, b_i\}$.

Let f and g be nonzero polynomials of $K[\mathbf{x}]$. Let c_f be the coefficient of $\mathrm{in}_<(f)$ in f and c_g that of $\mathrm{in}_<(g)$ in g. Then the polynomial

$$S(f, g) = \frac{\mathrm{lcm}(\mathrm{in}_<(f), \mathrm{in}_<(g))}{c_f \cdot \mathrm{in}_<(f)} f - \frac{\mathrm{lcm}(\mathrm{in}_<(f), \mathrm{in}_<(g))}{c_g \cdot \mathrm{in}_<(g)} g$$

is called the *S-polynomial* of f and g.

In other words, the S-polynomial of f and g can be obtained by canceling the initial monomials of f and g. For example, if $f = x_1 x_4 - x_2 x_3$ and $g = x_4 x_7 - x_5 x_6$, then with respect to $<_{\mathrm{lex}}$ one has

$$S(f, g) = x_7 f - x_1 g = x_1 x_5 x_6 - x_2 x_3 x_7,$$

and with respect to $<_{\mathrm{rev}}$ one has

$$S(f, g) = -x_5 x_6 f + x_2 x_3 g = x_2 x_3 x_4 x_7 - x_1 x_4 x_5 x_6.$$

We say that f *reduces to* 0 with respect to $g_1, g_2, \ldots, g_s$ if there is a standard expression (1.1) of f with respect to $g_1, g_2, \ldots, g_s$ with $f' = 0$.

Lemma 1.3.1. *Let f and g be nonzero polynomials of $K[\mathbf{x}]$ and suppose that $\mathrm{in}_<(f)$ and $\mathrm{in}_<(g)$ are relatively prime, i.e., $\mathrm{lcm}(\mathrm{in}_<(f), \mathrm{in}_<(g)) = \mathrm{in}_<(f)\mathrm{in}_<(g)$. Then $S(f, g)$ reduces to 0 with respect to f, g.*

Proof. To simplify the notation, we assume that each of the coefficients of $\mathrm{in}_<(f)$ in f and $\mathrm{in}_<(g)$ in g is 1. Let $f = \mathrm{in}_<(f) + f_1$ and $g = \mathrm{in}_<(g) + g_1$. Since $\mathrm{in}_<(f)$ and $\mathrm{in}_<(g)$ are relatively prime, it follows that

$$S(f, g) = \mathrm{in}_<(g) f - \mathrm{in}_<(f) g$$
$$= (g - g_1) f - (f - f_1) g$$
$$= f_1 g - g_1 f$$

We claim that $\text{in}_<(f_1)\text{in}_<(g)$ cannot coincides with $\text{in}_<(g_1)\text{in}_<(f)$. In fact, if $\text{in}_<(f_1)\text{in}_<(g) = \text{in}_<(g_1)\text{in}_<(f)$, then, since $\text{in}_<(f)$ and $\text{in}_<(g)$ are relatively prime, it follows that $\text{in}_<(f)$ divides $\text{in}_<(f_1)$. However, since $\text{in}_<(f_1) < \text{in}_<(f)$, this is impossible. Let, say, $\text{in}_<(f_1 g) < \text{in}_<(g_1 f)$. Then $\text{in}_<(S(f, g)) = \text{in}_<(g_1 f)$. Hence $S(f, g) = f_1 g - g_1 f$ is a standard expression of $S(f, g)$ with respect to f, g with a remainder 0. Thus $S(f, g)$ reduces to 0 with respect to f, g. $\square$

1.3.2   Buchberger Criterion

We now come to the most important theorem in the theory of Gröbner bases.

Lemma 1.3.2. *Let w be a monomial and $f_1, f_2, \ldots, f_s$ polynomials with $\text{in}_<(f_i) = w$ for all $1 \leq i \leq s$. Let $g = \sum_{i=1}^{s} b_i f_i$ with each $b_i \in K$ and suppose that $\text{in}_<(g) < w$. Then there exist $c_{jk} \in K$ with*

$$g = \sum_{1 \leq j, k \leq s} c_{jk} S(f_j, f_k).$$

Proof. Let c_i be the coefficient of $w = \text{in}_<(f_i)$ in f_i. Then $\sum_{i=1}^{s} b_i c_i = 0$. Let $g_i = (1/c_i) f_i$. Then

$$S(f_j, f_k) = g_j - g_k, \quad 1 \leq j, k \leq s.$$

Hence

$$\sum_{i=1}^{s} b_i f_i = \sum_{i=1}^{s} b_i c_i g_i$$

$$= b_1 c_1 (g_1 - g_2) + (b_1 c_1 + b_2 c_2)(g_2 - g_3)$$

$$+ (b_1 c_1 + b_2 c_2 + b_3 c_3)(g_3 - g_4)$$

$$+ \cdots + (b_1 c_1 + \cdots + b_{s-1} c_{s-1})(g_{s-1} - g_s)$$

$$+ (b_1 c_1 + \cdots + b_s c_s) g_s.$$

Since $\sum_{i=1}^{s} b_i c_i = 0$, it follows that

$$\sum_{i=1}^{s} b_i f_i = \sum_{i=2}^{s} (b_1 c_1 + \cdots + b_{i-1} c_{i-1}) S(f_{i-1}, f_i),$$

as desired. $\square$

Theorem 1.3.3 (Buchberger Criterion). *Let I be an ideal of the polynomial ring $K[\mathbf{x}]$ and $\mathscr{G} = \{g_1, g_2, \ldots, g_s\}$ a system of generators of I. Then $\mathscr{G}$ is a Gröbner basis of I if and only if the following condition is satisfied:*

($\star$)   *For all $i \neq j$, $S(g_i, g_j)$ reduces to 0 with respect to $g_1, g_2, \ldots, g_s$.*

Proof. ("Only If") Suppose that a system of generators $\mathscr{G} = \{g_1, g_2, \ldots, g_s\}$ is a Gröbner basis of I. Since the S-polynomial $S(g_i, g_j)$ of g_i and g_j belongs to the ideal $\langle g_i, g_j \rangle$. In particular, $S(g_i, g_j) \in I$. Since $\mathscr{G}$ is a Gröbner basis of I, Corollary 1.2.5 guarantees that $S(g_i, g_j)$ reduces to 0 with respect to $g_1, g_2, \ldots, g_s$, as required.

("If") Let $\mathscr{G} = \{g_1, g_2, \ldots, g_s\}$ be a system of generators of I which satisfies the condition $(\star)$.

(First Step) If a nonzero polynomial f belongs to I, then we write $\mathscr{H}_f$ for the set of sequences $(h_1, h_2, \ldots, h_s)$ with each $h_i \in K[\mathbf{x}]$ such that

$$f = \sum_{i=1}^{s} h_i g_i. \tag{1.2}$$

Since $\mathscr{G} = \{g_1, g_2, \ldots, g_s\}$ is a system of generators of I, it follows that $\mathscr{H}_f$ is nonempty. We associate each sequence $(h_1, h_2, \ldots, h_s) \in \mathscr{H}_f$ with the monomial

$$\delta_{(h_1, h_2, \ldots, h_s)} = \max\{\mathrm{in}_<(h_i g_i) : h_i g_i \neq 0\}.$$

Then

$$\mathrm{in}_<(f) \leq \delta_{(h_1, h_2, \ldots, h_s)}. \tag{1.3}$$

Now, among all of the monomials $\delta_{(h_1, h_2, \ldots, h_s)}$ with $(h_1, h_2, \ldots, h_s) \in \mathscr{H}_f$, we are especially interested in the monomial

$$\delta_f = \min_{(h_1, h_2, \ldots, h_s) \in \mathscr{H}_f} \delta_{(h_1, h_2, \ldots, h_s)}.$$

Then the inequality (1.3) says that

$$\mathrm{in}_<(f) \leq \delta_f.$$

In the following discussion, we will assume that the monomial $\delta_{(h_1, h_2, \ldots, h_s)}$ arising from the equality (1.2) coincides with δ_f.

(Second Step) Suppose for a while that $\mathrm{in}_<(f) = \delta_f$. Then, in the right-hand side of the equality (1.2), there is $h_i g_i \neq 0$ with $\mathrm{in}_<(f) = \mathrm{in}_<(h_i g_i)$. In particular $\mathrm{in}_<(f)$ belongs to the monomial ideal generated by $\mathrm{in}_<(g_1), \mathrm{in}_<(g_2), \ldots, \mathrm{in}_<(g_s)$.

Hence, if we can prove that $\mathrm{in}_<(f) = \delta_f$ for any nonzero polynomial $f \in I$, then

$$\mathrm{in}_<(I) = \langle \mathrm{in}_<(g_1), \mathrm{in}_<(g_2), \ldots, \mathrm{in}_<(g_s) \rangle$$

and $\mathscr{G}$ turns out to be a Gröbner basis of I.

(Third Step)   Now, suppose that there is a nonzero polynomial $f \in I$ with $\text{in}_<(f) < \delta_f$. If we can get a contradiction, then our proof finishes.

We rewrite the right-hand side of the equality (1.2) as

$$(\sharp) \quad f = \sum_{\text{in}_<(h_i g_i) = \delta_f} h_i g_i + \sum_{\text{in}_<(h_i g_i) < \delta_f} h_i g_i$$

$$= \sum_{\text{in}_<(h_i g_i) = \delta_f} c_i \cdot \text{in}_<(h_i) g_i + \sum_{\text{in}_<(h_i g_i) = \delta_f} (h_i - c_i \cdot \text{in}_<(h_i)) g_i$$

$$+ \sum_{\text{in}_<(h_i g_i) < \delta_f} h_i g_i,$$

where $c_i \in K$ is the coefficient of $\text{in}_<(h_i)$ in h_i. The first equality is clear. The second equality is the consequence of the simple rewriting

$$h_i = c_i \cdot \text{in}_<(h_i) + (h_i - c_i \cdot \text{in}_<(h_i)).$$

A crucial fact is that every monomial u belonging to the support of

$$\sum_{\text{in}_<(h_i g_i) = \delta_f} (h_i - c_i \cdot \text{in}_<(h_i)) g_i + \sum_{\text{in}_<(h_i g_i) < \delta_f} h_i g_i$$

satisfies $u < \delta_f$. Hence, the hypothesis that $\text{in}_<(f) < \delta_f$ guarantees that

$$\text{in}_< \left(\sum_{\text{in}_<(h_i g_i) = \delta_f} c_i \cdot \text{in}_<(h_i) g_i \right) < \delta_f.$$

However, since $\text{in}_<(h_i g_i) = \delta_f$, one has

$$\text{in}_<(\text{in}_<(h_i) g_i) = \delta_f.$$

It then follows from Lemma 1.3.2 that, by using those S-polynomials

$$S(\text{in}_<(h_j) g_j, \text{in}_<(h_k) g_k)$$

with $\text{in}_<(h_j g_j) = \text{in}_<(h_k g_k) = \delta_f$ and $c_{jk} \in K$, we can rewrite the first sum in the right-hand side of the second equality of $(\sharp)$ as follows:

$$\sum_{\text{in}_<(h_i g_i) = \delta_f} c_i \cdot \text{in}_<(h_i) g_i = \sum_{j,k} c_{jk} S(\text{in}_<(h_j) g_j, \text{in}_<(h_k) g_k). \qquad (1.4)$$

Since $\text{in}_<(h_j g_j) = \text{in}_<(h_k g_k) = \delta_f$, it follows that

$$S(\text{in}_<(h_j) g_j, \text{in}_<(h_k) g_k) = (1/b_j) \text{in}_<(h_j) g_j - (1/b_k) \text{in}_<(h_k) g_k,$$

where b_j is the coefficient of $\text{in}_<(g_j)$ in g_j. Here each monomial u belonging to the support of $S(\text{in}_<(h_j)g_j, \text{in}_<(h_k)g_k)$ satisfies $u < \delta_f$.

Let

$$u_{jk} = \delta_f / \text{lcm}(\text{in}_<(g_j), \text{in}_<(g_k)).$$

Then

$$u_{jk}S(g_j, g_k)$$

$$= u_{jk}\left[\frac{\text{lcm}(\text{in}_<(g_j), \text{in}_<(g_k))}{b_j \cdot \text{in}_<(g_j)}g_j - \frac{\text{lcm}(\text{in}_<(g_j), \text{in}_<(g_k))}{b_k \cdot \text{in}_<(g_k)}g_k\right]$$

$$= \delta_f\left[\frac{1}{b_j \cdot \text{in}_<(g_j)}g_j - \frac{1}{b_k \cdot \text{in}_<(g_k)}g_k\right]$$

$$= \frac{\text{in}_<(h_j)}{b_j}g_j - \frac{\text{in}_<(h_k)}{b_k}g_k$$

$$= S(\text{in}_<(h_j)g_j, \text{in}_<(h_k)g_k).$$

By using the equality (1.4), there exists an expression of the form

$$\sum_{\text{in}_<(h_l g_l)=\delta_f} c_i \cdot \text{in}_<(h_i)g_i = \sum_{j,k} c_{jk}u_{jk}S(g_j, g_k), \quad c_{jk} \in K \qquad (1.5)$$

with

$$\text{in}_<(u_{jk}S(g_j, g_k)) < \delta_f.$$

The condition $(\star)$ guarantees the existence of an expression of $S(g_j, g_k)$ of the form

$$S(g_j, g_k) = \sum_{i=1}^{s} p_i^{jk}g_i, \quad \text{in}_<(p_i^{jk}g_i) \le \text{in}_<(S(g_j, g_k)), \qquad (1.6)$$

where $p_i^{jk} \in K[\mathbf{x}]$. Combining (1.6) with (1.5) yields

$$\sum_{\text{in}_<(h_i g_i)=\delta_f} c_i \cdot \text{in}_<(h_i)g_i = \sum_{j,k} c_{jk}u_{jk}\left(\sum_{i=1}^{s} p_i^{jk}g_i\right). \qquad (1.7)$$

We rewrite the right-hand side of the equality (1.7) as $\sum_{i=1}^{s} h_i'g_i$. Then

$$\text{in}_<(h_i'g_i) < \delta_f.$$

Finally, by virtue of (1.7) together with the second equality of ($\sharp$), it turns out that there exists an expression of f of the form

$$f = \sum_{i=1}^{s} h_i'' g_i, \quad \text{in}_<(h_i'' g_i) < \delta_f.$$

The existence of such an expression contradicts the definition of δ_f, as desired. □

In applying Buchberger's criterion it is not always necessary to check whether *all* S-polynomials $S(g_i, g_j)$ with $i \neq j$ reduce to 0 with respect to $g_1, \ldots, g_s$. In fact, Lemma 1.3.1 says that if $\text{in}_<(g_i)$ and $\text{in}_<(g_j)$ are relatively prime, then $S(g_i, g_j)$ reduces to 0 with respect to g_i, g_j. Thus in particular $S(g_i, g_j)$ reduces to 0 with respect to $g_1, g_2, \ldots, g_s$. Hence we only check those S-polynomials $S(g_i, g_j)$ with $i \neq j$ such that $\text{in}_<(g_i)$ and $\text{in}_<(g_j)$ possess at least one common variable.

Corollary 1.3.4. *If $g_1, \ldots, g_s$ are nonzero polynomials belonging to $K[\mathbf{x}]$ such that $\text{in}_<(g_i)$ and $\text{in}_<(g_j)$ are relatively prime for all $i \neq j$, then $\{g_1, \ldots, g_s\}$ is a Gröbner basis of $I = \langle g_1, \ldots, g_s \rangle$.*

Example 1.3.5. Let $n = 7$ and consider the reverse lexicographic order $<_{\text{rev}}$. Let $f = x_1 x_4 - x_2 x_3$, $g = x_4 x_7 - x_5 x_6$ and $I = \langle f, g \rangle$. Then, since $\text{in}_{<_{\text{rev}}}(f) = x_2 x_3$ and $\text{in}_{<_{\text{rev}}}(g) = x_5 x_6$ are relatively prime, it follows that $\{f, g\}$ is a Gröbner basis of I with respect to $<_{\text{rev}}$.

Example 1.3.6. Let $f = x_1 x_4 - x_2 x_3$, $g = x_4 x_7 - x_5 x_6$ and $I = \langle f, g \rangle$. Example 1.1.19 shows that $\{f, g\}$ cannot be a Gröbner basis of I with respect to the lexicographic order $<_{\text{lex}}$. On the other hand, If $h = S(f, g) = x_1 x_5 x_6 - x_2 x_3 x_7$, then $\{f, g, h\}$ is a Gröbner basis of I with respect to $<_{\text{lex}}$. To see why this is true, we must check the criterion ($\star$) for $S(f, g)$, $S(g, h)$ and $S(f, h)$. First, $S(f, g) = h$ reduces to 0 with respect to h. Since $\text{in}_{<_{\text{lex}}}(g)$ and $\text{in}_{<_{\text{lex}}}(h)$ are relatively prime, $S(g, h)$ reduces to 0 with respect to g, h. Moreover, since

$$S(f, h) = x_5 x_6 f - x_4 h = x_2 x_3 x_4 x_7 - x_2 x_3 x_5 x_6 = x_2 x_3 g,$$

it follows that $S(f, h)$ reduces to 0 with respect to g.

1.3.3 Buchberger Algorithm

One of the advantages of Buchberger criterion is that it yields an algorithm, called Buchberger algorithm, which supplies a procedure for computing a Gröbner basis of an ideal I of $K[\mathbf{x}]$ from a system of generators of I.

- Let I be an ideal of the polynomial ring $K[\mathbf{x}]$ and $\mathscr{G} = \{g_1, g_2, \ldots, g_s\}$ its system of generators. If each S-polynomial $S(g_i, g_j)$, $1 \leq i < j \leq s$, reduces to 0 with respect to $g_1, g_2, \ldots, g_s$, then Buchberger criterion guarantees that $\mathscr{G}$ is a Gröbner basis of I.

- Otherwise there is $S(g_i, g_j)$ with nonzero remainder g_{s+1}. It follows from the definition of a remainder that none of $\text{in}_<(g_i)$ divides $\text{in}_<(g_{s+1})$. Hence the monomial ideal

$$\langle \text{in}_<(g_1), \text{in}_<(g_2), \ldots, \text{in}_<(g_s) \rangle$$

  is strictly contained in the monomial ideal

$$\langle \text{in}_<(g_1), \text{in}_<(g_2), \ldots, \text{in}_<(g_s), \text{in}_<(g_{s+1}) \rangle.$$

- Since $S(g_i, g_j) \in I$, it follows that $g_{s+1} \in I$. Now, replace $\mathscr{G}$ with

$$\mathscr{G}' = \mathscr{G} \cup \{g_{s+1}\},$$

  which is a system of generators of I with a redundant polynomial g_{s+1}. We then apply Buchberger criterion to $\mathscr{G}'$. If each $S(g_i, g_j)$, $1 \leq i < j \leq s + 1$, reduces to 0 with respect to $g_1, g_2, \ldots, g_s, g_{s+1}$, then Buchberger criterion guarantees that $\mathscr{G}'$ is a Gröbner basis of I.

- Otherwise there is $S(g_k, g_\ell)$ with nonzero remainder g_{s+2} and

$$\langle \text{in}_<(g_1), \text{in}_<(g_2), \ldots, \text{in}_<(g_s), \text{in}_<(g_{s+1}) \rangle$$

  is strictly contained in

$$\langle \text{in}_<(g_1), \text{in}_<(g_2), \ldots, \text{in}_<(g_s), \text{in}_<(g_{s+1}), \text{in}_<(g_{s+2}) \rangle.$$

- Again, the remainder g_{s+2} belongs to I. We thus apply Buchberger criterion to

$$\mathscr{G}'' = \mathscr{G}' \cup \{g_{s+2}\},$$

  which is a system of generators of I with redundant polynomials g_{s+1} and g_{s+2}.
- By virtue of Theorem 1.1.3, it follows that the above procedure will terminate after a finite number of steps, and a Gröbner basis of I can be obtained.
- In fact, if the above procedure will eternally persist, then there exists a strictly increasing infinite sequence of monomial ideals

$$\langle \text{in}_<(g_1), \ldots, \text{in}_<(g_s) \rangle \subset \langle \text{in}_<(g_1), \ldots, \text{in}_<(g_s), \text{in}_<(g_{s+1}) \rangle$$

$$\subset \cdots \subset \langle \text{in}_<(g_1), \ldots, \text{in}_<(g_s), \text{in}_<(g_{s+1}), \ldots, \text{in}_<(g_{s+k}) \rangle \subset \cdots$$

Theorem 1.1.3 says that the set of minimal elements of the set of monomials

$$\mathscr{M} = \{\text{in}_<(g_1), \ldots, \text{in}_<(g_s), \text{in}_<(g_{s+1}), \ldots\}$$

is finite. If

$$\text{in}_<(g_{i_1}), \text{in}_<(g_{i_2}), \ldots, \text{in}_<(g_{i_q}), \qquad i_1 < i_2 < \cdots < i_q,$$

are the minimal elements of $\mathcal{M}$, then for all $j > i_q$ one has

$$\langle \text{in}_<(g_{i_1}), \text{in}_<(g_{i_2}), \ldots, \text{in}_<(g_{i_q}) \rangle$$
$$= \langle \text{in}_<(g_1), \text{in}_<(g_2), \ldots, \text{in}_<(g_{i_q}), \text{in}_<(g_{i_q+1}), \ldots, \text{in}_<(g_j) \rangle,$$

which is a contradiction.

The reader may have observed that the basic fact which guarantees that the above procedure terminates after a finite number of steps is again Theorem 1.1.3. The above algorithm which, starting from a system of generators of I, enables us to find a Gröbner basis of I is said to be *Buchberger algorithm*.

Example 1.3.7. We follow Example 1.1.23. Let $n = 10$ and $I = \langle f_1, f_2, f_3, f_4, f_5 \rangle$ the ideal of $K[x_1, x_2, \ldots, x_{10}]$, where

$$f_1 = x_1 x_8 - x_2 x_6, \quad f_2 = x_2 x_9 - x_3 x_7, \quad f_3 = x_3 x_{10} - x_4 x_8,$$
$$f_4 = x_4 x_6 - x_5 x_9, \quad f_5 = x_5 x_7 - x_1 x_{10}.$$

In Example 1.1.23 it is shown that there exists no monomial order $<$ such that $\mathcal{F} = \{f_1, f_2, f_3, f_4, f_5\}$ is a Gröbner basis of I. In what follows, by using Buchberger algorithm, we compute a Gröbner basis of I with respect to the lexicographic order as well as that with respect to the reverse lexicographic order.

(lexicographic order) The initial monomials of f_1, f_2, f_3, f_4, f_5 are

$$x_1 x_8, \quad x_2 x_9, \quad x_3 x_{10}, \quad x_4 x_6, \quad x_1 x_{10},$$

respectively. Recall that if $\text{in}_{<_{\text{lex}}}(f_i)$ and $\text{in}_{<_{\text{lex}}}(f_j)$ with $i \neq j$ are relatively prime, then $S(f_i, f_j)$ reduces to 0. Thus the S-polynomials which we must check are

$$S(f_1, f_5) = x_{10} f_1 + x_8 f_5 = x_5 x_7 x_8 - x_2 x_6 x_{10},$$
$$S(f_3, f_5) = x_1 f_3 + x_3 f_5 = x_3 x_5 x_7 - x_1 x_4 x_8.$$

One has

$$S(f_3, f_5) = -x_4 f_1 - x_2 x_4 x_6 + x_3 x_5 x_7$$
$$= -x_4 f_1 - x_2 f_4 - x_2 x_5 x_9 + x_3 x_5 x_7$$
$$= -x_4 f_1 - x_2 f_4 - x_5 f_2,$$

which reduces to 0. On the other hand, $S(f_1, f_5)$ itself is a remainder with respect to f_1, f_2, f_3, f_4, f_5. Thus, letting

$$f_6 = x_5 x_7 x_8 - x_2 x_6 x_{10},$$

we consider $\mathscr{F}' = \{f_1, f_2, f_3, f_4, f_5, f_6\}$ to be a system of generators of I (with a redundant polynomial f_6) and apply Buchberger criterion to $\mathscr{F}'$. Since $\mathrm{in}_{<_{\mathrm{lex}}}(f_6) = x_2 x_6 x_{10}$, the S-polynomials which we must check are

$$S(f_2, f_6) = x_6 x_{10} f_2 + x_9 f_6 = x_5 x_7 x_8 x_9 - x_3 x_6 x_7 x_{10}$$
$$= x_7(x_5 x_8 x_9 - x_3 x_6 x_{10}) = x_7(-x_6 f_3 - x_4 x_6 x_8 + x_5 x_8 x_9)$$
$$= -x_7(x_6 f_3 + x_8 f_4),$$
$$S(f_3, f_6) = x_2 x_6 f_3 + x_3 f_6 = x_3 x_5 x_7 x_8 - x_2 x_4 x_6 x_8$$
$$= x_8(x_3 x_5 x_7 - x_2 x_4 x_6) = x_8(-x_2 f_4 - x_2 x_5 x_9 + x_3 x_5 x_7)$$
$$= -x_8(x_5 f_2 + x_2 f_4),$$
$$S(f_4, f_6) = x_2 x_{10} f_4 + x_4 f_6 = x_4 x_5 x_7 x_8 - x_2 x_5 x_9 x_{10}$$
$$= x_5(x_4 x_7 x_8 - x_2 x_9 x_{10}) = x_5(-x_{10} f_2 - x_3 x_7 x_{10} + x_4 x_7 x_8)$$
$$= -x_5(x_{10} f_2 + x_7 f_3),$$
$$S(f_5, f_6) = -x_2 x_6 f_5 + x_1 f_6 = x_1 x_5 x_7 x_8 - x_2 x_5 x_6 x_7$$
$$= x_5 x_7 f_1.$$

Each of them reduces to 0. Thus $\mathscr{F}'$ is a Gröbner basis of I with respect to the lexicographic order.

(reverse lexicographic order) The initial monomials of f_1, f_2, f_3, f_4, f_5 are

$$x_2 x_6, \quad x_3 x_7, \quad x_4 x_8, \quad x_4 x_6, \quad x_5 x_7,$$

respectively. Thus the S-polynomials which we must check are

$$S(f_1, f_4) = -x_4 f_1 - x_2 f_4 = x_2 x_5 x_9 - x_1 x_4 x_8,$$
$$S(f_2, f_5) = -x_5 f_2 - x_3 f_5 = x_1 x_3 x_{10} - x_2 x_5 x_9,$$
$$S(f_3, f_4) = -x_6 f_3 - x_8 f_4 = x_5 x_8 x_9 - x_3 x_6 x_{10}.$$

Since

$$S(f_1, f_4) = x_1 f_3 + x_2 x_5 x_9 - x_1 x_3 x_{10},$$

its remainder is $-S(f_2, f_5)$. Thus, letting

$$f_6 = x_2 x_5 x_9 - x_1 x_3 x_{10},$$
$$f_7 = x_5 x_8 x_9 - x_3 x_6 x_{10},$$

we consider $\mathscr{F}'' = \{f_1, f_2, f_3, f_4, f_5, f_6, f_7\}$ to be a system of generators of I and apply Buchberger criterion to $\mathscr{F}''$. The initial monomials of f_6 and f_7 are $x_2 x_5 x_9$ and $x_5 x_8 x_9$, respectively. Thus the S-polynomials which we must check are

$$S(f_1, f_6) = -x_5 x_9 f_1 - x_6 f_6 = x_1 x_3 x_6 x_{10} - x_1 x_5 x_8 x_9$$
$$= x_1(x_3 x_6 x_{10} - x_5 x_8 x_9) = -x_1 f_7,$$
$$S(f_3, f_7) = -x_5 x_9 f_3 - x_4 f_7 = x_3 x_4 x_6 x_{10} - x_3 x_5 x_9 x_{10}$$
$$= x_3 x_{10}(x_4 x_6 - x_5 x_9) = x_3 x_{10} f_4,$$
$$S(f_5, f_6) = x_2 x_9 f_5 - x_7 f_6 = x_1 x_3 x_7 x_{10} - x_1 x_2 x_9 x_{10}$$
$$= x_1 x_{10}(x_3 x_7 - x_2 x_9) = -x_1 x_{10} f_2,$$
$$S(f_5, f_7) = x_8 x_9 f_5 - x_7 f_7 = x_3 x_6 x_7 x_{10} - x_1 x_8 x_9 x_{10}$$
$$= x_{10}(x_3 x_6 x_7 - x_1 x_8 x_9) = x_{10}(-x_6 f_2 + x_2 x_6 x_9 - x_1 x_8 x_9)$$
$$= -x_{10}(x_6 f_2 + x_9 f_1).$$

Each of them reduces to 0. Thus $\mathscr{F}''$ is a Gröbner basis of I with respect to the reverse lexicographic order.

Problem 1.3.8. By using Buchberger algorithm, compute a Gröbner basis of the ideal of Problem 1.1.24 with respect to the lexicographic order as well as that with respect to the reverse lexicographic order.

1.4 Elimination Theory

The reader might be familiar with the elimination technique which can be used for solving simultaneous equations, say,

$$\begin{cases} 2x + y = 3 \\ x + 3y = 4 \end{cases}$$

The elimination theorem generalizes the above simple elimination technique. It is a fascinating result which demonstrates the power of Gröbner bases.

1.4.1 Elimination Theorem

Let $K[\mathbf{x}] = K[x_1, x_2, \ldots, x_n]$ be the polynomial ring and write $B_{i_1 i_2 \cdots i_m}$ for the subset of $K[\mathbf{x}]$ consisting of those $f \in K[\mathbf{x}]$ such that each monomial belonging to $\mathrm{supp}(f)$ is a monomial in the variables $x_{i_1}, x_{i_2}, \ldots, x_{i_m}$, where $1 \leq i_1 < i_2 < \cdots < i_m \leq n$. Thus

$$B_{i_1 i_2 \cdots i_m} = K[x_{i_1}, x_{i_2}, \ldots, x_{i_m}].$$

If f and g belong to $B_{i_1 i_2 \cdots i_m}$, then the sum and the product of f and g again belong to $B_{i_1 i_2 \cdots i_m}$. Thus $B_{i_1 i_2 \cdots i_m}$ itself is the polynomial ring.

A monomial order $<$ on $K[\mathbf{x}]$ can be naturally induce the monomial order $<'$ on $B_{i_1 i_2 \cdots i_m}$. More precisely, for monomials u and v belonging to $B_{i_1 i_2 \cdots i_m}$, one has $u <' v$ if and only if $u < v$ in $K[\mathbf{x}]$. Unless confusion arises, the monomial order $<'$ on $B_{i_1 i_2 \cdots i_m}$ induced by a monomial order $<$ on $K[\mathbf{x}]$ will be also written as $<$.

In general, if I is an ideal of $K[\mathbf{x}]$, then $I \cap B_{i_1 i_2 \cdots i_m}$ is an ideal of $B_{i_1 i_2 \cdots i_m}$. It is then natural to ask, for given a Gröbner basis $\mathscr{G}$ of I, whether $\mathscr{G} \cap B_{i_1 i_2 \cdots i_m}$ is a Gröbner basis of $I \cap B_{i_1 i_2 \cdots i_m}$ or not.

Theorem 1.4.1 (The Elimination Theorem). *Let $<$ be a monomial order on $K[\mathbf{x}]$ and $\mathscr{G}$ a Gröbner basis of an ideal I of $K[\mathbf{x}]$ with respect to $<$. Suppose that*

$$(\clubsuit) \quad \text{For each } g \in \mathscr{G}, \text{ one has } g \in B_{i_1 i_2 \cdots i_m} \text{ if } \mathrm{in}_<(g) \in B_{i_1 i_2 \cdots i_m}.$$

Then $\mathscr{G} \cap B_{i_1 i_2 \cdots i_m}$ is a Gröbner basis of $I \cap B_{i_1 i_2 \cdots i_m}$ with respect to $<$ on $B_{i_1 i_2 \cdots i_m}$.

Proof. What we must prove is that the initial ideal $\mathrm{in}_<(I \cap B_{i_1 i_2 \cdots i_m})$ of the ideal $I \cap B_{i_1 i_2 \cdots i_m}$ is generated by

$$\{\mathrm{in}_<(g) \ : \ g \in \mathscr{G} \cap B_{i_1 i_2 \cdots i_m}\}.$$

Let u be a monomial belonging to $\mathrm{in}_<(I \cap B_{i_1 i_2 \cdots i_m})$. Then there is $0 \neq f \in I \cap B_{i_1 i_2 \cdots i_m}$ with $\mathrm{in}_<(f) = u$. Since $f \in I$, one has $u \in \mathrm{in}_<(I)$. Now, since $\mathscr{G}$ is a Gröbner basis of I, there is $g \in \mathscr{G}$ such that $\mathrm{in}_<(g)$ divides u. Since $u \in B_{i_1 i_2 \cdots i_m}$ and since $\mathrm{in}_<(g)$ divides u, it follows that $\mathrm{in}_<(g) \in B_{i_1 i_2 \cdots i_m}$. Hence the condition $(\clubsuit)$ guarantees that g belongs to $B_{i_1 i_2 \cdots i_m}$. Consequently, for any monomial u belonging to the initial ideal $\mathrm{in}_<(I \cap B_{i_1 i_2 \cdots i_m})$, there is $g \in \mathscr{G} \cap B_{i_1 i_2 \cdots i_m}$ such that $\mathrm{in}_<(g)$ divides u. Hence $\mathrm{in}_<(I \cap B_{i_1 i_2 \cdots i_m})$ is generated by $\{\mathrm{in}_<(g) \ : \ g \in \mathscr{G} \cap B_{i_1 i_2 \cdots i_m}\}$, as desired. $\square$

Corollary 1.4.2. *Let $<_{\mathrm{purelex}}$ denote the pure lexicographic order on $K[\mathbf{x}]$ and*

$$B_{\geq p} = K[x_p, x_{p+1}, \ldots x_n].$$

Let $\mathscr{G}$ be a Gröbner basis of an ideal I of $K[\mathbf{x}]$ with respect to $<_{\mathrm{purelex}}$. Then $\mathscr{G} \cap B_{\geq p}$ is a Gröbner basis of $I \cap B_{\geq p}$ with respect to $<_{\mathrm{purelex}}$.

Proof. We must prove the condition $(\clubsuit)$ of Theorem 1.4.1 is satisfied. If $g \in \mathscr{G}$ and if its initial monomial $\mathrm{in}_{<_{\mathrm{purelex}}}(g)$ belongs to $B_{\geq p}$, then $\mathrm{in}_{<_{\mathrm{purelex}}}(g)$ is a monomial in the variables $x_p, x_{p+1}, \ldots x_n$. Hence by the definition of the pure lexicographic order $<_{\mathrm{purelex}}$ it follows that each monomial belonging to the support of g is a monomial in $x_p, x_{p+1}, \ldots x_n$. Thus $g \in B_{\geq p}$, as desired. $\square$

As one of the typical applications of Corollary 1.4.2, we discuss the problem of computing the intersection of ideals.

Let, in general, I and J be ideals of the polynomial ring $K[\mathbf{x}]$. Then the sum $I + J$ and the intersection $I \cap J$ are defined as follows:

$$I + J = \{f + h \, ; \, f \in I, h \in J\},$$
$$I \cap J = \{f \in K[\mathbf{x}] \, ; \, f \in I, f \in J\}.$$

Then both $I + J$ and $I \cap J$ are ideals of $K[\mathbf{x}]$. Let $\{f_1, f_2, \ldots\}$ be a system of generators of I and $\{h_1, h_2, \ldots\}$ that of J. Then

$$\{f_1, f_2, \ldots, h_1, h_2, \ldots\}$$

is a system of generators of $I + J$. However, to find a system of generators of $I \cap J$ is rather difficult.

With adding a new variable t to $K[\mathbf{x}]$, we consider the polynomial ring

$$K[t, \mathbf{x}] = K[t, x_1, x_2, \ldots, x_n]$$

in $n + 1$ variables. If I and J are ideals of $K[\mathbf{x}]$, then we introduce ideals tI and $(1 - t)J$ of $K[t, \mathbf{x}]$ as follows:

$$tI = \langle \{tf \, ; \, f \in I\} \rangle,$$
$$(1 - t)J = \langle \{(1 - t)f \, ; \, f \in J\} \rangle.$$

Then

Lemma 1.4.3. *As ideals of $K[\mathbf{x}]$ one has*

$$I \cap J = (tI + (1 - t)J) \cap K[\mathbf{x}].$$

Proof. Let $f \in K[\mathbf{x}]$ belong $I \cap J$. Since $f \in I$ one has $tf \in tI$, and since $f \in J$, one has $(1 - t)f \in (1 - t)J$. Hence $f = tf + (1 - t)f \in tI + (1 - t)J$.

On the other hand, if a polynomial $f(\mathbf{x}) \in K[\mathbf{x}]$ belongs to $tI + (1 - t)J$, then there exist $f_i \in I$, $f'_j \in J$ and $h_i, h'_j \in K[t, \mathbf{x}]$ such that

$$f(\mathbf{x}) = t \sum_i f_i(\mathbf{x})h_i(t, \mathbf{x}) + (1 - t) \sum_j f'_j(\mathbf{x})h'_j(t, \mathbf{x}).$$

Letting $t = 0$ one has $f = \sum_j f'_j(\mathbf{x})h'_j(0, \mathbf{x}) \in J$, and letting $t = 1$ one has $f = \sum_i f_i(\mathbf{x})h_i(1, \mathbf{x}) \in I$. Hence $f \in I \cap J$, as required. $\square$

Let $<_{\text{purelex}}$ be the pure lexicographic order on the polynomial ring $K[t, \mathbf{x}] = K[t, x_1, x_2, \ldots, x_n]$ induced by the ordering $t > x_1 > x_2 > \cdots > x_n$. Let I and J be ideal of $K[\mathbf{x}]$. If $\{f_1, f_2, \ldots\}$ is a system of generators of I and $\{h_1, h_2, \ldots\}$ that of J, then a system of generators of the ideal $tI + (1 - t)J$ of $K[t, \mathbf{x}]$ is

$$\{tf_1, tf_2, \ldots, (1 - t)h_1, (1 - t)h_2, \ldots\}.$$

Now Buchberger algorithm gives a Gröbner basis $\mathscr{G}$ of $tI + (1-t)J$ with respect to $<_{\text{purelex}}$. Corollary 1.4.2 then guarantees that

$$\mathscr{G}' = \{\, g \in \mathscr{G} \;;\; t \text{ does not appear in } g \,\}$$

is a Gröbner basis of $(tI + (1-t)J) \cap K[\mathbf{x}]$. Hence Lemma 1.4.3 says that $\mathscr{G}'$ is a Gröbner basis of $I \cap J$ with respect to the pure lexicographic order on $K[\mathbf{x}]$ induced by $x_1 > x_2 > \cdots > x_n$. Thus in particular $\mathscr{G}'$ is a system of generators of $I \cap J$.

Example 1.4.4. Let $n = 2$. Let $I = \langle x^2 \rangle$ and $J = \langle xy \rangle$ be ideals of $K[x, y]$. We compute $I \cap J$. We apply Buchberger algorithm to the system of generators $\{tx^2, (1-t)xy\}$ of the ideal $tI + (1-t)J$ of $K[t, x, y]$. The S-polynomial of tx^2 and $(1-t)xy$ is $x^2 y$. We then apply Buchberger criterion to the system of generators $\{tx^2, (1-t)xy, x^2 y\}$ of $tI + (1-t)J$. The S-polynomial of tx^2 and $x^2 y$ is 0. The S-polynomial of $(1-t)xy$ and $x^2 y$ is $x^2 y$. Thus $\{tx^2, (1-t)xy, x^2 y\}$ is a Gröbner basis of $tI + (1-t)J$. Hence $I \cap J = \langle x^2 y \rangle$.

Example 1.4.5. Let $n = 1$. Let $I = \langle x(x-1) \rangle$ and $J = \langle x^3 \rangle$ be ideals of $K[x]$. In order to compute $I \cap J$, Buchberger algorithm can be applied to the system of generators $\{tx(1-x), (1-t)x^3\}$ of the ideal $tI + (1-t)J$ of $K[t, x]$. A routine computation shows that

$$\{tx(1-x), (1-t)x^3, (t-x^2)x, x^5 - x^3, x^4 - x^3\}$$

is a Gröbner basis of $tI + (1-t)J$. In particular the initial ideal of $tI + (1-t)J$ is $\langle x^4, tx \rangle$. Hence the reduced Gröbner basis of $tI + (1-t)J$ is $\{(t-x^2)x, x^4 - x^3\}$. Thus $I \cap J = \langle x^4 - x^3 \rangle$.

Problem 1.4.6. Let $n = 3$. Compute the intersection $I \cap J$ of the ideals $I = \langle xy^2 \rangle$ and $J = \langle yz, z^3 \rangle$ of $K[x, y, z]$.

1.4.2   Solving Simultaneous Equations

Corollary 1.4.2 provides a powerful technique for solving simultaneous equations. Recall that a zero point of a polynomial $f \in K[\mathbf{x}] = K[x_1, x_2, \ldots, x_n]$ is a point $(a_1, a_2, \ldots, a_n)$ belonging to the space

$$K^n = \{(a_1, a_2, \ldots, a_n) : a_1, a_2, \ldots, a_n \in K\}$$

with

$$f(a_1, a_2, \ldots, a_n) = 0.$$

Let $V(f)$ denote the set of zero points of f. If I is an ideal of $K[\mathbf{x}]$, then $V(I) \subset K^n$ is defined by setting

$$V(I) = \bigcap_{f \in I} V(f).$$

Lemma 1.4.7. *Let $\{f_1, f_2, \ldots, f_s\}$ be a system of generators of an ideal I of $K[\mathbf{x}]$. Then*

$$V(I) = \bigcap_{i=1}^{s} V(f_i).$$

Proof. Clearly $V(I) \subset \bigcap_{i=1}^{s} V(f_i)$. Let $(a_1, a_2, \ldots, a_n) \in K^n$ belong to $\bigcap_{i=1}^{s} V(f_i)$. Let $f \in I$ and $f = \sum_{i=1}^{s} g_i f_i$, where each $g_i \in K[\mathbf{x}]$. Since $f_i(a_1, a_2, \ldots, a_n) = 0$, one has $f(a_1, a_2, \ldots, a_n) = 0$. Thus $(a_1, a_2, \ldots, a_n) \in V(f)$. This is true for all $f \in I$. It then follows that $(a_1, a_2, \ldots, a_n) \in V(I)$. $\square$

Corollary 1.4.8. *Let $\{f_1, f_2, \ldots, f_s\}$ and $\{g_1, g_2, \ldots, g_t\}$ be systems of generators of an ideal I of $K[\mathbf{x}]$. Then the set of solutions of the simultaneous equations*

$$f_1 = f_2 = \cdots = f_s = 0$$

and that of

$$g_1 = g_2 = \cdots = g_t = 0$$

coincide.

Example 1.4.9. We solve the simultaneous equations

$$\begin{cases} x + y = 3 \\ y + z = 5 \\ x + z = 4. \end{cases} \tag{1.8}$$

In the elementary algebra, adding three equations yields $x + y + z = 6$. Hence, say, $x = (x + y + z) - (y + z) = 1$. Thus the solution is $(x, y, z) = (1, 2, 3)$.

Now, we solve the simultaneous equations (1.8) in the language of Gröbner bases. Let $n = 3$ and $K[x_1, x_2, x_3] = K[x, y, z]$. Let

$$f_1 = x + y - 3, \quad f_2 = y + z - 5, \quad f_3 = x + z - 4$$

and

$$I = \langle f_1, f_2, f_3 \rangle.$$

By using Buchberger algorithm, we compute a Gröbner basis of I with respect to $<_{\text{purelex}}$. Since the initial ideals of f_1, f_2, f_3 are x, y, x, respectively, we only check $S(f_1, f_3) = f_1 - f_3 = y - z + 1$. Its remainder with respect to f_1, f_2, f_3 is $-2z + 6$. Let $f_4 = z - 3$ and apply Buchberger criterion to $\{f_1, f_2, f_3, f_4\}$. It then turns out that $\{f_1, f_2, f_3, f_4\}$ is a Gröbner basis of I. Since it is not minimal, after removing, say, f_3, one has a minimal Gröbner basis

$$\mathscr{G} = \{x + y - 3, \, y + z - 5, \, z - 3\}.$$

Now, in order to solve the simultaneous equations (1.8), Corollary 1.4.8 says that we may solve the simultaneous equations

$$\begin{cases} x + y - 3 = 0 \\ y + z - 5 = 0 \\ z - 3 = 0. \end{cases} \tag{1.9}$$

To solve the simultaneous equations (1.9) is easy. The third equation says that $z = 3$. Then the second equation gives $y = 2$. Finally, from the first equation, one has $x = 1$.

This is the advantage of Corollary 1.4.2. In fact, if $I \cap K[z] \neq \langle 0 \rangle$, then $\mathscr{G} \cap K[z]$ is a Gröbner basis of $I \cap K[z]$. In particular, since $\mathscr{G} \cap K[z] \neq \emptyset$, it follows that $\mathscr{G}$ contains a polynomial belonging to $K[z]$. It is $z - 3$. In other words, an equation in z can be obtained. In the next step, if $I \cap K[y, z] \neq \langle 0 \rangle$, then $\mathscr{G} \cap K[y, z]$ is a Gröbner basis of $I \cap K[y, z]$. In particular, since $\mathscr{G} \cap K[y, z] \neq \emptyset$, it follows that $\mathscr{G}$ contains a polynomial belonging to $K[y, z]$. It is $y + z - 5$. In other words, an equation in y and z is obtained.

Problem 1.4.10. Compute a Gröbner basis of the ideal

$$\langle 3y + 4z - 7, \, 3x + 5y - 7z - 10, \, -x - y + 2z + 3 \rangle$$

of $K[x, y, z]$ with respect to $<_{\mathrm{purelex}}$ and solve the simultaneous equations

$$\begin{cases} 3y + 4z = 7 \\ 3x + 5y - 7z = 10 \\ -x - y + 2z = -3. \end{cases}$$

Example 1.4.11. We compute a Gröbner basis of the ideal

$$I = \langle x^2 + y + z - 1, \, x + y^2 + z - 1, \, x + y + z^2 - 1 \rangle$$

of $K[x, y, z]$ with respect to $<_{\mathrm{purelex}}$ and solve the simultaneous equations

$$\begin{cases} x^2 + y + z = 1 \\ x + y^2 + z = 1 \\ x + y + z^2 = 1. \end{cases} \tag{1.10}$$

The reduced Gröbner basis $\mathscr{G}_{\mathrm{red}}(I; <_{\mathrm{purelex}})$ of I with respect to $<_{\mathrm{purelex}}$ is

$$\{x + y + z^2 - 1, \; yz^2 + \frac{z^4 - z^2}{2}, \; y^2 - y - z^2 + z, \; z^6 - 4z^4 + 4z^3 - z^2\}.$$

Instead of solving the simultaneous equations (1.10), we may solve the following simultaneous equations

$$\begin{cases} x + y + z^2 = 1 \\ yz^2 + \frac{z^4 - z^2}{2} = 0 \\ y^2 - y - z^2 + z = 0 \\ z^6 - 4z^4 + 4z^3 - z^2 = 0. \end{cases} \tag{1.11}$$

Corollary 1.4.2 says that $\mathscr{G}_{\mathrm{red}}(I; <_{\mathrm{purelex}}) \cap K[z]$ is the reduced Gröbner basis of $I \cap K[z]$. Thus, if $I \cap K[z] \neq \langle 0 \rangle$, then a polynomial in the variable z belongs to $\mathscr{G}_{\mathrm{red}}(I; <_{\mathrm{purelex}})$. Such a polynomial arises by eliminating x and y in the simultaneous equations (1.10). It is routine to solve the simultaneous equations (1.11). Since

$$z^6 - 4z^4 + 4z^3 - z^2 = z^2(z - 1)^2(z^2 + 2z - 1) = 0,$$

one has

$$z = 0, 1, -1 \pm \sqrt{2}.$$

Problem 1.4.12. Find all the solutions of the simultaneous equations (1.11).

Example 1.4.13. Computing the reduced Gröbner basis of the ideal

$$I = \langle x^2 + y^2 + z^2 - 4, \; x^2 + 2y^2 - 5, \; xz - 1 \rangle$$

of $K[x, y, z]$ with respect to $<_{\mathrm{purelex}}$, we solve the simultaneous equations

$$\begin{cases} x^2 + y^2 + z^2 = 4 \\ x^2 + 2y^2 = 5 \\ xz = 1. \end{cases} \tag{1.12}$$

The reduced Gröbner basis $\mathscr{G}_{\mathrm{red}}(I; <_{\mathrm{purelex}})$ is

$$\{x + 2z^3 - 3z, \; y^2 - z^2 - 1, \; 2z^4 - 3z^2 + 1\}.$$

Thus, in order to solve simultaneous equations (1.12), we may solve the following simultaneous equations

$$\begin{cases} x + 2z^3 - 3z = 0 \\ y^2 - z^2 = 1 \\ 2z^4 - 3z^2 = -1. \end{cases} \tag{1.13}$$

From the third equation, it follows that

$$z^2 = \frac{3 \pm \sqrt{9-8}}{4} = 1, \frac{1}{2}.$$

Hence

$$z = \pm 1, \pm 1/\sqrt{2}.$$

Problem 1.4.14. Find all the solutions of the simultaneous equations (1.13).

Problem 1.4.15. Compute the reduced Gröbner basis of the ideal

$$\langle x^2 + 2y^2 - y - 2z, \; x^2 - 8y^2 + 10z - 1, \; x^2 - 7yz \rangle$$

with respect to $<_{\text{purelex}}$ and find all the solutions of the simultaneous equations

$$\begin{cases} x^2 + 2y^2 - y - 2z = 0 \\ x^2 - 8y^2 + 10z = 1 \\ x^2 - 7yz = 0. \end{cases}$$

1.5   Toric Ideals

Toric ideals are indispensable for the application of Gröbner bases to, e.g., algebraic statistics (Chap. 4) and convex polytopes (Chap. 5). This section is devoted to the study on the foundation of toric ideals.

1.5.1   Configuration Matrices

Let $A = (a_{ij})_{\substack{1 \le i \le d \\ 1 \le j \le n}}$ be a $d \times n$ matrix and

$$\mathbf{a}_j = \begin{bmatrix} a_{1j} \\ a_{2j} \\ \vdots \\ a_{dj} \end{bmatrix}, \qquad 1 \le j \le n$$

the column vectors of A.

Let $\mathbb{Z}$ denote the set of integers and write $\mathbb{Z}^{d \times n}$ for the set of $d \times n$ matrices $A = (a_{ij})_{\substack{1 \le i \le d \\ 1 \le j \le n}}$ with each $a_{ij} \in \mathbb{Z}$.

The *inner product* of vectors $\mathbf{a} = [a_1, a_2, \ldots, a_d]^\mathsf{T}$ and $b = [b_1, b_2, \ldots, b_d]^\mathsf{T}$, where T stands for the transpose, belonging to $\mathbb{R}^d$ is defined to be

$$\mathbf{a} \cdot \mathbf{b} = \sum_{i=1}^{d} a_i b_i.$$

A matrix $A = (a_{ij})_{\substack{1 \le i \le d \\ 1 \le j \le n}} \in \mathbb{Z}^{d \times n}$ is called a *configuration matrix* if there exists $\mathbf{c} \in \mathbb{R}^d$ such that

$$\mathbf{a}_j \cdot \mathbf{c} = 1, \qquad 1 \le j \le n.$$

Example 1.5.1. (a) Given a matrix $A \in \mathbb{Z}^{(d-1) \times n}$, we write $A' \in \mathbb{Z}^{d \times n}$ for the matrix which is obtained by adding the raw vector $[1, 1, \ldots, 1]$ to A as the dth raw. Then A' is a configuration matrix. In fact, if $\mathbf{c} = [0, 0, \cdots, 0, 1]^\top \in \mathbb{R}^d$, then $\mathbf{a}_j \cdot \mathbf{c} = 1$ for each column vector $\mathbf{a}_j$ of A'.

(b) Suppose that the sum of each column of a matrix $A \in \mathbb{Z}^{d \times n}$ is constant, say, $\sum_{i=1}^{d} a_{ij} = k$. If $\mathbf{c} = [1/k, 1/k, \cdots, 1/k]^\top \in \mathbb{R}^d$, then $\mathbf{a}_j \cdot \mathbf{c} = 1$ for each column vector $\mathbf{a}_j$ of A. Thus A is a configuration matrix.

1.5.2   Binomial Ideals

A *binomial* belonging to $K[\mathbf{x}] = K[x_1, x_2, \ldots, x_n]$ is a polynomial of the form $u - v$, where u and v are monomials of the same degree belonging to $K[\mathbf{x}]$. A *binomial ideal* is an ideal of $K[\mathbf{x}]$ generated by binomials. Corollary 1.1.22 says that a binomial ideal possesses a system of generators consisting of a finite number of binomials.

Theorem 1.5.2. *Let I be a binomial ideal of $K[\mathbf{x}]$. Then the reduced Gröbner basis of I with respect to an arbitrary monomial order on $K[\mathbf{x}]$ consists of binomials.*

Proof. In general, if f and g are binomials, then their S-polynomial $S(f, g)$ is again a binomial. It then follows from the argument done in the proof of Theorem 1.2.1 that a remainder of a binomial with respect to several binomials can be chosen as a binomial. Thus, applying Buchberger algorithm to a system of generators of a binomial ideal I consisting of a finite number of binomials, a minimal Gröbner basis $\mathscr{G} = \{g_1, g_2, \ldots, g_s\}$, where each g_i is a binomial, of I can be obtained.

Let $g_i = u_i - v_i$, where u_i and v_i are monomials with $u_i = \mathrm{in}_<(g_i)$. Recall that $\mathscr{G}$ is reduced if v_i cannot be divided by u_j for $i \ne j$. Suppose that $\mathscr{G}$ is not reduced and, say, v_2 is divided by u_1. Let $v_2 = w u_1$, where w is a monomial. We then replace g_2 with $g_2' = g_2 + w g_1 (= u_2 - w v_1)$. Let $g_2' = u_2 - v_2'$. Then $\{g_1, g_2', g_3, \ldots, g_s\}$ is a minimal Gröbner basis of I consisting of binomials with $v_2' (= w v_1) < (w u_1 =) v_2$. Thus, after a finite number of steps, the reduced Gröbner basis of I consisting of binomials arises.   □

1.5.3 Toric Ideals

Given a configuration matrix $A \in \mathbb{Z}^{d \times n}$, write $\mathrm{Ker}_{\mathbb{Z}} A$ for the set of column vectors $\mathbf{b} \in \mathbb{Z}^n$ with $A\mathbf{b} = \mathbf{0}$, where $\mathbf{0}$ is the zero vector of $\mathbb{R}^d$. That is to say,

$$\mathrm{Ker}_{\mathbb{Z}} A = \{ \mathbf{b} \in \mathbb{Z}^n \ : \ A\mathbf{b} = \mathbf{0} \}.$$

Lemma 1.5.3. *If a column vector* $\mathbf{b} = [b_1, b_2, \ldots, b_n]^\top \in \mathbb{Z}^n$ *belongs to* $\mathrm{Ker}_{\mathbb{Z}} A$, *then*

$$b_1 + b_2 + \cdots + b_n = 0.$$

Proof. Since A is a configuration matrix, there is $\mathbf{c} \in \mathbb{R}^d$ with $\mathbf{a}_j \cdot \mathbf{c} = 1$ for all column vectors $\mathbf{a}_j$ of A. Since $A\mathbf{b} = \mathbf{0}$, one has $\sum_{j=1}^n b_j \mathbf{a}_j = \mathbf{0}$. Thus

$$\left(\sum_{j=1}^n b_j \mathbf{a}_j \right) \cdot \mathbf{c} = \sum_{j=1}^n b_j (\mathbf{a}_j \cdot \mathbf{c}) = \sum_{j=1}^n b_j = 0,$$

as desired. □

Now, given a column vector

$$\mathbf{b} = \begin{bmatrix} b_1 \\ b_2 \\ \vdots \\ b_n \end{bmatrix}$$

belonging to $\mathrm{Ker}_{\mathbb{Z}} A$, we introduce the binomial $f_{\mathbf{b}} \in K[\mathbf{x}]$ defined by

$$f_{\mathbf{b}} = \prod_{b_i > 0} x_i^{b_i} - \prod_{b_j < 0} x_j^{-b_j}.$$

Since Lemma 1.5.3 guarantees that the degree of $\prod_{b_i > 0} x_i^{b_i}$ coincides with that of $\prod_{b_j < 0} x_j^{-b_j}$, it follows that $f_{\mathbf{b}}$ is, in fact, a binomial. For example, if $\mathbf{b} = [2, -3, 0, 1]^\top$, then $f_{\mathbf{b}} = x_1^2 x_4 - x_2^3$.

Let $A \in \mathbb{Z}^{d \times n}$ be a configuration matrix. The binomial ideal

$$I_A = \langle \{ f_{\mathbf{b}} \ : \ \mathbf{b} \in \mathrm{Ker}_{\mathbb{Z}} A \} \rangle$$

of $K[\mathbf{x}]$ is called the *toric ideal* of A.

Corollary 1.5.4. *The reduced Gröbner basis* $\mathscr{G}_{\mathrm{red}}(I_A; <)$ *of a toric ideal* I_A *consists of binomials.*

Given a configuration matrix, which is even simple, to compute its toric ideal is, in general, rather difficult.

Example 1.5.5. We compute the toric ideal I_A of the configuration matrix

$$A = \begin{bmatrix} 0 & 1 & 1 & 0 & 1 \\ 0 & 1 & 0 & 1 & 1 \\ 0 & 0 & 1 & 1 & 1 \\ 1 & 1 & 1 & 1 & 1 \end{bmatrix}.$$

Since $\mathbf{b} = [1, -1, -1, -1, 2]^T$ belongs to $\mathrm{Ker}_{\mathbb{Z}} A$, the binomial $f_{\mathbf{b}} = x_1 x_5^2 - x_2 x_3 x_4$ belongs to I_A. In fact, it turns out that $I_A = \langle x_1 x_5^2 - x_2 x_3 x_4 \rangle$. To see why this is true, in the vector space $\mathbb{Q}^5$, we study the simultaneous linear equations

$$\begin{bmatrix} 0 & 1 & 1 & 0 & 1 \\ 0 & 1 & 0 & 1 & 1 \\ 0 & 0 & 1 & 1 & 1 \\ 1 & 1 & 1 & 1 & 1 \end{bmatrix} \begin{bmatrix} z_1 \\ z_2 \\ z_3 \\ z_4 \\ z_5 \end{bmatrix} = \begin{bmatrix} 0 \\ 0 \\ 0 \\ 0 \end{bmatrix}.$$

The space of its solution is

$$\{ r[1, -1, -1, -1, 2]^T : r \in \mathbb{Q} \}.$$

Hence

$$\mathrm{Ker}_{\mathbb{Z}} A = \{ m[1, -1, -1, -1, 2] : m \in \mathbb{Z} \}.$$

Thus

$$\{ (x_1 x_5^2)^m - (x_2 x_3 x_4)^m : m = 1, 2, \ldots \}$$

is a system of generators of I_A. However, since

$$(x_1 x_5^2)^m - (x_2 x_3 x_4)^m = (f_{\mathbf{b}} + x_2 x_3 x_4)^m - (x_2 x_3 x_4)^m \in \langle f_{\mathbf{b}} \rangle,$$

it follows that $I_A = \langle f_{\mathbf{b}} \rangle$.

Problem 1.5.6. Compute the toric ideal of the configuration matrix

$$\begin{bmatrix} 1 & 0 & 0 & 1 \\ 1 & 1 & 0 & 0 \\ 0 & 1 & 1 & 0 \\ 0 & 0 & 1 & 1 \end{bmatrix}.$$

1.5.4   Toric Rings

The definition of toric ideals in the previous subsection has an advantage which does require no special knowledge of commutative algebra. In this subsection, however, we study toric ideals in the language of commutative algebra.

Let $t_1, t_2, \ldots, t_d$ be variables. Let $A = (a_{ij})_{\substack{1 \leq i \leq d \\ 1 \leq j \leq n}} \in \mathbb{Z}^{d \times n}$ be a configuration matrix. To each column vector

$$\mathbf{a}_j = \begin{bmatrix} a_{1j} \\ a_{2j} \\ \vdots \\ a_{dj} \end{bmatrix},$$

we associate the monomial

$$\mathbf{t}^{\mathbf{a}_j} = t_1^{a_{1j}} t_2^{a_{2j}} \cdots t_d^{a_{dj}}$$

with allowing negative powers. If $f = f(x_1, x_2, \ldots, x_n) \in K[\mathbf{x}]$, then we define $\pi(f)$ by setting

$$\pi(f) = f(\mathbf{t}^{\mathbf{a}_1}, \mathbf{t}^{\mathbf{a}_2}, \ldots, \mathbf{t}^{\mathbf{a}_n}).$$

In other words, $\pi(f)$ is a rational function in $t_1, t_2, \ldots, t_d$ obtained by substituting $\mathbf{t}^{\mathbf{a}_i}$ for each x_i in f. Let

$$K[A] = \{ \pi(f) \, : \, f \in K[\mathbf{x}] \}.$$

Then in $K[A]$ the sum and the product can be naturally defined. We say that $K[A]$ is the *toric ring* of A.

Example 1.5.7.   The toric ring of the configuration matrix

$$\begin{bmatrix} 1 & 0 & -1 \\ 0 & 1 & -1 \\ 1 & 1 & 1 \end{bmatrix}$$

is the set of all rational functions obtained by replacing x_1, x_2, x_3 of $f = f(x_1, x_2, x_3)$ belonging to $K[x_1, x_2, x_3]$ with $t_1 t_3, t_2 t_3, t_1^{-1} t_2^{-1} t_3$, respectively.

Lemma 1.5.8.   *If monomials $u, v \in K[\mathbf{x}]$ satisfies $\pi(u) = \pi(v)$, then u and v has the same degree.*

Proof. Let $u = \prod_{j=1}^n x_j^{c_j}$ and $v = \prod_{j=1}^n x_j^{d_j}$. Then $\pi(u) = \prod_{j=1}^n \mathbf{t}^{c_j \mathbf{a}_j}$ and $\pi(v) = \prod_{j=1}^n \mathbf{t}^{d_j \mathbf{a}_j}$. Hence, if $\pi(u) = \pi(v)$, then $\sum_{j=1}^n c_j \mathbf{a}_j = \sum_{j=1}^n d_j \mathbf{a}_j$. Then, by using the technique appearing in the proof of Lemma 1.5.3, one has $\sum_{j=1}^n c_j = \sum_{j=1}^n d_j$.   □

Lemma 1.5.9. *The toric ideal $I_A \subset K[\mathbf{x}]$ of a configuration A coincides with*

$$\langle \{\, f \in K[\mathbf{x}] \,:\, \pi(f) = 0 \,\} \rangle.$$

Proof. **(First Step)** We show that the ideal $\langle \{\, f \in K[\mathbf{x}] \,:\, \pi(f) = 0 \,\} \rangle$ is a binomial ideal. Write $f \in K[\mathbf{x}]$ for $f = \sum_{i=1}^{t} f_i$, where $f_i \in K[\mathbf{x}]$, such that

- if monomials u and v belong in $\mathrm{supp}(f_i)$, then $\pi(u) = \pi(v)$;
- if $u \in \mathrm{supp}(f_i)$ and $v \in \mathrm{supp}(f_j)$ with $i \neq j$, then $\pi(u) \neq \pi(v)$.

Now, we write $f_i = \sum_{j=1}^{s_i} c_{ij} u_{ij}$, where $0 \neq c_{ij} \in K$ and where each u_{ij} is a monomial. Since $\pi(u_{ij}) = \pi(u_{ik})$ for all j and k, it follows that $\pi(f_i) = (\sum_{j=1}^{s_i} c_{ij}) \pi(u_{i1})$. Hence

$$\pi(f) = \sum_{i=1}^{t} \left(\sum_{j=1}^{s_i} c_{ij} \right) \pi(u_{i1}).$$

Recall that $\pi(u_{i1}) \neq \pi(u_{i'1})$ if $i \neq i'$. Thus, if $\pi(f) = 0$, then $\sum_{j=1}^{s_i} c_{ij} = 0$ for all $1 \leq i \leq t$. Since $c_{i1} = -\sum_{j=2}^{s_i} c_{ij}$, it follows that

$$f_i = \sum_{j=2}^{s_i} c_{ij} (u_{ij} - u_{i1}).$$

Since $\pi(u_{ij} - u_{i1}) = \pi(u_{ij}) - \pi(u_{i1}) = 0$, Lemma 1.5.8 says that the degree of u_{ij} and that of u_{i1} coincide. Hence the ideal $\langle \{\, f \in K[\mathbf{x}] \,:\, \pi(f) = 0 \,\} \rangle$ is generated by those binomials $u - v$ with $\pi(u) = \pi(v)$.

(Second Step) If $f = \prod_{j=1}^{n} x_j^{c_j} - \prod_{j=1}^{n} x_j^{d_j}$ is a binomial, then

$$\pi(f) = \prod_{j=1}^{n} \mathbf{t}^{c_j \mathbf{a}_j} - \prod_{j=1}^{n} \mathbf{t}^{d_j \mathbf{a}_j}.$$

Thus

$$\pi(f) = \mathbf{t}^{\sum_{j=1}^{n} c_j \mathbf{a}_j} - \mathbf{t}^{\sum_{j=1}^{n} d_j \mathbf{a}_j}.$$

Hence $\pi(f) = 0$ if and only if

$$\sum_{j=1}^{n} c_j \mathbf{a}_j - \sum_{j=1}^{n} d_j \mathbf{a}_j = 0.$$

In other words, one has $\pi(f) = 0$ if and only if

$$\begin{bmatrix} c_1 \\ c_2 \\ \vdots \\ c_n \end{bmatrix} - \begin{bmatrix} d_1 \\ d_2 \\ \vdots \\ d_n \end{bmatrix} \in \mathrm{Ker}_{\mathbb{Z}} A.$$

Hence a binomial belongs to the binomial ideal $\langle \{\, f \in K[\mathbf{x}] \,:\, \pi(f) = 0 \,\} \rangle$ if and only if it belongs to the toric ideal I_A.

$\square$

In order to define the toric ring, we employ monomials with allowing negative powers. However, we can avoid negative powers.

Lemma 1.5.10. *Given a configuration matrix* $A = [\mathbf{a}_1, \mathbf{a}_2, \dots, \mathbf{a}_n] \in \mathbb{Z}^{d \times n}$, *there is* $\mathbf{a} \in \mathbb{Z}^d$ *such that the matrix*

$$B = [\mathbf{a}_1 + \mathbf{a}, \mathbf{a}_2 + \mathbf{a}, \dots, \mathbf{a}_n + \mathbf{a}]$$

is a configuration matrix with nonnegative entries. Moreover, the toric ideal I_A *of* A *coincides with the toric ideal* I_B *of* B.

Proof. Since A is a configuration matrix, there is $\mathbf{c} \in \mathbb{R}^d$ with $\mathbf{a}_j \cdot \mathbf{c} = 1$ for each $1 \le j \le n$. Choose an adequate vector $\mathbf{a} \in \mathbb{Z}^d$ such that each entry of the matrix B is a nonnegative integer. We may assume that $\mathbf{a} \cdot \mathbf{c} \ne -1$. (In fact, if $\mathbf{a} \cdot \mathbf{c} = -1$, then, for a vector $\mathbf{a}' \in \mathbb{Z}^d$ with nonnegative components, one has $(\mathbf{a} + \mathbf{a}') \cdot \mathbf{c} \ne -1$.) Let

$$r = \frac{1}{1 + \mathbf{a} \cdot \mathbf{c}}.$$

Then

$$(\mathbf{a}_j + \mathbf{a}) \cdot r\mathbf{c} = 1, \quad 1 \le j \le n.$$

Thus the matrix $B \in \mathbb{Z}^{d \times n}$ is a configuration matrix.

Now, if a vector $\mathbf{b} = [b_1, b_2, \dots, b_n]^\top \in \mathbb{Z}^n$ satisfies $b_1 + b_2 + \cdots + b_n = 0$, then

$$B\mathbf{b} - A\mathbf{b} = (b_1 + b_2 + \cdots + b_n)\mathbf{a} = 0.$$

Thus Lemma 1.5.3 says that $I_A = I_B$, as required. $\square$

A technique based on Corollary 1.4.2 to compute the toric ideal of a configuration matrix is now introduced.

Let $A = [\mathbf{a}_1, \mathbf{a}_2, \dots, \mathbf{a}_n] \in \mathbb{Z}^{d \times n}$ be a configuration matrix each of whose entries is nonnegative. Let

$$K[\mathbf{x}, \mathbf{t}] = K[x_1, x_2, \dots, x_n, t_1, t_2, \dots, t_d]$$

be the polynomial ring in $(n + d)$ variables and define the ideal J_A of $K[x, t]$ by

$$J_A = \langle x_1 - \mathbf{t}^{\mathbf{a}_1}, x_2 - \mathbf{t}^{\mathbf{a}_2}, \ldots, x_n - \mathbf{t}^{\mathbf{a}_n} \rangle.$$

Lemma 1.5.11. *The toric ideal $I_A \subset K[\mathbf{x}]$ of A is equal to the intersection of the ideal $J_A \subset K[\mathbf{x}, \mathbf{t}]$ and $K[\mathbf{x}]$, i.e.,*

$$I_A = J_A \cap K[\mathbf{x}].$$

Proof. If a polynomial $f = f(x_1, x_2, \ldots, x_n) \in K[\mathbf{x}]$ belongs to I_A, then $\pi(f) = 0$. Thus $f(\mathbf{t}^{\mathbf{a}_1}, \mathbf{t}^{\mathbf{a}_2}, \ldots, \mathbf{t}^{\mathbf{a}_n}) = 0$. Then, since

$$f((x_1 - \mathbf{t}^{\mathbf{a}_1}) + \mathbf{t}^{\mathbf{a}_1}, (x_2 - \mathbf{t}^{\mathbf{a}_2}) + \mathbf{t}^{\mathbf{a}_2}, \ldots, (x_n - \mathbf{t}^{\mathbf{a}_n}) + \mathbf{t}^{\mathbf{a}_n}) \in J_A,$$

it follows that $f \in J_A \cap K[\mathbf{x}]$. Hence $I_A \subset J_A \cap K[\mathbf{x}]$.

On the other hand, if a polynomial $f = f(x_1, x_2, \ldots, x_n) \in K[\mathbf{x}]$ belongs to J_A, then there exist polynomials $g_1, g_2, \ldots, g_n$ belonging to $K[\mathbf{x}, \mathbf{t}]$ such that

$$f(x_1, x_2, \ldots, x_n) = g_1(\mathbf{x}, \mathbf{t})(x_1 - \mathbf{t}^{\mathbf{a}_1}) + \cdots + g_n(\mathbf{x}, \mathbf{t})(x_n - \mathbf{t}^{\mathbf{a}_n}).$$

Then $\pi(f) = f(\mathbf{t}^{\mathbf{a}_1}, \mathbf{t}^{\mathbf{a}_2}, \ldots, \mathbf{t}^{\mathbf{a}_n}) = 0$. Thus $f \in I_A$. Hence $J_A \cap K[\mathbf{x}] \subset I_A$. $\square$

Let $<_{\text{purelex}}$ denote the pure lexicographic order on $K[\mathbf{x}, \mathbf{t}]$ induced by the ordering

$$t_1 > t_2 > \cdots > t_d > x_1 > x_2 > \cdots > x_n$$

and compute the reduced Gröbner basis $\mathscr{G}_{\text{red}}(J_A; <_{\text{purelex}})$ of J_A with respect to $<_{\text{purelex}}$. Corollary 1.4.2 then guarantees that $\mathscr{G}_{\text{red}}(J_A; <_{\text{purelex}}) \cap K[\mathbf{x}]$ is the reduced Gröbner basis of I_A with respect to $<_{\text{purelex}}$. In particular $\mathscr{G}_{\text{red}}(J_A; <_{\text{purelex}}) \cap K[\mathbf{x}]$ is a system of generators of I_A.

Example 1.5.12. By using the above technique, we compute the toric ideal of the configuration matrix

$$A = \begin{bmatrix} 0 & 1 & 2 & 3 \\ 1 & 1 & 1 & 1 \end{bmatrix}.$$

Let $K[\mathbf{x}, \mathbf{t}]$ be the polynomial ring in the variables $x_1, x_2, x_3, x_4, t_1, t_2$ and $<_{\text{purelex}}$ the pure lexicographic order on $K[\mathbf{x}, \mathbf{t}]$ induced by the ordering

$$t_1 > t_2 > x_1 > x_2 > x_3 > x_4.$$

The reduced Gröbner basis $\mathscr{G}_{\text{red}}(J_A; <_{\text{purelex}})$ of the ideal

$$J_A = \langle x_1 - t_2, x_2 - t_1 t_2, x_3 - t_1^2 t_2, x_4 - t_1^3 t_2 \rangle$$

of $K[\mathbf{x}, t]$ with respect to $<_{purelex}$ is

$$\{x_2x_4 - x_3^2, x_1x_4 - x_2x_3, x_1x_3 - x_2^2, t_2 - x_1, t_1x_3 - x_4, t_1x_2 - x_3, t_1x_1 - x_2\}.$$

Thus it follows from Corollary 1.4.2 that

$$\{x_2x_4 - x_3^2, x_1x_4 - x_2x_3, x_1x_3 - x_2^2\}$$

is the reduced Gröbner basis of the toric ideal I_A with respect to $<_{purelex}$.

Problem 1.5.13. Imitating the discussion of Example 1.5.12, compute a system of generators of the toric ideal of the configuration matrix

$$A = \begin{bmatrix} 4 & 0 & 0 & 2 & 1 & 1 \\ 0 & 4 & 0 & 1 & 2 & 1 \\ 0 & 0 & 4 & 1 & 1 & 2 \end{bmatrix}.$$

1.6 Residue Class Rings and Hilbert Functions

After the foundation of residue class rings of the polynomial ring is studied, Macaulay's theorem on initial ideals follows and Hilbert functions are introduced.

1.6.1 Residue Classes and Residue Class Rings

Let $K[\mathbf{x}] = K[x_1, x_2, \ldots, x_n]$ be the polynomial ring in n variables over K and I an ideal of $K[\mathbf{x}]$ with $I \neq K[\mathbf{x}]$. Given a polynomial $f \in K[\mathbf{x}]$, we write $[f]$ for

$$f + I = \{f + g : g \in I\} \ (\subset K[\mathbf{x}]).$$

Since $0 \in I$, it follows that $f = f + 0 \in f + I$. Thus $f \in [f]$. We call $[f]$ a *residue class* of $K[\mathbf{x}]$ modulo I. In particular $I = [0]$ is a residue class. In addition, one has $[f] = I$ if and only if $f \in I$.

Lemma 1.6.1. *Let f and g be polynomials belonging to $K[\mathbf{x}]$. If $[f] \cap [g] \neq \emptyset$, then $[f] = [g]$.*

Proof. Let $h \in [f] \cap [g]$. Then there exist $f_1, g_1 \in I$ with $h = f + f_1 = g + g_1$. Since $f - g = g_1 - f_1 \in I$, one has $(g_1 - f_1) + I = I$. Hence

$$f + I = g + (g_1 - f_1) + I = g + ((g_1 - f_1) + I) = g + I,$$

as desired.   □

Lemma 1.6.2. *Let f and g be polynomials belonging to $K[\mathbf{x}]$. Then the following conditions are equivalent:*

(i) $[f] = [g]$;
(ii) $g \in [f]$;
(iii) $f - g \in I$.

Proof. Suppose (i). Since $g \in [g]$, one has $g \in [f]$. Thus (ii) follows. Suppose (ii). There is $h \in I$ with $g = f + h$. Hence $f - g = -h \in I$. Thus (iii) follows. Finally, suppose (iii). If $f - g \in I$, then

$$f + I = (g + (f - g)) + I = g + ((f - g) + I) = g + I.$$

Thus (i) follows.   □

Let $[f]$ and $[g]$ be residue classes. Then we define $[f] + [g]$ by setting

$$[f] + [g] = \{ p + q : p \in [f], q \in [g] \} \tag{1.14}$$

and $[f][g]$ by setting

$$[f][g] = \{ pq : p \in [f], q \in [g] \}. \tag{1.15}$$

Lemma 1.6.3. *If $[f]$ and $[g]$ are residue classes, then*

$$[f] + [g] = [f + g], \qquad [f][g] \subset [fg].$$

Proof. Since $I + I = I$, it follows that

$$[f] + [g] = (f + g) + (I + I) = (f + g) + I = [f + g].$$

Since $fI = \{ fh : h \in I \} \subset I$ and $I^2 = \{ pq : p, q \in I \} \subset I$, it follows that

$$[f][g] = fg + fI + gI + I^2 \subset fg + I = [fg],$$

as desired.   □

Corollary 1.6.4. *Let $[f]$, $[f_0]$, $[g]$ and $[g_0]$ be residue classes. If $[f] = [f_0]$ and $[g] = [g_0]$, then*

$$[f + g] = [f_0 + g_0], \qquad [fg] = [f_0 g_0].$$

Proof. The right-hand side of the definition (1.14) is the set of the sum of each element belonging to $[f]$ and each element belonging to $[g]$. The right-hand side of the definition (1.15) is the set of the product of each element belonging to $[f]$ and each element belonging to $[g]$. Hence if $[f] = [f_0]$, $[g] = [g_0]$, then it is clear that $[f] + [g] = [f_0] + [g_0]$ and $[f][g] = [f_0][g_0]$.

Lemma 1.6.3 says that $[f] + [g] = [f + g]$ and $[f_0] + [g_0] = [f_0 + g_0]$. Thus one has $[f + g] = [f_0 + g_0]$.

Lemma 1.6.3 says that $[f][g] \subset [fg]$ and $[f_0][g_0] \subset [f_0g_0]$. Thus in particular $[fg] \cap [f_0g_0] \neq \emptyset$. Hence, by using Lemma 1.6.1, one has $[fg] = [f_0g_0]$. □

Problem 1.6.5. Deduce Corollary 1.6.4 from Lemma 1.6.2.

The *residue class decomposition* of $K[\mathbf{x}]$ modulo I is a set of residue classes $\{T_\lambda : \lambda \in \Lambda\}$ of $K[\mathbf{x}]$ modulo I such that

- if $\lambda \neq \mu$, then $T_\lambda \cap T_\mu = \emptyset$;
- $K[\mathbf{x}] = \bigcup_{\lambda \in \Lambda} T_\lambda$.

Lemma 1.6.1 says that the residue class decomposition of $K[\mathbf{x}]$ modulo I exists uniquely. It is denoted by $K[\mathbf{x}]/I$. If we choose an element f_λ of each residue class T_λ, then Lemma 1.6.2 says that $T_\lambda = [f_\lambda]$. Hence

$$K[\mathbf{x}]/I = \{[f_\lambda] : \lambda \in \Lambda\}.$$

We then call $\{f_\lambda : \lambda \in \Lambda\}$ a *complete system of representatives* of the residue class decomposition of $K[\mathbf{x}]$ modulo I.

Lemma 1.6.3 guarantees that if $T_\lambda, T_\mu \in K[\mathbf{x}]/I$, then there exist a unique $\nu \in \Lambda$ and a unique $\xi \in \Lambda$ such that

$$T_\lambda + T_\mu = T_\nu, \quad T_\lambda T_\mu \subset T_\xi.$$

Thus we can define the sum $+_{K[\mathbf{x}]/I}$ and the product $\cdot_{K[\mathbf{x}]/I}$ in $K[\mathbf{x}]/I$ by setting

$$T_\lambda +_{K[\mathbf{x}]/I} T_\mu = T_\nu, \quad T_\lambda \cdot_{K[\mathbf{x}]/I} T_\mu = T_\xi.$$

Example 1.6.6. Let $I = \langle x^2 + x + 1 \rangle$ be an ideal of the polynomial ring $K[x]$ in one variable. Let $f(x) \in K[x]$ and divide $f(x)$ by $x^2 + x + 1$. Let $q(x)$ be its quotient and $ax + b$ its remainder, where $a, b \in K$. Thus $f(x) = (x^2 + x + 1)q(x) + (ax + b)$. Since $(x^2 + x + 1)q(x) \in I$, it follows that

$$f(x) + I = (ax + b) + ((x^2 + x + 1)q(x) + I) = (ax + b) + I$$

Thus $[f(x)] = [ax + b]$. Since the degree of each nonzero polynomial belonging to I is bigger than or equal to 2, for residue classes $[ax + b]$ and $[a'x + b']$, where $a, b, a', b' \in K$, one has $(ax + b) - (a'x + b') \in I$ if and only if $ax + b = a'x + b'$. It then follows from Lemma 1.6.2 that if $ax + b \neq a'x + b'$, then $[ax + b] \neq [a'x + b']$. Let $T_{a,b} = [ax + b]$. Then $\{T_{a,b} : a, b \in K\}$ is the residue class decomposition of $K[x]$ modulo I. In other words, $\{ax + b : a, b \in K\}$ is a complete system of representatives of the residue class decomposition of $K[x]$ modulo I. In order to compute the sum $+_{K[x]/I}$ and the product $\cdot_{K[x]/I}$, we must find residue classes $T_{c,d}$ and $T_{c',d'}$ such that

$$T_{a,b} + T_{a',b'} = T_{c,d}, \quad T_{a,b} \cdot T_{a',b'} \subset T_{c',d'}.$$

Lemma 1.6.3 says that

$$T_{c,d} = [(a + a')x + (b + b')] = T_{a+a',b+b'}, \quad T_{c',d'} = [(ax + b)(a'x + b')].$$

Thus the remainder obtained by dividing

$$(ax + b)(a'x + b') = aa'x^2 + (ab' + a'b)x + bb'$$

by $x^2 + x + 1$ is $c'x + d'$. For example, in order to compute $T_{1,-1} \cdot_{K[x]/I} T_{2,3}$, since the remainder obtained by dividing $(x - 1)(2x + 3) = 2x^2 + x - 3$ by $x^2 + x + 1$ is $-x - 5$, one has $T_{1,-1} \cdot_{K[x]/I} T_{2,3} = T_{-1,-5}$.

Example 1.6.7. Apart from the discussion of the residue class decomposition of the polynomial ring, we briefly explain the residue class decomposition of the set of integers. Fix an integer $n > 0$ and write T_i, where $i = 0, 1, \ldots, n - 1$, for the set of those integers q such that i is the remainder obtained by dividing q by n. Thus

$$T_i = \{i + na : a \in \mathbb{Z}\}.$$

If $i \neq j$, then $T_i \cap T_j = \emptyset$. Moreover,

$$\mathbb{Z} = \bigcup_{i=0}^{n-1} T_i$$

Now, we define $T_i + T_j$ and $T_i \cdot T_j$ by setting

$$T_i + T_j = \{a + b : a \in T_i, b \in T_j\},$$

$$T_i \cdot T_j = \{ab : a \in T_i, b \in T_j\}.$$

It then follows that there exist a unique k and a unique ℓ such that

$$T_i + T_j = T_k, \quad T_i \cdot T_j \subset T_\ell.$$

In fact, k is the remainder obtained by dividing $i + j$ by n and ℓ is the remainder obtained by dividing ij by n. Thus in the finite set $\mathbb{Z}_n = \{T_0, T_1, \ldots, T_{n-1}\}$, we can define the sum $+_{\mathbb{Z}_n}$ and the product $\cdot_{\mathbb{Z}_n}$ by setting

$$T_i +_{\mathbb{Z}_n} T_j = T_k, \quad T_i \cdot_{\mathbb{Z}_n} T_j = T_\ell.$$

The reader can understand the similarities between the residue class decomposition of $K[x]$, which is explained in Example 1.6.6, and that of $\mathbb{Z}$.

Let I be an ideal of $K[\mathbf{x}]$ and $\{g_1, g_2, \ldots, g_s\}$ the reduced Gröbner basis of I with respect to a monomial order on $K[\mathbf{x}]$. Let $f \in K[\mathbf{x}]$ and f' a unique remainder of f with respect to $g_1, g_2, \ldots, g_s$. Since $f - f' \in I$, it follows that the residue class $[f]$ coincides with $[f']$. If g' is the remainder of $g \in K[\mathbf{x}]$ and if $f' \neq g'$, then $f' - g' \notin I$. Thus $[f'] \neq [g']$. In fact, since $f - f' \in I$ and $g - g' \in I$, it follows that $(f - g) - (f' - g') \in I$. Thus if $f' - g' \in I$, then $f - g \in I$. Corollary 1.2.5 then says that the remainder of $f - g$ is 0. Since $f' - g'$ is the remainder of $f - g$, one has $f' - g' = 0$.

In other words, if $\{f_\lambda : \lambda \in \Lambda\}$ is the set of those $f' \in K[\mathbf{x}]$ such that f' can be obtained as the remainder of a polynomial belonging to $K[\mathbf{x}]$ with respect to $g_1, g_2, \ldots, g_s$, then $\{f_\lambda : \lambda \in \Lambda\}$ is a complete system of representatives of the residue class decomposition of $K[\mathbf{x}]/I$.

In $K[\mathbf{x}]/I = \{[f_\lambda] : \lambda \in \Lambda\}$, the computation of the sum $+_{K[\mathbf{x}]/I}$ and the product $\cdot_{K[\mathbf{x}]/I}$ can be achieved with imitating the technique explained in Example 1.6.6. In general, if $\lambda, \nu \in \Lambda$, then $f_\lambda + f_\mu$ can be obtained as a remainder, but $f_\lambda f_\mu$ is not necessarily obtained as a remainder. Thus

$$[f_\lambda] +_{K[\mathbf{x}]/I} [f_\mu] = [f_\lambda + f_\mu],$$

$$[f_\lambda] \cdot_{K[\mathbf{x}]/I} [f_\mu] = [(f_\lambda f_\mu)'],$$

where $(f_\lambda f_\mu)'$ is the remainder of $f_\lambda f_\mu$.

Example 1.6.8. Let $n = 7$. The residue class decomposition of $K[x_1, x_2, \ldots, x_7]$ by the ideal $I = \langle f, g \rangle$, where $f = x_1 x_4 - x_2 x_3$ and $g = x_4 x_7 - x_5 x_6$, is studied. Recall from Example 1.3.5 together with Example 1.3.6 that $\{f, g\}$ is the reduced Gröbner basis of I with respect to the reverse lexicographic order $<_{rev}$ and that $\{f, g, h\}$, where $h = x_1 x_5 x_6 - x_2 x_3 x_7$, is the reduced Gröbner basis of I with respect to the lexicographic order $<_{lex}$. With using each of $<_{rev}$ and $<_{lex}$, each of the monomials $x_1 x_7$ and x_4^2 can be obtained as a remainder. Now, $(x_1 x_7)(x_4^2)$ can be a remainder with using $<_{rev}$, but cannot be a remainder with using $<_{lex}$. In fact, since

$$(x_1 x_7)(x_4^2) = (f - x_2 x_3)(g - x_5 x_6) = (g - x_5 x_6)f - x_2 x_3 g + x_2 x_3 x_5 x_6,$$

the remainder of $(x_1 x_7)(x_4^2)$ is $x_2 x_3 x_5 x_6$. Hence if we use $<_{rev}$, then

$$[x_1 x_7] \cdot_{K[\mathbf{x}]/I} [x_4^2] = [x_1 x_4^2 x_7].$$

If we use $<_{lex}$, then

$$[x_1 x_7] \cdot_{K[\mathbf{x}]/I} [x_4^2] = [x_2 x_3 x_5 x_6].$$

However, the product $\cdot_{K[\mathbf{x}]/I}$ depends only on I and is independent from a monomial order. In fact, since

$$x_1 x_4^2 x_7 - x_2 x_3 x_5 x_6 = x_1 x_4 g + x_5 x_6 f \in I,$$

one has $[x_1 x_4^2 x_7] = [x_2 x_3 x_5 x_6]$. Consequently, if we employ the set of remainders with respect to (the polynomials belonging to) the reduced Gröbner basis as a complete system of representatives of the residue class decomposition, then it occurs that different monomial orders yield different complete systems of representatives.

The sum $+_{K[\mathbf{x}]/I}$ and the product $\cdot_{K[\mathbf{x}]/I}$ in $K[\mathbf{x}]/I$ are analogues of those in $K[\mathbf{x}]$. With emphasizing the algebraic structure with the sum and the product, we call $K[\mathbf{x}]/I$ the *residue class ring* of $K[\mathbf{x}]$ modulo I.

1.6.2 Macaulay's Theorem

In this subsection, with assuming that the reader is familiar with linear algebra, we introduce *Macaulay's theorem* on initial ideals.

Fix a monomial order $<$ on $K[\mathbf{x}] = K[x_1, x_2, \ldots, x_n]$. Let I be an ideal of $K[\mathbf{x}]$ with $I \neq K[\mathbf{x}]$ and $\mathrm{in}_<(I)$ the initial ideal of I with respect to $<$. A monomial $w \in K[\mathbf{x}]$ is called *standard* with respect to $\mathrm{in}_<(I)$ if $w \notin \mathrm{in}_<(I)$.

Let $\{T_\lambda : \lambda \in \Lambda\}$ be the residue class decomposition of $K[\mathbf{x}]$ modulo I. If $a \in K$ and T_λ is a residue class, then the scalar product $a\, T_\lambda$ is defined by $[a] \cdot_{K[\mathbf{x}]/I} T_\lambda$. Then the residue class ring $K[\mathbf{x}]/I$ is a vector space over K with the sum $+_{K[\mathbf{x}]/I}$ and the scalar product.

Theorem 1.6.9 (Macaulay). *Fix a monomial order $<$ on $K[\mathbf{x}] = K[x_1, x_2, \ldots, x_n]$. Let I be an ideal of $K[\mathbf{x}]$ with $I \neq K[\mathbf{x}]$ and $\mathrm{in}_<(I)$ the initial ideal of I with respect to $<$. Then*

$$\mathscr{B} = \{[w] \in K[\mathbf{x}]/I : w \text{ is a monomial with } w \notin \mathrm{in}_<(I)\}$$

is a K-basis of the vector space $K[\mathbf{x}]/I$ over K.

Proof. Let $\mathscr{G}_{\mathrm{red}}(I; <) = \{g_1, g_2, \ldots, g_s\}$ denote the reduced Göbner basis of I with respect to $<$ and $\{f_\lambda : \lambda \in \Lambda\}$ the set of those polynomials $f' \in K[\mathbf{x}]$ such that f' can be obtained as the remainder of a polynomial f with respect to $g_1, g_2, \ldots, g_s$. Let $K[\mathbf{x}]/I = \{[f_\lambda] : \lambda \in \Lambda\}$. Since f_λ is a remainder, it follows that each monomial belonging to the support of $f_\lambda \neq 0$ is standard. Let $f_\lambda = a_1 w_1 + a_2 w_2 + \cdots + a_t w_t$, where each w_i is a standard monomial and each $a_i \in K$. Then in $K[\mathbf{x}]/I$ one has

$$[f_\lambda] = a_1[w_1] + a_2[w_2] + \cdots + a_f[w_t].$$

Hence $K[\mathbf{x}]/I$ is spanned by $\mathscr{B}$ over K.

Now, we show that $\mathscr{B}$ is linearly independent. Let $u_1, u_2, \ldots, u_\ell$ be standard monomials with $u_1 < u_2 < \cdots < u_\ell$ and $b_1, b_2, \ldots, b_\ell$ nonzero elements belonging to K. Suppose that

$$b_1[u_1] + b_2[u_2] + \cdots + b_\ell[u_\ell] = [0].$$

Then $[\sum_{i=1}^{\ell} b_i u_i] = [0]$. Thus $\sum_{i=1}^{\ell} b_i u_i \in I$. Hence the initial monomial u_ℓ of $\sum_{i=1}^{\ell} b_i u_i$ belongs to $\mathrm{in}_<(I)$. This contradict the fact that u_ℓ is standards.   □

In general, an ideal I of $K[\mathbf{x}]$ is called a 0-*dimensional ideal* if the dimension of the vector space $K[\mathbf{x}]/I$ over K is finite. By virtue of Theorem 1.6.9 an ideal $I \subset K[\mathbf{x}]$ is 0-dimensional if and only if the number of standard monomials with respect to $\mathrm{in}_<(I)$ is finite.

1.6.3   Hilbert Functions

We now turn to the discussion of the residue class ring of $K[\mathbf{x}] = K[x_1, x_2, \ldots, x_n]$ modulo an ideal generated by homogeneous polynomials. A *homogeneous ideal* is an ideal which is generated by homogeneous polynomials.

Lemma 1.6.10. *Let* $f \in K[\mathbf{x}]$ *be a polynomial of degree d and*

$$f = f^{(0)} + f^{(1)} + \cdots + f^{(d)},$$

where each $f^{(j)} \in K[\mathbf{x}]$ *is a homogeneous polynomial of degree j. Suppose that f belongs to a homogeneous ideal I of* $K[\mathbf{x}]$. *Then each* $f^{(j)}$ *belongs to I.*

Proof. Suppose that I is generated by homogeneous polynomials $g_1, g_2, \ldots, g_s$. Let d_i be the degree of g_i. If $f \in I$, then there exist polynomials $h_1, h_2, \ldots, h_s$ with

$$f = h_1 g_1 + h_2 g_2 + \cdots + h_s g_s. \tag{1.16}$$

Each h_k can be expressed as

$$h_k = h_k^{(0)} + h_k^{(1)} + \cdots + h_k^{(e_k)},$$

where e_k is the degree of h_k and $h_k^{(j)}$ is a homogeneous polynomial of degree j. Comparing the homogeneous polynomials of degree j appearing in the both sides of the equality (1.16), it follows that

$$f^{(j)} = \sum_{i=1}^{s} h_i^{(j-d_i)} g_i.$$

Hence $f^{(j)} \in I$, as desired.   □

Lemma 1.6.11. *The reduced Gröbner basis of a homogeneous ideals of* $K[\mathbf{x}]$ *with respect to a monomial order consists of homogeneous polynomials.*

Proof. If f and g are homogeneous polynomials, then its S-polynomial $S(f, g)$ is again a homogeneous polynomial. On the other hand, the division algorithm says that a remainder f' of a homogeneous polynomial f with respect to homogeneous polynomials $g_1, g_2, \ldots, g_s$ is again a homogeneous polynomial. Then Buchberger algorithm guarantees that a homogeneous ideal possesses a Gröbner basis consisting of homogeneous polynomials. Now, the required result follows immediately from the proof of Theorem 1.2.6.   □

Let I be a homogeneous ideal of $K[\mathbf{x}] = K[x_1, x_2, \ldots, x_n]$ and $\{g_1, g_2, \ldots, g_s\}$ the reduced Gröbner basis of I with respect to a monomial order. Let, as before, $\{ f_\lambda : \lambda \in \Lambda \}$ denote the set of those polynomials $f' \in K[\mathbf{x}]$ such that f' can be obtained as the remainder of a polynomial $f \in K[\mathbf{x}]$ with respect to $g_1, g_2, \ldots, g_s$. We recall that $\{ f_\lambda : \lambda \in \Lambda \}$ is a complete system of representatives of the residue class decomposition of $K[\mathbf{x}]$ modulo I.

Given an integer $j \geq 0$, we write $(K[\mathbf{x}]/I)_j$ for the subspace

$$\{ [f_\lambda] \in K[\mathbf{x}]/I : f_\lambda \text{ is a homogeneous polynomial of degree } j \}$$

of the vector space $K[\mathbf{x}]/I$.

Lemma 1.6.12. *If $i \neq j$, then*

$$(K[\mathbf{x}]/I)_i \cap (K[\mathbf{x}]/I)_j = \{[0]\}.$$

Proof. Let $\lambda, \nu \in \Lambda$. Let f_λ be a homogeneous polynomial of degree i and f_ν that of degree j. If $[f_\lambda] = [f_\nu]$, then $f_\lambda - f_\nu \in I$. Since I is a homogeneous ideal, it follows from Lemma 1.6.10 that $f_\lambda \in I$ and $f_\nu \in I$. Hence $[f_\lambda] = [f_\nu] = [0]$, as desired.   □

Lemma 1.6.13. *The residue class ring $K[\mathbf{x}]/I$ is the direct sum*

$$K[\mathbf{x}]/I = \bigoplus_{j=0}^{\infty} (K[\mathbf{x}]/I)_j$$

of the subspaces $(K[\mathbf{x}]/I)_j$ with $j = 0, 1, 2, \ldots$ as a vector space over K. In other words, each element $[f_\lambda]$ of $K[\mathbf{x}]/I$ can be expressed uniquely of the form

$$[f_\lambda] = \sum_{j=0}^{\infty} [f_{\nu_j}],$$

where $[f_{\nu_j}] \in (K[\mathbf{x}]/I)_j$ is $[0]$ except for a finite number of j's.

Proof. Write f_λ of the form

$$f_\lambda = f_\lambda^{(j_1)} + f_\lambda^{(j_2)} + \cdots + f_\lambda^{(j_d)}, \qquad j_1 < j_2 < \cdots < j_d,$$

where $f_\lambda^{(j_i)}$ is a homogeneous polynomial of degree j_i. Since f_λ is a remainder with respect to $g_1, g_2, \ldots, g_s$, it follows that each $f_\lambda^{(j_i)}$ is again a remainder. Thus $[f_\lambda^{(j_i)}] \in (K[\mathbf{x}]/I)_{j_i}$. Hence in $K[\mathbf{x}]/I$ one has

$$[f_\lambda] = [f_\lambda^{(j_1)}] + [f_\lambda^{(j_2)}] + \cdots + [f_\lambda^{(j_d)}].$$

Suppose that $[f_\lambda]$ possesses another expression

$$[f_\lambda] = [f_{\lambda_{k_1}}] + [f_{\lambda_{k_2}}] + \cdots + [f_{\lambda_{k_e}}], \qquad k_1 < k_2 < \cdots < k_e,$$

where $[f_{\lambda_{k_\ell}}] \in (K[\mathbf{x}]/I)_{k_\ell}$. Then in $K[\mathbf{x}]/I$ one has

$$[f_\lambda^{(j_1)} + f_\lambda^{(j_2)} + \cdots + f_\lambda^{(j_d)}] = [f_{\lambda_{k_1}} + f_{\lambda_{k_2}} + \cdots + f_{\lambda_{k_e}}].$$

It then follows from Lemma 1.6.2 that

$$(f_\lambda^{(j_1)} + f_\lambda^{(j_2)} + \cdots + f_\lambda^{(j_d)}) - (f_{\lambda_{k_1}} + f_{\lambda_{k_2}} + \cdots + f_{\lambda_{k_e}})$$

belongs I. Since I is a homogeneous ideal, by using Lemma 1.6.10, if $j_i = k_\ell$, then $f_\lambda^{(j_i)} - f_{\lambda_{k_\ell}} \in I$. Then in $K[\mathbf{x}]/I$ one has $[f_\lambda^{(j_i)}] = [f_{\lambda_{k_\ell}}]$. In particular if $j_i \notin \{k_1, k_2, \ldots, k_e\}$, then $f_\lambda^{(j_i)} \in I$. Hence in $K[\mathbf{x}]/I$ one has $[f_\lambda^{(j_i)}] = [0]$. □

Lemma 1.6.14. *As a K-basis of $(K[\mathbf{x}]/I)_j$, we can choose*

$$\mathscr{B}_j = \{[w] \in K[\mathbf{x}]/I \ : \ w \text{ is a monomial of degree } j \text{ with } w \notin \text{in}_<(I)\}.$$

In particular, the vector space $(K[\mathbf{x}]/I)_j$ is of finite dimension.

Proof. In the proof of Theorem 1.6.9, assuming f_λ is a homogeneous polynomial of degree j, it follows that the subspace $(K[\mathbf{x}]/I)_j$ is spanned by $\mathscr{B}_j$. Since $\mathscr{B}$ is linearly independent, its subset $\mathscr{B}_j$ is again linearly independent. □

Let $H(K[\mathbf{x}]/I; j)$ denote the dimension of the vector space $(K[\mathbf{x}]/I)_j$ over K. We say that the function $H(K[\mathbf{x}]/I; j)$, $j = 0, 1, 2, \ldots$, is the *Hilbert function* of $K[\mathbf{x}]/I$. In particular, since $I \neq K[\mathbf{x}]$, one has $H(K[\mathbf{x}]/I; 0) = 1$.

Theorem 1.6.15. *Let $<$ be a monomial order on $K[\mathbf{x}]$ and $I \neq K[\mathbf{x}]$ a homogeneous ideal of $K[\mathbf{x}]$. Then the Hilbert function $H(K[\mathbf{x}]/I; j)$ of $K[\mathbf{x}]/I$ is equal to the number of standard monomials of degree j with respect to $\text{in}_<(I)$.*

Proof. The result follows immediately from Lemma 1.6.14. □

Example 1.6.16. Let $n = 2$. We compute the Hilbert function of the residue class ring $K[x, y]/I$ of $K[x, y]$ modulo the ideal $I = \langle x^2 + y^2, x^3 \rangle$. Since $\text{in}_{<_{\text{lex}}}(I) = \langle x^2, xy^2, y^4 \rangle$, the standard monomials with respect to $\text{in}_{<_{\text{lex}}}(I)$ are

$$1, x, y, xy, y^2, y^3.$$

Thus the sequence $\{H(K[x, y]/I; i)\}_{i=0}^{\infty}$ is

$$1, 2, 2, 1, 0, 0, 0, \ldots$$

and I is a 0-dimensional ideal.

Corollary 1.6.17. *Let $<$ be a monomial order on $K[\mathbf{x}]$ and $I \neq K[\mathbf{x}]$ a homogeneous ideal of $K[\mathbf{x}]$. Then the Hilbert function of $K[\mathbf{x}]/I$ coincides with that of $K[\mathbf{x}]/\mathrm{in}_<(I)$.*

Proof. In general, since $\mathrm{in}_<(I) = \mathrm{in}_<(\mathrm{in}_<(I))$, it follows that a monomial $w \in K[\mathbf{x}]$ is standard with respect to $\mathrm{in}_<(I)$ if and only if w is standard with respect to $\mathrm{in}_<(\mathrm{in}_<(I))$. Theorem 1.6.15 says that $H(K[\mathbf{x}]/I; j)$ coincides with the number of standard monomials of degree j with respect to $\mathrm{in}_<(I)$. Hence $H(K[\mathbf{x}]/I; j) = H(K[\mathbf{x}]/\mathrm{in}_<(I); j)$ for all j, as desired.   □

Example 1.6.18. Again, we study the ideal $I = \langle f, g \rangle$, where $f = x_1 x_4 - x_2 x_3$ and $g = x_4 x_7 - x_5 x_6$, of $K[x_1, x_2, \ldots, x_7]$ and compute the Hilbert function of $K[\mathbf{x}]/I$. By virtue of Theorem 1.6.15, the computation of the Hilbert function results in enumerating the standard monomials. A monomial order can be chosen arbitrarily. We work with a reverse lexicographic order. Its advantage is that the initial ideal is relatively simple. Now, since $\mathrm{in}_{<_{\mathrm{rev}}}(I) = \langle x_2 x_3, x_5 x_6 \rangle$, a monomial of degree j

$$w = x_1^{a_1} x_2^{a_2} \cdots x_7^{a_7}, \qquad a_1 + a_2 + \cdots + a_7 = j$$

is standard with respect to $\mathrm{in}_{<_{\mathrm{rev}}}(I)$ if and only if w can be divided by neither $x_2 x_3$ nor $x_5 x_6$. Thus the standard monomials of degree j are

$$x_1^{a_1} x_3^{1+a_3} x_4^{a_4} x_6^{1+a_6} x_7^{a_7}, \qquad a_1 + a_3 + a_4 + a_6 + a_7 = j - 2,$$

$$x_1^{a_1} x_3^{1+a_3} x_4^{a_4} x_5^{1+a_5} x_7^{a_7}, \qquad a_1 + a_3 + a_4 + a_5 + a_7 = j - 2,$$

$$x_1^{a_1} x_2^{1+a_2} x_4^{a_4} x_6^{1+a_6} x_7^{a_7}, \qquad a_1 + a_2 + a_4 + a_6 + a_7 = j - 2,$$

$$x_1^{a_1} x_2^{1+a_2} x_4^{a_4} x_5^{1+a_5} x_7^{a_7}, \qquad a_1 + a_2 + a_4 + a_5 + a_7 = j - 2,$$

$$x_1^{a_1} x_4^{a_4} x_6^{1+a_6} x_7^{a_7}, \qquad a_1 + a_4 + a_6 + a_7 = j - 1,$$

$$x_1^{a_1} x_4^{a_4} x_5^{1+a_5} x_7^{a_7}, \qquad a_1 + a_4 + a_5 + a_7 = j - 1,$$

$$x_1^{a_1} x_3^{1+a_3} x_4^{a_4} x_7^{a_7}, \qquad a_1 + a_3 + a_4 + a_7 = j - 1,$$

$$x_1^{a_1} x_2^{1+a_2} x_4^{a_4} x_7^{a_7}, \qquad a_1 + a_2 + a_4 + a_7 = j - 1,$$

$$x_1^{a_1} x_4^{a_4} x_7^{a_7}, \qquad a_1 + a_4 + a_7 = j.$$

A routine work shows that the number of these monomials is

$$4\binom{j+2}{j-2} + 4\binom{j+2}{j-1} + \binom{j+2}{j}. \tag{1.17}$$

Hence the Hilbert function $H(K[\mathbf{x}]/I; n)$ coincides with the formula (1.17).

Problem 1.6.19. By enumerating the standard monomials with respect to the initial ideal in$_{<_{\mathrm{lex}}}(I)$, compute the Hilbert function $H(K[\mathbf{x}]/I;n)$ of Example 1.6.18.

1.7   Historical Background

In this section, we will briefly survey the historical background of Gröbner bases and provide some fundamental references.

In the middle of 1960s, the Gröbner basis was introduced independently by Hironaka [14] and Buchberger [5]. Hironaka came up with the idea of standard bases in the process of solving the outstanding problem in algebraic geometry, the resolution of singularities of algebraic varieties. On the other hand, Buchberger created Gröbner bases for his dissertation for which the research topic had been given by his advisor Wolfgang Gröbner. The topic seems to be the problem of creating a technique which enables us to find explicitly a set of monomials which is a K-basis of the residue class ring of the polynomial modulo a 0-dimensional ideal. Hironaka's standard bases work in the local ring, while Buchberger's Gröbner bases work in the polynomial ring. There is no essential difference between the idea of standard bases and that of Gröbner bases. However, it must be an obvious fact that Buchberger criterion and Buchberger algorithm opened the fascinating research area called computer algebra in the modern algebra.

Looking further back in the history, the idea of Gröbner bases with respect to the reverse lexicographic order already appeared in the famous paper [15] by Macaulay in 1927. Macaulay studied the problem of finding a characterization of the Hilbert functions of residue class rings of the polynomial ring modulo homogeneous ideals. Macaulay discovered the fundamental fact that the Hilbert function of the residue class ring of a homogeneous ideal coincides with that of its initial ideal (Corollary 1.6.17). It then follows that his original problem resulted in the enumerative problem on counting monomials, which we discussed in Example 1.6.18. Macaulay's work stimulated the study on enumerative combinatorics on monomials and promoted the birth of the active area called commutative algebra and combinatorics, which originated in Stanley's work [24]. See [13, 25].

After the pioneer works of Hironaka and Buchberger, Gröbner bases had not been in the limelight for about 20 years. However, a turning point occurred in the middle of 1980s, when David Bayer and Michael Stillman developed the computer software, named Macaulay, which has a great influence on commutative algebra and algebraic geometry. Since Gröbner bases were indispensable for developing the software Macaulay, Gröbner bases became common knowledge for researchers on commutative algebra and algebraic geometry. This can be the first breakthrough in the progress of Gröbner bases.

The entry of Gröbner bases into the world of applied mathematics was achieved by Conti and Traverso [6], where an algebraic algorithm to solve problems of integer programming by means of Gröbner bases is proposed. Conti–Travelso algorithm is studied in, e.g., [8]. We briefly discuss Conti–Travelso algorithm by using a simple example. Let $\mathbb{Z}_{\geq 0}$ denote the set of nonnegative integers. We consider the problem on integer programming of the standard type

$$\min\{\mathbf{c} \cdot \mathbf{z} \ : \ A\mathbf{z} = \mathbf{b}, \ \mathbf{z} \in \mathbb{Z}_{\geq 0}^5\},$$

where $\mathbf{z} = [z_1, z_2, z_3, z_4, z_5]^\top$, $\mathbf{c} = [0, 1, 0, 1, 1]$, $\mathbf{b} = [25, 34, 18]^\top$, and where

$$A = \begin{bmatrix} 1 & 1 & 1 & 1 & 1 \\ 0 & 1 & 2 & 1 & 0 \\ 0 & 0 & 1 & 2 & 1 \end{bmatrix}.$$

In other words, the problem is to find a vector $\mathbf{z} \in \mathbb{Z}_{\geq 0}^5$ with $A\mathbf{z} = \mathbf{b}$ which minimizes $\mathbf{c} \cdot \mathbf{z}$. The vector $\mathbf{c}$ is called a cost vector and a vector $\mathbf{z}$ satisfying the condition $A\mathbf{z} = \mathbf{b}$ is called a feasible solution. A feasible solution $\mathbf{z}$ which minimizes $\mathbf{c} \cdot \mathbf{z}$ is called an optimal solution. For the sake of convenience we assume that the matrix A is a configuration matrix and each component of $\mathbf{c}$ is nonnegative. The toric ideal of A is

$$I_A = \langle x_2 x_5^2 - x_1^2 x_4, x_2 x_4 - x_3 x_5, x_2^2 x_5 - x_1^2 x_3 \rangle.$$

By using the cost vector $\mathbf{c}$ we introduce the monomial order $<_\mathbf{c}$, which is discussed in Example 1.1.13. The reduced Gröbner basis of I_A with respect to $<_\mathbf{c}$ is

$$\mathscr{G} = \{x_3 x_5^3 - x_1^2 x_4^2, x_2 x_5^2 - x_1^2 x_4, x_2 x_4 - x_3 x_5, x_2^2 x_5 - x_1^2 x_3\}.$$

We then choose an arbitrary feasible solution, say, $[1, 10, 10, 4, 0]^\top$ and associate it with the monomial $w = x_1 x_2^{10} x_3^{10} x_4^4$. The remainder of w with respect to the binomials belonging to $\mathscr{G}$ is $x_1^7 x_3^{17} x_5$. It then turns out that the vector $[7, 0, 17, 0, 1]^\top$ associated with $x_1^7 x_3^{17} x_5$ is one of the optimal solutions.

Toric ideals were spread rapidly under the great influence of Sturmfels [26]. As one of the effective techniques to compute the dimension of the solution space of a hypergeometric equation, Gel'fand et al. [11] introduced the notion of regular triangulations. In [26] it is shown that the Stanley–Reisner ring of a regular triangulation is just the radical ideal of the initial ideal of a toric ideal. As a result, toric ideals turned out to be a bridge between the theory of monomial ideals [12] and the theory of triangulations of convex polytopes, and the algebraic theory of triangulations of convex polytopes was developed quickly. This can be the second breakthrough in the progress of Gröbner bases. In this algebraic frame of convex polytopes, one of the most important results is the discovery of an example of a convex polytope for which neither any triangulation the number of whose simplices is smallest nor any triangulation the number of whose simplices is biggest is regular [18]. We refer the reader to Chap. 5 for the topics on toric ideals, Gröbner bases and convex polytopes.

In commutative algebra, in connection with so-called Koszul algebras, a toric ideal generated by quadratic binomials is important. If a toric ring is Koszul, then its toric ideal is generated by quadratic binomials. In addition, if the toric ideal of a toric ring possesses a Gröbner basis consisting of quadratic binomials, then the toric ring is Koszul. With considering this background, to find a non-Koszul toric ring whose toric ideal is generated by quadratic binomials as well as to find a Koszul toric ring whose toric ideal possesses no Gröbner basis consisting of quadratic binomials had been a pending problem. Both examples were constructed independently by Ohsugi and Hibi [19] and Roos and Sturmfels [22].

The study of Gröbner bases in the ring of differential operators started gradually in the 1980s. A dramatic breakthrough was done by Oaku [16, 17], where, based on Buchberger algorithm, new and effective algorithms on D-modules were created. Since regular triangulations originated in the study of hypergeometric equations, the algebraic development of toric ideals naturally had the great influence on the study of hypergeometric equations. The textbook [23] published in 2000, in which the authors focus distinguished results on algorithms on D-modules together with computational and algebraic results on toric ideals, established a new method on the study of hypergeometric equations and succeeded in making a new trend for further research on hypergeometric equations.

An epoch-making application of Gröbner bases to statistics is originated in the paper [9] by Diaconis and Sturmfels. In the examination of a statistical model, when we achieve Markov chain Monte Carlo method, to find a Markov basis is required. In [9], it is shown that a Markov basis corresponds to a system of generators of the toric ideal arising from a statistical model. Hence the technique explained in Example 1.5.12 enables us to find a Markov basis of a statistical model. Later the new and exciting research area called algebraic statistics was born and it has been developing rapidly. This can be the third breakthrough in the progress of Gröbner bases. Algebraic statistics supplies commutative algebra with new problems [2, 21]. Conversely, toric ideals studied in commutative algebra supply algebraic statistics with new statistical models [3, 20]. The interrelationship between algebraic statistics and commutative algebra is worth studying hardly. We refer the reader to Chap. 4 for the detailed study on algebraic statistics.

As standard textbooks on Gröbner bases, we recommend [1, 4, 7]. In the frame of commutative algebra, the generic initial ideal is indispensable, which is extensively studied in [10, 12].

References

1. W.W. Adams, P. Loustaunau, *An Introduction to Gröbner Bases* (American Mathematical Society, Providence, 1994)
2. S. Aoki, T. Hibi, H. Ohsugi, A. Takemura, Gröbner bases of nested configurations. J. Algebra **320**, 2583–2593 (2008)
3. S. Aoki, T. Hibi, H. Ohsugi, A. Takemura, Markov basis and Gröbner basis of Segre–Veronese configuration for testing independence in group-wise selections. Ann. Inst. Stat. Math. **62**, 299–321 (2010)

4. T. Becker, W. Weispfenning, *Gröbner Bases* (Springer, New York, 1993)
5. B. Buchberger, An algorithm for finding the basis elements of the residue class ring of a zero dimensional polynomial ideal, Ph.D. Dissertation (University of Innsbruck, 1965)
6. P. Conti, C. Traverso, Buchberger algorithm and integer progamming, in *Applied Algebra, Algebraic Algorithms and Error Correcting Codes*, ed. by H. Mattson, T. Mora, T. Rao. Lecture Notes in Computer Science, vol. 539 (Springer, Berlin, 1991), pp. 130–139
7. D. Cox, J. Little, D. O'Shea, *Ideals, Varieties, and Algorithms* (Springer, Berlin, 1992)
8. D. Cox, J. Little, D. O'Shea, *Using Algebraic Geometry*. Graduate Texts in Mathematics, vol. 185 (Springer, Berlin, 1998)
9. P. Diaconis, B. Sturmfels, Algebraic algorithms for sampling from conditional distributions. Ann. Stat. **26**, 363–397 (1998)
10. D. Eisenbud, *Commutative Algebra with a View Towards Algebraic Geometry*. Graduate Texts in Mathematics, vol. 150 (Springer, Berlin, 1995)
11. I.M. Gel'fand, M.M. Kapranov, A.V. Zelevinsky, *Discriminants, Resultants, and Multidimensional Determinants* (Birkhäuser, Boston, 1994)
12. J. Herzog, T. Hibi, *Monomial Ideals*. Graduate Texts in Mathematics, vol. 260 (Springer, Berlin, 2010)
13. T. Hibi, *Algebraic Combinatorics on Convex Polytopes* (Carslaw Publications, Glebe, 1992)
14. H. Hironaka, Resolution of singularities of an algebraic variety over a field of characteristic zero. Ann. Math. **79**, 109–203, 205–326 (1964)
15. F.S. Macaulay, Some properties of enumeration in the theory of modular systems. Proc. Lond. Math. Soc. **26**, 531–555 (1927)
16. T. Oaku, An algorithm of computing b-functions. Duke Math. J. **87**, 115–132 (1997)
17. T. Oaku, Algorithms for *b*-functions, restrictions, and algebraic local cohomology groups of *D*-modules. Adv. Appl. Math. **19**, 61–105 (1997)
18. H. Ohsugi, T. Hibi, A normal $(0, 1)$-polytope none of whose regular triangulations is unimodular. Discrete Comput. Geom. **21**, 201–204 (1999)
19. H. Ohsugi, T. Hibi, Toric ideals generated by quadratic binomials. J. Algebra **218**, 509–527 (1999)
20. H. Ohsugi, T. Hibi, Compressed polytopes, initial ideals and complete multipartite graphs. Ill. J. Math. **44**, 391–406 (2000)
21. H. Ohsugi, T. Hibi, Toric ideals arising from contingency tables, in *Commutative Algebra and Combinatorics*. Ramanujan Mathematical Society Lecture Notes Series, vol. 4 (Ramanujan Mathematical Society, Mysore, 2007), pp. 91–115
22. J.-E. Roos, B. Sturmfels, A toric ring with irrational Poincaré–Betti series. C. R. Acad. Sci. Paris Ser. I Math. **326**, 141–146 (1998)
23. M. Saito, B. Sturmfels, N. Takayama, *Gröbner Deformations of Hypergeometric Differential Equations* (Springer, Berlin, 2000)
24. R.P. Stanley, The upper bound conjecture and Cohen–Macaulay rings. Stud. Appl. Math. **54**, 135–142 (1975)
25. R.P. Stanley, *Combinatorics and Commutative Algebra*, 2nd edn. (Birkhäuser, Boston, 1996)
26. B. Sturmfels, *Gröbner Bases and Convex Polytopes* (American Mathematical Society, Providence, 1996)

Chapter 2
Warm-Up Drills and Tips for Mathematical Software

Tatsuyoshi Hamada

Abstract In Chap. 1, we studied the basic theory of Gröbner bases. Our goal is to use mathematical software to further our research. In this chapter, we will begin with warm-up drills in order to learn the basic ideas necessary for using mathematical software. We will use MathLibre, a mathematical software environment. It is a collection of mathematical software and free documents which form a kind of Live Linux system. The Linux operating system is compatible with UNIX, and many mathematical research systems have been developed on a UNIX system. It is thus important to know the command line interface, Emacs editor, and the fundamental ideas of the UNIX environment. If you are already familiar with this environment, you can skip this chapter; otherwise, please try and enjoy the world of MathLibre.

2.1 Using MathLibre

We now introduce *MathLibre*, a mathematical software execution environment. MathLibre is a kind of *Linux* operating system that boots from a DVD. Linux is a *UNIX*-compatible computer environment used for education and research. Math-Libre includes over 100 mathematical software systems that have been developed all over the world. Once MathLibre has been booted, the mathematical software is immediately available for anyone to try.

T. Hamada (✉)
Department of Applied Mathematics, Fukuoka University, Nanakuma, Fukuoka 814-0180, Japan
e-mail: hamada@fukuoka-u.ac.jp

T. Hibi (ed.), *Gröbner Bases: Statistics and Software Systems*,
DOI 10.1007/978-4-431-54574-3_2, © Springer Japan 2013

2.1.1 How to Get MathLibre

In the following, we will assume the computer environment is a PC with a Microsoft Windows operating system. MathLibre [4] is an open-source project that can be downloaded from the Internet.[1] The DVD-R has about four gigabytes of data, and it will take over 30 min to download it. It contains the ISO image file, which, when burned to a DVD, will produce a DVD-bootable version of MathLibre. On your computer, please double-click the icon for the DVD drive; we will find some folders and files in the Explorer window. If there is only one file on the DVD, then it is necessary to reconfigure the DVD burning software and rewrite the ISO image.

2.1.2 How to Boot and Shut Down MathLibre

After successfully making and rebooting the MathLibre DVD, we can find the penguin icon. Press the Enter key and, after a display of the boot-sequence messages, the desktop environment of MathLibre will be displayed. In some cases, it will reboot Windows; if this is the case, the *BIOS* settings need to be reconfigured. When a PC is rebooted, the message "BIOS Setup" is briefly displayed. After pressing the correct function key, usually <F2> or <F8>, we can find the "Boot" menu in the BIOS configuration. By changing the order of booting, we can boot from a DVD or a USB storage device. If you are not familiar with computers, consult a specialist for help.

If you are using an Apple MacOS X computer with an Intel CPU, you can boot from the DVD by using the "C" key.

2.1.3 Various Mathematical Software Packages

MathLibre includes many mathematical software packages, such as *CoCoA*, *GeoGebra*, *gfan*, *KSEG*, *Macaulay2*, *Maxima*, *Octave*, *Polymake*, *R*, *Risa/Asir*, *Singular*, *surfex*, *Sage*, and others. The applications introduced in this book show only a subset of the mathematical abilities of MathLibre.

Select the Math menu from the start menu at the bottom left-hand side of the screen, as you would do for Windows Start. Alternatively, double click the "Math Software" icon; there is a collection of start-up icons for mathematical software and a "MathLibre Start" button, which leads to an HTML file that contains short introductions and links to the developers of the various software packages.

---

[1] http://www.mathlibre.org/.

2.2 File Manager

In this section, we introduce some of the basic management of files in MathLibre. If you click on the *PCManFM* icon ▪ in the bottom panel (the *lxpanel*), a list of files and folders will be displayed. This is the file manager, *PCManFM*. It is a tool for moving, removing, and duplicating files and folders. It can also be used to start applications (Fig. 2.1).

In the MathLibre Live system, /home/user is our *home directory* and user is our username. The home directory is a special folder in which we can freely make files and folders. Linux's *directory* and Windows' *folder* have almost the same meaning. In a Linux environment such as MathLibre, the files and folders are in a *tree structure*, such as the one shown in Fig. 2.2. In the figure, the folders are represented by circles. The root of the tree structure is the *root directory*, and it is represented by a slash mark /. Unlike Windows, there is no concept of C

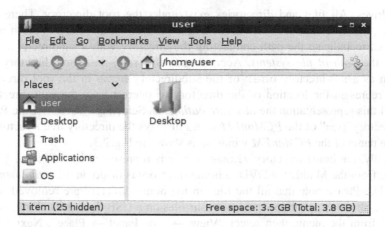

Fig. 2.1 PCManFM

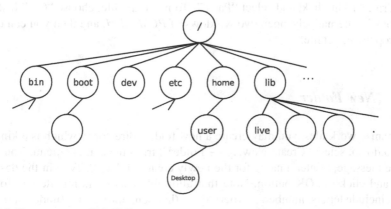

Fig. 2.2 Tree structure of MathLibre

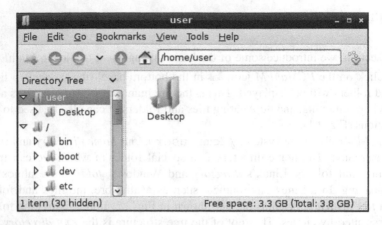

Fig. 2.3 Tree structure of MathLibre with *PCManFM*

or D drives. All files and directories exist under the root directory. There are, however, multiple *subdirectories* branching from the root directory. A slash mark / is used to indicate the path connecting the directory with a subdirectory. This creates the *path of file systems*. According to this rule, the home directory can be seen as a subdirectory user of the subdirectory home in the root directory /. We represent the location of the directory of interest from the root directory. We call this representation the *absolute pathname*. Selecting "View → Side Panel → Directory Tree" of the *PCManFM* menu displays the directory tree structure in the side panel of the *PCManFM* window, as shown in Fig. 2.3.

Usually, the home directory /home/user is represented by a tilde ~. When booting from the MathLibre DVD, a home directory is made on the *main memory* of the PC. Please note that all the files in the home directory are removed when the PC is shut down, but you can save your files on a USB flash drive. First, select "Copy" from the menu, then select "View → Side Panel → Place". Next, select the location where you wish to save the file (either the hard disk drive or a USB flash drive), right click, and select "Paste". To move the file, choose "Cut" instead of "Copy". Alternatively, open two windows of *PCManFM*, and then you can drag and drop the target file.

2.2.1 New Folder

To organize work files, you may create a new folder (directory), which is a kind of file. To do so, select "Create New... → Folder:" from the context menu. You will see the message, "Enter a name for the newly created folder:". Type in the desired name and click the OK button. Note that allowable characters for file and folder names include letters, numbers, . (period), - (hyphenation), and _ (underline).

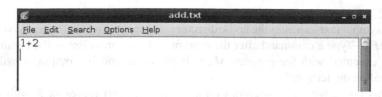

Fig. 2.4 *Leafpad*

2.2.2   New Text File

When using mathematical software, it is often convenient to save commands or scripts in a *text file*. A text file is structured as a sequence of lines of electronic text.[2] To create a new text file, in the context menu of *PCManFM*, select "Create New... → Blank File". The message "Enter a name for the newly created file:" will be displayed. Type in the name for the new file and click the OK button; a blank text file with 0 bytes of data will be created.

Double click on the newly created file to launch MathLibre's default *text editor*, *Leafpad*. *Leafpad* is similar to *Notepad* of Windows; it is very simple and can be used easily by anyone. Besides *Leafpad*, MathLibre contains other editors. *Emacs* and *vim* are popular text editors for UNIX users. *Emacs* is not only an editor, but also an environment for developing and computing. A lot of mathematical software uses the *Emacs* interface, and in Sect. 2.6, we will introduce the *Emacs* with MathLibre.

Exercise 2.2.1. Make a text file add.txt. As shown in Fig. 2.4, type the characters $1 + 2$ and then click on the "Enter" key. Save this file to the home directory.

2.3   Terminal

Using MathLibre, we can take advantage of mathematical software that has been developed around the world. In order to explore special research software, it is helpful to know how to operate Linux, one of the UNIX-derived systems. It uses a traditional method in which a command is input to the *terminal* by using a keyboard. We will use the *GNOME Terminal* application. To begin, we launch the terminal by single-clicking the third icon. ■ When we open the terminal, we will see the following:

```
user@debian:~$
```

---

[2]Text files have an important role in UNIX. If you want to learn more about it, we recommend [2, 3].

We call the phrase "`user@debian:~$`" the *prompt*. In the prompt, "~" is a special symbol that indicates the home directory /home/user that we created with *PCManFM*. Type a command after the prompt and then press Enter; the command will be executed with the program *shell*. If the command has output, it will be displayed on the terminal.

In this book, we will sometimes omit the prompt and represent it with only $ or #. These prompts represent the general user mode and the system administration mode, respectively. Since the system administration mode is unnecessary for using mathematical software, please work in the general user mode.

2.3.1 Files and Directories

To obtain information about files and directories, use the command ls, which is an abbreviation of the word "list". In the following, we can see the directory Desktop and text file add.txt which were made in an earlier exercise.

```
user@debian:~$ ls
Desktop add.txt
```

On the actual screen, directories are represented in blue and files are shown in white. If we want to know more about a file, for example, its date or size, enter the command with the "long" option: "-l".

```
user@debian:~$ ls -l
drwxr-xr-x 2 user user 4096 2011-01-29 11:14 Desktop
-rw-r--r-- 2 user user    4 2011-01-29 11:14 add.txt
```

To facilitate our work, we will make the working directory in our home directory. The command for making a directory is mkdir, which is an abbreviation of "make directory".

```
user@debian:~$ mkdir tutorial
```

After the command mkdir, type in the new directory name; in this case, it is tutorial. The new name is the *argument* of the command mkdir. It is a distinctive feature of the terminal that there is no message when we have made a new directory. To verify the existence of a directory, use the command ls or *PCManFM*.

```
user@debian:~$ ls
Desktop add.txt tutorial
```

At any time, the directory that we are working with is called the *current working directory*. The current working directory is represented with one period " . ". When we launch the terminal, the current working directory is the home directory. The command for changing the current working directory is cd, which is an abbreviation

Fig. 2.5 Parent and child
directories

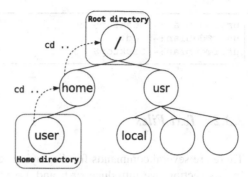

of "change directory". After typing in the command, type in the argument (the name
of the directory to which you want to move), and then execute the command.

```
user@debian:~$ cd tutorial
user@debian:~/tutorial$
```

It should be noted that the current working directory ~/tutorial is included in
the prompt. ~/tutorial is a subdirectory of our home directory, and it is also
called a *child directory*. If a directory contains a child directory, we call it a *parent
directory*, and we represent it with a double period "..". Therefore, this command
will move from a child directory to its parent directory:

```
user@debian:~/tutorial$ cd ..
user@debian:~$
```

For example, if the starting point, the current working directory, is our home
directory /home/user, then using the absolute pathname will move us to the
directory /usr/local:

```
user@debian:~$ cd /usr/local
user@debian:/usr/local$
```

On the other hand, using the symbols for parent directory ".." and path /, we can
represent it like this (Fig. 2.5):

```
user@debian:~$ cd ../../usr/local
user@debian:/usr/local$
```

The location relative to our current working directory is called the
../../usr/local *relative pathname*. Whether to use the relative pathname or
the absolute pathname depends on the situation.

Using the command cd with no argument is a way to return quickly to our home
directory.

```
user@debian:/usr/local$ cd
user@debian:~$
```

And with "cd - ", we can quickly go back to the previous directory.

```
user@debian:~/Desktop$ cd
user@debian:~$ cd -
user@debian:~/Desktop$
```

2.3.2 Text Files

There are several commands for displaying text files: cat, more, less, and lv. In this section, we introduce cat and less. If we set the text file name as the argument of cat, the contents of the text file are displayed on the terminal. For example, we will display the file which we made in Exercise 2.2.1.

```
user@debian:~$ cat add.txt
1+2
```

There is no problem when displaying a small file like add.txt, but if we want to display a large file on the terminal, we will not be able to read it because the contents will be streaming past. It is better to use the command less for displaying large files. As an example, here is the file /etc/passwd.

```
user@debian:~$ less /etc/passwd
root:x:0:0:root:/root:/bin/bash
daemon:x:1:1:daemon:/usr/sbin:/bin/sh
bin:x:2:2:bin:/bin:/bin/sh
...
```

The command less is useful, and it allows us to choose our position in the contents by using the space key and the cursor. We can search forward with / and backward with ?. To end the display, enter q. We can read manuals with the man command; this is important when learning Linux commands.

　　For example, to read the manual description of the command less, enter the following:

```
user@debian:~$ man less
```

2.3.3 Input and Output

The command bc is a standard calculation tool in Linux. By using bc and the text file "add.txt", we can calculate the sum:

```
user@debian:~$ bc < add.txt
3
```

In this situation, the less-than sign < is called a *redirection*. In UNIX systems, a redirection allows us to choose a text file as input. Using the greater-than sign as a redirection > allows us to output the result to a text file.

```
user@debian:~$ bc < add.txt > answer.txt
user@debian:~$ ls
Desktop answer.txt add.txt
user@debian:~$ cat answer.txt
3
```

With the output redirection >, we can use cat to make a new file. Note that ^D means to simultaneously press the D key and the Ctrl key, Ctrl+D . It indicates the end of the input data.

```
user@debian:~$ cat > multi.txt
3*4
^D
user@debian:~$ ls
Desktop answer.txt add.txt multi.txt
user@debian:~$ cat multi.txt
3*4
user@debian:~$ bc < multi.txt
12
```

2.3.4   Character Codes

In the previous subsection, when we input characters to text files, it is recognized as bit data and interpreted using *The American Standard Code for Information Interchange (ASCII)*, a character encoding system that was originally based on the English alphabet. Using a standard tool of UNIX, od, we can examine how the characters are treated in a computer. The command od is an abbreviation of "octal dump". Octal is the base-8 number system, which uses the digits 0 to 7; a dump is an exact copy of the data as it is held in the computer. In the following table, we list the binary, octal, decimal, and hexadecimal equivalents for the numbers one to sixteen. The binary number system uses only two symbols, $\{0, 1\}$. The octal system uses $\{0, 1, 2, 3, 4, 5, 6, 7\}$, decimal uses $\{0, 1, 2, 3, 4, 5, 6, 7, 8, 9\}$, and hexadecimal uses $\{0, 1, 2, 3, 4, 5, 6, 7, 8, 9, a, b, c, d, e, f\}$.

Using the command od with a qualifier, we can also display the contents of the file in hexadecimal. For example, suppose we wish to build a text file ABC.txt that contains only the three characters "ABC" and then observe it with a hexadecimal dump. We can do so, as follows. First, use cat to create a text file, using the Enter key and Ctrl+D to finish editing.

Binary	Octal	Decimal	Hexadecimal
0	0	0	0
1	1	1	1
10	2	2	2
11	3	3	3
100	4	4	4
101	5	5	5
110	6	6	6
111	7	7	7
1000	10	8	8
1001	11	9	9
1010	12	10	a
1011	13	11	b
1100	14	12	c
1101	15	13	d
1110	16	14	e
1111	17	15	f
10000	20	16	10
…	…	…	…

```
user@debian:~$ cat > ABC.txt
ABC
^D
```

We can then display the hexadecimal dump by using the command od.

```
user@debian:~$ od -Ad -tx1 ABC.txt
000000 41 42 43 0a
000004
```

What is listed here is the information recorded as ASCII code in the storage device. Note that the file name "ABC.txt" is not included in the file itself. The six digits on the left side provide the decimal address of one byte of the stored data. Memory addresses from 000000 up to 000003 are allocated for the data in the file ABC.txt. Since the data is represented in hexadecimal notation, that corresponds to 'A'=41, 'B'=42, 'C'=43. A single two-digit number written in hexadecimal equals 1 byte (= 8 bits). The last two-digit number, 0a, is the control code for LF, which stands for LineFeed. The letters of the English alphabet, numerals, and symbols are each represented by 1 byte in binary. For example, 'A' is 01000001; it is represented by eight digits in binary. If we use hexadecimal, 'A' is 41. Binary notation is cumbersome for humans to read, so in many cases, data is represented in hexadecimal. One obvious convenience of using hexadecimal is that four digits of binary correspond to a single digit of hexadecimal.

Exercise 2.3.1. Use the command man ascii on the terminal to find the ASCII code.

2.4   How to Write Mathematical Documents

When using MathLibre to create a document (such as a research paper) that contains mathematical formulas, we use the *TEX* system. TEX is a typesetting system designed and mostly written by Donald E. Knuth, who is a famous mathematician and computer scientist.

In this book, we will introduce LATEX, which is widely used in mathematical communities. It was originally written by Leslie Lamport, and the current version is LATEX 2_ε. We can make a PDF file from TEX source code by using the command pdflatex. Because it is suitable for the construction of mathematical documents and for structural descriptions, and is not limited to mathematics, it is widely used for writing papers and books. For example, this book was written using LATEX 2_ε.

In order to create a PDF file from LATEX 2_ε source code, we need the following typesetting process:

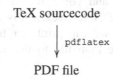

In this section, we will use the terminal.

2.4.1   Writing a TEX Document

We need a tool with which to create and edit TEX source code. In this case, we will use the *Emacs* text editor, and will launch it from the terminal. If you are not familiar with *Emacs*, you can use another text editor, such as *Leafpad*.

First, we will create a file of TEX source code with the name "sample.tex". Adding & at the end of command allows us to execute *Emacs* as a background job. That is, we can continue to use the terminal while *Emacs* is executing.

```
user@debian:~$ emacs sample.tex &
```

We next write four lines in sample.tex,

```
\documentclass{article}
\begin{document}
Hello LaTeX
\end{document}
```

LATEX commands always start with backslash \. After editing the file sample.tex, we save the file and exit *Emacs*. For more information on how to use *Emacs*, see Sect. 2.6.

2.4.2 Making a PDF File

A PDF file can be created by typesetting the source file sample.tex with the command pdflatex.

```
user@debian:~$ pdflatex sample.tex
This is pdfTeX, Version 3.1415926-2.4-1.40.13
(TeX Live 2012/Debian) restricted \write18 enabled.
...
```

The pdflatex command creates three new files: sample.aux, sample.log, and sample.pdf.

To view the PDF file, sample.pdf, we use the command evince.

```
user@debian:~$ evince sample.pdf &
```

Only the sentence "Hello LaTeX" will be displayed because it is the only thing between \begin{document} and \end{document}. We embed the special TEX command in our source code. Note that LATEX is a kind of markup language.

After setting up the printer, we can print our file by using the File menu. Alternatively, you can move the PDF file to the another environment, such as Windows or MacOS.

2.4.3 Brief Introduction to TEX Source Code

The first line of TEX source code, \documentclass{}, is the command that reads the settings for the file. For writing a LATEX document, we start with the following command.

Listing 2.1 TEX source code

```
\documentclass[options]{class}
```

In the sample, the document class is *article* and the default paper size is *letterpaper*. We can change the options; the paper can be other sizes, such as *legalpaper* or *a4paper*, and there are other document classes, such as the standard one, *report*, as well as *book*, *letter*, and *slides*. To create a presentation file, use *beamer*, and for a poster session, use *a0poster*. With the appropriate LATEX options, we can create many styles of documents. The lines starting and ending points of the document are \begin{document} and \end{document}, respectively, and the lines in between are called the document environment.

Listing 2.2 TEX source code

```
\documentclass[options]{class}
\begin{document}
............................
.....creating documents.....
............................
\end{document}
```

This is the basic style of LaTeX 2_ε source code.

With \begin{ } and \end{ }, we can create various typesettings.

Listing 2.3 TeX source code

```
\begin{environment name}
...
\end{environment name}
```

There are environment names for many common usages, such as "equation" for mathematical formulas and "itemize" for list structures. For more detail, please refer, for example, to Wikibooks.[3]

2.4.4 Math Formulas

LaTeX 2_ε is good for describing mathematical formulas. When we create a mathematical object in TeX source code, we set the beginning and end points in our document. For example, we want to create the following:

Listing 2.4 PDF view

$$x^2 + y^2 + z^2 - 4 = 0 \qquad\qquad (2.1)$$

We can produce this by using the equation environment, as follows.

Listing 2.5 TeX source code

```
\begin{equation}
  x^2+y^2+z^2-4=0
\end{equation}
```

The equation number is automatically added, unless we add an asterisk *, as shown.

Listing 2.6 TeX source code

```
\begin{equation*}
  x^2+y^2+z^2-4=0
\end{equation*}
```

Or we can abbreviate it:

Listing 2.7 TeX source code

```
\[
  x^2+y^2+z^2-4=0
\]
```

We can use $$...$$ instead of \[...\], but this is not recommended because it can make it difficult to see the start and end points. Similarly, to write mathematical characters and equations in our documents, we can use \(and \) or $...$, but for the same reason as above, it is better to use \(and \).

---

[3]http://en.wikibooks.org/wiki/LaTeX/.

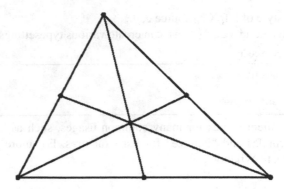

Fig. 2.6 Centroid of a triangle

Listing 2.8 PDF view

Using x_1, x_2, and x_3, we consider the polynomial $x_1^2 + 3x_1 x_2 - x_3^2$.

Listing 2.9 TeX source code

```
Using \(x_{1}, x_{2}\), and \(x_{3}\),
we consider the polynomial \(x_{1}^2+3x_{1}x_{2}-x_{3}^2\).
```

Exercise 2.4.1. Edit sample.tex with *Emacs*. Write a formula in the equation environment between \begin{document} and \end{document}. Next, typeset it with pdflatex and then check the PDF file with the evince viewer.

2.4.5 graphicx Package

There are some additional packages for LaTeX 2_ε, for example, the graphicx package, which lets us embed into our document a graphic file with a PDF, PNG (Portable Network Graphics), or JPEG (Joint Photographic Experts Group) format. As an example, use the following TeX source code Listing 2.10 to embed the graphics file centroid.png.

The command for embedding a graphics file is \includegraphics[options]{graphics filename}, and the command \caption{} is for naming the graphic file in the document. The *figure* environment is for determining the location of the figure (Fig. 2.6).

Listing 2.10 TeX source code

```
\documentclass{jsarticle}
\usepackage{graphicx}
\begin{document}
......
\begin{figure}[htbp]
\centering
\includegraphics[height=4cm]{centroid.png}
\caption{Centroid of a triangle}
```

```
\end{figure}
......
\end{document}
```

In this section, we presented an introduction to the basic idea of typesetting on the terminal; it is very similar to compiling the source code of a program. In MathLibre, there are many TEX editing environments. We can select from *Kile*,[4] *TeXstudio*,[5] *TeXworks*,[6] and *Texmaker*.[7] There are advantages and disadvantages to each of these environments.

2.5 Various Math Software Systems

There are so many mathematical software systems in MathLibre that it can be confusing. In this section, we deviate slightly from the primary topic of this book because we want you to enjoy mathematical software. We therefore introduce dynamic geometry software, which allows us to create and manipulate geometric constructions with a simulated compass and ruler. It is very popular for educational use.

When I first encountered this, I misunderstood and thought that it was a tool for only elementary geometry. However, as I used the software to create geometrical objects, I began to see interesting applications for the function of drawing trajectories and the construction of recursive methods. I believe that this software has potential for helping us visualize various mathematical ideas. In MathLibre, there is a lot of dynamic geometry software because I like it. One of them includes the automatic proof assistant system that uses the method of Wu and Gröbner bases. This is not covered in this book, but if you are interested in this topic, you can refer to [1].

2.5.1 KSEG

One of the basic dynamic geometry software systems is *KSEG*.[8] This open-source software was written in the C++ programming language by Ilya Baran. Using *KSEG*, we can deform, rotate, and move geometrical objects while maintaining their properties. We can measure the distance between two points and the angles of a triangle, and then perform calculations with them. With a function for creating the locus of restricted objects, we can draw various geometric curves. We can make Koch and dragon curves by using a recursive method.

---

[4]http://kile.sourceforge.net/.

[5]http://texstudio.sourceforge.net.

[6]http://www.tug.org/texworks/.

[7]http://www.xmlmath.net/texmaker/.

[8]http://www.mit.edu/~ibaran/kseg.html.

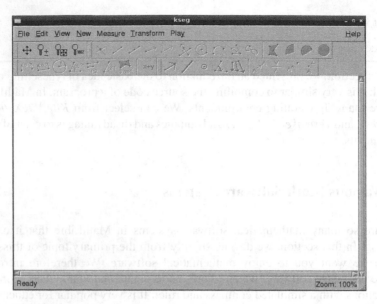

Fig. 2.7 *KSEG* window

Fig. 2.8 *KSEG* menu

2.5.1.1 How to Start KSEG

From the start menu, click on the Math software submenu and select KSeg .
Alternatively, to launch from the terminal, enter kseg. The following *KSEG*
window will be displayed.

There are menus at the top of the window, and button icons right below them.
The pictures on the icon buttons indicate their functions (Figs. 2.7 and 2.8).

KSEG has only four main types of function.

1. Draw a point by right clicking.
2. Select points, lines, and circles by left clicking. (They can be selected by using
   the shift key and rectangle selection.)
3. Create a geometrical object by using a menu or button.
4. Delete geometrical objects using the Ctrl+Del keys.

Fig. 2.9 Right clicking the appropriate position

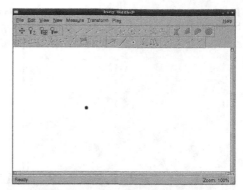

Fig. 2.10 Right clicking the other place

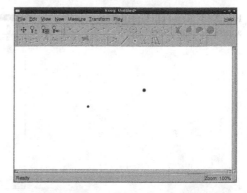

Fig. 2.11 Selecting two points with Shift key

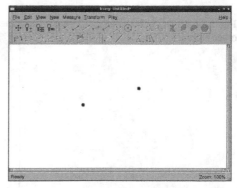

For example, we need two points in order to draw a segment. To create them, right click on two places on the *KSEG* screen. After selecting these two points, create a segment using the menu "New → Segment" or the button of "Segment", Figs. 2.9–2.12.

In a similar way, we can create a "Line" or a "Half line". If we select two points and click the "Circle" button, we create a circle centered at the first point and going through the second point, Figs. 2.13–2.16.

Fig. 2.12 Selecting "New →
Segment"

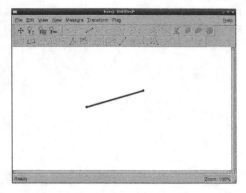

Fig. 2.13 Right clicking the
appropriate position

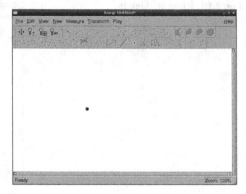

Fig. 2.14 Right clicking the
other place

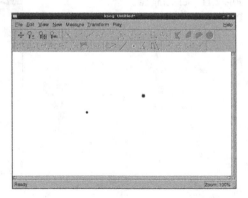

2.5.1.2 Creating a Triangle

Creating three points and using rectangle selection is a convenient way to create a
triangle, Figs. 2.17–2.20.

Fig. 2.15 Selecting two points with the Shift key

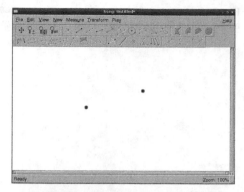

Fig. 2.16 Selecting "New → Circle"

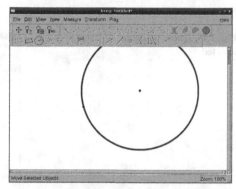

Fig. 2.17 Drawing three points for creating a triangle

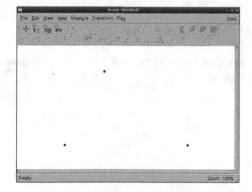

2.5.1.3 Centroid of a Triangle

We can find the locations of the various centers of a triangle with *KSEG*. In this subsection, we will explain how to draw the centroid. After we draw the centroid, we can drag the points of the triangle to see its dynamic deformation, Figs. 2.21–2.36. Determining the location of the circumcenter, the orthocenter, and the centers of the incircle and excircle of a triangle are left as exercises for the reader.

Fig. 2.18 Dragging the
mouse around the three points

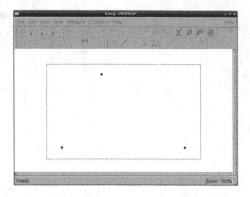

Fig. 2.19 Three points are
selected

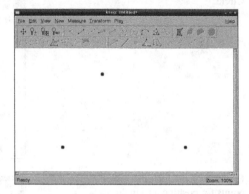

Fig. 2.20 Selecting "New →
Segment"

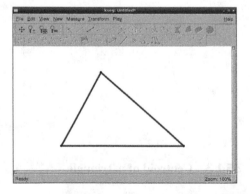

Fig. 2.21 Drawing a triangle

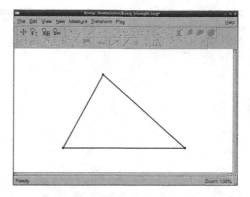

Fig. 2.22 Selecting an edge of the triangle

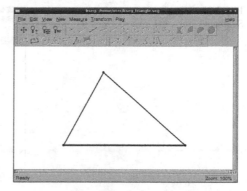

Fig. 2.23 "New → Midpoint"

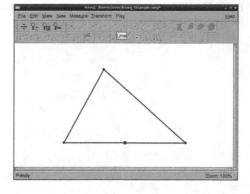

Fig. 2.24 Selecting a vertex and the midpoint with the Shift key

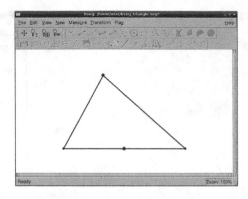

Fig. 2.25 Selecting "New → Segment"

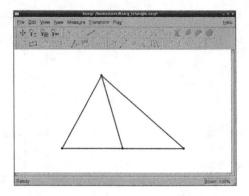

Fig. 2.26 After the same step for the other two points

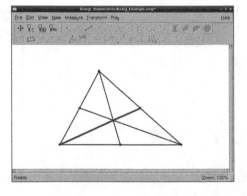

Fig. 2.27 Selecting two
medians in the triangle

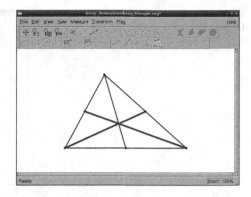

Fig. 2.28 Selecting "New →
Crosspoint"

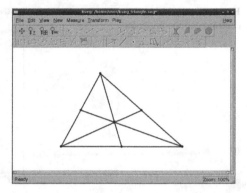

Fig. 2.29 Selecting a vertex
and centroid

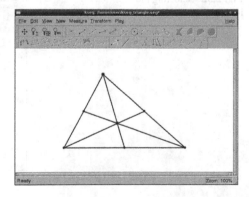

Fig. 2.30 "Measure →
Distance"

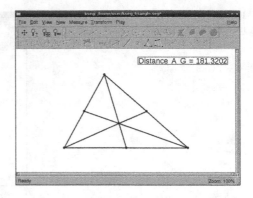

Fig. 2.31 Measuring the
centroid and the midpoint

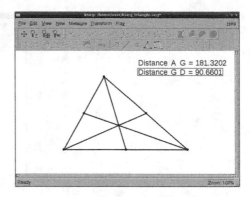

Fig. 2.32 Selecting the
longer one

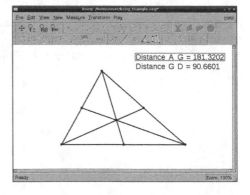

Fig. 2.33 "Measure →
Calculate"

Fig. 2.34 Moving the cursor
to the end of line, pushing $\frac{x}{y}$

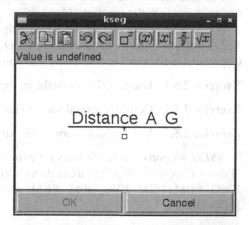

Fig. 2.35 Selecting the
shorter one

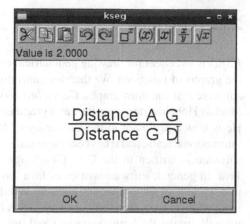

Fig. 2.36 Clicking "OK"

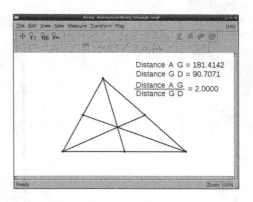

The centroid is exactly two-thirds of the way along each median. We can check this property with the function "Measure".

For more information, see the *KSEG* help document. There are samples on the *KSEG* web page and on the MathLibre DVD. In the *KSEG* help file, go to "File → Copy as Construction" to find out how to construct geometrical objects.

Exercise 2.5.1. Use *KSEG* to draw the centers of a triangle.

Exercise 2.5.2. Consider several ways to use *KSEG* to draw conic sections.

Exercise 2.5.3. Use *KSEG* to draw Koch curves.

KSEG supports various formats for exporting graphics. From the menu, choose "File → Export to Image"; you can then choose one of the following formats: BMP, JPEG, PBM, PGM, PNG, PPM, XBM, and XPM. PDFLᴬTᴇX supports PNG and JPEG formats.

2.5.2   GeoGebra

KSEG is excellent for drawing geometrical objects, but it does not support drawing the graphs of functions. We therefore introduce *GeoGebra*, dynamic mathematical software that can draw graphs. *GeoGebra* is open source and was first developed by Markus Hohenwarter when he was a graduate student at the University of Salzburg. He is now a professor at the University of Linz. *GeoGebra* was developed by an international team, and it has became popular all over the world. Much mathematical software is written in the C or C++ language, but *GeoGebra* was developed in Java. In general, software written in Java has the advantage of being easy to move to a variety of environments, such as Windows, Mac, and UNIX. *GeoGebra* is mathematical education software, and so this flexibility is important.[9] In this section, we will discuss the basic operations and investigate the trajectories of the vertex of a parabola.

---

[9]A disadvantage is that software developed in Java may be slower than that developed in C or C++.

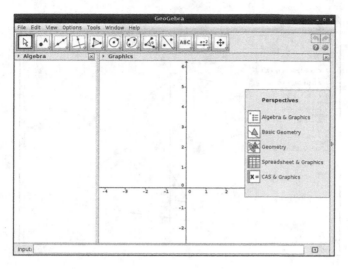

Fig. 2.37 *GeoGebra* window

2.5.2.1   GeoGebra Basics

To launch *GeoGebra*, select the icon ⭕ from the start menu. To execute it from the terminal, enter the command `geogebra`. With *KSEG*, we first create the points and the select the geometrical object to be created from a menu. With *GeoGebra*; however, we first select the geometrical object and then change to drawing mode. For example, if we want to draw a point in the *GeoGebra* window, we first click the icon for Point ●^{A}. We then create a new point by clicking the appropriate position on the window. To create a circle, first chose the Circle icon ⨀. There are four methods for drawing a circle: "Circle with Center through Point" (the default), "Circle with Center and Radius", "Compass", and "Circle through Three Points". Select one of these methods by clicking the small triangle in the right bottom of the icon. There are some differences between the interfaces of *KSEG* and *GeoGebra*, but both of them are good software systems for dynamic geometry (Fig. 2.37).

We can enter a command to the Input Bar, draw the graph of a function, draw the tangent line, or calculate the integral.

First, though, we will draw some points, lines, triangles, and circles. Click the "Move" icon, and then you can use the cursor to freely move the geometrical objects. When we move a circle, the changes in the equation of the circle are displayed on the left side of the window Fig. 2.38.

2.5.2.2   Graph of a Function

We can draw a graph by typing the function into the Input bar of *GeoGebra*. Type in `y=x^2` and press the Enter key to draw the parabola $y = x^2$.

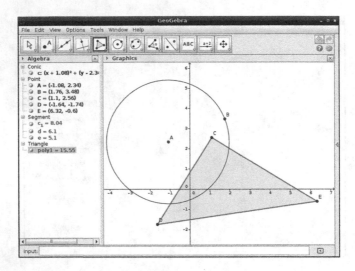

Fig. 2.38 Geometrical object with *GeoGebra*

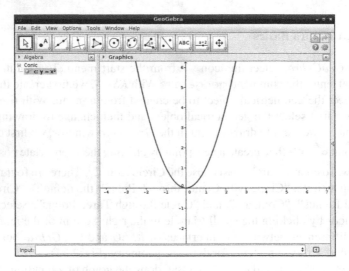

Fig. 2.39 Parabola with *GeoGebra*

The parabola is automatically named c by *GeoGebra*. The function `Vertex[]` will show the vertex of the conic. The argument of the `Vertex[]` function is the name of the conic. In this case, we input `Vertex[c]` and press Enter. The vertex named A is on the origin $(0, 0)$, and by clicking the cursor icon, we can move the parabola. As we move it, we can observe the changes in the vertex A and the equation c of the parabola (Fig. 2.39).

2.5.2.3 Slider of GeoGebra

Suppose we have the following problem: "What is the trajectory of the vertex A of the parabola $y = x^2 - 2ax + 1$ when we change the value of the constant a?"
   Slider is a convenient tool for solving this problem.

1. Click the icon of "Slider".
2. Click the appropriate point of the graphics area.
3. The small new window shows the default data of the Slider function; click the Apply button.
4. Type y=x^2-2*a*x+1 in the Input Bar and press Enter.
5. Type Vertex[c] in the Input Bar and again press Enter.
6. Click the Move icon and change the value of the slider.
7. Observe the vertex.

2.5.2.4 Trace On

To follow the changes in the vertex A, we can use a feature called "Trace On".

1. Right click the vertex A.
2. Select the check box of "Trace On".
3. Move the slider and observe the state of the vertex A.

Note that the trace is also a parabola.

2.5.2.5 Creating a Graphics File

GeoGebra also supports exporting to various file formats. As an example, this is how to export to a PNG file (Fig. 2.40):

1. Create a figure with *GeoGebra*.
2. Select from the menu File → Export → Graphic View as Picture (png, eps)...
3. Select Portable Network Graphics (png).
4. Press the Save button.
5. Save the file with a suitable name to an appropriate directory.

GeoGebra supports exporting as a dynamic web page in Java Applet and Javascript. There is a community site for collecting educational materials for *GeoGebra*: GeoGebraTube.[10] It is easy to upload files from the application to this archive. *GeoGebra* also interfaces with the TEX system. It is able to export PGF/Tikz source code; this is a TEX macro package for embedding graphics into TEX source code.

---

[10]http://www.geogebratube.org/.

Fig. 2.40 Taylor Polynomial
with *GeoGebra*

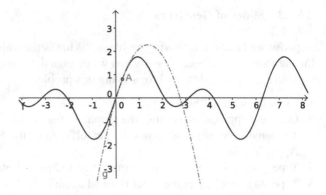

By typesetting with the command `pdflatex`, we can produce a document with embedded graphics.

Exercise 2.5.4. Using *GeoGebra*, draw a graph of the function $f(x) = \sin(x) + \sin(2x)$.

Exercise 2.5.5. Select a point A on the graph of the function $f(x) = \sin(x) + \sin(2x)$, and draw a graph of the second-degree Taylor polynomial on A.

Exercise 2.5.6. Make a slider of $1 \leq a \leq 5$ with the increment equal to 1. Draw a graph of the Taylor polynomial on A with degree a.

2.5.2.6   GeoGebra with Risa/Asir

By combining *GeoGebra* with other software, we can display the result of a Gröbner basis. Consider the polynomials $x^2 + y^2 - a$ and $xy - b$ of the two variables x and y. In this case, $a, b \in R$ are constants. We can draw the graph of the implicit function. We can create a and b with Slider, and we can choose the range of default values of a and b. After making the two sliders, input the equation `x^2+y^2-a=0`. It is a circle, and we can change its radius. We can input `x*y-b=0`, a hyperbola. Click on the icon for Intersect Two Objects, and select the circle and the hyperbola. When the default value of the sliders are $a = 1, b = 1$, the circle and the hyperbola do not intersect. The result of this operation is "A undefined" because the solution of this system of equations is a complex number. If we change the value so that $a = 2$, there are two intersects, and when a is greater than 2, then are four cross points, which are displayed as real solutions.

Set $a = 4$. *GeoGebra* supports rounding, so select "Option → Rounding → 4 Decimal Places". The coordinates of the four intersection points are $(0.5176, 1.9319)$, $(-0.5176, -1.9319)$, $(-1.9319, -0.5176)$, and $(1.9319, 0.5176)$ (Fig. 2.41).

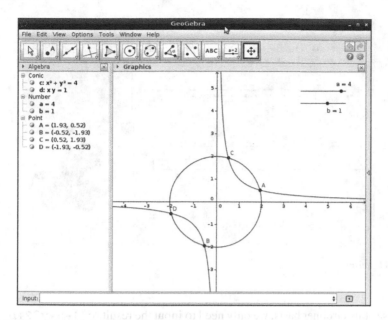

Fig. 2.41 Four intersection points with *GeoGebra*

We will calculate the Gröbner basis by using a computer algebra system, *Risa/Asir*.[11] It is launched in a similar way to the previous programs. To execute it from a terminal, enter openxm fep asir. To learn more about the uses of *Risa/Asir*, please see the "Risa/Asir Drill Book"[6].

As an example, here are the commands nd_gr and pari, in *Risa/Asir*.

```
[1371] G=nd_gr([x^2+y^2-a,x*y-b],[x,y],0,2);
[y^4-a*y^2+b^2,-b*x-y^3+a*y]
[1372] pari(roots,base_replace(G[0],[[a,4],[b,1]]));
[ -1.931851652578136573  -0.5176380902050415246
0.5176380902050415246  1.931851652578136573 ]
```

Using the command

nd_gr(Polynomial List, Variable List, P, Order),

we can compute the Gröbner basis of G[0] =y^4-a*y^2+b^2,
G[1]=-b*x-y^3+a*y. In this example, the argument for Order is 2, which is the lexicographical order (refer Corollary 1.4.2). With the command
pari(Roots, polynomial), we can calculate the roots of the polynomial. In this case, we solved the polynomial G[0] with $a = 4, b = 1$. This result gives the y coordinates of the four intersects. We see it nearly coincides with the results of *GeoGebra*.

---

[11]http://www.math.kobe-u.ac.jp/Asir/asir.html.

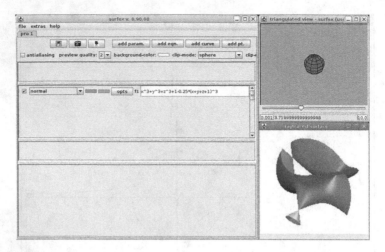

Fig. 2.42 *surfex*

To see this Gröbner basis, we only need to input the result, `y^4-a*y^2+b^2=0` and `-b*x-y^3+a*y=0`, to the Input Bar of *GeoGebra*.

We can draw it with the function `ifplot` of *Risa/Asir*. The commands *surf* and *Sage* are able to draw implicit functions.

Exercise 2.5.7. Let $f(x, y)$ be a polynomial of degree n, and let $g(x, y)$ be a polynomial of degree m. Assume that f and g are relatively prime. At most, how many intersections of $f = g = 0$ are there? Try to determine the number with the help of Geogebra.[12]

2.5.3 Surf Family

surf was written by Stephan Endrass. It is a tool for drawing real algebraic geometry. With it, we can create beautiful graphics of plane algebraic curves, algebraic surfaces, and hyperplane sections of surfaces. *surf* supports a macro language that is very similar to the C language, but *surfex* and *surfer* have been released, and they have interactive interfaces. Using *surfex* and *surfer*, we can observe the graphics of surfaces from a dynamically changing viewpoint. All of these are included in MathLibre, but we will introduce *surfex*. When *surfex* is launched, these windows are displayed: The main window has the following four buttons (Fig. 2.42):

---

[12]Answer: mn (Bezout theorem).

Fig. 2.43
Figure 2.45 + Fig. 2.46

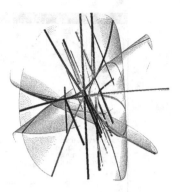

add param.   creating a parameter slider,
add eqn.   creating a surface that represents an implicit function,
add curve   creating a curve as an intersection of surfaces,
add pt.   creating a point.

There is already an example polynomial $x^3 + y^3 + z^3 + 1 - 0.25(x + y + z + 1)^3$, which is called the Cayley cubic, and its surface is displayed in the small window. There are two small windows, one is the "triangulated view" and the other is the "raytraced surface". To change the viewpoint of the display, drag on image in the "triangulated view" window. We can change the colors of surfaces by using the "opts" button. There is also a transparency mode, which is very helpful for observing the intersection of surfaces. The resolution of the graphic can be changed by configuring the parameter of "preview quality" in the main window; the best graphics are when the value is set to 1. There are some buttons in the main window, one of which is the camera icon, which captures the graphics in JPEG format. Using *surfex* with Example 1.4.11 from the first section, we create the following pictures. Because of the difference in the dimensions, it is little bit difficult to see with printed pictures, but we note Fig. 2.43 is similar to Fig. 2.44. All the surfaces in the figure are created by the original polynomials; to reduce complexity, the surfaces from the Gröbner basis have been set to transparent (Figs. 2.45 and 2.46).

2.5.4 *Maxima*

In this section, we introduce a general purpose system for computer algebra, *Maxima*. *Maxima* is the descendant of *MACSYMA*, which has a long history and is written in the Lisp language. MACSYMA was developed at MIT for a research project on artificial intelligence. In this book, we also introduce other research systems: *CoCoA*, *Macaulay2*, *Risa/Asir*, and *Singular*. These recent systems have been developed mainly for mathematics research. In the 1970s and 1980s, *MAC-SYMA* supported the Risch algorithm for indefinite integrals; it was commercialized by the company Symbolics [5]. *Maxima* was developed by William Schelter and

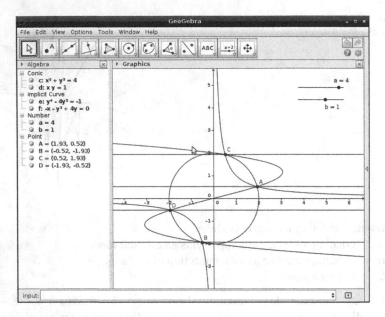

Fig. 2.44 Implicit functions with *GeoGebra*

Fig. 2.45 Intersection curves
with the original polynomials

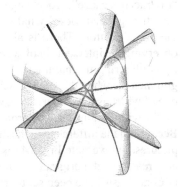

is based on a 1982 version of *MACSYMA*; it later became open source. In 2001, Schelter passed away while traveling in Russia, but *Maxima* is now continuously maintained by a team of developers.[13] *Maxima* supports many operations, including factoring; solving algebraic, differential, and integral equations; manipulating limits, series, matrices; and drawing graphs.

We can launch *Maxima* either from the menu or by entering maxima, xmaxima, or wxmaxima to the terminal. As shown here, we launch *Maxima* from the terminal.

---

[13]http://maxima.sourceforge.net/.

Fig. 2.46 Intersection curves
with Gröbner basis

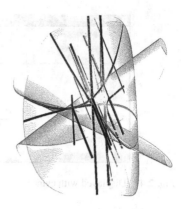

```
user@debian:~$ maxima

Maxima 5.27.0 \url{http://maxima.sourceforge.net}
using Lisp GNU Common Lisp (GCL) GCL 2.6.7 (a.k.a. GCL)
Distributed under the GNU Public License. See the file COPYING.
Dedicated to the memory of William Schelter.
The function bug_report() provides bug reporting information.
(%i1)
```

(`%i1`) is the interactive prompt for *Maxima*, after which a command or function
can be entered, followed by a semicolon. To quit the program, enter `quit();`

When learning *Maxima*, `describe()` is an important function. For
example, entering the command `describe(factor);` will display the help
documents for the function `factor()`. The command `describe(string)`
is equivalent to `describe(string, exact)`. If such an item exists, it will
find one with the exact same title (case-insensitive) as `string`. The command
`describe(string, inexact)` finds all help documents for items that
contain `string` in their titles. Note that following the interactive prompt with
`? foo` (with a space between `?` and `foo`) is equivalent to `describe(foo,
exact)`, and `?? foo` is equivalent to `describe(foo, inexact)`. There is
a lot of documentation for *Maxima* in MathLibre and on the Internet.

2.5.4.1   Output of TEX Source Code

TEX is an excellent typesetting system, but it can be difficult for complicated
mathematical formulas. Thus, it would be helpful to have mathematical software
produce output that is formatted as TEX source code. In fact, there are several
mathematical software systems that can do this; in this section, we will show how
to do it using the function `tex()` in *Maxima* (Fig. 2.47).

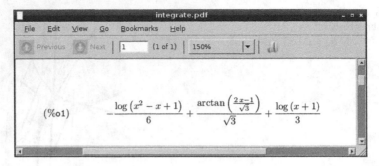

Fig. 2.47 Displayed with evince

Listing 2.11 tex() of *Maxima*

```
(%i1) integrate(1/(x^3+1),x);
                                            2 x - 1
                        2              atan(-------)
                 log(x  - x + 1)           sqrt(3)       log(x + 1)
(%o1)            - --------------- + ------------- + ----------
                         6               sqrt(3)          3
(%i2) tex(%o1);
$$-{{\log \left(x^2-x+1\right)}\over{6}}+{{\arctan \left({{2\,x-1
 }\over{\sqrt{3}}}\right)}\over{\sqrt{3}}}+{{\log \left(x+1\right)
 }\over{3}}\leqno{\tt (\%o1)}$$
(%o2)                                                    (\%o1)
(%i3)
```

Inputting the command after *Maxima*'s interactive prompt (%i1), allows us to calculate the indefinite integral of the function $f(x) = 1/(x^3 + 1)$. The result of the computation can be referenced to as (%o number).

2.5.4.2 Working Record of a Calculation

We can save the result of a computation by using redirection, but we can also save the entire work record to a text file by using the UNIX command script on the terminal. In the following example, we use *Maxima*, but the method is the same for any other mathematical software which is executed from the terminal.

```
user@debian:~$ script
Script started, file is typescript
user@debian:~$ maxima

Maxima 5.27.0 \url{http://maxima.sourceforge.net}
using Lisp GNU Common Lisp (GCL) GCL 2.6.7 (a.k.a. GCL)
Distributed under the GNU Public License. See the file COPYING.
Dedicated to the memory of William Schelter.
The function bug_report() provides bug reporting information.
(%i1) integrate(1/(x^3+1),x);
                                            2 x - 1
                        2              atan(-------)
```

```
                    log(x - x + 1)                 sqrt(3)      log(x + 1)
(%o1)             - --------------- + ------------- + ----------
                          6                   sqrt(3)          3
(%i2) quit();
user@debian:~$ exit
Script done, file is typescript
user@debian:~$ ls
Desktop typescript
user@debian:~$
```

When we execute the command `script` with no arguments, the working record
is saved to the file `typescript`. If we specify the file name in the argument as
follows, it will be saved in the specified file.

```
user@debian:~$ script logfile.txt
Script started, file is logfile.txt
```

The text editor *Emacs* can be used to view the working record and to alter it to try
different approaches (ref. Sect. 2.6).

2.5.5 *R*

R is a programming language and environment for statistics and graphics.[14] It is
very similar to the *S* language and statistical calculation environment, although *R*
and *S* were developed independently. *R* is open source and has grammar similar to
the *S* language. To execute *R*, enter R to the terminal.

```
user@debian:~$ R

R version 2.15.1 (2012-06-22) -- "Roasted Marshmallows"
Copyright (C) 2012 The R Foundation for Statistical Computing
ISBN 3-900051-07-0
Platform: x86_64-pc-linux-gnu (64-bit)

R is free software and comes with ABSOLUTELY NO WARRANTY.
We are welcome to redistribute it under certain conditions.
Type 'license()' or 'licence()' for distribution details.

R is a collaborative project with many contributors.
Type 'contributors()' for more information and
'citation()' on how to cite R or R packages in publications.

Type 'demo()' for some demos, 'help()' for on-line help, or
'help.start()' for an HTML browser interface to help.
Type 'q()' to quit R.

>
```

---

[14]http://www.r-project.org/.

The interactive prompt for *R* is >. There are various ways to execute *R*, for example *Rcommander*, *Rkward*, and *RStudio*. They can be found on the start menu or by entering the command Rcmdr, rkward, or rstudio. *R* can also be executed in the text editor *Emacs*, as with other mathematical software systems.

2.5.6 Sage

Sage is a free open-source mathematics software system. It combines the power of many existing open-source packages, such as *Maxima*, *PARI*, *R*, *Singular*, and *surf*, into a common Python-based interface. The lead developer of *Sage* is William Stein, a professor at the University of Washington. There are large communities of *Sage* developers and users all over the world. This software supports a huge range of mathematics, including basic algebra, calculus, from elementary to very advanced number theory, cryptography, numerical computation, commutative algebra, group theory, combinatorics, graph theory, exact linear algebra, and much more. To run *Sage* on MathLibre, select "Math → SAGE" from the start menu. The message shown below will appear in the new terminal window. The *notebook* is a browser-based interface for *Sage*, and in this system, it will start automatically.

```
----------------------------------------------------------------------
| Sage Version 5.7, Release Date: 2013-02-19                          |
| Type "notebook()" for the browser-based notebook interface.         |
| Type "help()" for help.                                             |
----------------------------------------------------------------------

Please wait while the {\it Sage Notebook\/} server starts...
Setting permissions of DOT_SAGE directory so only you can read and write it.
The {\it notebook\/} files are stored in: sage_notebook.sagenb
```

Before proceeding further, it is necessary to enter and confirm a password.

```
Please choose a new password for the {\it Sage}\/ Notebook 'admin' user.
Do _not_ choose a stupid password, since anybody who could guess our password
and connect to our machine could access or delete our files.
NOTE: Only the hash of the password you type is stored by {\it Sage}.
You can change our password by typing notebook(reset=True).

Enter new password:
Retype new password:
```

A new window of the *Iceweasel*[15] web browser will open with *Sage notebook* (Fig. 2.48). We can create a new worksheet by clicking on "New Worksheet" (Fig. 2.49).

For example, we can change the name of a worksheet; here, it is "ex1". We can then integrate a function. When we click the check box of "Typeset", the typeset formula in Fig. 2.50 will be displayed.

To see how to draw the curve of an implicit function, we can find an example in the Help file. We can use *Singular* and *surf* for to plot a curve using *Sage* (Fig. 2.51).

---

[15]Iceweasel is *Firefox*, rebranded.

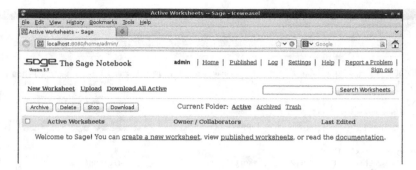

Fig. 2.48 *Sage* notebook

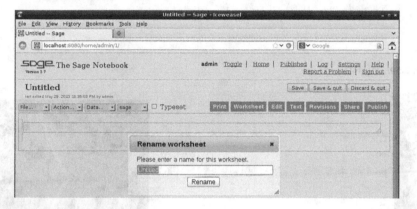

Fig. 2.49 New Worksheet

```
s = singular.eval
s('LIB "surf.lib";')
s("ring rr0 = 0,(x1,x2),dp;")
s("ideal I = x1^3 - x2^2;")
s("plot(I);")\\
```

2.6 Emacs

Knowledge of the text editor *Emacs* can be considered a fundamental skill that
will be useful in many ways. *Emacs* has a long history, but it is still popular with
professional users. It is unusual in that someone who learned how to use *Emacs* 30
years ago can still use it in the same way now. TEX and UNIX are also classics in
the same way. *Emacs* can be hard to learn at first, but once learned, it is an excellent
and indispensable tool. Many software systems incorporate aspects of *Emacs*; for
example, some of *Emacs* key bindings are used for terminals.

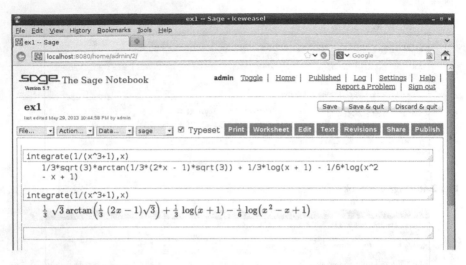

Fig. 2.50 integrate() in *Sage*

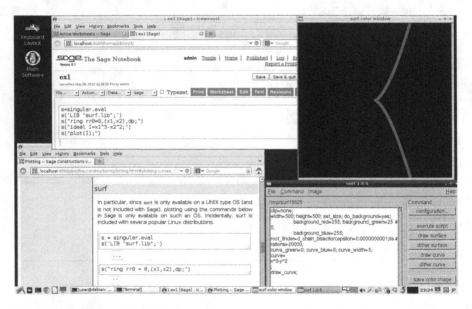

Fig. 2.51 *Singular* and *surf* in *Sage*

In this book, we perform computations using text files that contain commands or scripts of mathematical software. We need a text editor for these files, and *Emacs*, once learned, allows our work to proceed smoothly. For using some mathematical software, such as the programming language Lisp, we can use *Emacs* as an environment.

When we are using computers, inputting characters with a keyboard or drawing with a mouse, we typically are not aware that these manipulations involve reading data from, or writing data to, a buffer on the memory drive.

The official name of *Emacs* is "GNU Emacs". It was developed by Richard Stallman, who proposed the idea of free software. It was developed and published by the FSF (Free Software Foundation) and is available for free.

2.6.1 Starting Emacs

In this section, we will introduce only a small subset of the many features of *Emacs*. If you are already familiar with *Emacs*, you can skip this section.

If you want to learn *Emacs* thoroughly, there are many specialized books to help you do so. For an introduction, however, open the terminal, move to the directory that contains a tutorial that we made, ~/tutorial, and execute *Emacs*.

```
user@debian:~$ cd tutorial
user@debian:~/tutorial$ emacs
```

In this example, we call *Emacs* without arguments.

When we execute *Emacs*, a welcome message is displayed. The most important thing is how to quit *Emacs*, which can be done by using the mouse to select "File → Quit" from the menu. It is also possible to quit by typing and entering C-x C-c. This is the abbreviation for typing "x" and "c" while pressing the "Ctrl" key.

After booting *Emacs*, press the "q" key, and the following will be displayed.

```
;; This buffer is for notes you don't want to save, and for Lisp evaluation.
;; If you want to create a file, visit that file with C-x C-f,
;; then enter the text in that file's own buffer.
```

This is an area called a *buffer*, which is used for short memos and executing Lisp code. It shows that the command for saving a file is C-x C-f. If we want to use a mouse, we can select "File → Visit New File...". If we are familiar with the text editor of Windows or MacOS, we can use *Emacs* by clicking the menu interface. In order to demonstrate the full capability of the original *Emacs*, however, we should use only the keyboard. At first, it is sufficient to use *Emacs* with only the graphical user interface, but it is best to become familiar with other capabilities presented in textbooks and trying them for yourself (Fig. 2.52).[16]

Here we show how to use *Emacs*. We will create a file named "emacs.txt". First, we need to prepare the buffer for the file; the command for this is C-x C-f.

```
Find file: ~/tutorial/
```

---

[16] *Emacs* always uses the Ctrl key, and so it would be better to change the keyboard layout and set the key to the immediate left of "A" as a Ctrl key.

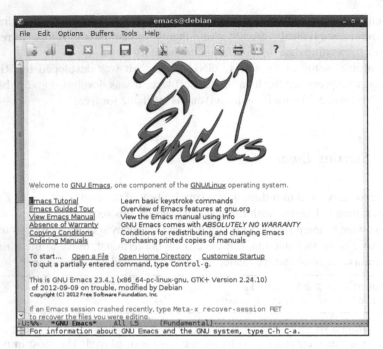

Fig. 2.52 Initial message of *Emacs*

This is called the *echo area*. In the echo area, type the following and press Enter.

```
Find file: ~/tutorial/emacs.txt
```

We have now prepared the buffer for creating the file ~/tutorial/emacs.txt. We can find New file in the echo area, and please input the following.

```
Emacs is a screen editor.
```

After typing the sentence, make a new line by pressing the Enter key. We operate *Emacs* with the "Ctrl" and "Meta" keys. The abbreviation C-x means to press the "x" key along with the "Ctrl" key. For example, the command C-x o means press "x" with the "Ctrl" key and release the finger from the "Ctrl" key and press "o". The command M-x means press the "x" key along with the "Meta" key. "Meta" keys, however, were found on earlier computers but not on modern PCs; instead, use the "Alt" key. Alternatively, press the "Esc" key, release it, and press the "x" key.

We now want to save the data of the buffer named "emacs.txt" to a file. The command to save a file is C-x C-s. When we enter this command, the following message is displayed in the echo area.

```
Wrote /home/user/tutorial/emacs.txt
```

We have now saved the data from the buffer to the directory "tutorial" in our home directory. To quit *Emacs*, input the command C-x C-c.

We can check for the created file as follows.

```
user@debian:~/tutorial$ ls emacs.txt
emacs.txt
```

We can find out the size of this file as follows.

```
user@debian:~/tutorial$ ls -l emacs.txt
-rw-r--r-- 1 user user 26 2009-09-14 10:34 emacs.txt
```

We can view the contents of this file as follows.

```
user@debian:~/tutorial$ cat emacs.txt
Emacs is a screen editor.
user@debian:~/tutorial$
```

Finally, using the command emacs with the file name as an argument, we can edit this file.

```
user@debian:~/tutorial$ emacs emacs.txt
```

When we open this file with *Emacs*, the contents will be displayed.

2.6.2   Cut and Paste

As an example, we will now edit the buffer and change the word "screen" to "text". When we open the file, the character "E" at the beginning of the line should be blinking. If we input a character, it will be displayed in this location. This blinking part indicates the location of the cursor, which we can move by using the arrow keys. We can also use the Ctrl key to move the cursor: Using C-f to move the cursor to the character "n" of the word "screen", remove the word "screen" with the BS key, then enter "text". The buffer is now edited, and we will see the following.

```
Emacs is a text editor.
```

At this point, the cursor is in the position of the character "t" at the end of the word "text".

2.6.3   Editing Multiple Lines

We can edit multiple lines, as follows. Move the cursor to the next line with the command C-n. After the word "editor", there will be a control character indicating a new line. After moving the cursor key, we can now edit the next line.

```
Emacs is a text editor.
Emacs is a computer environment.
```

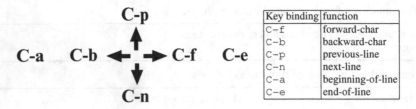

Fig. 2.53 Cursor moving of *Emacs*

The features of *Emacs* are implemented using the *Emacs* Lisp programming language. The input of characters and moving of the cursor are realized by calling Lisp functions. To explicitly call a Lisp function using its name, input M-x, which opens a mini buffer in the echo area. Enter the function name.

```
M-x
```

For example, if we input M-x forward-char, the cursor moves to the next character. Just to move the cursor, it is bothersome to input a function name; thus, these functions are usually assigned to shortcut keys. We call this binding the keys. At first, we may use the cursor keys, but it is more efficient to keep our hands on the keyboard. Below is a table of *Emacs* key bindings for moving the cursor (Fig. 2.53).

All of the file operations that were introduced in this section have key bindings with Lisp functions.

Key binding	Function
C-x C-c	Save-buffers-kill-emacs
C-x C-f	Find-file
C-x C-s	Save-buffer
M-x	Execute-extended-command

Exercise 2.6.1. Input another sentence.

```
Emacs is a text editor.
Emacs is a computer environment.

At its core is an interpreter for Emacs Lisp,
a dialect of the Lisp programming language
with extensions to support text editing.
```

After completing changes, we can save the buffer to a file. The command is C-x C-s.

Exercise 2.6.2. With C-p, move the cursor to the third line and insert a new line.

```
Emacs is a text editor.
Emacs is a computer environment.
Emacs is the extensible, customizable editor.
At its core is an interpreter for Emacs Lisp,
a dialect of the Lisp programming language
with extensions to support text editing.
```

We will focus on the mode line, which corresponds to the current line number.

Exercise 2.6.3. Move the cursor, and check the line number.

Not all editing operations are assigned to key bindings. For example, the function for displaying the line number is "what-line"; input M-x what-line, and the line number is displayed in the echo area. In order to move to a specified line number, there is a Lisp function "goto-line". When we execute the following command,

```
M-x goto-line
```

we will see

```
Goto line:
```

in the echo area; input the desired number and press Enter to move the cursor to that line.

Exercise 2.6.4. Using the Lisp function "goto-line", move to the fifth line. Now, using the Lisp function "what-line", display the current line number in the echo area.

Exercise 2.6.5. Execute the command C-x C-c without saving a file. What is the message in the echo area?

In this situation, the following message may be displayed.

```
Save file /home/user/tutorial/emacs.txt? (y, n, !, ., q, C-r or C-h)
```

This happens when there is an edited buffer which has not yet been saved. If you want to save it, enter y; if you do not want to save it, enter n. For now, save this file by entering y.

2.6.4   Remove Again

Again use *Emacs* to open the text file "emacs.txt".

```
user@debian:~/tutorial$ emacs emacs.txt &
```

The position of the cursor is on the first character of the first line, E. To remove the character under the cursor, we can use the *Emacs* command C-d. Enter C-d, to remove the character E.

Exercise 2.6.6. Remove the word "Emacs" with `C-d`.

To remove a single line, use the command `C-k`. This key binding will remove all the text from the position of the cursor to the end of the line. When used with the key binding `C-a`, which moves the cursor to the beginning of the line, we can quickly remove a line.

Key binding	Function
BS	Backward-delete-char
C-d	Delete-char
C-k	Kill-line

What should you do to remove multiple lines?

2.6.5 *Point, Mark, and Region*

In *Emacs*, the words "point", "mark", and "region" have special meanings.

Concept	Meaning
Point	The position between the previous character and the current position of the cursor
Mark	a Unique point that the user can save to the buffer
Region	The area between the positions of point and mark

For example, here the cursor is located on the indefinite article a,

```
Emacs is a computer environment.
Emacs is the extensible, customizable, self-documenting, real-time editor.
```

The point is located between the blank space " " and a. We will set the mark at this point; the command for setting the mark is `C-SPC`, where `SPC` means the space bar. We now move the cursor to the character "t" of the definite article `the`.

```
Emacs is a computer environment.
Emacs is t he extensible, customizable, self-documenting, real-time editor.
```

The region is now as shown here.

```
         a computer environment.
Emacs is
```

In MathLibre, the specified region is displayed with a yellow background. In another environment, the mark and region may be invisible. In that case, input `C-x C-x` twice to find the position of mark. There is another choice, the LISP function "transient-mark-mode", which changes the mode of the region displayed.

When you have verified the region, now try to remove it. The command for removing it is C-w. The removed region will be saved to a buffer called the "kill ring". We can now reinsert the region that was saved to the kill ring by using the command C-y. There is also a command for duplicating the region, M-w. Generally, deleting a region or line C-w and C-k is called "cut"; duplicating the characters with M-w is called "copy", and inserting characters with C-y is called "paste". Moving text is called "cut & paste", and duplicating text is called "copy & paste".

By the way, the kill ring is in the shape of a ring. Using M-y after the command C-y obtains the previous elements in order.

Key binding	Function
C-w	Kill-region
M-w	Kill-ring-save
C-y	Yank
M-y	Yank-pop

2.6.6   Undo, Redo, and Etc.

If a mistake is made when entering a key or calling a function, it can be canceled with the command C-g. This is a very important and useful command.

In the following table, we summarize the *Emacs* Lisp functions which will be necessary in addition to the features described in this section. *Emacs* can be operated with a mouse, but it is faster and easier if only the keyboard is used.

Key binding	Function
C-g	Quit
C-x i	Insert-buffer
C-x C-w	Write-file
C-s	isearch-forward
C-r	isearch-backward
M-%	Query-replace
C-x o	Other-window
C-x u	Advertised-undo
C-_	Undo

When the command M-x help-with-tutorial is entered, a basic introduction of *Emacs* will be displayed. You should read it.

2.6.7   Command and Shell

We introduced *Emacs* as a text editor. At this point, we will introduce the extended features that use the Lisp language and show how it can be used as a computer environment.

In Sect. 2.3, we entered various commands. We can also use a variety of commands with *Emacs*. If we enter M- !, "Shell command:" will be displayed in the echo area. We can now execute a command. For example, after executing M- ! and entering ls, a list of the files in the current working directory will be displayed.

If we want to input commands exclusively, we can execute the *shell* in *Emacs*. The command to change to the *shell* mode is M-x shell. We will then see the prompt of the *shell* and can use it in a way similar to the terminal. For example, we can move the cursor and copy and paste regions.

In addition, MathLibre also contains a file manager (M-x dired), the Tower of Hanoi (M-x hanoi), Goban (M-x gomoku), and Tetris (M-x tetris). All of these are implemented using the *Emacs* Lisp language.

2.6.8   Math Software Environment

Some mathematical software systems can be used with *Emacs*. We summarize the commands in the following table. When we use software in *Emacs*, it is very similar to using the *shell* on *Emacs*. Executing mathematical software on the terminal is easy and simple, but using a system on *Emacs* may be convenient because of the various features of *Emacs*. In particular, the computer algebra system *Maxima* contains a Lisp package named *imaxima*; this package allows us to elegantly typeset our results, Fig. 2.54.

Maxima	M-x imaxima
Macaulay2	M-x M2
Singular	M-x singular
R	M-x R

2.7   Other Ways of Booting MathLibre

When we boot MathLibre from a DVD, some think it is too slow and noisy. We would like to recommend the use of a virtual machine, which is a software-implemented version of hardware that runs at the applications layer of the operating system. A virtual machine can emulate the computer architecture of Windows, MacOS X, or Linux.

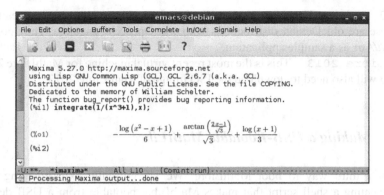

Fig. 2.54 M-x imaxima

Here are two the virtual machines which can be freely downloaded: *VMware Player* of VMware, Inc. and *VirtualBox* of the Oracle Corporation. *VMware Player* is an application for Windows and Linux; MacOS X users can buy a commercial product, *VMware Fusion. VirtualBox* is an open-source product that can be used on Windows, MacOS X, and Linux.

Currently, there are some virtual machine image files of MathLibre for *VMware Player* that can be downloaded from http://www.math.kobe-u.ac.jp/vmkm/. These commands shown in the image files support a persistent home directory, which can be used for daily work. In September 2009, we hosted a JST CREST Gröbner school at Kobe University; it was for graduate students who wanted to learn about Gröbner bases and mathematical software. We used a virtual machine as the standard computer environment and can thus recommend it as an everyday environment for research.

2.7.1   Various Virtual Machines

Here are some virtual machines for related project. KNOPPIX/Math was a project for archiving mathematical software and free documents in KNOPPIX, another Live Linux. MathLibre is a direct descendant of KNOPPIX/Math. All of them are for *VMware Player*. If you use Windows, you will have to install the VMware Player; for MacOS, install VMware Fusion. These virtual machines are compressed and must be extracted.

VMware/KNOPPIX/Math 2008   This is a virtual machine for "knxm 2008-kobe.iso". It is the latest version of KNOPPIX/Math with a KDE desktop and a multi-lingual environment.

VMware/Knoppix/Math 2010(en)   This is a virtual machine for KNOP-PIX/Math 2010. You will also need the ISO image of "knoppix_v6.2.1-math-dvd-icms2010-20100730-en.iso"

Small VM/KM These are the small virtual machines lenny-ox and etch-ox. These are old Debian distributions with the OpenXM package. They contain *Risa/Asir* as a sample application.

MathLibre 2013 This is the most recent virtual machine for MathLibre 2013. You will also need to download the ISO image of MathLibre 2013.

2.7.2 Making a USB-Bootable MathLibre

This is a third way to boot MathLibre. We are currently experimenting with implementing a shell script that makes MathLibre bootable from a USB device. A USB flash drive is a very convenient and economic device. With a USB-bootable version, when we want to create a mathematical software environment for our students, we can copy it to bootable devices and redistribute them. It takes about 20 min to copy the operating system and applications onto a USB device.

1. Boot with MathLibre DVD.
2. Connect a USB flash drive that has over 8 GB of memory.
3. Execute the following command.

```
user@debian:~$ sudo mkusbmath
```

4. View the list of USB devices.

```
-------------------------------------------------------------
 mkusbmath: shell script for making USB bootable MathLibre
-------------------------------------------------------------

Please select the target device from the following list:

/dev/sdb usb-Generic-_SD_MMC_058F63626420-0:0
/dev/sdc usb-Generic-_Compact_Flash_058F63626420-0:1
/dev/sdd usb-Generic-_SM_xD_Picture_058F63626420-0:2
/dev/sde usb-Generic-_MS_MS-Pro_058F63626420-0:3
/dev/sdf usb-ELECOM_MF-LSU2_eb972e623b2081-0:0

Please input your target device (ex. /dev/sdc) :
```

5. Chose /dev/sdf, input "/dev/sdf", and press Enter.

```
/dev/sdf
Unmounting the mounted partitions.

We've selected the device: /dev/sdf

By the following operation,
all files in /dev/sdf will be !!!REMOVED!!!.

It takes over 20 minutes for making,
would you start this operation? (y/n)
```

6. If you input y, then all the files in the flash drive will be removed. If that is OK, please input "y" and press Enter.

```
y
You're  copying OS image to /dev/sdf
   1.3GB at  112.1MB/s  eta: 0:00:22 34 [======                    ]
```

7. The process will take more than 15 min. You will see a progress bar; it may take a few minutes after reaching 100 %.
8. After the copying is completed, a persistent home directory is automatically created.

```
We've finished copying OS image.
We're making persistent volume.
```

9. The following message is displayed when the procedure is completed.

```
We've finished making persistent volume.
You've got an USB bootable MathLibre.
```

MathLibre was developed using the Debian Live system.[17] Debian GNU/Linux is a Linux distribution, and it uses the APT system to manage packages. We can install additional Debian packages with the command apt-get. To install additional packages, we have to update the resource database. To do this, we have to change to administration mode by using the command sudo.

```
user@debian:~$ sudo apt-get update
```

The command sudo executes a command as another user. For example, if we want to add another general purpose computer algebra system, "axiom", we input the following command.

```
user@debian:~$ sudo apt-get install axiom
```

Note that all packages that are required by the specified package will also be retrieved and installed.

2.7.3 How to Install MathLibre to an Internal Hard Disk

Beginning with MathLibre 2013, the installation of MathLibre to an internal hard disk is supported. The install menu can be found while booting, Fig. 2.55; when the "Install" or "Graphical install" menus are selected, we can install MathLibre as well as ordinary Debian distributions. There is an manual for installing the Debian Project.[18]

---

[17]http://live.debian.net/.

[18]http://www.debian.org/releases/stable/installmanual.

Fig. 2.55 MathLibre boot menu

References

1. D. Cox, J. Little, D. O'Shea, *Ideals, Varieties, and Algorithms: An Introduction to Computational Algebraic Geometry and Commutative Algebra* (Springer, Berlin, 2006)
2. M. Gancarz, *The UNIX Philosophy* (Digital Press, Bedford, 1994)
3. B.W. Kernighan, R. Pike, *The Practice of Programming* (Addison-Wesley, Reading, 1999)
4. MathLibre Project, http://www.mathlibre.org/
5. J. Moses, Macsyma: A Personal History (May, 2008), http://esd.mit.edu/Faculty_Pages/moses/Macsyma.pdf
6. Risa/Asir Drill Book, http://www.math.kobe-u.ac.jp/Asir/asir.html

Chapter 3
Computation of Gröbner Bases

Masayuki Noro

Abstract In Chap. 1, we presented the theoretical foundation of Gröbner bases, and many of our computations were carried out by hand. When we want to apply Gröbner bases to practical problems, however, in most cases, we will need the help of computers. There are many mathematical software systems which support the computation of Gröbner bases, but we will often encounter cases which require careful settings or preprocessing in order to be efficient. In this chapter, we explain various methods to efficiently use a computer to compute Gröbner bases. We also present some algorithms for performing operations on the ideals realized by Gröbner bases. These operations are implemented in several mathematical software systems: Singular, Macaulay2, CoCoA, and Risa/Asir. We will illustrate the usage of these systems mainly by example.

In this chapter, we explain how to compute various objects related to Gröbner bases, which were explained in Chap. 1. In Sect. 3.1, we explain the fundamental tools necessary for the efficient computation of Gröbner bases. You do not have to understand the details of the proofs in order to use Gröbner bases as a computational tool. However, homogenization is useful in various situations, and it is helpful to understand the principles behind it. In Sect. 3.2, we introduce the most popular software systems for computing Gröbner bases: Macaulay2, Singular, and CoCoA. Sections 3.3–3.5 describe various applications of Gröbner bases, and for each topic, we give examples using the above three systems. Section 3.6 describes how to use Risa/Asir for computations related to Gröbner bases. Finally, in Sect. 3.7, we introduce a new primary ideal decomposition algorithm, and we present its implementation in Macaulay2.

M. Noro (✉)
Department of Mathematics, Graduate School of Science, Kobe University,
1-1 Rokkodai, Nada-ku, Kobe 657-8501, Japan
e-mail: noro@math.kobe-u.ac.jp

T. Hibi (ed.), *Gröbner Bases: Statistics and Software Systems*,
DOI 10.1007/978-4-431-54574-3_3, © Springer Japan 2013

Most of the theoretical background for understanding this chapter was presented in Chap. 1. Using the functions related to Gröbner bases for practical applications, however, also requires knowledge of the use of finite fields. A finite field is a finite set in which addition, subtraction, multiplication, and division by a nonzero element are defined, and they are often used as coefficient fields for polynomial rings. A typical example is the set of all residue classes modulo a prime p, in other words, the set of all the remainders modulo p. This finite field is denoted by $\mathbb{F}_p$. The Buchberger algorithm works for ideals in polynomial rings over finite fields. A Gröbner basis computation over $\mathbb{F}_p$ does not cause intermediate coefficient swells, which often happens with computations over $\mathbb{Q}$. It is instructive to observe the behavior of an algorithm by computing various examples.

Lists are another important concept when using mathematical software systems. A list is a sequence of data, and since a list itself is considered to be data, a list can be recursively structured data. Results returned by computations related to Gröbner bases are often structured data represented by lists. If a list is returned, then it must be processed in order to extract the information contained. Usually the method for processing a list depends on the system being used; consult the appropriate manual.

3.1 Improving the Efficiency of the Buchberger Algorithm

The Buchberger algorithm introduced in Sect. 1.3.3 can be described as the following procedure.

Algorithm 3.1.1 $(Buchberger0(F))$.

Input: a set of polynomials $F = \{f_1, \ldots, f_l\}$
Output: a Gröbner basis G of $\langle F \rangle$
$D \leftarrow \{\{f, g\} \mid f, g \in F; f \neq g\}$
$G \leftarrow F$
while $(D \neq \emptyset)$ do
$\qquad C = \{f, g\} \leftarrow$ an element of D
$\qquad D \leftarrow D \setminus \{C\}$
$\qquad h \leftarrow$ a remainder of $S(f, g)$ on division by G
$\qquad$ if $h \neq 0$ then
$\qquad\qquad D \leftarrow D \cup \{\{f, h\} \mid f \in G\}$
$\qquad\qquad G \leftarrow G \cup \{h\}$
$\qquad$ endif
end while
return G

In this algorithm, if a nonzero remainder h is generated, then pairs of h and all elements of G are added to D, and h is added to G. Therefore you might expect that D would continue to grow with each iteration. But in reality, the remainders all eventually become 0, and the execution terminates. When the algorithm terminates, all the S polynomials are reduced to 0 and G is a Gröbner basis of $\langle F \rangle$. The output G is constructed by adding many polynomials to the input F, and, in general, it will be redundant. The reduced Gröbner basis defined in Sect. 1.2.2 can be obtained from a minimal Gröbner basis constructed from G.

Algorithm 3.1.2 $(MinimalGB(G))$.

Input: a Gröbner basis G of $\langle G \rangle$
Output: a minimal Gröbner basis of $\langle G \rangle$
$G_m \leftarrow \emptyset$
while $G \neq \emptyset$ do
    $g \leftarrow$ an element of G
    $G \leftarrow G \setminus \{h \in G \mid in_<(g) \mid in_<(h)\}$
    if $\{h \in G \mid in_<(h) \mid in_<(g)\} = \emptyset$ then $G_m \leftarrow G_m \cup \{g\}$
end while
return G_m

Algorithm 3.1.3 $(ReducedGB(G))$.

Input : a minimal Gröbner basis G of $\langle G \rangle$
Output : the reduced Gröbner basis of $\langle G \rangle$
$R \leftarrow \emptyset$
while $G \neq \emptyset$ do
    $g \leftarrow$ an element of G
    $G \leftarrow G \setminus \{g\}$
    $r \leftarrow$ a remainder of g on division by $R \cup G$
    $c \leftarrow$ the coefficient of $in_<(r)$ in r
    $R \leftarrow R \cup \{r/c\}$
end while
return R

Algorithm 3.1.1 is the most primitive form of the Buchberger algorithm, and it is not efficient on computers. In particular, it has the following serious drawbacks:

1. If the remainder of an S polynomial on division by G is not 0, then the number of elements in D increases by the number of elements in G.
2. The selection of the pair C affects the computational efficiency.

In order to overcome these difficulties, several criteria for eliminating unnecessary S polynomials and several strategies for selecting the S-pairs have been proposed. We will explain the most popular strategy for each of these. In the following section, $R = K[x_1, \ldots, x_n]$.

3.1.1 Elimination of Unnecessary S-Pairs

Definition 3.1.4. For monic polynomials $f_1, \ldots, f_m \in R$, we set $T_i = \text{in}_<(f_i)$, $T_{ij} = \text{LCM}(T_i, T_j)$ $(i \neq j)$. Let $(e_1, \ldots, e_m)$ be the standard basis of R^m, and set $S_{ij} = \frac{T_{ij}}{T_i} e_i - \frac{T_{ij}}{T_j} e_j$ $(i \neq j)$. For monomials $t_1, \ldots, t_m$, the expression $t_1 e_1 + \cdots + t_m e_m$ is said to be T-homogeneous if there exists a monomial t such that $t = t_i T_i$ for $i = 1, \ldots, m$.

Remark 3.1.5. $S = \{S_{ij} \mid 1 \leq i < j \leq m\}$ is a T-homogeneous basis of $\text{syz}(T_1, \ldots, T_m)$ (see Sect. 3.5.3).

Under this definition, the Buchberger criterion is refined as follows:

Theorem 3.1.6. *Suppose that $f_1, \ldots, f_m$ are monic. Then $F = \{f_1, \ldots, f_m\}$ is a Gröbner basis of $\langle F \rangle$ if and only if*

$$S_{ij} \in S' \Rightarrow S(f_i, f_j) \bmod F = 0$$

for a subset S' of $S = \{S_{ij} \mid 1 \leq i < j \leq m\}$ such that S' is a basis of $\text{syz}(T_1, \ldots, T_m)$.

Problem 3.1.7. Prove Theorem 3.1.6 by modifying the proof of the Buchberger criterion in Sect. 1.3.3.

This theorem states that it is not necessary to reduce all the S polynomials in order to check that a polynomial set F is a Gröbner basis. Instead, it is only necessary to reduce the S polynomials created from a basis S' of $\text{syz}(T_1, \ldots, T_m)$. We now introduce a well-known method for choosing such a basis.

Lemma 3.1.8. *For $T_{ijk} = \text{LCM}(T_i, T_j, T_k)$ $(1 \leq i < j < k \leq m)$,*

$$\frac{T_{ijk}}{T_{ij}} S_{ij} + \frac{T_{ijk}}{T_{jk}} S_{jk} + \frac{T_{ijk}}{T_{ki}} S_{ki} = 0.$$

Corollary 3.1.9. *If there exists k such that $T_{ij} = T_{ijk}$, then $\langle S \rangle = \langle S \setminus \{S_{ij}\} \rangle$.*

In order to use this corollary to eliminate unnecessary S_{ij}'s, we define an appropriate total ordering in S. We then eliminate S_{ij} if S_{ij} is the largest among S_{ij}, S_{jk}, S_{ki} with respect to this total ordering and $T_{ij} = T_{ijk}$ holds. A possible such ordering is the order of processing S_{ij}, that is, $S_{ij} > S_{kl}$ if S_{ij} is processed after S_{kl}. This is the one proposed by Buchberger. A more systematic method was proposed by Gebauer and Möller [4] and is widely used. In this method, the following total ordering $\prec$ is defined.

$$S_{ij} \prec S_{kl} \Leftrightarrow T_{ij} < T_{kl} \text{ or } (T_{ij} = T_{kl} \text{ and } (j < l \text{ or } (j = l \text{ and } i < k)))$$

It is convenient to introduce the following three properties of (f_i, f_j, f_k) $(i < j)$.

Definition 3.1.10.

$$F_k(i, j) \Leftrightarrow k < i \text{ and } T_{jk} = T_{ij}$$

$$M_k(i, j) \Leftrightarrow k < j \text{ and } T_k \mid T_{ij} \text{ and } T_{jk} \neq T_{ij}$$

$$B_k(i, j) \Leftrightarrow k > j \text{ and } T_k \mid T_{ij} \text{ and } T_{ik} \neq T_{ij} \text{ and } T_{jk} \neq T_{ij}$$

Problem 3.1.11. Check that these properties give the condition that S_{ij} is the largest among S_{ij}, S_{jk}, S_{ki} with respect to $\prec$. (We note that $t \mid s$ implies $t \prec s$.)

Theorem 3.1.12. *We define a subset S' of S by*

$$S' = \{S_{ij} \mid F_k(i, j), M_k(i, j), B_k(i, j) \text{ do not hold for any } k\}.$$

Then S' is a basis of $\mathrm{syz}(T_1, \ldots, T_m)$.

According to this theorem, we can eliminate S_{ij} from D if one of $F_k(i, j)$, $M_k(i, j)$, or $B_k(i, j)$ holds for some k. Furthermore, according to Lemma 1.3.1, we can eliminate S_{ij} such that $T_{ij} = T_i T_j$ or equivalently $\mathrm{GCD}(T_i, T_j) = 1$.

3.1.2 Strategies for Selecting S-Pairs

We now give an example which shows that, in the Buchberger algorithm, the strategy of selecting the S pairs affects the computational efficiency.

Example 3.1.13. Consider an ideal in $\mathbb{F}_{32003}[x, y, z, u]$:

$$I = \langle 3zx^3 + x^2 + 3y - 2, x^4 - yx - y, 3x - 2zy^2 + uz - 2, -2x^2 + uy^2 \rangle.$$

If we compute a Gröbner basis with respect to the lexicographic ordering such that $x > y > z > u$, one selection strategy (strategy 1) gives only 169 intermediate bases but another selection strategy (strategy 2) gives 1,465 intermediate bases.

Buchberger proposed the following strategy.

Definition 3.1.14 (Normal Selection Strategy). Select the pair from D which has the smallest T_{ij} with respect the term ordering. This strategy is called the *normal (selection) strategy*.

We can use the normal strategy to efficiently compute a Gröbner basis with respect to a graded term ordering. However, we note that the computation using the normal strategy is often inefficient if the term ordering is not graded. In fact, strategy 2 in Example 3.1.13 is the normal strategy. In this case, it was known that the homogenization, which will be explained in Sect. 3.1.3, is useful for the efficient computation of Gröbner bases. On the other hand, it is possible that homogenization may increase the computational costs because it increases the number of variables. In order to achieve the effect of homogenization without actual homogenization, Giovini et al. proposed the *sugar selection strategy*.[1]

Definition 3.1.15 (Sugar Selection Strategy). Let F be a set of input polynomials. We define the *sugar $s(f)$* of a polynomial f which is created from F as follows:

1. For $f \in F$, $s(f) = \text{tdeg}(f)$, where $\text{tdeg}(f)$ denotes the *total degree* of f.
2. $s(f + g) = \text{MAX}(s(f), s(g))$.
3. $s(mf) = \text{tdeg}(m) + s(f)$.

An S pair is then selected by the normal strategy from those S pairs whose sugar is the smallest.

The sugar strategy is used in most implementations of the Buchberger algorithm, and it behaves well with both graded and nongraded orderings. We note that, in the example above, strategy 1 is the sugar strategy.

3.1.3 Homogenization

Sometimes the Buchberger algorithm with the sugar strategy causes unnecessary intermediate coefficient swells over $\mathbb{Q}$ (see Sect. 3.2.4). In this case, it may be possible to avoid this by performing an actual homogenization.

Definition 3.1.16 (Homogenization). For $f \in R$, $f \neq 0$, we define the *homogenization $f^h \in R[x_0] = K[x_0, \ldots, x_n]$* by

$$f^h = x_0^{\text{tdeg}(f)} f(x_1/x_0, \ldots, x_n/x_0).$$

For a term ordering $<$ in R, we define the homogenization $<^h$ to be the term ordering in $R[x_0]$ such that for $t, s \in R$

$$x_0^i t <^h x_0^j s \Leftrightarrow i + \text{tdeg}(t) < j + \text{tdeg}(s) \text{ or } (i + \text{tdeg}(t) = j + \text{tdeg}(s) \text{ and } t < s).$$

For a homogeneous polynomial $h \in R[x_0]$, we define the *dehomogenization* by $h|_{x_0=1} = h(1, x_1, \ldots, x_n)$.

The following is clear from the definition of $<^h$.

Proposition 3.1.17. *For $f = c_1 t_1 + \cdots + c_m t_m \in R$ ($t_m < \cdots < t_1$), if $f^h = c_1 x_0^{i_1} t_1 + \cdots + c_m x_0^{i_m} t_m$ then $x_0^{i_m} t_m^h < \cdots < x_0^{i_1} t_1$. For a homogeneous polynomial $h = c_1 x_0^{i_1} t_1 + \cdots + c_m x_0^{i_m} t_m \in R[x_0]$ such that $x_0^{i_m} t_m^h < \cdots <^h x_0^{i_1} t_1$, we have $t_m < \cdots < t_1$.*

---

[1] The name "sugar" comes from [5]. The sugar strategy was first implemented in CoCoA.

Theorem 3.1.18. *For $F = \{f_1 \ldots, f_m\} \subset R$, let $G^h = \{g_1, \ldots, g_l\}$ be a Gröbner basis of $\langle F^h \rangle$ ($F^h = \{f_1^h, \ldots, f_m^h\}$) with respect to $<^h$ consisting of homogeneous polynomials. Then $G = \{g_1|_{x_0=1}, \ldots, g_l|_{x_0=1}\}$ is a Gröbner basis of $\langle F \rangle$ with respect to $<$.*

Problem 3.1.19. Prove Theorem 3.1.18.

3.1.4 Buchberger Algorithm (an Improved Version)

We now present a version of the Buchberger algorithm which incorporates the improvements described above.

Algorithm 3.1.20 ($BuchbergerCore(F)$).

Input: a sequence of polynomials $F = (f_1, \ldots, f_l)$
Output: a Gröbner basis of $\langle F \rangle$
$D \leftarrow \emptyset$
for $m = 1$ to l do
   $D \leftarrow UpdatePairs(D, F, m)$
end for
$m \leftarrow l$
while ($D \neq \emptyset$) do
      $D_{min} = \{\{i, j\} \in D \mid s(S(f_i, f_j))$ is minimal $\}$
      $C = \{i, j\}$ ←an element of D_{min} such that T_{ij} is minimal with respect to
               the term ordering
      $D \leftarrow D \setminus \{C\}$
      $h \leftarrow$ a remainder of $S(f_i, f_j)$ on division by $\{f_1, \ldots, f_m\}$
      if $h \neq 0$ then
         $F \leftarrow (f_1, \ldots, f_m, h)$
         $m \leftarrow m + 1$
         $D \leftarrow UpdatePairs(D, F, m)$
      endif
end while
return F

Algorithm 3.1.21 ($UpdatePairs(D, F, m)$).

Input: a set of pairs of indices D, a sequence of polynomials $F = (f_1, \ldots, f_m)$
Output: a set of pairs created from F after eliminating unnecessary pairs
$D \leftarrow \{\{i, j\} \in D \mid B_m(i, j)$ does not hold$\}$
$N \leftarrow \{\{i, m\} \mid i = 1, \ldots, m - 1, F_k(i, m)$ does not hold for any $k = 1, \ldots, i-1 \}$
$N \leftarrow \{\{i, m\} \in N \mid M_k(i, m)$ does not hold for any $k = 1, \ldots, m - 1$ $(k \neq i) \}$
$N \leftarrow \{\{i, m\} \in N \mid GCD(T_i, T_m) \neq 1\}$
return $D \cup N$

Algorithm 3.1.22 $(Buchberger(F, H))$.

if $H = 1$ then
  $G^h \leftarrow BuchbergerCore(F^h)$
  $G \leftarrow G^h|_{x_0=1}$
else
  $G \leftarrow BuchbergerCore(F)$
endif
$G \leftarrow MinimalGB(G)$
$G \leftarrow ReducedGB(G)$
return G

An ideal generated by homogeneous polynomials is called a homogeneous ideal, so F^h in Algorithm 3.1.22 is a homogeneous ideal. If a set of input polynomials consists of homogeneous polynomials, then all the S polynomials and their remainders are homogeneous and the sugar coincides with the total degree. In this case, we may assume that the specified term ordering is graded and the sugar strategy coincides with the normal strategy. Suppose that we execute Algorithm 3.1.20 for a set of homogeneous polynomials.

Definition 3.1.23. In Algorithm 3.1.20, we set

$$D_d = \{\{i, j\} \in D \mid \text{tdeg}(T_{ij}) = d\}, \quad F_d = \{f_i \in F \mid \text{tdeg}(f_i) \le d\}.$$

Proposition 3.1.24. *Let G be the output of Algorithm 3.1.20. If $D_d = \emptyset$ holds in Algorithm 3.1.20, then $F_d = \{g \in G \mid \text{tdeg}(g) \le d\}$.*

Proof. If $D_d = \emptyset$, then, for each of the S polynomials and their remainders created after that point, its total degree is greater than d. Therefore F_d is stable until the termination of the algorithm.

Proposition 3.1.25. *Let G be the output of Algorithm 3.1.20. We set*

$$S_d = \{a \text{ remainder of } S(f_i, f_j) \text{ on division by } F_d \mid \{i, j\} \in D_d\}$$

for D_d at the point of $D_{d-1} = \emptyset$. We take the set of all monomials $(t_D, \ldots, t_1)$ of total degree d ($D = \dim_K R_d$, $t_D > t_{D-1} > \cdots > t_1$) as an ordered K-basis of $R_d = \{f \in R \mid f \text{ is homogeneous and } \text{tdeg}(f) = d\}$. Suppose that we obtain a basis S'_d of $\text{Span}_K(S_d)$ by applying Gaussian elimination to S_d with respect to this K-basis. Then $S'_d \cup F_d = \{g \in G \mid \text{tdeg}(g) \le d\}$.

Proof. We show that a remainder of $S(f_i, f_j)$ on division by $S'_d \cup F_d$ is 0 for $\{i, j\} \in D_d$. Let $r \in S_d$ be a remainder of $S(f_i, f_j)$ on division by F_d. Since $S'_d = \{g_1, \ldots, g_k\}$ $(\text{in}_<(g_1) > \cdots > \text{in}_<(g_k))$ is a basis of $\text{Span}_K(S_d)$, r can be written as $r = c_1 g_1 + \cdots + c_k g_k$ $(c_1, \ldots, c_k \in K)$. Then $r - c_1 g_1 - \cdots - c_k g_k$ is a remainder of r on division by S'_d, and it is equal to 0.

Based on this proposition, we can replace old basis elements with their remainders on division by new basis elements when we execute Algorithm 3.1.20 for a homogeneous ideal. F_d becomes a reduced polynomial set by this procedure, and it may reduce the computational costs in the subsequent remainder computations.

3.2   Using Macaulay2, SINGULAR, and CoCoA

In this section, we illustrate the fundamental use of Macaulay2[6], SINGULAR[3], and CoCoA[2], which are all available in KNOPPIX/Math.[2] We recommend running Macaulay2 and SINGULAR in Emacs. See Chap. 2 for details of Emacs.

3.2.1   Getting Started

In KNOPPIX/Math, all systems can be started from the Math Software icon 🐧 in the panel, the $\sqrt{x}$ Math submenu in the main menu, or a terminal emulator.

3.2.1.1   Macaulay2

If you start Macaulay2 from the Math menu, it runs in a buffer of Emacs (Fig. 3.1). If you type M2 in a terminal emulator, then Macaulay2 runs in the terminal emulator. In this case, various facilities provided by the Emacs environment will not be available.[3] Therefore, it is convenient to run Macaulay2 in Emacs if you are familiar with Emacs. You can also use Macaulay2 in GNU TEXmacs. The viewHelp command invokes a web browser, and the documentation page will then be displayed (Fig. 3.2). You will find useful information at

Macaulay 2->getting started->a first Macaulay 2 session.

Information on individual commands is available from the index.

3.2.1.2   SINGULAR

If you start ESingular from the Math menu or type ESingular into a terminal emulator, SINGULAR will run in a buffer of Emacs (Fig. 3.3). If you type help in

---

[2]The versions discussed here of Macaulay2, Singular, and CoCoA are 1.4, 3.1.2, and 4.7.5, respectively. They were all executed on a KNOPPIX/Math virtual machine, and the computing time is not accurate.

[3]Only command line editing is available.

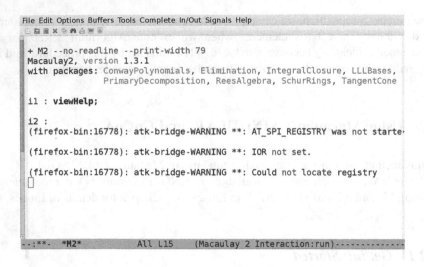

Fig. 3.1 Macaulay2

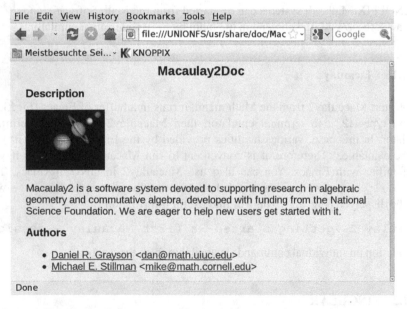

Fig. 3.2 Macaulay2 help browser

ESingular, then the manual is displayed as an Emacs info screen (Fig. 3.4). If you run `Singular` in a terminal emulator, the `help` command invokes a web browser, and the manual is displayed in the browser.

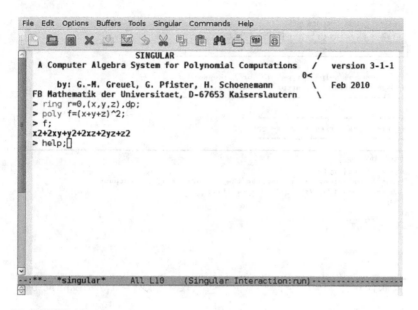

Fig. 3.3 SINGULAR

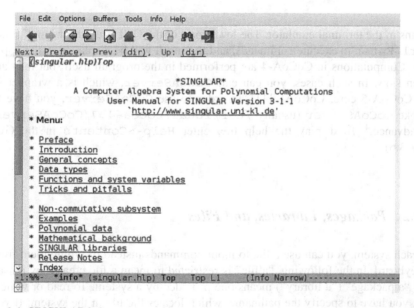

Fig. 3.4 SINGULAR help

3.2.1.3 CoCoA

If you start CoCoA from the Math menu or type xcocoa in a terminal emulator, CoCoA runs in its own GUI (Fig. 3.5). If you type cocoa in a terminal emulator,

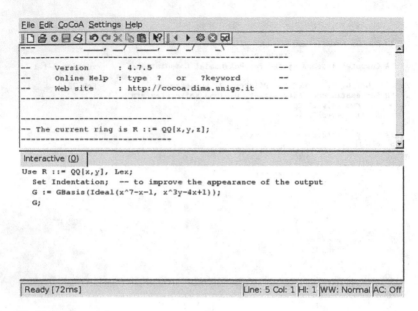

Fig. 3.5 CoCoA (xcocoa)

it runs in the terminal emulator. The lower part of the GUI is used for input. Enter
`Ctrl+Enter` to execute commands, and the result will be displayed in the upper
part. Computations in CoCoA-4 are performed in the program itself, but they are
often slow. In such cases, you can call `CoCoAServer`, which is a wrapper of
the CoCoA-5 core, CoCoALib[1]. In order to use `CoCoAServer`, you have to
invoke `CoCoAServer` (usually `/usr/local/cocoa-4.7/CoCoAServer`)
in advance.[4] To display the help file, enter `Help->Contents` in the GUI
(Fig. 3.6).

3.2.2 Packages, Libraries, and Files

In each system, you can use a file to input commands instead of typing them from
a keyboard. In the following, "a file" is restricted to mean a file created by a user,
and "a package" ("a library") means one provided by a system. To read or write a
file, you have to specify the pathname, which locates the file in the system. If you
start a system from a terminal emulator, the pathname may be relative to the current
directory at the time when the system was started. For instance, the file abc in the
current directory would be specified by `"abc"`, and the file abc in the subdirectory

---

[4] If you are using CoCoA-5, you do not have to do this.

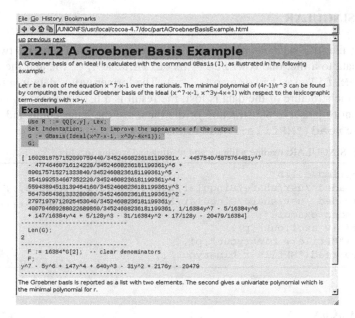

Fig. 3.6 CoCoA help

pqr of the current directory would be specified by `"pqr/abc"`. If a system is invoked from the menu, then the current directory is set to the home directory of the user.

3.2.2.1 Macaulay2

To read commands from a file, use `load`. To read a package, use `loadPackage`. To write data to a file, use *filename*`<<`. In the following example, a new file `out` is created, a list `L` is converted to a character string, the string with a newline character is written to the file, and the file is closed.

Listing 3.1 Macaulay2: reading a package and writing to a file

```
i1 : loadPackage "Normaliz";
i2 : load "normaliz-example";
i3 : L

                 2          2    3     2      2
o3 = {ideal (x , x*y, y , z , x*z , y*z ),...}
o3 : List
i4 : "out"<<toString(L)<<endl<<close;
```

3.2.2.2 SINGULAR

To read a file, use < *filename*. To read a library, use `LIB`. There are two ways
to write data to a file: one is `write(":w filename",...)`, and the other is
`write("MPfile:w filename",...)`. The former writes data in a readable
form, but the structure of the data is lost; the result is a list separated by commas.
If you want to preserve the structure of the data, use the latter command; then to
read it, use `read("MPfile:r filename")`.

Listing 3.2 SINGULAR: reading a library and reading and writing to a file

```
> LIB "primdec.lib";
// ** loaded /usr/share/Singular/LIB/primdec.lib (12508,...)
// ** loaded /usr/share/Singular/LIB/general.lib (12231,...)
> <"primdec-example";
> write(":w asciiout",p);
> write("MPfile:w binaryout",p);
> def pp=read("MPfile:r binaryout");
```

3.2.2.3 CoCoA

To read a file or a package, use `Source`. In order to read a package in the CoCoA
package directory, enter `CocoaPackagePath()` followed by/before the file
name of the package. Note that a package is loaded automatically when calling a
function in the package provided by CoCoA. If you are using the GUI, the content of
a file can be read in the buffer by entering `File->Open`, and you can edit it before
executing it by entering `Ctrl+Enter`. To write data to a file, use `OpenOFile`
and `Print On`.

Listing 3.3 CoCoA: reading a package and reading and writing to a file

```
Source "gb-example";
G;
D:=OpenOFile("out");
Print G On D;
Close(D);
```

3.2.3 Rings, Term Orderings, and Polynomials

In Macaulay2, SINGULAR, and CoCoA, it is necessary to declare a base ring before
starting a computation. A term ordering is specified with the base ring. Below, we
explain how to define the typical term orderings: a graded reverse lexicographic
ordering (grevlex), a lexicographic ordering (lex), and a block (or elimination)
ordering (see Sect. 3.3.1).

3.2.3.1 Macaulay2

QQ and ZZ/p denote $\mathbb{Q}$ and $\mathbb{F}_p = \mathbb{Z}/p\mathbb{Z}$, respectively, and these can be set as coefficient fields. The default term ordering is grevlex, and MonomialOrder is used to specify a different term ordering. An example is QQ[x,y,z,Monomial Order=>Lex], which declares a polynomial ring with a lex ordering such that $x > y > z$. Another example is ZZ/37[x,y,z,u,v,MonomialOrder=>{2,3}], which declares a polynomial ring with a block ordering such that $\{x,y\} \gg \{z,u,v\}$ and the grevlex ordering is applied to each block (see Definition 3.3.1). Indeterminates, which are not declared in the base ring, are not allowed. Each polynomial belongs to a base ring, and if operands of a computation belong to different base rings, it is necessary to apply map. We note that in Macaulay2, variables do not have a defined type and can hold any object.

Listing 3.4 Macaulay2: declaration of a base ring and input of polynomials

```
i1 : R=QQ[x,y,z];
i2 : f=(x+y+z)^2
        2              2              2
o2 = x  + 2x*y + y  + 2x*z + 2y*z + z
o2 : R
i3 : g=y+u
stdio:3:4:(1):[0]: error: no method for binary operator
+ applied to objects:
--              y (of class R)
--      +       u (of class Symbol)

i4 : S=QQ[x,y,z,u]
i5 : f+u
stdio:5:2:(1):[0]: error: expected pair to have a method
for '+'
i6 : h=(map(S,R))(f);
i7 : h+u
        2              2                  2
o7 = x  + 2x*y + y  + 2x*z + 2y*z + z  + u
o7 : S
```

3.2.3.2 SINGULAR

Use ring to declare a base ring. The coefficient field is specified by the characteristic. That is, 0 denotes the field of rationals, and a prime p ($p < 32767$) denotes $\mathbb{F}_p$. Some examples of ring declarations are shown below.

- grevlex ordering
      ring r=0,(x,y,z),dp;
- lex ordering
      ring r=0,(a,b,c),lp;

- block ordering
    ```
    ring r=32003,(u,v,w,x),(dp(2),dp(2));
    ```
    $(dp(2),dp(2))$ means that $\{u, v\} \gg \{w, x\}$ and grevlex is applied to each block.

Listing 3.5 SINGULAR: declaration of a base ring and input of polynomials

```
> ring r=0,(x,y,z),dp;
> poly f=(x+y+z)^2;
> poly g=f+u;
  ? `u` is not defined
  ? error occurred in or before STDIN line 4: `poly g=f+u;`
  ? expected poly-expression. type 'help poly;'
> ring s=0,(x,y,z,u),dp;
> poly h=imap(r,f);
> h+u;
x2+2xy+y2+2xz+2yz+z2+u
```

As in Macaulay2, imap must be applied if the operands of a computation belong to different base rings. In SINGULAR, the type of a variable must be declared before it is used.

3.2.3.3 CoCoA

A base ring is declared by Use, QQ denotes the field of rationals, and ZZ/(p) denotes $\mathbb{F}_p$ (p is a prime less than 32767). The default term ordering is grevlex. Other term orderings can be specified as follows.

- lex ordering
    ```
    Use QQ[x,y], Lex;
    ```
- elimination ordering
    ```
    Use QQ[x,y,z,u], Elim(x..y);
    ```

The latter example declares an elimination ordering that eliminates $\{x, y\}$. Enter the Ord command to display the details of the term ordering.

Listing 3.6 CoCoA: declaration of a base ring and input of polynomials

```
Use R::=QQ[x,y,z];
F:=(x+y+z)^2;
F+u;
ERROR: Undefined indeterminate u
CONTEXT: F + u
-------------------------------
Use S::=QQ[x,y,z,u];
B:=BringIn(F);
B+u;
x^2 + 2xy + y^2 + 2xz + 2yz + z^2 + u
-------------------------------
Use U::=QQ[x[1..5]],Elim(x[1]..x[3]);
Ord(U);
```

```
Mat([
  [1,  1,  1,  0,  0],
  [0,  0,  0,  1,  1],
  [0,  0,  0,  0, -1],
  [0,  0, -1,  0,  0],
  [0, -1,  0,  0,  0]
])
```

In CoCoA, an indeterminate is represented by a single lowercase letter. It can be followed by indices. A variable begins with an uppercase letter. BringIn corresponds to map in Macaulay2 and imap in SINGULAR. In the example above, we see from the matrix defining the elimination ordering that it is not a block ordering.

3.2.4 Computation of Gröbner Bases

We now explain how to compute Gröbner bases in each of these systems. The following is the cyclic-7 (C_7) ideal, which is often used as a benchmark for Gröbner basis computations. It sometimes causes intermediate coefficient swells over $\mathbb{Q}$.

Example 3.2.1 (cyclic-7).

$$C_7 = \langle c_0 + c_1 + c_2 + c_3 + c_4 + c_5 + c_6,$$

$$c_0c_1 + c_1c_2 + c_2c_3 + c_3c_4 + c_4c_5 + c_5c_6 + c_6c_0,$$

$$c_0c_1c_2 + c_1c_2c_3 + c_2c_3c_4 + c_3c_4c_5 + c_4c_5c_6 + c_5c_6c_0 + c_6c_0c_1,$$

$$c_0c_1c_2c_3 + c_1c_2c_3c_4 + c_2c_3c_4c_5 + c_3c_4c_5c_6 + c_4c_5c_6c_0 + c_5c_6c_0c_1 + c_6c_0c_1c_2,$$

$$c_0c_1c_2c_3c_4 + c_1c_2c_3c_4c_5 + c_2c_3c_4c_5c_6 + c_3c_4c_5c_6c_0 + c_4c_5c_6c_0c_1 + c_5c_6c_0c_1c_2$$

$$+ c_6c_0c_1c_2c_3,$$

$$c_0c_1c_2c_3c_4c_5 + c_1c_2c_3c_4c_5c_6 + c_2c_3c_4c_5c_6c_0 + c_3c_4c_5c_6c_0c_1 + c_4c_5c_6c_0c_1c_2$$

$$+ c_5c_6c_0c_1c_2c_3 + c_6c_0c_1c_2c_3c_4,$$

$$c_0c_1c_2c_3c_4c_5c_6 - 1 \rangle$$

3.2.4.1 Macaulay2

In Macaulay2, an ideal is generated by the command ideal. Use the command gb to compute a Gröbner basis. The result is returned as a Gröbner basis object. The generating set can be obtained as a matrix by using the command gens. In the following example, we compute the Gröbner basis for cyclic-7 with respect to a grevlex ordering. Many possible switches are available for controlling the computations, but the computation finished quickly without requiring them. In this

example, the generating set was obtained as a 1×209 matrix by using the command
gens. The i-th element can be obtained by the command g_i. Note that the index
i starts at 0.

Listing 3.7 Macaulay2: Gröbner basis computation

```
i1 : R=QQ[c0,c1,c2,c3,c4,c5,c6];
i2 : I=ideal(c0+c1+c2+c3+c4+c5+c6,...);
o2 : Ideal of R
i3 : G=gb I;
i4 : g=gens G;
               1        209
o4 : Matrix R  <--- R
i5 : g_0
o5 = | c0+c1+c2+c3+c4+c5+c6 |
       1
o5 : R
```

3.2.4.2 SINGULAR

In SINGULAR, an ideal is generated by substituting a list of polynomials for a
variable that is declared to be an ideal. Use the command groebner to compute
a Gröbner basis. The result can be obtained by declaring an ideal variable and
then substituting the result for the variable. The generators are obtained by entering
the variable followed by an index; the index starts at 1. In the following example,
we compute a Gröbner basis for cyclic-7 over $\mathbb{F}_{32003}$. SINGULAR quickly computes
Gröbner bases over finite fields, but it is often slower over the field of rationals. This
is thought to be caused by the unnecessary coefficient swells.

Listing 3.8 SINGULAR: Gröbner basis computation

```
> ring r=32003,(c0,c1,c2,c3,c4,c5,c6),dp;
> ideal i=c0+c1+c2+c3+c4+c5+c6,...;
> ideal g=groebner(i);
> g[1];
c0+c1+c2+c3+c4+c5+c6
> g[2];
c1^2+c1*c3-c2*c3+c1*c4-c3*c4+c1*c5-c4*c5+2*c1*c6+...
```

3.2.4.3 CoCoA

In CoCoA, an ideal is generated by the command Ideal, and a Gröbner basis
is computed by the command GBasis (the reduced Gröbner basis is computed
by ReducedGBasis). Computing Gröbner bases is very slow in general, and
we recommend the use of the command GBasis5 (or ReducedGBasis5) after
starting CoCoAServer.

Listing 3.9 CoCoA: starting `CoCoAServer`

```
noro@ubuntu:~ /usr/local/cocoa-4.7/CoCoAServer
------[   Starting CoCoAServer on port 49344 (0xc0c0)   ]------

Provides operations defined in the following libraries:
  CoCoALib-0.9931 (frobby)
  CoCoALib-0.9931 (groebner)
  CoCoALib-0.9931 (combinatorics)
  CoCoALib-0.9931 (approx)
```

Even when using `GBasis5`, coefficient swells may occur when the input ideals are nonhomogeneous. For example, it takes a very long time to compute the Gröbner basis for cyclic-7, but `GBasis5` runs very fast if the input polynomials are homogenized in advance. To homogenize, use the command `Homogenized`.

Listing 3.10 CoCoA: Gröbner basis computation

```
Use QQ[a,b,c,d,e,f,g,t];
I := Ideal(
abcdefg-1,
abcdef+bcdefg+cdefga+defgab+efgabc+fgabcd+gabcde,
abcde+bcdef+cdefg+defga+efgab+fgabc+gabcd,
abcd+bcde+cdef+defg+efga+fgab+gabc,
abc+bcd+cde+def+efg+fga+gab,
ab+bc+cd+de+ef+fg+ga,
a+b+c+d+e+f+g
);
H:=Homogenized(t,Gens(I));
HG:=ReducedGBasis5(Ideal(H));
-- CoCoAServer: computing Cpu Time = 16.9891
DHG := Subst(HG,t,1);
G  := ReducedGBasis5(Ideal(DHG));
-- CoCoAServer: computing Cpu Time = 149.265
```

For the Gröbner basis `HG` obtained for the homogenized input, a (nonreduced) Gröbner basis `DHG` of `I` is obtained by substituting 1 for t in `DHG`. Finally, we obtain the reduced Gröbner basis of `I` by applying the command `ReducedGBasis5` to `DHG`.[5]

3.2.5 Computation of Initial Ideals

The set of initial monomials of the elements in a Gröbner basis G of an ideal $I \subset R$ ($R = K[x_1, \ldots, x_n]$) generates the initial ideal $\mathrm{in}_<(I)$ of I.

---

[5]This takes a very long time.

3.2.5.1 Macaulay2

In Macaulay2, the initial monomial of a polynomial is called the lead monomial, and it is extracted by using the command `leadMonomial`. To extract the coefficient and term of the lead monomial, use the commands `leadCoefficient` and `leadTerm`, respectively. The initial ideal is obtained if we apply the command `leadTerm` to an ideal. In this case, the Gröbner basis is computed automatically.

Listing 3.11 Macaulay2: computation of initial ideal

```
i1 : R=QQ[x,y,z];
i2 : I=ideal(x^2*y^2-z^2,x^3-y*z^2,x^2*z^4-y^2);
o2 : Ideal of R
i3 : J=ideal leadTerm I
              3  2 2  3 2  5  6   2 4
o3 = ideal (x , x y , y z , y , z , x z )
o3 : Ideal of R
```

3.2.5.2 SINGULAR

When the command `lead` is applied to an ideal, it simply returns the ideal generated by the initial monomials of the generators of the ideal. Therefore, it is necessary to explicitly compute a Gröbner basis of the ideal before applying `lead`.

Listing 3.12 SINGULAR: computation of initial ideal

```
> ring r=0,(x,y,z),dp;
> ideal i=x^2*y^2-z^2,x^3-y*z^2,x^2*z^4-y^2;
> lead(i);
_[1]=x2y2
_[2]=x3
_[3]=x2z4
> ideal g=groebner(i);
> lead(g);
_[1]=x3
...
_[6]=x2z4
```

3.2.5.3 CoCoA

The command `LT` automatically computes a Gröbner basis before finding the initial monomials.

Listing 3.13 CoCoA: computation of initial ideal

```
Use R::= QQ[x,y,z];
I:=Ideal(x^2*y^2-z^2,x^3-y*z^2,x^2*z^4-y^2);
LT(I);
Ideal(x^3, x^2y^2, x^2z^4, y^3z^2, z^6, y^5)
```

3.2.6 Computation of Quotient and Remainder

An immediate application of the remainder computation is to test whether a polynomial f belongs to an ideal I. This can be done by checking the remainder of f on division by a Gröbner basis of I with respect to any term ordering. If we have a Gröbner basis of an ideal J, then we can test whether $I \subset J$ holds for the two ideals. In the examples below, we compute the remainder to check whether some power of f belongs to an ideal I. Note that we cannot use this method to show that no power of f belongs to I. See Sect. 3.3.3 for how to check this by computing a Gröbner basis.

3.2.6.1 Macaulay2

Using Macaulay2, it is possible to compute the quotient and the remainder when a polynomial is divided by a Gröbner basis or a matrix.

- remainder(f, g) returns the remainder r of f on division by g.
- quotient(f, g) returns the quotient r of f on division by g.
- quotientRemainder(f, g) returns (q, r) for the quotient q and the remainder of f on division by g.

The input f is a matrix, and g is a Gröbner basis or a matrix. If g is a Gröbner basis, then 0 is returned as the quotient. If g is a matrix, then q and r such that $gq + r = f$ are returned. For example, if g is a row vector $(g_0, \ldots, g_l)$, then q is a column vector ${}^t(q_0, \ldots, q_l)$ such that $q_0 g_0 + \cdots + q_l g_l + r = f$, where the remainder is r.

Listing 3.14 Macaulay2: computation of quotient and remainder

```
i1 : R=QQ[x,y,z];
i2 : I=ideal(x^4*y^2+z^2-4*x*y^3*z-2*y^5*z,x^2+2*x*y^2+y^4);
o2 : Ideal of R
i3 : G=gb I;
i4 : g=gens G;
             1        3
o4 : Matrix R  <--- R
i5 : f=y*z-x^3;
i6 : remainder(matrix{{f}},G)
o6 = | -x3+yz |
             1        1
o6 : Matrix R  <--- R
i7 : remainder(matrix{{f^2}},G)
o7 = | 2x2y3z+2x3yz+2y2z2+2xz2 |
             1        1
o7 : Matrix R  <--- R
```

Listing 3.15 Macaulay2: computation of quotient and remainder (continued)

```
i8  : remainder(matrix{{f^3}},G)
o8 = 0
                 1        1
o8  : Matrix R   <--- R
i9  : qr=quotientRemainder(matrix{{f^3}},g);
o9  : Sequence
i10 : q=qr_0;
                 3        1
o10 : Matrix R   <--- R
i11 : g*q
o11 = | -x9+3x6yz-3x3y2z2+y3z3 |
                 1        1
o11 : Matrix R   <--- R
i12 : g*q-f^3
o12 = 0
```

In this example, $f^3 \in I$ is obtained by computing the remainders of f, f^2, and f^3 from division by a set of generators of I.

We can check whether two ideals are identical by using the command `remainder`. For this purpose, we can also use the operator `==`. When `J==I` is executed, Gröbner bases of both ideals are automatically computed, and their equivalence is checked. To check for ideal membership or ideal inclusion, use the command `isSubset`.

Listing 3.16 Macaulay2: ideal inclusion

```
i1  : R=QQ[x,y,z];
i2  : I=ideal(x*y^2-z^2,x^2*z-x^2,y^2*z^2-x^3);
o2  : Ideal of R
i3  : isSubset(ideal(x^2*z-x^2),I)
o3 = true
i4  : J=ideal(y^4,z^2*y^2,z^4,-y^2*x+z^2,z^2*x,x^2);
o4  : Ideal of R
i5  : isSubset(I,J)
o5 = true
```

In this example, the first `isSubset` checks $x^2z - x^2 \in I$. Since `isSubset` requires ideals for its arguments, the polynomial $x^2z - x^2$ is passed as a principal ideal. The second `isSubset` is a test of $I \subset J$. We note that it is not necessary to explicitly compute a Gröbner basis in order to test for inclusion.

3.2.6.2 SINGULAR

- The command `reduce(f, I)` returns the remainder of f on division by the generating set of an ideal I.

    If the given generating set is not a Gröbner basis, the remainder may not be 0 even if $f \in I$.
- The command `division(f, I)` returns the quotient and the remainder modulo an ideal I.

This command automatically computes a Gröbner basis of I, and the remainder is computed by using the Gröbner basis. The quotient is computed for the original generating set of I. The third element of the returned result is the multiplier for computing a remainder for the case that the ordering is not a term ordering, and this is equal to 1 for a term ordering.

Listing 3.17 SINGULAR: computation of quotient and remainder

```
> ring r=0,(x,y,z),dp;
> ideal i=x^4*y^2+z^2-4*x*y^3*z-2*y^5*z,x^2+2*x*y^2+y^4;
> poly f=y*z-x^3;
> reduce(f,i);
// ** i is no standard basis
-x3+yz
> reduce(f^3,i);
// ** i is no standard basis
-x9+3x6yz-3x3y2z2+y3z3
> division(f^3,i);
[1]:
   _[1,1]=x3y2+2x4+y3z
   _[2,1]=-x7+2x3y3z+3x4yz+2y4z2-2x2z2
[2]:
   _[1]=0
[3]:
   _[1,1]=1
```

The command `reduce` can be applied to an ideal, and we can check $I = J$ by checking that both `reduce(I, J)` and `reduce(J, I)` consist of only 0.

3.2.6.3 CoCoA

- The command $NF(f, I)$ returns the remainder of f modulo an ideal I.

  This command automatically computes a Gröbner basis and returns the remainder on division by the Gröbner basis.
- The command $DivAlg(f, l)$ returns the quotient and the remainder of f on division by a list of polynomials l. The quotient is returned as a list. If the polynomial list l is not a Gröbner basis, the remainder may not be 0 even if $f \in I$.

Listing 3.18 CoCoA: computation of quotient and remainder

```
Use R::=QQ[x,y,z];
I:=Ideal(x^4*y^2+z^2-4*x*y^3*z-2*y^5*z,x^2+2*x*y^2+y^4);
F:=y*z-x^3;
NF(F^3,I);
0
-------------------------------
QR:=DivAlg(F^3,Gens(I));
QR;
Record[Quotients := [0, 0],
Remainder := -x^9 + 3x^6yz - 3x^3y^2z^2 + y^3z^3]
```

```
-----------------------------------
QR:=DivAlg(F^3,GBasis(I));
QR;
Record[Quotients := [-4xy^2z^2 - 2x^2z^2, -2xyz, -x^3 - yz],
Remainder := 0]
```

3.3 Operations on Ideals by Using Gröbner Bases

In this section, we introduce algorithms for realizing various operations for ideals by using Gröbner bases. In many systems, most of these operations are provided as a single command. Understanding of the methods, however, may allow you to perform this more efficiently. We set $R = K[x_1, \ldots, x_n]$.

3.3.1 Elimination Ordering

In many operations on ideals, it is necessary to eliminate variables. A term ordering satisfying the condition in Theorem 1.4.1 is called an elimination ordering, and it can be used to eliminate variables. A lex ordering is a typical elimination ordering, but because it is not efficient, its use should be avoided as far as possible. In many cases, a term ordering called a block ordering (a product ordering) is sufficient.

Definition 3.3.1 (A Block Ordering). Suppose that term orderings $<_Y$ and $<_Z$ are given on $K[Y]$ and $K[Z]$, respectively, for two sets of indeterminates Y, Z ($Y \cap Z = \emptyset$). We define a term ordering $<$ on $K[X]$ ($X = Y \cup Z$) by

$$t_Y t_Z < s_Y s_Z \Leftrightarrow t_Y <_Y s_Y \text{ or } (t_Y = s_Y \text{ and } t_Z <_Z s_Z)$$

for $t_Y, s_Y \in K[Y]$ and $t_Z, s_Z \in K[Z]$. Then $<$ is an elimination ordering on $K[X]$. We call $<$ the *block ordering* (product ordering) such that $Y \gg Z$.

For an ideal I in $K[X]$ ($X = Y \cup Z$, $Y \cap Z = \emptyset$), a generating set of $I_Z = I \cap K[Z]$ is given by $G_Z = G \cap K[Z]$ for a Gröbner basis G with respect to any elimination ordering $<$ such that $Y \gg Z$. Furthermore G_Z is a Gröbner basis of I_Z with respect to $<'$ which is the restriction of $<$ to $K[Z]$. For this purpose, we recommend the use of a block ordering or an elimination ordering similar to a block ordering (see Sect. 3.2.3.3). As an example, we will compute $I \cap \mathbb{Q}[z]$ for an ideal $I = \langle x^2 - z, xy - 1, x^3 - x^2 y - x^2 - 1 \rangle \subset \mathbb{Q}[x, y, z]$.

3.3.1.1 Macaulay2

We compute a Gröbner basis G of I with respect to a block ordering such that $\{x, y\} \gg \{z\}$ and then apply a grevlex ordering to each block. Then Gz, a

Gröbner basis of $I Âě\ cap\ Q[z]$, is extracted from G. In order to compute Gz, use the command selectInSubring. The command selectInSubring(i, m) returns the matrix that consists of those columns of m whose variables do not belong to the first i blocks of the block ordering.

Listing 3.19 Macaulay2: Gröbner basis computation of an elimination ideal

```
i1 : R=QQ[x,y,z,MonomialOrder=>{2,1}];

i2 : I=ideal(x^2-z,x*y-1,x^3-x^2*y-x^2-1);
o2 : Ideal of R
i3 : G=gens gb I;
                1          3
o3 : Matrix R   <--- R
i4 : Gz=selectInSubring(1,G)
o4 = | z3-3z2-z-1 |
```

3.3.1.2 SINGULAR

Let m be the product of variables in a variable set Y. A generating set of $I \cap K[X \setminus Y]$ can be computed by using the command eliminate(I, m). Any term ordering can be set for the base ring.

Listing 3.20 SINGULAR: Gröbner basis computation of an elimination ideal

```
> ring r=0,(x,y,z),dp;
> ideal i=x^2-z,x*y-1,x^3-x^2*y-x^2-1;
> eliminate(i,x*y);
_[1]=z3-3z2-z-1
```

3.3.1.3 CoCoA

The command Elim(v, I) computes the ideal $I \cap K[X \setminus v]$ for the set of variables X of the base ring. Any term ordering can be set. The input v is a variable or a list of variables, such as $[x, y]$, $x..z$.

Listing 3.21 CoCoA: Gröbner basis computation of an elimination ideal

```
Use R::=QQ[x,y,z];
I:=Ideal(x^2-z,x*y-1,x^3-x^2*y-x^2-1);
Elim(x..y,I);
Ideal(-1/2z^3 + 3/2z^2 + 1/2z + 1/2)
```

In each system, a Gröbner basis computation with respect to an elimination ordering is used to compute an elimination ideal. This is done according to a setting determined by each system, and it may cause a difficulty with the computations. Frequently, homogenization can be applied to overcome this difficulty. As an example, we compute $I \cap \mathbb{Q}[d, e]$ in CoCoA for

$$I = \langle 4a^3 + 3cb^2a^2 - a + 2, -3ca^3 + 3c^2a^2 - a + 3cb^4 + c^3,$$
$$-6a^2 + (-3c^3 - 4)a + 3, (4cb^2 - 1)a^2 + (db + c)a + b^4,$$
$$2a^3 - 3a^2 + (-d^2 - 2e)b + 3 \rangle.$$

If we apply the built-in command Elim5, it will continue to run for several minutes; so instead, we will try homogenization.

Listing 3.22 CoCoA: Computation of an elimination ideal via homogenization

```
Use R::=QQ[a,b,c,d,e,t];
I:=Ideal(4*a^3+3*c*b^2*a^2-a+2,-3*c*a^3+3*c^2*a^2-a+3*c*b^4+c^3,
-6*a^2+(-3*c^3-4)*a+3,(4*c*b^2-1)*a^2+(d*b+c)*a+b^4,
2*a^3-3*a^2+(-d^2-2*e)*b+3);
H:=Homogenized(t,Gens(I));
J:=Ideal(H);
E:=Elim5(a..c,J);
-- CoCoAServer: computing Cpu Time = 34.3644
--------------------------------
DE:=Subst(E,t,1);
G:=ReducedGBasis5(Ideal(DE));
-- CoCoAServer: computing Cpu Time = 0.5099
```

The computation successfully terminates within 35 s.

Gröbner bases with respect to an elimination ordering can be applied to compute Gröbner bases over rational function fields.

Theorem 3.3.2. *Let* $\{f_1, \ldots, f_l\} \subset K[X, Y]$ *be a set of generators of an ideal* $I \subset K(Y)[X]$. *For* $J = \langle f_1, \ldots, f_l \rangle \subset K[X, Y]$, *a Gröbner basis of* J *with respect to an elimination ordering* $<$ *such that* $X \gg Y$ *is a Gröbner basis of* I *with respect to a term ordering* $<'$ *which is the restriction of* $<$ *to* $K(Y)[X]$.

Problem 3.3.3. Show Theorem 3.3.2.

A Gröbner basis over a rational function field can be computed by using the Buchberger algorithm, but this requires polynomial GCD computations in order to reduce fractions; this can be expensive. If we apply Theorem 3.3.2, it is not necessary to compute polynomial GCDs. Instead, the number of S polynomials will increase. It is hard to predict which method will be most efficient for a particular input ideal.

3.3.2 Sum, Product, and Intersection of Ideals

For ideals $I = \langle A \rangle$ and $J = \langle B \rangle$ in R, the sets of generators of the sum $I + J = \{f + g \mid f \in I, g \in J\}$ and the product $IJ = \langle \{fg \mid f \in I, g \in J\} \rangle$ are $A \cup B$ and $\{ab \mid a \in A, b \in B\}$, respectively. But these sets of generators are not necessarily Gröbner bases even if A and B are Gröbner bases. In order to determine the Gröbner bases of these ideals, it is necessary to compute them.

Lemma 1.4.3 shows that the intersection of ideals can be obtained by computing an elimination ideal. The intersection of ideals $I_1, I_2, \ldots$ is obtained by executing the command intersect$(I_1, I_2, \ldots)$ in Macaulay2 and SINGULAR and by intersection$(I_1, I_2, \ldots)$ in CoCoA. Here we show an example in Macaulay2. In this example, we compute the primary decomposition (see Sect. 3.7) of a binomial ideal I, and we then check that the intersection of all components coincides with I.

Listing 3.23 Macaulay2: intersection of ideals

```
i1 : R=QQ[a,b,c,d,e,f,g,h,i];
i2 : I=ideal(e*a-d*b,f*b-e*c,h*d-g*e,i*e-h*f);
o2 : Ideal of R
i3 : PD=primaryDecomposition(I);
i4 : J=intersect(PD)
o4 = ideal (f*h - e*i, e*g - d*h, c*e - b*f, b*d - a*e)
o4 : Ideal of R
i5 : I==J
o5 = true
```

3.3.3   Radical Membership Test

Let I be an ideal, and let f be a polynomial. If $f \in I$, then f vanishes on $V(I)$, the *zero set* of I, but the converse does not hold in general. If K is an algebraically closed field, then f vanishes on $V(I)$ if and only if $f \in \sqrt{I}$ (Hilbert's Nullstellensatz).

Definition 3.3.4. For an ideal $I \subset R$, we define the *radical* $\sqrt{I}$ of I by

$$\sqrt{I} = \{f \in R \mid f^m \in I \text{ for some positive integer } m\}$$

$\sqrt{I}$ is an ideal in R.

Theorem 3.3.5. *For an ideal $I \subset R$ and $f \in R$, the following are equivalent.*

1. *$f \in \sqrt{I}$.*
2. *$R[t]I + \langle tf - 1 \rangle = K[x_1, \ldots, x_n, t]$.*
3. *The reduced Gröbner basis of $R[t]I + \langle tf - 1 \rangle$ with respect to any term ordering is equal to $\{1\}$.*

Problem 3.3.6. Show Theorem 3.3.5. (It is easy to show 1. $\Rightarrow$ 2. To show 2. $\Rightarrow$ 1., write 1 as an element of $R[t]I + \langle tf - 1 \rangle$ and substitute $1/f$ for t.)

Using this theorem, $f \in \sqrt{I}$ can be checked by computing a Gröbner basis of $R[t]I + \langle tf - 1 \rangle$. Note that we can apply any term ordering, but usually grevlex is used.

In the following example, we check $f \in \sqrt{I}$ in Sect. 3.2.6 by using the method explained here.

3.3.3.1 Macaulay2

Listing 3.24 Macaulay2: radical membership test

```
i1 : R=QQ[t,x,y,z];
i2 : I=ideal(x^4*y^2+z^2-4*x*y^3*z-2*y^5*z,x^2+2*x*y^2+y^4);
o2 : Ideal of R
i3 : f=y*z-x^3;
i4 : gens gb (I+ideal(t*f-1))
o4 = | 1 |
```

Since the reduced Gröbner basis of $R[t]I + \langle tf - 1 \rangle$ is $\{1\}$, $f \in \sqrt{I}$ holds.

3.3.3.2 SINGULAR

Listing 3.25 SINGULAR: radical membership test

```
> ring r=0,(t,x,y,z),dp;
> ideal i=x^4*y^2+z^2-4*x*y^3*z-2*y^5*z,x^2+2*x*y^2+y^4;
> poly f=y*z-x^3;
> ideal j=t*f-1,i;
> groebner(j);
_[1]=1
```

Since the reduced Gröbner basis of $R[t]I + \langle tf - 1 \rangle$ is $\{1\}$, $f \in \sqrt{I}$ holds.

3.3.3.3 CoCoA

Listing 3.26 CoCoA: radical membership test

```
Use R::=QQ[x,y,z];
I:=Ideal(x^4*y^2+z^2-4*x*y^3*z-2*y^5*z,x^2+2*x*y^2+y^4);
F:=y*z-x^3;
IsInRadical(F,I);
True
-------------------------------
MinPowerInIdeal(F,I);
3
```

A built-in function IsInRadical is available. Furthermore, for a polynomial $f \in \sqrt{I}$, the command MinPowerInIdeal computes the minimal m such that $f^m \in I$.

3.3.4 Ideal Quotient and Saturation

Definition 3.3.7 (Ideal Quotient). For ideals $I, J \subset R$, we define the *ideal quotient* $I : J$ by $I : J = \{f \mid fJ \subset I\}$. For $J = \langle f \rangle$, we write $I : J$ as $I : f$.

The following theorem is an easy consequence of the definition.

Theorem 3.3.8. *1. If $J = \langle g_1, \ldots, g_l \rangle$, then $I : J = \bigcap_{i=1}^{l} (I : g_i)$.*

2. $I : g = (I \cap \langle g \rangle)/g$.

$(I \cap \langle g \rangle)/g$ is an ideal generated by the quotients of the generators of $I \cap \langle g \rangle$ on division by g. Therefore, $I : J$ is obtained by computing the intersections of the ideals.

Definition 3.3.9 (Saturation). For ideals $I, J \subset R$, the *saturation* $I : J^\infty$ is defined by $I : J^\infty = \bigcup_{m=1}^{\infty} (I : J^m)$. For $J = \langle f \rangle$, we write $I : J^\infty$ as $I : f^\infty$.

Theorem 3.3.10. *1. If $J = \langle g_1, \ldots, g_l \rangle$, then $I : J = \bigcap_{i=1}^{l} (I : g_i^\infty)$.*

2. $I : g^\infty = (I + \langle tg - 1 \rangle) \cap R$.

Problem 3.3.11. Show Theorem 3.3.10 by using Sect. 3.3.3.

According to this theorem, $I : J^\infty$ can be computed from the intersection of the ideals followed by elimination. In the next example, we will compute $I : x^k = (I : x^{k-1}) : x$ recursively for $I = \langle x^4 - y^5, x^3 - y^7 \rangle \subset \mathbb{Q}[x, y]$ until $I : x^k$ becomes stable; we will then check that it coincides with $I : x^\infty$.

3.3.4.1 Macaulay2

The ideal quotient $I : J$ and the saturation $I : J^\infty$ for ideals I, J are computed by the commands $\text{quotient}(I, J)$ and $\text{saturate}(I, J)$, respectively.

Listing 3.27 Macaulay2: ideal quotient and saturation

```
i1 : R=QQ[x,y];
i2 : I=ideal(x^4-y^5,x^3-y^7);
o2 : Ideal of R
i3 : I1=quotient(I,x);
o3 : Ideal of R
i4 : I2=quotient(I1,x);
o4 : Ideal of R
i5 : I1==I2
o5 = false
i6 : I3=quotient(I2,x);
o6 : Ideal of R
i7 : I2==I3
o7 = false
i8 : I4=quotient(I3,x);
o8 : Ideal of R
i9 : I3==I4
o9 = true
```

```
i10 : J=saturate(I,x);
o10 : Ideal of R
i11 : I3==J;
o11 = true
```

3.3.4.2 SINGULAR

$I : J$ is computed by the command $\text{quotient}(I, J)$. $I : J^\infty$ is computed by $\text{sat}(I, J)$, which is defined in elim.lib. $\text{sat}(I, J)$, and which returns a pair $I : J^\infty$ and the smallest integer m such that $I : J^\infty = I : J^m$.

Listing 3.28 SINGULAR: ideal quotient and saturation

```
> ring r=0,(x,y),dp;
> ideal i=x^4-y^5,x^3-y^7;
> LIB "elim.lib";
...
> list s=sat(i,x);
> s;
[1]:
   _[1]=xy2-1
   _[2]=y5-x4
   _[3]=x5-y3
[2]:
   3
> ideal g3=quotient(i,x^3);
> g3=groebner(g3);
> g3;
g3[1]=xy2-1
g3[2]=y5-x4
g3[3]=x5-y3
> size(reduce(g3,s[1]));
0
> size(reduce(s[1],g3));
0
```

The command size returns the number of nonzero elements in a list. The above result shows that g3 coincides with s[1].

3.3.4.3 CoCoA

$I : J$ and $I : J^\infty$ are computed by the commands $\text{Colon}(I, J)$ and I:J, $\text{Saturation}(I, J)$, respectively. Both functions accept only an ideal as the second argument.

Listing 3.29 CoCoA: ideal quotient and saturation

```
Use R::=QQ[x,y];
I:=Ideal(x^4-y^5,x^3-y^7);
I:Ideal(x^3);
```

```
Ideal(xy^2 - 1, y^5 - x^4, x^5 - y^3)
-------------------------------
Colon(I,Ideal(x^3));
Ideal(xy^2 - 1, y^5 - x^4, x^5 - y^3)
-------------------------------
Saturation(I,Ideal(x));
Ideal(xy^2 - 1, y^5 - x^4, x^5 - y^3)
```

3.3.5 Computation of a Radical

As seen in Sect. 3.3.3, the test for radical membership is performed by computing a Gröbner basis. However, the computation of the radical of an ideal is not an easy task. In this section, as an application of an operation on ideals, we outline an algorithm for computing radicals.

Definition 3.3.12 (The Squarefree Part of a Polynomial). Let $f = f_1^{n_1} \cdots f_l^{n_l}$ $(n_1, \ldots, n_l \geq 1)$ be the irreducible factorization of a polynomial f. We call the product $f_1 \cdots f_l$ the *squarefree part* of f.

Theorem 3.3.13 (Seidenberg). *Let K be a perfect field, and let I be a zero-dimensional ideal in $K[x_1, \ldots, x_n]$ (see Sect. 3.4.1). Let f_i $(i = 1, \ldots, n)$ be the squarefree part of the generator of $I \cap K[x_i]$. Then $\sqrt{I} = I + \langle f_1, \ldots, f_n \rangle$.*

Definition 3.3.14 (Extension and Contraction). Let I be an ideal in $K[X]$. For $U \subset X$, an ideal in $K(U)[X \setminus U]$ generated by I is denoted by I^e and is called the *extension* of I to $K(U)[X \setminus U]$. For an ideal J in $K(U)[X \setminus U]$, an ideal $J \cap K[X]$ in $K[X]$ is denoted by J^c and is called the *contraction* of J to $K[X]$.

Theorem 3.3.15. *Let J be an ideal in $K(U)[X \setminus U]$. Suppose that a Gröbner basis $G = \{g_1, \ldots, g_l\} \subset K[U][X \setminus U] = K[X]$ of J with respect to a term ordering $<_1$ on $X \setminus U$ is given. Let $h_i \in K[U]$ be the leading coefficient of g_i with respect to $<_1$, and let f be the squarefree part of $\mathrm{LCM}(h_1, \ldots, h_l)$. Then $J^c = \tilde{J} : f^\infty$ for $\tilde{J} = \langle G \rangle \subset K[X]$.*

Problem 3.3.16. Show Theorem 3.3.15.

Corollary 3.3.17. *Let $<_1$, $<_2$ be term orderings on $X \setminus U$, U, respectively, and let $<$ be the block ordering defined by $<_1$ and $<_2$ such that $(X \setminus U) \gg U$. Let $G = \{g_1, \ldots, g_l\}$ be a Gröbner basis of an ideal $I \subset K[X]$ with respect to $<$. Let $h_i \in K[U]$ be the leading coefficient of g_i with respect to $<_1$, and let f be the squarefree part of $\mathrm{LCM}(h_1, \ldots, h_l)$. Then $I^{ec} = I : f^\infty$.*

Definition 3.3.18 (Maximal Independent Set). Let I be an ideal in $K[X]$. A subset $U \subset X$ satisfying the following conditions is called a *maximal independent set* of I.

1. U is an independent set, that is, $K[U] \cap I = \{0\}$.
2. $K[U \cup \{x\}] \cap I \neq \{0\}$ for any $x \in X \setminus U$.

Remark 3.3.19. 1. The largest number of elements of a maximal independent set of I is equal to the Krull dimension $\dim I$.

2. A subset $U \subset X$ such that $\text{in}_<(I) \cap K[U] = \{0\}$ (strongly independent set) is an independent set. If the term ordering is graded, then the largest number of elements of a strongly independent set is equal to $\dim I$. Thus a maximal independent set is computed by using $\text{in}_<(I)$.

Theorem 3.3.20. *If* $I : f^\infty = I : f^s$ *then* $I = (I : f^s) \cap (I + \langle f^s \rangle)$.

Problem 3.3.21. Show Theorem 3.3.20.

Theorem 3.3.22. *Let* I *be an ideal in* $K[X]$, *and let* $U \subset X$ *be a maximal independent set for* I. *Then* f *in Corollary 3.3.17 satisfies* $f \notin I$ *and* $\sqrt{I} = \sqrt{I^{ec}} \cap \sqrt{I + \langle f \rangle}$.

Problem 3.3.23. Show Theorem 3.3.22.

Algorithm 3.3.24 ($Radical(I)$).

Input: an ideal I in $K[X]$
Output: the radical of I
if $I = K[X]$ then return $K[X]$
$U \leftarrow$ a maximal independent set of I
$f \leftarrow$ a polynomial such that $I^{ec} = I : f^\infty$ (computed by Corollary 3.3.17)
$\tilde{J} \leftarrow \sqrt{I^e}$ (computed by Theorem 3.3.13; I^e has dimension zero)
$J \leftarrow \tilde{J}^c$ (computed by Theorem 3.3.15)
$J' \leftarrow Radical(I + \langle f \rangle)$
return $J \cap J'$

Since $f \notin I$, $I + \langle f \rangle$ is strictly larger than I and this algorithm terminates by the Noetherian property. The output is equal to $\sqrt{I}$ by Theorem 3.3.22.

3.4 Change of Ordering

When using the Buchberger algorithm to compute a Gröbner basis with respect to a term ordering, the computation often gets stuck because there are too many intermediate basis elements or because their coefficients are too large. This phenomenon typically happens when using the Buchberger algorithm to compute a lex Gröbner basis. To avoid this difficulty, we can apply a *change of ordering* (basis conversion). In this method, we first compute a Gröbner basis for an input ideal and with respect to some other term ordering. Then use the computed Gröbner basis to compute one with respect to the target ordering. The fact that the input is also a Gröbner basis can be used to speed up the computation of the subsequent Gröbner basis. There are several methods for changing the ordering. Of these, we will introduce the FGLM algorithm for zero-dimensional ideals and the Hilbert-driven algorithm for homogeneous ideals.

3.4.1 FGLM Algorithm

Theorem 3.4.1. *For an ideal $I \subset R = K[X]$, the following are equivalent:*

1. $\dim I = 0$.
2. R/I *is finite dimensional as a K-vector space.*
3. *A Gröbner basis G of I contains g_i such that $\text{in}_<(g_i) = x_i^{m_i}$ for each $x_i(i = 1, \ldots, n)$;*
4. $V_{\overline{K}}(I)$ *is a finite set for the algebraic closure $\overline{K}$ of K.*
5. *For each x_i $(i = 1, \ldots, n)$, I contains a univariate polynomial of x_i.*

Definition 3.4.2 (Zero-Dimensional Ideal). An ideal $I \subset R$ is called a *zero-dimensional ideal* if it satisfies the conditions in Theorem 3.4.1.

A zero-dimensional ideal represents a system of algebraic equations for which the zero set is finite. If we have a lex Gröbner basis of a zero-dimensional ideal, then we can solve the corresponding system of equations by using the method in Sect. 1.4.2. However, it is often difficult to compute a lex Gröbner basis by directly applying the Buchberger algorithm. The *FGLM algorithm* uses linear algebra to compute a Gröbner basis G_1 of a zero-dimensional ideal I with respect to the target ordering (e.g., lex ordering) starting from a Gröbner basis G_0 of I with respect to a term ordering (e.g., grevlex ordering).

For a term ordering $<$, the set of all monomials outside $\text{in}_<(I)$ is called the *standard monomial set* and is denoted by $SM_<(I)$. By the definition of a Gröbner basis, $SM_<(I)$ forms a K-basis of R/I. In particular, if I is zero-dimensional, then $SM_<(I)$ is a finite set, because R/I is a finite-dimensional K-vector space. In the FGLM algorithm, we seek the elements in the standard monomial set $SM_{<_1}(I)$ with respect to the target term ordering $<_1$ in increasing order with respect to $<_1$. In order to do this, we need two operations:

1. Find the minimal monomial with respect to $<_1$ in a set of monomials.
2. Decide whether I contains a K-linear sum of given monomials. Compute it if it exists.

The algorithm is shown below. In the algorithm, $\text{NF}_<(f, F)$ denotes the remainder of f on division by a Gröbner basis F.

Algorithm 3.4.3 (the FGLM algorithm).

Input: a Gröbner basis F of a zero-dimensional ideal I with respect to $<$
Output: a Gröbner basis of I with respect to $<_1$
$G \leftarrow \emptyset; h \leftarrow 1; B \leftarrow \{h\}; H \leftarrow \emptyset$
do
    $N \leftarrow \{u \mid h <_1 u \text{ and } m \nmid u \text{ for all } m \in H\}$
    if $N = \emptyset$ then return G
(1) $h_1 \leftarrow$ the minimal monomial in N with respect to $<_1$
(2) $E \leftarrow \text{NF}_<(h_1, F) + \sum_{t \in B} a_t \text{NF}_<(t, F)$

if there exist $a_t \in K(t \in B)$ such that $E = 0$ then

$$G \leftarrow G \cup \{h_1 + \sum_{t \in B} a_t t\}; \; H \leftarrow H \cup \{h_1\}$$

else $B \leftarrow \{h_1\} \cup B$

$h \leftarrow h_1$

end do

Steps (1) and (2) in the algorithm correspond to the above operations 1 and 2, respectively. In Step (2), we obtain a system of linear equations of a_t by equating the coefficient of each monomial in E to 0. A solution of the system of equations gives a K-linear sum of $\{h_1\} \cup B$ in I. In Step (1), we know that the number of candidates for h_1 is finite, because of the following proposition.

Proposition 3.4.4. *In (1),* $h_1 \in x_1 B \cup \cdots \cup x_n B$.

We will not prove Proposition 3.4.4 or the correctness of Algorithm 3.4.3. The FGLM algorithm is implemented in most systems. The algorithm requires only computing the remainder and solving systems of linear equations. Therefore, its proof is left as an instructive exercise. We note that we have presented Algorithm 3.4.3 primarily to aid understanding of the essence of the FGLM algorithm. From the viewpoint of efficiency, some improvements are necessary.

We now apply the FGLM algorithm to the example in Sect. 3.3.1.3 in SINGULAR. Some coefficients in the lex Gröbner basis of this example are more than 10^{5000}, and it is difficult to use the Buchberger algorithm because of the coefficient swells.

Listing 3.30 SINGULAR: FGLM

```
> timer=1;
> option(redSB);
> ring r=0,(a,b,c,d,e),dp;
> ideal i= 4*a^3+3*c*b^2*a^2-a+2,-3*c*a^3+3*c^2*a^2-a
+3*c*b^4+c^3,-6*a^2+(-3*c^3-4)*a+3,(4*c*b^2-1)*a^2
+(d*b+c)*a+b^4,2*a^3-3*a^2+(-d^2-2*e)*b+3;
> ideal g=groebner(i);
//used time: 57.31 sec
> ring s=0,(a,b,c,d,e),lp;
> ideal j=fglm(r,g);
//used time: 32.55 sec
```

To compute a grevlex Gröbner basis, the option redSB is specified, because the command fglm in SINGULAR requires a reduced Gröbner basis as an input. The computing time of the lex Gröbner basis is shorter than that for the grevlex basis. But if we try to compute the lex basis by using the Buchberger algorithm, it will get stuck because of the coefficient swells. We note that the command stdfglm in the library standard.lib will execute this entire example.

3.4.2 Hilbert-Driven Algorithm

The FGLM algorithm can only be applied to zero-dimensional ideals. Here, we introduce the *Hilbert-driven algorithm*, which can be applied to any homogeneous ideal. If we apply the homogenization, the Hilbert-driven algorithm can be applied to any ideal. See [8] Sect. 5.3 for details.

We set $R_d = \{f \in R \mid f \text{ is homogeneous and tdeg}(f) = d\}$. As explained in Sect. 1.6, the Hilbert function $H(R/I; d)$ for a homogeneous ideal I is given by a Gröbner basis with respect to any term ordering. Section 3.1 tells us that too many unnecessary S-pairs may make the Buchberger algorithm inefficient. Suppose that we apply the Buchberger algorithm to a homogeneous ideal in increasing order with respect to the total degree of the S-polynomials (see Sect. 3.1.3). If the number of those monomials in R_d which are not divisible by the initial monomial of any intermediate basis element is equal to the value of $H(R/I; d)$, then it means that we have obtained a K-basis of $(R/I)_d$ and that the remaining S polynomials of the total degree d will be reduced to 0. We can eliminate unnecessary S-pairs in this way, which is called the Hilbert-driven algorithm.

All three of the systems discussed here implement the Hilbert-driven algorithm. In SINGULAR, the command `stdhilb` (defined in `standard.lib`) computes the Hilbert function of an input ideal and executes the Hilbert-driven algorithm for computing a Gröbner basis with respect to the target ordering. If the input ideal is not homogeneous, it is automatically homogenized.

Listing 3.31 SINGULAR: the Hilbert-driven algorithm

```
> ring r=0,(a,b,c,d,e,f,g),(dp(3),dp(4));
> timer=1;
> option(prot);
> ideal i=-3*a^2+2*f*b+3*f*d,(3*g*b+3*g*e)*a-3*f*c*b,
-3*g^2*a^2-c*b^2*a-g^2*f*e-g^4,e*a-f*b-d*c;
> ideal j=stdhilb(i);
compute hilbert series with slimgb in ring (0),
(a,b,c,d,e,f,g,@),(dp(8),C)
weights used for hilbert series: 1,1,1,1,1,1,1,1
slimgb in ring (0),(a,b,c,d,e,f,g,@),(dp(8),C)
CC2M[1,1](2)C3M[1,1](2)4M[2,2](5)C5M[5,4](14)C6M[11,5](19)...
NF:118 product criterion:36, ext_product criterion:11
std with hilb in (0),(a,b,c,d,e,f,g,@),(dp(3),dp(4),dp(1),C)
[255:5]2(34)s(33)s3s(34)s4(36)s(38)ss(39)s(42)--s5(44)s(45)...
...
26(10)---shhhhhh27(5)shhhhh28(4)-shhh29-shhh30-shhhhhhh
product criterion:453 chain criterion:41711
hilbert series criterion:912
//used time: 63.30 sec
dehomogenization
simplification
imap to ring (0),(a,b,c,d,e,f,g),(dp(3),dp(4),C)
```

In this example, a Gröbner basis of an elimination ideal is computed by using the Hilbert-driven algorithm. Information about the computation will be displayed if the option `prot` is specified: the grevlex Gröbner basis is computed by `slimgb`, a variant of the Buchberger algorithm, then the Hilbert function is computed and unnecessary S-pairs are removed by using the Hilbert function. In the displayed information, `h` means that an S-pair was removed by the Hilbert function. For example, `30-shhhhhhh` at following execution of `std with hilb` shows that all remaining S-pairs of sugar = 30 are known to be unnecessary after obtaining the basis element indicated by `s`. If we apply the usual Buchberger algorithm to this example, we will find that it takes a very long time to reduce the S-polynomials removed in the Hilbert-driven algorithm.

3.5   Computation of Gröbner Bases for Modules

The ideas of a Gröbner basis and the Buchberger algorithm are easily extended to modules over a polynomial ring. In this section, we introduce term orderings and extend the Buchberger algorithm for modules. We set $R = K[x_1, \ldots, x_n]$.

3.5.1   Term Orderings for Modules

We consider a free module R^m and its submodule. Let $e_1, \ldots, e_m$ be the standard bases of R^m. Then an element $f \in R^m$ is written as

$$f = c_1 t_1 e_{i_1} + \cdots + c_l t_l e_{i_l}, \tag{3.1}$$

where $t_1, \ldots, t_l$ are monomials in R and $c_1, \ldots, c_l \in K \setminus \{0\}$. We call $t e_i$ a monomial in R^m for a monomial $t \in R$.

Definition 3.5.1 (Term Ordering in R^m).   A term ordering in R^m is a total ordering $<$ in the set of all monomials in R^m satisfying the following conditions:

1. For all monomials $u, v \in R^m$ and $t \in R$, $u < v$ implies $tu < tv$.
2. For all monomials $t \in R$ and $u \in R^m$, $u \leq tu$.

Remark 3.5.2.   The condition 2 in Definition 3.5.1 can be replaced by

$$\text{For all monomials } t \in R \text{ and all } i = 1, \ldots, m, \ e_i \leq t e_i.$$

Definition 3.5.3 (Typical Term Orderings for Modules).   Let $<$ be a term ordering in R. We define two term orderings in R^m extending $<$.

1. *TOP (Term Over Position) extension* of $<$
    $$t e_i <_{\text{TOP}} s e_j \Leftrightarrow t < s \text{ or } (t = s \text{ and } i > j).$$

2. *POT (Position Over Term) extension* of $<$
   $te_i <_{POT} se_j \Leftrightarrow i > j$ or $(i = j$ and $t < s)$.

When we write $f \in R^m$ $(f \neq 0)$ as (3.1) and $t_1 e_{i_1} > t_2 e_{i_2} > \cdots > t_l e_{i_l}$, we call
$t_1 e_{i_1}$ the *initial monomial* of f and denote it by $\text{in}_<(f)$.
Dickson's lemma also holds for sets of monomials in R^m. Thus any submodule of
R^m generated by a set of monomials is finitely generated.

Definition 3.5.4. For a submodule M of R^m, $\{g_1, \ldots, g_k\} \subset M$ is called a Gröbner
basis of M with respect to a term ordering $<$ if it satisfies $\langle \{\text{in}_<(f) \mid f \in M\} \rangle =$
$\langle \text{in}_<(g_1), \ldots, \text{in}_<(g_k) \rangle$.

Since division also terminates in R^m, the remainder of an element in M on division
by a Gröbner basis of M is equal to 0. Therefore, a Gröbner basis of M generates
M over R.

Problem 3.5.5. For ideals $I = \langle f_1, \ldots, f_l \rangle$ and $J = \langle g_1, \ldots, g_m \rangle$ in R, we set

$$M = \left\langle \begin{pmatrix} f_1 \\ 0 \end{pmatrix}, \ldots \begin{pmatrix} f_l \\ 0 \end{pmatrix}, \begin{pmatrix} g_1 \\ g_1 \end{pmatrix}, \ldots \begin{pmatrix} g_m \\ g_m \end{pmatrix} \right\rangle \subset R^2.$$

Let G be a Gröbner basis of M with respect to the POT extension $<_{POT}$ of a term
ordering $<$ in R. Then $G_0 = \{g \in R \mid (0, g) \in G\}$ is a Gröbner basis of $I \cap J$ with
respect to $<$.

3.5.2 Buchberger Algorithm for Modules

In order to compute a Gröbner basis of a submodule of R^m, we extend the idea of
an S polynomial to a pair of elements in R^m.

Definition 3.5.6 (S Polynomial in a Module). For $f, g \in R^m \setminus \{0\}$, let $\text{in}_<(f) =$
te_i and $\text{in}_<(g) = se_j$ be the initials of f and g, respectively, and let c_f and c_g be
their coefficients in f and g, respectively. We define $S(f, g)$ as follows:

1. If $i \neq j$, then $S(f, g) = 0$;
2. If $i = j$, then $S(f, g) = (\text{LCM}(t, s)/c_f t) \cdot f - (\text{LCM}(t, s)/c_g s) \cdot g$.

Theorem 3.5.7. *For a submodule M of R^m, $G = \{g_1, \ldots, g_k\} \subset M \setminus \{0\}$ is a
Gröbner basis of M if and only if the remainder of $S(g_i, g_j)$ on division by G is
equal to 0 for all i, j.*

By this theorem, a Gröbner basis of a submodule M can be computed by the
Buchberger algorithm. We note that it is possible to apply the argument in Sect. 3.1.1
to eliminate the unnecessary S pairs.

3.5.3   Computation of Syzygy

The most fundamental operation on modules is to compute their syzygies.

Definition 3.5.8. Let M be an R module. For $f_1, \ldots, f_m \in M$,

$$\{(h_1, \ldots, h_m) \in R^m \mid h_1 f_1 + \cdots + h_m f_m = 0\}$$

is called the *syzygy* module of $(f_1, \ldots, f_m)$, and it is denoted by $\mathrm{syz}(f_1, \ldots, f_m)$.

For $f_1, \ldots, f_m \in R^l$, we set $M = \langle f_1, \ldots, f_m \rangle$. It is not easy to compute a set of generators of the syzygy module for a general input set. But if $\{f_1, \ldots, f_m\}$ is a Gröbner basis of M, then a set of generators of $\mathrm{syz}(M)$ is obtained by computing the remainder.

Theorem 3.5.9. *Suppose that* $G = \{f_1, \ldots, f_m\} \subset R^l$ *is a Gröbner basis of* $M = \langle G \rangle$, *and* $S(f_i, f_j) = u_i f_i - u_j f_j \ (u_i, u_j \in R)$ *is written as*

$$S(f_i, f_j) = \sum_{k=1}^m h_{ijk} f_k, \quad \mathrm{in}_<(h_{ijk} f_k) \leq \mathrm{in}_<(S(f_i, f_j))$$

$(k = 1, \ldots, m)$. *If we define* $s_{ij} \in R^m$ *by*

$$s_{ij} = u_i e_i - u_j e_j - \sum_{k=1}^m h_{ijk} e_k,$$

then $S = \{s_{ij} \mid 1 \leq i, j \leq m, \ S(f_i, f_j) \neq 0\}$ *is a set of generators of* $\mathrm{syz}(G)$.

Remark 3.5.10. The above S is a Gröbner basis of $\mathrm{syz}(G)$ with respect to a special term ordering called the Schreyer ordering in R^m (Schreyer's theorem).

Two methods are known for computing the syzygy module for a general vector $(f_1, \ldots, f_m) \in R^l$.

Algorithm 3.5.11 (cf. [5] Chap. 5, Sect. 3).

Input: $F = (f_1, \ldots, f_m)$, $f_i \in R^l$ $(i = 1, \ldots, m)$
Output: a set of generators of $\mathrm{syz}(F)$
$G = (g_1, \ldots, g_t) \leftarrow$ a Gröbner basis of $\langle F \rangle$
$C \leftarrow$ a (t, m) matrix such that $^t G = C \cdot {}^t F$
$D \leftarrow$ an (m, t) matrix such that $^t F = D \cdot {}^t G$
$S = \{s_1, \ldots, s_u\} \leftarrow$ a generating set of $\mathrm{syz}(G)$
$\{r_1, \ldots, r_m\} \leftarrow$ the set of row vectors of $I_m - DC$,
                where I_m is the identity matrix of size m
return $\{s_1 C, \ldots, s_u C, r_1, \ldots, r_m\}$

A set of generators of syz(G) can be computed by Theorem 3.5.9. D consists of the quotients of the elements in F on division by G. To compute C, it is necessary in the Buchberger algorithm to maintain not only the remainders but also the quotients, and the computational cost will increase. We note that the output of Algorithm 3.5.11 is not always a Gröbner basis of syz(F).

Algorithm 3.5.12 (cf. [5] Exercise 15 in Chap. 5, Sect. 3).

Input: $F = (f_1, \ldots, f_m)$, $f_i \in R^l$ $(i = 1, \ldots, m)$, a term ordering $<$
Output: a Gröbner basis of syz(F) with respect to $<_{\text{POT}}$
$(e_1, \ldots, e_m) \leftarrow$ the standard basis of R^m
$u_i \leftarrow (f_i, e_i) \in R^l \oplus R^m = R^{l+m}$ $(i = 1, \ldots, m)$
$\tilde{G} \leftarrow$ a Gröbner basis of $\langle u_1, \ldots, u_m \rangle$ with respect to $<_{\text{POT}}$
$S \leftarrow \{h \in R^m \mid (0, h) \in \tilde{G}\}$
return S

Problem 3.5.13. Show that Algorithm 3.5.12 outputs a Gröbner basis of syz(F) with respect to $<_{\text{POT}}$.

In Algorithm 3.5.12, it is necessary to compute a Gröbner basis for a module even if F is a set of generators of an ideal, and the cost of computing the Gröbner basis may be large. However, it is not necessary to maintain the quotients, and the output will be a Gröbner basis with respect to $<_{\text{POT}}$. Furthermore, if we set

$$G = \{g \in R^l \mid g \neq 0 \text{ and } (g, h) \in \tilde{G} \text{ for some } h\} = \{g_1, \ldots, g_t\},$$

$$C = \text{a } (t, m) \text{ matrix whose } i\text{-th row is } h_i \text{ for } (g_i, h_i) \in \tilde{G},$$

then G is a Gröbner basis of $\langle F \rangle$ and $^tG = C \cdot {}^tF$.

3.6 Computation in Risa/Asir

Macaulay2, SINGULAR, and CoCoA have been developed with the same design, a base ring with a term ordering is explicitly set, and the algorithms are applied to objects belonging to the base ring. The design of the software system *Risa/Asir*[9] is different from these three systems, and its usage is also different. Therefore, we explain it in a separate section.

3.6.1 Starting Risa/Asir

If you start Risa/Asir from the KNOPPIX/Math menu, then a terminal emulator is opened and `openxm fep asir` is executed. If you start Risa/Asir from a terminal emulator, then execute `openxm fep asir` because `asir` itself does not provide line editing.

3.6.2 Help Files and Manuals

To display the help files for a command, enter help("function"). To search
the manual, enter Math-Doc-Search on the desktop. Enter helph() to open a web
browser and display the manual.

3.6.3 Reading and Writing Files

In order to read a user program file or a library file, use load. This function searches
the directories listed in the environment variable ASIRLOADPATH. The value of
this variable is displayed by executing openxm env from a shell. The command
output("file") redirects the output to a file until output() is executed.
These inputs and outputs are all in human-readable format but another format is
possible. With bsave(Data,"file"), the output data can be saved to a binary
file which can then be read by using the command bload("file"). Although
only one dataset can be specified, it is possible to encapsulate multiple datasets in
a list.

3.6.4 Polynomials

In Asir, an indeterminate is represented by a string that begins with a lowercase letter
and is followed by letters, numbers, and the symbol _. If a polynomial contains an
indeterminate, it is converted to an internal recursive form by using the ordering of
indeterminates maintained in the system. A newly introduced indeterminate is put at
the end of the list. Coefficients are assumed to be rational numbers. In this system,
polynomials can be added, subtracted, and multiplied, provided that the ordering
of the indeterminates is unchanged. When executing a computation related to a
Gröbner basis, the term ordering should be specified. This feature may be annoying
to users who are accustomed to the other three systems, but it is convenient for
changing term orderings for the same inputs or for introducing new variables.

Listing 3.32 Input of polynomials

```
[1518]  F=(x+y+z)^2;
x^2+(2*y+2*z)*x+y^2+2*z*y+z^2
[1519]  G=F+u;
x^2+(2*y+2*z)*x+y^2+2*z*y+z^2+u
```

In Asir, polynomials are maintained in a *recursive representation*. In this representation, a polynomial is represented as a univariate polynomial with respect to a
main variable, and its coefficients are polynomials which do not contain the main
variable. When we execute a computation related to a Gröbner basis, it is convenient
to represent a polynomial as a sum of monomials. This representation is called a
distributed representation. In Asir, the conversion between two representations may
be done implicitly or explicitly.

3.6.5 Term Orderings

In Asir, a term ordering is specified by a pair: a *variable ordering* and *term-ordering type*. A variable ordering is a list of indeterminates, and it determines the index of each variable in the exponent vector of a monomial. For example, if a variable ordering is given by $[x, y, z, u, v, w]$, $x^a y^b z^c u^d v^e w^f$ is represented by the vector (a, b, c, d, e, f). For a variable ordering for n variables, we can set the following term-ordering types.

- Simple Term-ordering type
    This is given by an integer. 0, 1, and 2 represent grevlex, glex, and lex, respectively.
- Block-ordering type
    This is given by a list $[[O_1, n_1], [O_2, n_2], \ldots, [O_l, n_l]]$. The variable list is divided into l blocks $(n_1 + \cdots + n_l = n)$, and a simple term-ordering type O_i is applied to the i-th block. A typical one is $[[0, n_1], [0, n_2]]$, which is used to eliminate the first n_1 variables.
- Matrix-ordering type
    This is given by an $m \times n$ integer matrix M. For nonnegative integer vectors $e = (e_1, \ldots, e_n)$, $f = (f_1, \ldots, f_n)$, the ordering is defined by

$$e > f \Leftrightarrow \text{the topmost nonzero element of } M(e - f) \text{ is positive.}$$

In order for M to define a term ordering, it must satisfy the following conditions:

- $Me = 0 \Leftrightarrow e = 0$ for all integer vectors e.
- The topmost nonzero element of each column of M is positive.

The term-ordering type can be set by the command dp_ord. It can be also specified as an argument of certain functions.

Listing 3.33 Conversion to distributed representation

```
[1532]  F=x^2*y+y^3*z+x*z+x+1;
y*x^2+(z+1)*x+z*y^3+1
[1533]  dp_ord(0)$
[1534]  DF0=dp_ptod(F,[x,y,z]);
(1)*<<0,3,1>>+(1)*<<2,1,0>>+(1)*<<1,0,1>>+(1)*<<1,0,0>>
+(1)*<<0,0,0>>
[1535]  dp_ord(2)$
[1536]  DF2=dp_ptod(F,[x,y,z]);
(1)*<<2,1,0>>+(1)*<<1,0,1>>+(1)*<<1,0,0>>+(1)*<<0,3,1>>
+(1)*<<0,0,0>>
[1537]  G=F+u;
y*x^2+(z+1)*x+z*y^3+u+1
[1538]  DG=dp_ptod(G,[u,x,y,z]);
(1)*<<1,0,0,0>>+(1)*<<0,2,1,0>>+(1)*<<0,1,0,1>>+(1)*<<0,1,0,0>>
+(1)*<<0,0,3,1>>+(1)*<<0,0,0,0>>
[1539]  dp_ht(DG);
(1)*<<1,0,0,0>>
```

In this example, polynomials F, G are explicitly converted to distributed representations by the command dp_ptod. DF0 is sorted according to a grevlex ordering by dp_ord(0), and DF2 is sorted according to a lex ordering by dp_ord(2). Since DG is sorted according to a lex ordering in which u is the largest, the initial monomial of DG is <<1,0,0,0>>. Commands dp_ht, dp_hc, and dp_hm return the initial monomial, the coefficient of the initial monomial, and the initial monomial with its coefficient, respectively. Arithmetic operations between polynomials with distributed representations only produce correct results if they were created according to the same term ordering.

3.6.6   Computation of Gröbner Bases

In this section, we explain how to compute Gröbner bases in Asir. Functions related to Gröbner basis are defined in the libraries gr and noro_pd.rr, which must be loaded before these functions can be used. In Asir, an ideal is represented by a list of polynomials. The base ring for computing a Gröbner basis is determined by the arguments given to the following functions:

- nd_gr($Plist, Vlist, Char, Ord$)
  $Plist$ is a list of polynomials representing an ideal. This function executes the Buchberger algorithm over a polynomial ring $K[Vlist]$ with the term ordering specified by a variable ordering $Vlist$ and an ordering type Ord, where $K = \mathbb{Q}$ if $Char = 0$ and $K = \mathbb{F}_{Char}$ if $Char$ is a prime. It returns the reduced Gröbner basis of $\langle Plist \rangle$. The output is a list of polynomials.
- nd_gr_trace($Plist, Vlist, Homo, Prime, Ord$)
  This is a function for efficient Gröbner basis computation over the rationals. Shortcuts that use a finite field are applied. Set $Prime$ to 1 (see the manual for other settings). If $Homo$ is set to 1, the homogenization explained in Sect. 3.1.3 is applied. The output is the same as that for $Homo = 0$, but the computation with $Homo = 1$ avoids intermediate coefficient swells.[6]

Listing 3.34 Gröbner basis computation in Asir

```
[1517]  load("cyclic")$
[1527]  C=cyclic(7);
[c6*c5*c4*c3*c2*c1*c0-1,...]
[1528]  V=vars(C);
[c0,c1,c2,c3,c4,c5,c6]
[1529]  nd_gr(C,V,31991,0)$
...
2.303sec + gc : 0.07sec(2.429sec)
[1530]  nd_gr(C,V,0,0)$
(stopped after 5 minutes)
```

---

[6]If it is known that no coefficient swells will occur, then the homogenization is unnecessary.

```
[1530]  G=nd_gr_trace(C,V,1,1,0)$
...
25.84sec + gc : 7.833sec(34.56sec)
[1531]  G[0];
(((23853922665902000713066̲2*c6*c4-...
[1532]  length(G);
209
```

In this example, the reduced Gröbner basis of cyclic-7 (Example 3.2.1) was computed over $\mathbb{F}_{31991}$ and $\mathbb{Q}$. The first nd_gr was executed over a finite field, and it terminated in 2 s. The second nd_gr was executed over $\mathbb{Q}$, and it became stuck due to coefficient swells. When nd_gr_trace with *Homo* = 1 is applied, it terminated in 25 s. The output is a list of polynomials which generates the input ideal. The *i*-th element of a list *G* can be obtained by *G*[*i*] (*i* starts with 0).

3.6.7 Computation of Initial Ideals

To compute initial ideals in Asir, it is first necessary to compute a Gröbner basis. To take the initial of a polynomial, it is necessary to convert it to a distributed representation and then apply dp_ht. Next, apply dp_dtop to convert it to a recursive representation. If the ideal is zero-dimensional, the standard monomial set can be obtained by dp_mbase.

Listing 3.35 Computation of initial ideal in Asir

```
[1517]  B=[x^2*y^2-z^2,x^3-y*z^2,x^2*z^4-y^2];
[y^2*x^2-z^2,x^3-z^2*y,z^4*x^2-y^2]
[1518]  V=[x,y,z]$
[1519]  G=nd_gr(B,V,0,0);
[z^4*x^2-y^2,-y^4+z^6,-y^2*x+y^5,-z^2*x+z^2*y^3,
y^2*x^2-z^2,x^3-z^2*y]
[1520]  D=map(dp_ptod,G,V)$ H=map(dp_ht,D)$
[1521]  [1522]  map(dp_dtop,H,V);
[z^4*x^2,z^6,y^5,z^2*y^3,y^2*x^2,x^3]
[1523]  map(dp_dtop,dp_mbase(H),V);
[z^5*y^2*x,z^4*y^2*x,z^5*y*x,z^5*y^2,z*y^4*x,z^3*y*x^2,...]
[1524]  length(@@);
52
```

3.6.8 Computation of the Remainder

Use p_nf or p_true_nf to compute remainders. The former returns a polynomial with integer coefficients which is an integer multiple of the remainder. This is used for checking whether the remainder is 0. The latter returns a list [*num*, *den*] such that *num* is the polynomial returned by p_nf and *num/den* is the true remainder.

Listing 3.36 Computation of remainders

```
[1517]  B=[u2*u0-2*u2+3,(2*u1-1)*u0^2-u0-2*u2,2*u1^3+u2+4]$
[1518]  V=[u0,u1,u2]$
[1519]  G=nd_gr(B,V,0,0);
[10*u2^4+126*u2^3+637*u2^2+(586*u1-907)*u2-816*u0^2-...]
[1520]  Q=p_nf(u0^5+u1^5+u2^5,G,V,0);
2851262910*u2^3+30078832770*u2^2+(22194374760*u1-...
[1521]  QR=p_true_nf(u0^5+u1^5+u2^5,G,V,0);
[2851262910*u2^3+30078832770*u2^2+...,35373600]
```

3.6.9 Elimination

As stated in Sect. 3.3.1, the elimination ideal $I_Y = I \cap K[Y]$ for an ideal $I \subset K[Z]$ ($Z = X \cup Y, X \cap Y = \emptyset$) can be computed from a Gröbner basis with respect to an elimination ordering. We recommend using nd_gr_trace with homogenization for computations over the rationals. Use noro_pd.elimination to extract a Gröbner basis of I_Y from a Gröbner basis of I, with respect to an elimination ordering.

In the following example, we compute a Gröbner basis of an ideal generated by B with respect to an elimination ordering such that $\{u0, u1\} \gg \{u2\}$, and we use noro_pd.elimination to extract a univariate polynomial of $u2$ in the Gröbner basis.

Listing 3.37 Computation of an elimination ideal

```
[1664]  B=[u2*u0-2*u2+3,(2*u1-1)*u0^2-u0-2*u2,2*u1^3+u2+4]$
[1665]  V=[u0,u1,u2]$
[1666]  G1=nd_gr_trace(B,V,1,1,[[0,2],[0,1]])$
[1667]  noro_pd.elimination(G1,[u2]);
[8*u2^9+72*u2^8+292*u2^7-2036*u2^6-198*u2^5+20682*u2^4-...]
```

3.6.10 Computation of Minimal Polynomials

For an ideal I, the computation of $I \cap K[z]$ is a special case of elimination. The general method can be applied to this computation, but another efficient method is applicable if I is a zero-dimensional ideal. The command minipoly defined in gr computes a nonzero polynomial $m(t) \in \mathbb{Q}[t]$ of the minimal degree, such that $m(f) \in I$ for a zero-dimensional ideal $I \subset \mathbb{Q}[X]$ and $f \in \mathbb{Q}[X]$. The polynomial $m(t)$ is a generator of an ideal $(\mathbb{Q}[X, t]I + \mathbb{Q}[X, t](f - t)) \cap \mathbb{Q}[t]$, and it is unique up to a constant factor. We call $m(t)$ the minimal polynomial of f modulo I.

In the following example, we use minipoly to compute the minimal polynomial of $u7$ modulo an ideal katsura-7. The argument G is a Gröbner basis of the ideal with respect to a term ordering represented by $(V, 0)$. The result is returned as a univariate polynomial of the last argument.

Listing 3.38 Computation of minimal polynomial

```
[1518]  load("katsura")$
[1522]  B=katsura(7)$
[1523]  V=[u0,u1,u2,u3,u4,u5,u6,u7]$
[1524]  G=nd_gr_trace(B,V,1,1,0)$
[1525]  minipoly(G,V,0,u7,t)$
[1526]  deg(@@,t);
128
```

3.6.11 Change of Orderings for Zero-Dimensional Ideals

The command `tolex` defined in `gr` computes the reduced Gröbner basis of an ideal with respect to a lex ordering. The input ideal has to be given by a Gröbner basis with respect to some term ordering. A modified version of Algorithm 3.4.3 is implemented. The target variable ordering can be specified in the last argument.

Listing 3.39 Computation of a lex Gröbner basis by change of ordering

```
[1524]  G=nd_gr_trace(katsura(7),V=[u0,u1,u2,u3,u4,u5,u6,u7]
,1,1,0)$
3.27sec + gc : 1.067sec(4.524sec)
[1525]  G2=tolex(G,V,0,V)$
316.4sec + gc : 98.52sec(442.8sec)
```

3.6.12 Ideal Operations

We introduce the operations on ideals that are defined in `noro_pd.rr`. All commands assume that the inputs are ideals over the rationals. If an option `mod=p` is given, the computation is performed over $\mathbb{F}_p$.

3.6.12.1 Intersection of Ideals

Use `noro_pd.ideal_intersection` to compute the intersection of two ideals. Use `noro_pd.ideal_list_intersection` to compute the intersection of ideals given in a list.

Listing 3.40 Intersection of ideals

```
[1640]  B=[g*a-f*b,h*b-g*c,i*c-h*d,j*d-i*e,l*f-k*g,m*g-l*h,
n*h-m*i,o*i-n*j]$
[1708]  V=[a,b,c,d,e,f,g,h,i,j,k,l,m,n,o]$
[1709]  G=nd_gr(B,V,0,0)$
[1710]  PD=noro_pd.syci_dec(B,V)$
```

```
[1711] length(PD);
4
[1712] map(length,PD);
[10,5,3,1]
[1713] for(I=0,T=[1];I<4;I++)
  for(J=0,L=length(PD[I]);J<L;J++)
    T=noro_pd.ideal_intersection(T,PD[I][J][0],V,0);
[1649] gb_comp(T,G);
1
```

In this example, noro_pd.syci_dec computes the primary decomposition of an ideal and it is checked that the intersection of the computed components coincides with the initial ideal (see Sect. 3.7). The result of the decomposition is given as a list $[PD_0, PD_1, PD_2, PD_3]$, and each PD_i is a list consisting of $[Q_{ij}, P_{ij}]$ such that Q_{ij} is a P_{ij}-primary component of the ideal that was input.

3.6.12.2 Radical Membership Test

As explained in Sect. 3.3.3, radical membership can be tested by computing a Gröbner basis of $R[t]I + \langle tf - 1 \rangle$ or by using noro_pd.radical_membership. If $f \in \sqrt{I}$, the latter command returns 0; otherwise, it returns a list consisting of a Gröbner basis of $R[t]I + \langle tf - 1 \rangle$ with respect to a grevlex ordering and an indeterminate representing t.

Listing 3.41 Radical membership test

```
[1665] B=[(x+y+z)^50,(x-y+z)^50]$
[1666] V=[x,y,z]$
[1667] F=y$
[1668] cputime(1)$
0sec(1.907e-06sec)
[1669] noro_pd.radical_membership(F,B,V);
0
0.2267sec(0.2502sec)
[1670] nd_gr(cons(t*F-1,B),cons(t,V),0,0);
[1]
0.21sec(0.285sec)
```

3.6.12.3 Ideal Quotient and Saturation

To compute an ideal quotient by a polynomial, use noro_pd.colon. To compute an ideal quotient by an ideal, use noro_pd.ideal_colon. To compute a saturation by a polynomial, use noro_pd.sat. To compute a saturation by an ideal, use noro_pd.ideal_sat.

Listing 3.42 Ideal quotient and saturation

```
[1640]  B=[(x+y+z)^50,(x-y+z)^50]$
[1641]  V=[x,y,z]$
[1642]  noro_pd.sat(B,y,V);
[1]
[1643]  noro_pd.colon(B,y^98,V);
[-x-z,-y]
[1644]  noro_pd.ideal_colon(B,[(x+y+z)^49,(x-y+z)^49],V);
[-y^49*x-z*y^49,-y^50,-x^2-2*z*x+y^2-z^2]
[1645]  noro_pd.ideal_sat(B,[(x+y+z)^49,(x-y+z)^49],V);
[1]
```

3.7   An Example of Programming in Macaulay2

So far, we have illustrated fundamental computations related to Gröbner basis using
Macaulay2, SINGULAR, CoCoA, and Risa/Asir. In this section, as a practical application, we implement a new primary decomposition algorithm using Macaulay2.

3.7.1   Primary Decomposition of Ideals

Let $R = K[X] = K[x_1, \ldots, x_n]$ be an n-variate polynomial ring over a field K.

Definition 3.7.1. 1. A proper ideal P of R is called a *prime ideal* if

$$fg \in P \text{ implies } f \in P \text{ or } g \in P.$$

2. A proper ideal Q of R is called a *primary ideal* if

$$fg \in Q \text{ implies } f \in Q \text{ or } g \in \sqrt{Q}.$$

3. If Q is a primary ideal, then $P = \sqrt{Q}$ is a prime ideal. In this case, Q is called
   a P-primary ideal, and P is called the *associated prime ideal* of Q.

Theorem 3.7.2. *1. A proper ideal I of R is represented as $I = Q_1 \cap \cdots \cap Q_l$ by
a finite number of primary ideals $Q_1, \ldots, Q_l$.*
2. *If the $\sqrt{Q_i}$'s are distinct and $\bigcap_{j \neq i} Q_j \not\subset Q_i$ $(i = 1, \ldots, l)$, then*

$$\{\sqrt{Q_1}, \ldots, \sqrt{Q_l}\} = \{\sqrt{I : f} \mid f \in R, \sqrt{I : f} \text{ is a prime ideal}\}.$$

In particular, the $\sqrt{Q_i}$'s are uniquely determined by I.

Definition 3.7.3. The expression of I by the Q_i's in Theorem 3.7.2 is called a *primary decomposition* of I, and each Q_i is called a *primary component*. If the condition in statement 2 is satisfied, the decomposition is called minimal and the prime ideal $\sqrt{Q_i}$ is called an *associated prime* of I. The set of all associated primes of I is denoted by $\mathrm{Ass}(I)$. If an associated prime is minimal with respect to inclusion, it is called a *minimal associated prime*. A primary component corresponding to a minimal associated prime is called an *isolated primary component*. A non-isolated primary component is called an *embedded primary component*.

If Q_1 and Q_2 are P-primary, then $Q_1 \cap Q_2$ is also P-primary. Thus we obtain a minimal primary decomposition from any primary decomposition by combining components having the same associated prime into one component and eliminating redundant components with respect to inclusion. If an ideal I is a radical ideal, that is, $\sqrt{I} = I$, then its minimal primary decomposition consists of prime ideals.

Theorem 3.7.4.
In a minimal primary decomposition of a proper ideal I of R, the set of all isolated components of I is uniquely determined.

Several algorithms are known for computing a primary decomposition of an ideal I. Recently, we have found another algorithm (Kawazoe-Noro 2011) [7], and it can decompose several examples which are difficult to decompose when using the existing algorithms. We will explain the new algorithm briefly and implement it in Macaulay2.

3.7.2 SYCI Algorithm

Assuming that we are given an algorithm $MinimalAssociatedPrimes(I)$ for computing the prime decomposition of $\sqrt{I}$ for an ideal I, we present the SYCI algorithm for computing the minimal primary decomposition.

$MinimalAssociatedPrimes(I)$ can be realized by replacing the computation of the radical of a zero-dimensional ideal by the prime decomposition of the radical in Algorithm 3.3.24. The prime decomposition of a zero-dimensional radical ideal can be realized by the irreducible factorization of multivariate polynomials.

Definition 3.7.5. For ideals I, Q such that $I \subset Q$, an ideal J is called a *saturated separating ideal* for (I, Q) if $I = Q \cap (I + J)$ and $\sqrt{I : Q} = \sqrt{I + J}$.

If $\sqrt{I : Q} = \langle f_1, \ldots, f_l \rangle$, then, for a sufficiently large m, $J = (\sqrt{I : Q})^m$ or $J = \langle f_1^m, \ldots, f_l^m \rangle$ are saturated separating ideals. In the latter case, we can choose the power m individually for each f_i as follows:

Algorithm 3.7.6 ($SaturatedSeparatingIdeal(C, I, Q)$)**.**

Input: ideals I, Q such that $I \subset Q$
Output: a saturated separating ideal J for (I, Q)

$S \leftarrow$ a set of generators of $\sqrt{I} : Q$
$J = \{0\}$
for each $f \in S \setminus \sqrt{I}$ do
    $j \leftarrow 0$
    do $j \leftarrow j + 1$ while $Q \cap (I + J + \langle f^j \rangle) \neq I$
    $J \leftarrow J + \langle f^j \rangle$
end for
return J

Algorithm 3.7.6 terminates and outputs a saturated separating ideal.

We write the operation I^{ec} in Corollary 3.3.17 as I_U^{ec} to indicate explicitly a subset $U \subset X$.

Algorithm 3.7.7 ($PrimaryDecompositionSYCI(I)$).

Input: an ideal $I \subset R$
Output: a minimal primary decomposition $(QL_1, QL_2, \ldots, QL_l)$ of I
            $QL_i = (Q_{i1}, \ldots, Q_{in_i})$; Q_{ij} is a primary component of I
$i \leftarrow 1$; $Q_0 \leftarrow R$
do
  $PL_i = \{P_{i1}, \ldots, P_{in_i}\} \leftarrow MinimalAssociatedPrimes(I : Q_{i-1})$
  for $j = 1$ to n_i do
    $U_{ij} \leftarrow$ a maximal independent set for P_{ij}
    $f_{ij} \leftarrow$ an element of $(\bigcap_{k \neq j} P_{ik}) \setminus P_{ij}$
    $R_{ij} \leftarrow Q_{i-1} \cap (I : f_{ij}^\infty)_{U_{ij}}^{ec}$
    $J_{ij} \leftarrow SaturatedSeparatingIdeal(R_{ij}, Q_{i-1}, P_{ij})$
    $Q_{ij} \leftarrow (R_{ij} + J_{ij})_{U_{ij}}^{ec}$
  end for
  $QL_i = \{Q_{i1}, \ldots, Q_{in_i}\}$
  $Q_i \leftarrow R_{i1} \cap \cdots \cap R_{in_i}$
  If $Q_i = I$ then return $(QL_1, \ldots, QL_i)$
  $i \leftarrow i + 1$
end do

In Algorithm 3.7.7, the following statements hold.

1. PL_i consists of all minimal elements with respect to inclusion in $\text{Ass}(I) \setminus (PL_1 \cup \cdots \cup PL_{i-1})$. An element $P \in PL_i$ is called an associated prime of *level i*.
2. Q_{ij} is a P_{ij}-primary component of I, and the output is a minimal primary decomposition.
3. Q_i is the intersection of all primary components Q such that the level of $\sqrt{Q}$ is not greater than i. Q_i is independent of a minimal primary decomposition.
4. For any P_{ij}-primary component Q, $R_{ij} = Q_{i-1} \cap Q$, and R_{ij} is determined only by P_{ij}.
5. PL_i can be computed without computing J_{ij} and Q_{ij}.

3.7.3 Implementation in Macaulay2

Below are the functions necessary for implementing Algorithm 3.7.7.

Listing 3.43 Find a polynomial of the smallest total degree

```
mindeg = (L) -> (
  f := L#0; df := degree f;
  scan(L, g -> if degree g < df then (f = g; df = degree g));
  return f
)
```

This function returns a polynomial of the smallest total degree in a given list L. In Macaulay2, a function is defined by $(arg_1, \ldots, arg_l) \rightarrow (e_1; \ldots; e_m)$, and its name is given by =. In this form, e_i denotes an expression, and the value of the last expression gives the value of the function. The command return completes the execution of the function and returns the value. The command scan($list, function$) applies $function$ to each element of $list$; it corresponds to the "for" statement in the C language. A substitution can be performed by writing $a=b$ or $a:=b$. In the former case, the variable a is regarded as a global variable, and if it has not yet been declared before the substitution, it is generated. The latter statement initializes a local variable. To simply declare a local variable, use local. The i-th element of a list L is specified by L#i, where the index i starts with 0. The length of a list is given by #L.

Listing 3.44 Compute the set of all generators of P_1 not belonging to P_2

```
nonmember = (P1,P2) -> (
return select(first entries gens P1,f->not isSubset(ideal f,P2))
)
```

This function returns the set of all generators of P_1 not belonging to P_2. The function entries returns a list consisting of the rows of a matrix. In the output, each row is also converted to a list. The command gens applied to an ideal returns a row vector ($1 \times m$ matrix). Thus, entries in the above function returns a list with one element, and first returns the element, that is, the list of generators. Finally, the function select($list, function$) returns the subset of $list$ which consists of the elements that yield true when $function$ is applied.

Listing 3.45 Compute the squarefree part of a polynomial

```
squarefree = (f) -> value apply(factor f, i -> i#0)
```

The command factor returns the irreducible factorization as a $Product$ object. This is a kind of list consisting of ($factor, multiplicity$). The command apply($list, function$) applies $function$ to each element of $list$ and returns a list of the results. The operation i->i#0 outputs the first element of a list. In the above function, the input list of apply is a $Product$ object, and the output is also returned as a $Product$ object. Thus, the squarefree part of the input polynomial is returned in a factored form. Finally, the function value expands the polynomial.

Listing 3.46 Computation of a set of separators

```
separator = (I,PP) -> (
  R := ring I;
  S := new MutableList;
  scan(toList(0..#PP-1),i->S#i=1_R);
  scan(toList(0..#PP-1),i->
   scan(toList(0..#PP-1),j->
    if i != j then
      S#j =  lcm(S#j,squarefree(mindeg(nonmember(PP#i,PP#j)))))));
  );
  return S
)
```

This function outputs a set $\{f_1, \ldots, f_l\}$ such that $f_i \in (\bigcap_{j \neq i} P_j) \setminus P_i$ $(i = 1, \ldots, l)$
for a list of mutually disjoint primes $PP = \{P_1, \ldots, P_l\}$. Each f_i is called a
separator. If $s_{ij} \in P_i \setminus P_j$ for i, j $(i \neq j)$, then the squarefree part of $\prod_{j \neq i} s_{ij}$
can be used as a separator f_i. A *MutableList* is used for the S which holds the
intermediate values of the separators. This type of list is convenient when rewriting
its elements.

Listing 3.47 Computation of ideals which are minimal with respect to inclusion

```
removeRedundantComps = (L) -> (
  if #L == 1 then return L;
  S := new MutableList from L;
  for i from 0 to #L-1 do (
    if S#i == 0 then continue;
    for j from 0 to #L-1 do (
        if j == i or S#j == 0 then continue;
        if S#i == S#j or isSubset(S#i,S#j) then S#j = 0
    )
  );
  return toList(select(S,i -> i!=0))
)
```

This function outputs the list of those ideals in an input list which are minimal with
respect to inclusion. We use the input list to initialize a *MutableList* S , and an
ideal in S is replaced by 0 if it contains another ideal. The remaining ideals are
then packed into a list. The function toList converts any type of list into the most
generic type.

Listing 3.48 Prime decomposition of $\sqrt{I : J}$

```
colonMinimalPrimes = (I,J) -> (
  local K,PL,S;
  R := ring I; L := {};
  for f in first entries mingens J do (
    if f==1 then K=I else K=I:f;
    if K != ideal(1_R) then L = append(L,K)
  );
  L = removeRedundantComps(L); P := {};
```

```
    for K in L do (
      S = apply(first entries gens K,f->squarefree(f));
      PL = minimalPrimes ideal mingens gb ideal S;
      P = join(PL,P)
    );
    return removeRedundantComps(P)
)
```

If $J = \langle f_1, \ldots, f_l \rangle$, then $I : J = \bigcap_{i=1}^{l} (I : f_i)$. Since any set of generators

of J suffices to compute $\sqrt{I : J}$, we reduce the number of generators of J by applying mingens to $\{f_1, \ldots, f_l\}$. After we apply removeRedundantComps to the list of $I : f_i$'s to reduce the number of ideals to decompose, we apply minimalPrimes to each ideal in L, and we obtain a list P of prime components of $\sqrt{I : J}$. Finally, we apply removeRedundantComps to P to remove any redundant components.

Listing 3.49 Computation of a saturated separating ideal

```
saturatedSeparatingIdeal = (C,I,Q,Rad) -> (
  local fi;
  S := nonmember(C,Rad);
  if intersect(I+ideal S,Q) == I then return S;
  I1 := I;
  SSI := {};
  for f in S do (
    fi = f;
    while (intersect(Q,I1+ideal fi) != I) do (fi=fi*f);
    I1 = I1+ideal fi;
    SSI=append(SSI,fi)
  );
  return ideal SSI
)
```

This function executes Algorithm 3.7.6 if $\sqrt{I}$ is passed as the argument Rad.

Listing 3.50 Computation of I_U^{ec}

```
load "PrimaryDecomposition/Shimoyama-Yokoyama.m2";
myextract = (I,Y) -> (
  R := ring I;
  if #Y == 0 then f = 1_R
  else f := flattener(I,Y#0);
  if f != 1_R then return saturate(I,f)
  else return I
)
```

To compute I_U^{ec}, we need the LCM f of the leading coefficients of the elements in a Gröbner basis $G \subset K[U][X \setminus U] = K[X]$ of an ideal I in $K(U)[X \setminus U]$. This can be computed by the command flattener, which is defined in Shimoyama-Yokoyama.m2. I_U^{ec} is equal to $I : f^\infty$.

Listing 3.51 Primary ideal decomposition

```
sycidec = (I) -> (
  local PLi,QLi,RLi,Si,Ci,Yi,Ti;
  R := ring I; Qi := ideal(1_R); QL := {};
  for i from 1 do (
    PLi = colonMinimalPrimes(I,Qi);
    Si = separator(I,PLi);
    Ci = apply(Si,f->saturate(I,f));
    Yi = apply(PLi,P->independentSets(P,Limit=>1));
    RLi = apply(Ci,Yi,
      (c,y)->intersect(Qi,ideal gens gb myextract(c,y)));
    if i == 1 then ( Rad := intersect(PLi); QLi = RLi )
    else (
      Ti = apply(PLi,RLi,
        (p,r)->r+saturatedSeparatingIdeal(p,r,Qi,Rad));
      QLi = apply(Ti,Yi,(t,y)->ideal gens gb myextract(t,y))
    );
    QL = append(QL,QLi); Qi = intersect(RLi);
    if Qi == I then return QL
  )
)
```

This function implements Algorithm 3.7.7. In the program, `Qi` corresponds to Q_{i-1} in the algorithm. The function `independentSets` outputs a list of all the maximal independent sets for the input ideal. If the option `Limit=>1` is given, a list with a single element is returned. The element is a product of indeterminates and it represents a maximal independent set. It is necessary to pass $\sqrt{R_{ij}}$ to `saturatedSeparatingIdeal`, but it is always equal to $\sqrt{I}$, and we can use $Rad = \sqrt{I}$ computed at $i = 0$.

After reading the file `syci.m2`, which contains all the functions defined here by `load`, we are ready to compute the primary ideal decomposition.

Listing 3.52 An example of primary decomposition

```
i1 : load "syci.m2";
i2 : R=QQ[a,b,c,d,e,f,g,h,i,j,k,l,m,n,o];
i3 : I=ideal(g*a-f*b,h*b-g*c,i*c-h*d,j*d-i*e,l*f-k*g,m*g-l*h,
       n*h-m*i,o*i-n*j);
o3 : Ideal of R
i4 : timing (p=sycidec(I);)
o4 = -- 1.95582 seconds
i5 : #p
o5 = 4
i6 : apply(p,i->#i)
o6 = {10, 5, 3, 1}
o6 : List
i7 : intersect(apply(p,intersect))==I
o7 = true
i8 : apply(join(p#1,p#2,p#3),isPrimary)
o8 = {true, true, true, true, true, true, true, true, true}
o8 : List
```

In this example, we obtained ten isolated components and nine embedded components. We can check that the result actually gave a decomposition of I by applying intersect. By applying isPrimary, we can check that each component is a primary ideal. In this example, the execution of sycidec terminated in 2 s, but it would take 20 h if we had used the built-in function primaryDecomposition.

3.8 Additional Problems

Problem 3.8.1. Let $H = \langle 2zx + 3zy, 2x^2u + zx^2 + zx, zxu - 2zx \rangle$, $I = \langle 2zx^2 + 3zyx, 4zx^2 - 9zy^2, 2x^3u + zx^3 + zx^2, -zx^2u + 2zx^2, 2x^3u + 3yx^2u \rangle$, and $J = \langle y^2, x^3, 4x^2u - 3zy, zx^2, yx^2, zxu + 3zy, z^2u \rangle$ be ideals in $\mathbb{Q}[x, y, z, u]$.

1. Examine the inclusion relations for H, I, and J.
2. Examine the inclusion relations for $V(H)$, $V(I)$, and $V(J)$.

Problem 3.8.2. Let $I = \langle 3x^2yz^2 + 3z + (-2x + 2)y + 2x, 3yz^5 + (-xy^2 + 2)z - 2y^4 + 2y, xy^3z^3 - 2yz^2 - z - 2y + x^2 \rangle$ be an ideal in $\mathbb{Q}[x, y, z]$.

1. Show that I is zero-dimensional.
2. Compute $\dim_{\mathbb{Q}} \mathbb{Q}[x, y, z]/I$.
3. Show that the reduced Gröbner basis of I with respect to the lex order $x > y > z$ is of the form $\{g_0(z), x - g_1(z), y - g_2(z)\}$.

Problem 3.8.3. Set $\alpha = 3^{\frac{1}{3}}$ and $\beta = 5^{\frac{1}{3}}$.

1. Compute the minimal polynomial of $\alpha + \beta$ over $\mathbb{Q}$. (Compute $I \cap \mathbb{Q}[z]$ for an ideal $I = \langle x^5 - 3, y^3 - 5, z - (x + y) \rangle \subset \mathbb{Q}[x, y, z]$.)
2. Represent $\frac{1}{\alpha + \beta}$ by a polynomial of α, β over $\mathbb{Q}$. (Compute a Gröbner basis of an ideal $J = \langle (x + y)t - 1, x^5 - 3, y^3 - 5 \rangle$ in $\mathbb{Q}[x, y, t]$ with respect to an elimination ordering such that $\{t\} \gg \{x, y\}$.)

3.9 Answers to Problems

Problem 3.1.7 $S(g_j, g_k)$ in the proof of Theorem 1.3.3 can be replaced by a linear sum of the S-polynomials created from $S_{pq} \in S'$ which have monomial coefficients. Then the proof can be applied as is by using (1.6) for the S-polynomials.

Problem 3.1.11   Suppose that $F_k(i, j)$ holds for some $k < i$. Then we have $T_{jk} = T_{ij}$ and $T_{ijk} = T_{ij}$, which imply $S_{ij} = S_{jk} + \frac{T_{ij}}{T_{ki}} S_{ki}$. We can thus conclude that $S_{ki} \prec S_{ij}$ and $S_{jk} \prec S_{ij}$ by the definition of $\prec$. The other cases can be checked similarly.

Problem 3.3.3 Let $G \subset K[X, Y]$ be a Gröbner basis of J with respect to $<$. Then clearly $G \subset I$. If $f \in I$, then $hf \in J$ for some $h \in K[Y]$. Then there exists $g \in G$ such that $\text{in}_<(g) \mid \text{in}_<(hf)$. Since $<$ is an elimination order such that $X \gg Y$, if we set $\text{in}_<(g) = t_X t_Y$ and $\text{in}_<(hf) = s_X s_Y$ ($t_X, s_X \in K[X]$, $t_Y, s_Y \in K[Y]$), then t_X and s_X are the initials of g and f with respect to $<'$ in $K(Y)[X]$, respectively, and $t_X \mid s_X$. Hence G is a Gröbner basis of I with respect to $<'$.

Problem 3.3.6 If $f \in \sqrt{I}$, there exists a positive integer m such that $f^m \in I$. Then we have $1 = (1 - t^m f^m) + t^m f^m = (1 - tf)(t^{m-1} f^{m-1} + \cdots + 1) + t^m f^m \in \langle 1 - tf \rangle + R[t]I$. Conversely, if $1 \in R[t]I + \langle 1 - tf \rangle$, there exist $f_1, \ldots f_l \in I$, $a_1(t), \ldots, a_l(t) \in R[t]$, and $b(t) \in R[t]$ such that $1 = a_1(t) f_1 + \cdots + a_l(t) f_l + b(t)(1 - tf)$. By setting $t = 1/f$, we have an equation $1 = a_1(1/f) f_1 + \cdots + a_l(1/f) f_l$ in the quotient field of R. If we clear the denominators by multiplying by f^m such that $f^m a_1(1/f), \ldots, f^m a_l(1/f) \in R$, we have $f^m = (f^m a_1(1/f)) f_1 + \cdots + (f^m a_l(1/f)) f_l$, which implies $f^m \in I$.

Problem 3.3.16 If $h \in J^c$, then the remainder of h on division by G is equal to 0. Let $(a_1, \ldots, a_l)$ ($a_i \in K(U)[X \setminus U]$) be the quotient of the division. Then we have $h = a_1 g_1 + \cdots + a_l g_l$, and the denominators of the a_i's are all power products of $h_1, \ldots, h_l$. Thus $f^m h \in \tilde{J}$ for a sufficiently large m, and $h \in \tilde{J} : f^\infty$. Conversely, if $h \in \tilde{J} : f^\infty$, there exists a positive integer m such that $f^m h \in \tilde{J}$, which implies $h \in J \cap K[X] = H^c$.

Problem 3.3.21 It is clear that $I \subset (I : f^s) \cap (I + \langle f^s \rangle)$. If $h \in I : f^s$ and $h \in I + \langle f^s \rangle$, $hf^s \in I$ and $h = a + bf^s$ for some $a \in I, b \in R$. Then we have $hf^s = af^s + bf^{2s} \in I$ and $bf^{2s} \in I$. Then $I : f^{2s} = I : f^\infty = I : f^s$ implies $b \in I : f^s$. Thus we obtain $bf^s \in I$ and $h = a + bf^s \in I$.

Problem 3.3.23 If we take an integer s such that $I^{ec} = I : f^\infty = I : f^s$, we have $I = (I : f^\infty) \cap (I + f^s) = I^{ec} \cap (I + f^s)$ by Theorem 3.3.20, thus $\sqrt{I} = \sqrt{I^{ec}} \cap \sqrt{I + f^s}$. Then $\sqrt{I^{ec}} = \sqrt{I^e \cap R} = \sqrt{I^e}^c$ and $\sqrt{I + f^s} = \sqrt{I + \langle f \rangle}$ imply $\sqrt{I^{ec}} \cap \sqrt{I + f^s} = \sqrt{I^e}^c \cap \sqrt{I + \langle f \rangle}$.

Problem 3.5.5 We first show $G_0 \subset I \cap J$. For $h \in G_0$, $\begin{pmatrix} 0 \\ h \end{pmatrix} \in M$ implies that there exist $c_i, d_j \in R$ such that $\begin{pmatrix} 0 \\ h \end{pmatrix} = \sum_i c_i \begin{pmatrix} f_i \\ 0 \end{pmatrix} + \sum_j d_j \begin{pmatrix} g_j \\ g_j \end{pmatrix}$. Since $\sum_i c_i f_i = -\sum_j d_j g_j$ and $h = \sum_j d_j g_j$, we have $h \in J$ and $h \in I$. We next show that G_0 is a Gröbner basis of $I \cap J$ with respect to $<$. If $h \in I \cap J$, then there exist $c_i, d_j \in R$ such that $h = \sum_i c_i f_i = \sum_j d_j g_j$. Then $\sum_i c_i \begin{pmatrix} f_i \\ 0 \end{pmatrix} - \sum_j d_j \begin{pmatrix} g_j \\ g_j \end{pmatrix} = \begin{pmatrix} 0 \\ -\sum_j d_j g_j \end{pmatrix} = \begin{pmatrix} 0 \\ -h \end{pmatrix}$ implies $v = \begin{pmatrix} 0 \\ h \end{pmatrix} \in M$. Thus there exists $w = \begin{pmatrix} h_1 \\ h_2 \end{pmatrix} \in G$ such that $\text{in}_{<_{POT}}(w) \mid \text{in}_{<_{POT}}(v)$. By the definition of $<_{POT}$, $h_1 = 0$, and we have $h_2 \in G_0$. Since $\text{in}_<(h_2) \mid \text{in}_<(h)$, G_0 is a Gröbner basis of $I \cap J$.

Problem 3.8.1 1. By computing the Gröbner bases of the given ideals and checking the inclusion, it is proved that $I \subset J$, $I \neq J$, $I \subset H$, $I \neq H$, and there is no inclusion relation between J and H.

2. The result of part 1 implies $V(J) \subset V(I)$, $V(H) \subset V(I)$. Furthermore, we conclude $V(I) = V(H)$, $V(H) \neq V(J)$ from the radical membership test.

Problem 3.8.2 1. A Gröbner basis of I with respect to a grevlex ordering $<$ such that $x > y > z$ contains elements whose initials are x^7, y^7, and z^7. Thus I is zero-dimensional.

2. The initial ideal $\text{in}_<(I)$ is $\langle x^7, y^6x, y^7, zy^6, z^2y^5, z^3x^4, z^4x^3, z^7, yx^5, y^2x^4,$
$y^4x^2, zx^5, zyx^4, z^3y^2x, z^3y^3, z^4yx, z^5x, z^5y, zy^3x, z^2yx^2 \rangle$, and the standard
monomial set is $\{x^6, y^3x^3, y^5x, y^6, zy^2x^3, zy^5, z^2x^4, z^2y^4, z^3x^3, z^4x^2, z^4y^2, z^6,$
$x^5, yx^4, y^2x^3, y^3x^2, y^4x, y^5, zx^4, zyx^3, zy^2x^2, zy^4, z^2x^3, z^2y^2x, z^2y^3, z^3x^2, z^3$
$yx, z^3y^2, z^4x, z^4y, z^5, x^4, yx^3, y^2x^2, y^3x, y^4, zx^3, zyx^2, zy^2x, zy^3, z^2x^2, z^2yx,$
$z^2y^2, z^3x, z^3y, z^4, x^3, yx^2, y^2x, y^3, zx^2, zyx, zy^2, z^2x, z^2y, z^3, x^2, yx, y^2,$
$zx, zy, z^2, x, y, z, 1\}$. The number of elements in this set is 66. Thus
$\dim_\mathbb{Q} \mathbb{Q}[x, y, z]/I = 66$.

3. The result of computation shows that the reduced Gröbner basis is of the form $\{g_0(z), x - g_1(z), y - g_2(z)\}$. We note that the coefficients in g_1, g_2 are rather large, and it may take a long time, depending on the method of computation.

Problem 3.8.3 1. The minimal polynomial $m(z)$ of $\alpha + \beta$ is obtained as $m(z) \in \mathbb{Q}[z]$ such that $I \cap \mathbb{Q}[z] = \langle m(z) \rangle$ for $I = \langle x^5 - 3, y^3 - 5, z - (x + y) \rangle$. The result is $m(z) = z^{15} - 25z^{12} - 9z^{10} + 250z^9 - 1350z^7 - 1250z^6 + 27z^5 - 10125z^4 + 3125z^3 - 1350z^2 - 5625z - 3152$.

2. $\mathbb{Q}(\alpha, \beta) = \mathbb{Q}[\alpha, \beta]$ implies that $\frac{1}{\alpha + \beta}$ can be represented by a polynomial $g(\alpha, \beta)$ of α, β over $\mathbb{Q}$. Then we have $(x+y)g(x, y) \equiv 1 \mod \langle x^5-3, y^3-5 \rangle$. Since $t - g(x, y) \in J$ for $J = \langle (x + y)t - 1, x^5 - 3, y^3 - 5 \rangle$, the reduced Gröbner basis of J with respect to an elimination order such that $\{t\} \gg \{x, y\}$ contains an element $t - g(x, y)$. The result is $g(x, y) = \frac{1}{3152}((-15y^2 + 125y + 9)x^4 + (-125y^2 - 9y + 75)x^3 + (9y^2 - 75y + 625)x^2 + (75y^2 - 625y - 45)x + 625y^2 + 45y - 375)$.

References

1. J. Abbott, A.M. Bigatti, CoCoALib: a C++ library for doing computations in commutative algebra. http://cocoa.dima.unige.it/cocoalib
2. CoCoA Team, A system for doing computations in commutative algebra. http://cocoa.dima.unige.it
3. W. Decker, G.-M. Greuel, G. Pfister, H. Schönemann, Singular 3-1-2—a computer algebra system for polynomial computations (2010). http://www.singular.uni-kl.de/
4. R. Gebauer, H.M. Möller, On an installation of Buchberger's algorithm. J. Symbolic Comput. **6**, 275–286 (1988)
5. A. Giovini, T. Mora, G. Niesi, L. Robbiano, C. Traverso, "One sugar cube, please" or selection strategies in the Buchberger algorithm, in *Proceedings of the ISSAC 1991* (ACM, New York, 1991), pp. 49–54

6. D.R. Grayson, M.E. Stillman, Macaulay2, a software system for research in algebraic geometry. http://www.math.uiuc.edu/Macaulay2/
7. T. Kawazoe, M. Noro, Algorithms for computing a primary ideal decomposition without producing intermediate redundant components. J. Symbolic Comput. **46**, 1158–1172 (2011)
8. M. Kreuzer, L. Robbiano, *Computational Commutative Algebra 1* (Springer, Berlin, 2000); *Computational Commutative Algebra 2* (Springer, Berlin, 2005)
9. M. Noro, N. Takayama, H. Nakayama, K. Nishiyama, K. Ohara, Risa/Asir: a computer algebra system. http://www.math.kobe-u.ac.jp/Asir/asir.html

Chapter 4
Markov Bases and Designed Experiments

Satoshi Aoki and Akimichi Takemura

Abstract Markov bases first appeared in a 1998 work by Diaconis and Sturmfels (Ann Stat 26:363–397, 1998). In this paper, they considered the problem of estimating the p values for conditional tests for data summarized in contingency tables by Markov chain Monte Carlo methods; this is one of the fundamental problems in applied statistics. In this setting, it is necessary to have an appropriate connected Markov chain over the given finite sample space. Diaconis and Sturmfels formulated this problem with the idea of a Markov basis, and they showed that it corresponds to the set of generators of a well-specified toric ideal. Their work is very attractive because the theory of a Gröbner basis, a concept of pure mathematics, can be used in actual problems in applied statistics. In fact, their work became one of the origins of the relatively new field, computational algebraic statistics. In this chapter, we first introduce their work along with the necessary background in statistics. After that, we use the theory of Gröbner bases to solve actual applied statistical problems in experimental design.

S. Aoki (✉)
Department of Mathematics and Computer Science, Kagoshima University,
Korimoto, Kagoshima 890-0065, Japan
e-mail: aoki@sci.kagoshima-u.ac.jp

A. Takemura
Department of Mathematical Informatics, Graduate School of Information Science
and Technology, University of Tokyo, Bunkyo, Tokyo 113-8656, Japan
e-mail: takemura@stat.t.u-tokyo.ac.jp

T. Hibi (ed.), *Gröbner Bases: Statistics and Software Systems*,
DOI 10.1007/978-4-431-54574-3_4, © Springer Japan 2013

4.1   Conditional Tests of Contingency Tables

4.1.1   Sufficient Statistics

In this chapter, we consider qualitative data analyses. Qualitative data can be expressed in categories such as {Yes, No}, and they commonly result from such things as questionnaires in which respondents choose an answer from several options, or clinical trials where the responses are difficult to quantify. The random variables in this chapter are discrete, and for the variables and their frequencies, we will consider various statistical models. We will use capital letters (such as X) to represent random variables and lower-case letters (such as x) to represent observations. Qualitative data are often summarized in *contingency tables*. To match the context of contingency tables, we will consider only random variables that are limited to nonnegative integers values $\{0, 1, 2, \ldots\}$. We will use a bold letter to represent a vector or a multidimensional variable, and an index to denote each variable. For example, to denote a three-dimensional random variable, we will write $X = (X_1, X_2, X_3)$. We will write

$$X = (X_{ij}) = (X_{11}, \ldots, X_{1J}, \ldots, X_{I1}, \ldots, X_{IJ})$$

to represent IJ random variables in an $I \times J$ table (i.e., an $I \times J$ contingency table). These are common notational conventions in the context of contingency tables.

The joint probability function of X is written as $p(x) = \Pr(X = x)$, and $p(x)$ is usually characterized by a *parameter*. The aim of statistical inference is to use observations to estimate or to test a hypothesis about this parameter. In statistical inference, the concept of *sufficient statistics* plays an important role.

Definition 4.1.1 (Sufficient Statistic). Let $T(X) = (T_1(X), \ldots, T_k(X))$ be a function of X, i.e., a statistic, and let $\theta = (\theta_1, \ldots, \theta_v)$ be a parameter. T is called a sufficient statistic for θ if the conditional probability function of X for a given T,

$$p(x \mid t) = \Pr(X = x \mid T(X) = t),$$

does not depend on θ.

To better understand the concept of a sufficient statistic, consider the following simple example.

Example 4.1.2 (Estimating and Testing a Hypothesis Using a Sufficient Statistic). It is well known that the center of gravity of a common die is slightly biased, and the probabilities of the faces are not equal. To verify this, we roll a die n times and count the number of times a particular face (say, the face 1) occurs.

This experiment is described as follows. We are interested in the true probability of the face 1 of a particular die, which we denote by the parameter p. The experiment of rolling this die n times can be represented by random variables, as follows:

$$X_i = \begin{cases} 1, & \text{if 1 appears in the } i\text{th rolling,} \\ 0, & \text{if 1 does not appear in the } i\text{th rolling} \end{cases}$$

for $i = 1, \ldots, n$. The result of this experiment is that we obtain values for the random variables $X_1, \ldots, X_n$. Finally, we assume that the n rolls of this die are conducted independently. The above assumption is summarized as "$X_1, \ldots, X_n$ are independent Bernoulli random variables with success probability p". The joint probability function of $X = (X_1, \ldots, X_n)$ is

$$p(x) = \prod_{i=1}^{n} p^{x_i}(1-p)^{1-x_i} = p^t(1-p)^{n-t},$$

where $t = x_1 + \cdots + x_n$. On the other hand, $T = X_1 + \cdots + X_n$, the random variable that represents the total number of times the face 1 resulted in the n rolls, follows the binomial distribution $\mathrm{Bin}(n, p)$ with the probability function

$$p(t) = \binom{n}{t} p^t(1-p)^{n-t}, \quad t = 0, 1, \ldots, n.$$

We see that the joint conditional probability function of X for fixed $T = t$ is given as

$$p(x \mid t) = \frac{p(x)}{p(t)} = \frac{1}{\binom{n}{t}},$$

which does not depend on the parameter p. Thus, T is a sufficient statistic for p, by definition.

The meaning of a sufficient statistic is explained as follows. If we know the value of T, then knowing X provides no further information about p; therefore, knowing T is sufficient for the inference of p. In the above experiment, we lose the information about the sequence of the t occurrences of 1 and the $n - t$ occurrences of $2, \ldots, 6$ if we obtain only t instead of $x = (x_1, \ldots, x_n)$. We can see, however, that information about the sequence is not necessary for the inference of p. Some frequently used statistical methods for the inference of p are as follows.

- The point estimation (maximum likelihood estimation) of p is given by $\hat{p} = \dfrac{t}{n}$.
- The conventional 95% confidence interval of p is given by

$$\hat{p} - 1.96\sqrt{\frac{\hat{p}(1-\hat{p})}{n}} \le p \le \hat{p} + 1.96\sqrt{\frac{\hat{p}(1-\hat{p})}{n}}.$$

- One of the test statistics for the hypothesis

$$H_0 : p = p_0 \ (p_0 = 1/6, \text{for example})$$
$$H_1 : p \neq p_0$$

is given by $(\hat{p} - p_0) \Big/ \sqrt{\dfrac{p_0(1 - p_0)}{n}}$. (The two-sided test of the level $100\alpha\%$

can be conducted by comparing this value to the two-sided $100(1 - \alpha/2)\%$ point of the standard normal distribution.)

We can see from the confidence interval above, for example, that we would need to roll a die more than a million times in order to obtain an estimate of the third decimal place of $\hat{p}$. Although the design of the sample size is another important topic in statistical theory, we will not give it further consideration here. From the above, we see that the natural estimates and test statistics of p are constructed only in terms of the observed value of the sufficient statistic t. It is important to note that estimates and test statistics based on sufficient statistic are *optimal* in the appropriate ways. Although we have omitted the theoretical background for these, it can be found in [18, 19].

To obtain a sufficient statistic for each problem, we must derive a conditional probability function from the definition. Fortunately, there is a theorem that allows us to avoid this somewhat cumbersome calculation.

Theorem 4.1.3 (Factorization Theorem). *T is a sufficient statistic for θ if and only if the probability distribution of X is factored as*

$$p(x, \theta) = h(x)g(T(x), \theta). \tag{4.1}$$

Proof. We only consider the case of discrete random variables. Suppose that the probability function $p(x, \theta)$ is factored as (4.1). Then the probability function of $T(x)$ is written as

$$\Pr(T = t) = \sum_{x:T(x)=t} p(x, \theta) = \sum_{x:T(x)=t} h(x)g(T(x), \theta) = g(t, \theta) \sum_{x:T(x)=t} h(x).$$

Therefore, we have

$$\Pr(X = x \mid T = t) = \frac{\Pr(X = x, T = t)}{\Pr(T = t)} = \frac{g(t, \theta)h(x)}{g(t, \theta) \displaystyle\sum_{y:T(y)=t} h(y)} = \frac{h(x)}{\displaystyle\sum_{y:T(y)=t} h(y)},$$

and we see that T is a sufficient statistic for θ. Conversely, suppose that T is a sufficient statistic for θ. Because $\Pr(T = t)$ is a function of t and θ, we write $\Pr(T = t) = g(t, \theta)$. From the definition of a sufficient statistic, we can also write $\Pr(X = x \mid T = t) = h(x)$. Then we have

$$p(x, \theta) = \Pr(T = t)\Pr(X = x \mid T = t) = g(t, \theta)h(x),$$

which is the factorization (4.1). $\square$

From the factorization theorem, we can directly derive a sufficient statistic from the probability function. Note that the factorization theorem also holds for continuous cases. For the proof, see Corollary 1 in Sect. 2.6 of [18].

We now give another definition.

Definition 4.1.4 (Exponential Family). $p(x, \theta)$ belongs to a (k-parameter) exponential family if

$$p(x, \theta) = h(x) \exp\left(\sum_{j=1}^{k} T_j(x) \phi_j(\theta) - c(\theta) \right). \tag{4.2}$$

For a probability function in the exponential family, we see from the factorization theorem that $(T_1, \ldots, T_k)$ is a k-dimensional sufficient statistic. In fact, the distributions treated in this chapter belong to the exponential family. Specifically, the problem considered in this chapter is a testing hypothesis

$$H_0 : \lambda = (0, \ldots, 0)$$
$$H_1 : \lambda \neq (0, \ldots, 0)$$

for the parameter θ in the exponential family, where we consider the transform $\theta \leftrightarrow (\lambda, \psi)$. Here, our concern is focused on λ, and we are not interested in ψ, which is called a *nuisance parameter*.

We have now given the necessary definitions, and in the next section, we will consider examples with contingency tables.

4.1.2 2 × 2 Contingency Tables

A contingency table is a cross-classified table of frequencies. For example, consider a questionnaire with ten questions. If there are I_i choices for question i, $i = 1, \ldots, 10$, the reply for each respondent is one of the $I_1 \times \cdots \times I_{10}$ combinations of choices. If we count the frequency for each of these combinations of choices, we can produce a $I_1 \times \cdots \times I_{10}$ table of frequencies. This is an example of a 10-way contingency table. In the first half of this chapter, we consider data that is summarized in contingency tables. Contingency tables with 2 axes, called two-way contingency tables, can be described by simple statistical models. In this section, we consider the simplest two-way case, 2×2, i.e., there are only two levels for each axis.

In the following, we introduce three typical examples of 2×2 contingency tables and consider natural statistical models. The aim of this section is to show that the sufficient statistics for the three models are written in the same form.

Example 4.1.5 (The Case of Independent Binomial Distributions). To investigate the contribution of smoking to the risk for stomach cancer, 20 patients with stomach

Table 4.1 Smoking experiences of stomach cancer patients and healthy people (imaginary data)

	Smoking	Nonsmoking	Total
Cases	14	6	20
Controls	56	44	100

cancer (cases) and 100 healthy people (controls) were asked whether they had ever smoked. The result of this research is summarized in Table 4.1. We then ask, is there a relation between smoking and stomach cancer?

Table 4.1 is an example of a 2×2 contingency table. A natural model for these data is

$$X_1 \sim \text{Bin}(n_1, p_1), \quad X_2 \sim \text{Bin}(n_2, p_2), \quad X_1 \perp\!\!\!\perp X_2,$$

where X_1 and X_2 are random variables that represent smoking cases and smoking controls, respectively. Here Bin denotes a binomial distribution, and $\perp\!\!\!\perp$ denotes independence. The parameters p_1 and p_2 are the probability of smoking of cases and controls, respectively. Note that the sample sizes n_1 and n_2 are fixed.

The observations and corresponding probabilities are summarized as follows.

Observations	Smoking	Nonsmoking	Total
Cases	x_1	$n_1 - x_1$	n_1
Controls	x_2	$n_2 - x_2$	n_2

(4.3)

Probabilities	Smoking	Nonsmoking	Total
Cases	p_1	$1 - p_1$	1
Controls	p_2	$1 - p_2$	1

The joint probability function of $X = (X_1, X_2)$ is the product of two independent binomial distributions given by

$$p(x) = \binom{n_1}{x_1}\binom{n_2}{x_2} p_1^{x_1}(1-p_1)^{n_1-x_1} p_2^{x_2}(1-p_2)^{n_2-x_2},$$

$$x_1 = 0, 1, \ldots, n_1, \quad x_2 = 0, 1, \ldots, n_2.$$

(4.4)

Now consider the natural statistical model for this experiment. The aim of this research is to reveal whether there is a significant difference in the smoking experience of the cases and controls. (The relation between lung cancer and smoking may be obvious. However, the relation between stomach cancer and smoking is not so clear.) From Table 4.1, we see that the smoking ratio of cases ($14/20$) is higher than that of the controls ($56/100$), and thus we might want to conclude that "there is a significant difference". However, there is another possibility that we need to consider: there is no true relation and the (false-positive) result was observed by chance. We can answer this question by a *testing hypothesis*. In the testing hypothesis, the assertion that we want to reject is treated as the *null hypothesis* (or *null model*). In this example, the null hypothesis is "there is no true relation between the stomach cancer and smoking"; this is also the natural statistical model in which we are interested.

In general, the null model is expressed as a manifold in the parameter space. The parameter space in this example is the two-dimensional square

$$\{(p_1, p_2) \mid 0 \leq p_1, p_2 \leq 1\},$$

and the null model is the line segment $p_1 = p_2$ within it. Following the conventions for a testing hypothesis, we write the null model as H_0 and write the alternative model as H_1. Then we have

$$
\begin{aligned}
H_0 &: p_1 = p_2 \\
H_1 &: p_1 \neq p_2.
\end{aligned}
\tag{4.5}
$$

Expression (4.5) is simple. However, we will prepare another expression for later use. Consider the parameter transform

$$\psi = \log \frac{p_2}{1 - p_2}, \quad \lambda = \log \frac{p_1(1 - p_2)}{p_2(1 - p_1)} \tag{4.6}$$

for a two-dimensional parameter (p_1, p_2), where we assume $0 < p_1, p_2 < 1$. This is a one-to-one transform. Then the null model and the alternative model can be written in terms of (ψ, λ) as

$$
\begin{aligned}
H_0 &: \lambda = 0 \\
H_1 &: \lambda \neq 0.
\end{aligned}
\tag{4.7}
$$

Note that expressions (4.5) and (4.7) are equivalent. We also note that although (4.7) looks more complicated than (4.5), it is a natural form in the exponential family setting or when this is extended to higher-dimensional problems. In fact, λ is called a *log odds ratio* of stomach cancer and smoking, and it is frequently used in medical statistics. Some readers who like to gamble, such as betting on horse racing, may be familiar with the term *"odds"*. In this problem, $p_1/(1 - p_1)$ and $p_2/(1 - p_2)$ are the odds of smoking for stomach cancer patients and healthy people, respectively.

Now we investigate the relation between the new parameter and the exponential family. From the inverse transform of (4.6),

$$p_1 = \frac{e^{\lambda + \psi}}{1 + e^{\lambda + \psi}}, \quad p_2 = \frac{e^{\psi}}{1 + e^{\psi}},$$

we have

$$p(x) = \binom{n_1}{x_1}\binom{n_2}{x_2} \exp\left((x_1 + x_2)\psi + x_1\lambda - n_1 \log(1 + e^{\psi + \lambda}) - n_2 \log(1 + e^{\psi})\right).$$

$$\tag{4.8}$$

Table 4.2 Which party is
supported and voting
intention (imaginary data)

	Ruling	Opposition	Total
Voting	22	43	65
Nonvoting	14	21	35
Total	36	64	100

The expression (4.8) can also be obtained by setting

$$x = (x_1, x_2), \; \theta = (\psi, \lambda), \; h(x) = \binom{n_1}{x_1}\binom{n_2}{x_2},$$
$$T_1(x) = x_1 + x_2, \; T_2(x) = x_1, \; \phi_1(\theta) = \psi, \; \phi_2(\theta) = \lambda,$$
$$c(\theta) = n_1 \log(1 + e^{\psi+\lambda}) + n_2 \log(1 + e^\psi)$$

in the general form of an exponential family (4.2), i.e., we see that the model of independent binomial distributions belongs to an exponential family. Note also that $x_1 + x_2$ and x_1 are sufficient statistics for the parameters ψ and λ, respectively, and under the null model $\lambda = 0$, ψ is the only parameter for which the sufficient statistic is $x_1 + x_2$. Recall that n_1 and n_2 are fixed. In the 2×2 table of observations (upper table of (4.3)), the sum of the first column, $x_1 + x_2$, is the sum of smokers. If we fix $x_1 + x_2$, the sum of the second column (the sum of nonsmokers) is also fixed.

Example 4.1.6 (The Case of a Multinomial Distribution). Table 4.2 is a result of a survey of the political views of 100 young people. They were asked

(1) Which of parties (ruling or opposition) do you support?
(2) Will you vote in the next election?

We wish to know, is there a relation between which party an individual supports and their intent to vote?

Table 4.2 is another example of a 2×2 contingency table, and it has a different structure from the one in Example 4.1.5. In Table 4.2, only the total sum 100 is fixed, whereas two row sums are fixed in Example 4.1.5. In this situation, it is natural to assume a four-dimensional random variable $X = (X_{11}, X_{12}, X_{21}, X_{22})$ that follows a *multinomial distribution* $M(n, (p_{11}, p_{12}, p_{21}, p_{22}))$. Here, X_{ij} and p_{ij} are the random variable and the probability, respectively, for the reply of which party is supported and if the person intends to vote. (We treat $i = 1, 2$ as (Voting, Nonvoting) and $j = 1, 2$ as (Ruling, Opposition).) When placed into 2×2 tables, we have the following.

Observation	Ruling	Opposition	Total
Voting	x_{11}	x_{12}	
Nonvoting	x_{21}	x_{22}	
Total			n

Probability	Ruling	Opposition	Total
Voting	p_{11}	p_{12}	
Nonvoting	p_{21}	p_{22}	
Total			1

The joint probability function of $X = (X_{11}, X_{12}, X_{21}, X_{22})$ is

$$p(x) = \frac{n!}{x_{11}!x_{12}!x_{21}!x_{22}!} p_{11}^{x_{11}} p_{12}^{x_{12}} p_{21}^{x_{21}} p_{22}^{x_{22}},$$

$$p_{11} + p_{12} + p_{21} + p_{22} = 1, \quad x_{11} + x_{12} + x_{21} + x_{22} = n. \tag{4.9}$$

As we did in Example 4.1.5, we consider a natural statistical model. In this case, our aim is to consider the influence of which party is supported on the intention to vote. Therefore, we are interested in the null model, "there is no relation between which party is supported and voting intention". In the context of two-way contingency tables, this model is called an *independence model* or *independence hypothesis* between the rows and columns. One of the common expressions of this null model is as follows. As in Example 4.1.5, we consider a one-to-one parameter transformation between $p = (p_{11}, p_{12}, p_{21}, p_{22})$ and $(\psi_1, \psi_2, \lambda)$ in the region that $p_{ij} > 0$ for all i, j:

$$\psi_1 = \log \frac{p_{12}}{p_{22}}, \quad \psi_2 = \log \frac{p_{21}}{p_{22}}, \quad \lambda = \log \frac{p_{11} p_{22}}{p_{12} p_{21}}. \tag{4.10}$$

We can then express the model as

$$H_0 : \lambda = 0$$
$$H_1 : \lambda \neq 0.$$

This is the same expression as (4.7) in Example 4.1.5. Under the null hypothesis $\lambda = 0$, the ratios of which party is supported is the same for both the voter and the nonvoter (p_{11}/p_{12} and p_{21}/p_{22}, respectively), for example, and the rows (voting intention) and columns (which party is supported) are selected independently. We can now write the joint probability function based on the new parameter. The inverse transform of (4.10) is

$$p_{11} = \frac{e^{\psi_1 + \psi_2 + \lambda}}{1 + e^{\psi_1} + e^{\psi_2} + e^{\psi_1 + \psi_2 + \lambda}}$$

$$p_{12} = \frac{e^{\psi_1}}{1 + e^{\psi_1} + e^{\psi_2} + e^{\psi_1 + \psi_2 + \lambda}}$$

$$p_{21} = \frac{e^{\psi_2}}{1 + e^{\psi_1} + e^{\psi_2} + e^{\psi_1 + \psi_2 + \lambda}}$$

$$p_{22} = \frac{1}{1 + e^{\psi_1} + e^{\psi_2} + e^{\psi_1 + \psi_2 + \lambda}}.$$

Substituting this into (4.9), we have

$$p(x) = \frac{n!}{x_{11}!x_{12}!x_{21}!x_{22}!} \exp\left((x_{11} + x_{12})\psi_1 + (x_{11} + x_{21})\psi_2 + x_{11}\lambda \right.$$
$$\left. -n\log(1 + e^{\psi_1} + e^{\psi_2} + e^{\psi_1 + \psi_2 + \lambda})\right). \tag{4.11}$$

Table 4.3 Numbers of defective products
under various conditions of heating time and
catalyst (imaginary data)

	Catalyst A	Catalyst B
Long	5	12
Short	7	6

We see that the multinomial distribution again belongs to the exponential family.
From (4.11), we also see that $(X_{11} + X_{12}, X_{11} + X_{21})$ is a sufficient statistic for
the parameter (ψ_1, ψ_2) under the null hypothesis $\lambda = 0$. Because the total sum n
is fixed in this example, fixing the sufficient statistic under the null hypothesis is
equivalent to fixing the row sums and the column sums in the 2×2 table.

Example 4.1.7 (The Case of a Poisson Distribution). Consider a manufacturing
process in a factory for which you wish to determine how to reduce the number
of defective products. To investigate this, you could use various combinations of
heating time (long or short) and catalyst (A or B), and then count the number
of defectives manufactured under each condition. The results are summarized in
Table 4.3. Which length of heating time and which catalyst is most desirable?

Table 4.3 is another type of 2×2 table. Unlike Examples 4.1.5 and 4.1.6, no column
sums or row sums are fixed at the start. We may simply choose (Long, Catalyst A)
as the desirable combination since it has the smallest number of defective products.
However, another interpretation of Table 4.3 leads to choosing (Short, Catalyst A)
if we ignore an interaction effect of (Long, Catalyst A), because the total number
of defective products for the Short heating time $(7 + 6 = 13)$ is smaller than that
for the Long heating time $(5 + 12 = 17)$. Such a consideration leads to a statistical
model with the null hypothesis, "there is no interaction effect between the heating
time and which catalyst is used". A natural model for Table 4.3 is the independent
Poisson distribution:

$$X_{ij} \sim \text{Po}(\mu_{ij}), \ i, j = 1, 2, \ X_{ij} \text{ are independent}, \ \mu_{ij} > 0,$$

where X_{ij} is a random variable for the number of defective products for the level
(i, j). Here we write $i = 1, 2$ as heating time (Long, Short) and $j = 1, 2$ as catalyst
(Catalyst A, Catalyst B). μ_{ij} is the expected value of the number of defective
products for the level (i, j), i.e., $E(X_{ij}) = \mu_{ij}$. We summarize this notation in
the following 2×2 table.

Observation

	Catalyst A	Catalyst B
Long	x_{11}	x_{12}
Short	x_{21}	x_{22}

Expected value

	Catalyst A	Catalyst B
Long	μ_{11}	μ_{12}
Short	μ_{21}	μ_{22}

The joint probability function of $X = (X_{11}, X_{12}, X_{21}, X_{22})$ is

$$p(x) = \prod_{i=1}^{2} \prod_{j=1}^{2} \frac{\mu_{ij}^{x_{ij}} e^{-\mu_{ij}}}{x_{ij}!}, \quad x_{ij} = 0, 1, 2, \ldots. \tag{4.12}$$

In this example, the parameter $(\mu_{11}, \mu_{12}, \mu_{21}, \mu_{22})$ is four-dimensional. Here, we define a transform

$$\psi_0 = \log \mu_{22}, \ \psi_1 = \log \frac{\mu_{12}}{\mu_{22}}, \ \psi_2 = \log \frac{\mu_{21}}{\mu_{22}}, \ \lambda = \log \frac{\mu_{11}\mu_{22}}{\mu_{12}\mu_{21}}.$$

The meaning of this new parameter is seen from the inverse transform

$$\mu_{11} = e^{\psi_0+\psi_1+\psi_2+\lambda}, \ \mu_{12} = e^{\psi_0+\psi_1}, \ \mu_{21} = e^{\psi_0+\psi_2}, \ \mu_{22} = e^{\psi_0}. \tag{4.13}$$

First, similar to Examples 4.1.5 and 4.1.6, the null hypothesis, "there is no interaction effect between the heating time and the catalyst", is expressed as $\lambda = 0$. Under this null hypothesis, ψ_1 can be written as

$$\psi_1 = \log \frac{\mu_{12}}{\mu_{22}} = \log \frac{\mu_{11}}{\mu_{21}},$$

which represents the common (i.e., independent of catalyst) effect of setting the heating time to Long. Similarly, ψ_2 represents the common (i.e., independent of the heating time) effect of using Catalyst A. These effects are called the main effects for the heating time and the catalyst, respectively. λ is called a two-factor interaction effect between the heating time and the catalyst.

Substituting (4.13) into (4.12), we have

$$p(x) = \frac{1}{x_{11}!x_{12}!x_{21}!x_{22}!} \exp \left((x_{11} + x_{12} + x_{21} + x_{22})\psi_0 \right.$$
$$+ (x_{11} + x_{12})\psi_1 + (x_{11} + x_{21})\psi_2 + x_{11}\lambda$$
$$\left. - (e^{\psi_0+\psi_1+\psi_2+\lambda} + e^{\psi_0+\psi_1} + e^{\psi_0+\psi_2} + e^{\psi_0}) \right). \tag{4.14}$$

We see that (4.14) is an exponential family, and

$$(X_{11} + X_{12} + X_{21} + X_{22}, X_{11} + X_{12}, X_{11} + X_{21})$$

is a sufficient statistic for the parameter (ψ_0, ψ_1, ψ_2) under the null model H_0 : $\lambda = 0$. As in Examples 4.1.5 and 4.1.6, fixing a value to the sufficient statistic is equivalent to fixing a value to the row sums and the column sums.

Now we have seen three examples of 2×2 tables. In these examples, the assumed probability functions and null hypotheses differ according to the sampling schemes, as follows.

	Probability functions	Null models of interest
Example 4.1.5	Independent Binomial	Common proportions
Example 4.1.6	Multinomial	Independent rows and columns
Example 4.1.7	Independent Poisson	No two-factor interaction

However, with appropriate transformations of the parameters, the null hypotheses can all be written as

$$H_0 : \lambda = 0. \tag{4.15}$$

Also note that the sufficient statistic for the parameter under the null hypothesis coincides with the row sums and column sums of the 2×2 tables. In the next section, we consider tests of the null hypothesis (4.15) that are based on the sufficient statistic for the parameter of "no interest". We note that there is an advantage in that we need not consider differences of sampling schemes. Such an advantage is valid not only for general two-dimensional contingency tables but also for higher-dimensional contingency tables.

As the last topic of this section, we give another definition.

Definition 4.1.8 (Saturated Models). A model for which the number of parameters and the dimensionality of the data are the same is called a saturated model.

There are no assumptions about the probability structure of a saturated model. In the three examples in this section, the dimensions of a saturated model would be $\dim(p_1, p_2) = 2$ for Example 4.1.5, $\dim(p_{11}, p_{12}, p_{21}, p_{22}) = 3$ for Example 4.1.6 (because of the constraint $\sum p_{ij} = 1$), and $\dim(\mu_{11}, \mu_{12}, \mu_{21}, \mu_{22}) = 4$ for Example 4.1.7. In this chapter, we focus on testing a hypothesis of the form

$$H_0 : \quad \text{Submodel of the saturated model}$$
$$H_1 : \quad \text{Saturated model.}$$

4.1.3 Similar Tests

In this section, we consider testing the hypothesis

$$H_0 : \lambda = (0, \ldots, 0), \tag{4.16}$$

where the parameter is expressed as $\theta = (\psi, \lambda)$ by some transform. In the three examples of 2×2 tables in the previous section, the dimension of λ was one, and the dimensions of ψ was different for each problem. ψ, is called a nuisance parameter.

Generally, the testing procedure

$$T(x) \geq c \implies H_0 \text{ is rejected}$$

for some test statistics $T(X)$ of significance level α must satisfy some condition for the type I error, i.e.,

$$\Pr(T(X) \geq c \mid H_0 \text{ is true}) \leq \alpha \qquad (4.17)$$

for a given level of significance α. For the null hypothesis (4.16), (4.17) is written as

$$\sup_{\psi} \Pr(T(X) \geq c \mid \theta = (\psi, (0, \ldots, 0))) \leq \alpha. \qquad (4.18)$$

However, in general, it is difficult to evaluate the left-hand side of (4.18) or to seek tests that are powerful under (4.18). Therefore, it is a common approach to consider a class of similar tests.

Definition 4.1.9 (Similar Tests). A test is similar if the type I error $\Pr(T(X) \geq c \mid H_0$ is true) does not depend on the nuisance parameter.

To construct similar tests, one approach is to use the conditional distribution that fixes the sufficient statistic for the nuisance parameter. We will explain this approach for the case of 2×2 tables.

Similar Tests for 2×2 Contingency Tables (Fisher's Exact Tests)

We can write the three 2×2 tables from the previous section in the form

x_{11}	x_{12}	$x_{1\cdot}$
x_{21}	x_{22}	$x_{2\cdot}$
$x_{\cdot 1}$	$x_{\cdot 2}$	$x_{\cdot\cdot}$

where we use the dot notation ($\cdot$) to represent the sum with respect to the index. Because we have seen that the sufficient statistics for the nuisance parameter are the row sums and column sums, we will consider the conditional probability function that fixes them. We write it as the one variable function of x_{11} as

$$p(x_{11} \mid x_{\cdot\cdot}, x_{1\cdot}, x_{\cdot 1}, \lambda) = \frac{1}{C(\lambda)} \frac{\exp(\lambda x_{11})}{x_{11}!(x_{1\cdot} - x_{11})!(x_{\cdot 1} - x_{11})!(x_{\cdot\cdot} - x_{\cdot 1} - x_{1\cdot} + x_{11})!}, \qquad (4.19)$$

where $C(\lambda)$ is the normalizing constant defined by

$$C(\lambda) = \sum_{y=\max(0, x_{1\cdot}+x_{\cdot 1}-x_{\cdot\cdot})}^{\min(x_{1\cdot}, x_{\cdot 1})} \frac{\exp(\lambda y)}{y!(x_{1\cdot} - y)!(x_{\cdot 1} - y)!(x_{\cdot\cdot} - x_{\cdot 1} - x_{1\cdot} + y)!}. \qquad (4.20)$$

The summation in (4.20) is the sum with respect to all the values of x_{11} for fixed row sums and column sums. Equation (4.19) is called a *generalized hypergeometric distribution.* Note that some textbooks call the conditional distribution for general $I \times J$ tables (cf. (4.30) of Sect. 4.1.4, for example) the generalized hypergeometric distribution. In this book, we call the conditional probability function (4.19) the generalized hypergeometric distribution. The normalizing constant of the generalized hypergeometric distribution does not have a closed-form expression. However, to evaluate the significance probability in an actual test, it is sufficient to use the conditional probability function under the null model. Substituting $\lambda = 0$ into (4.19), we have

$$p(x_{11} \mid x.., x_{1.}, x._{1}, \lambda = 0) = \frac{\binom{x_{1.}}{x_{11}}\binom{x.. - x_{1.}}{x._{1} - x_{11}}}{\binom{x..}{x._{1}}}. \tag{4.21}$$

Equation (4.21) is called a *hypergeometric distribution.*

Proposition 4.1.10. *Substituting $\lambda = 0$ into (4.19) yields (4.21).*

Proof. Comparing (4.20) and (4.21), we want to show the relation

$$\sum_{y=\max(0,x_{1.}+x._{1}-x..)}^{\min(x_{1.},x._{1})} \binom{x_{1.}}{y}\binom{x.. - x_{1.}}{x._{1} - y} = \binom{x..}{x._{1}}. \tag{4.22}$$

The right-hand side is the number of ways to choose $x._{1}$ balls from among $x..$ balls. Imagine these $x..$ balls are painted in two colors, i.e., $x_{1.}$ red balls and $x.. - x_{1.}$ white balls. Then the number of red balls y in the selected $x._{1}$ balls satisfies the condition

$$\max(0, x_{1.} + x._{1} - x..) \leq y \leq \min(x_{1.}, x._{1}).$$

Because $\binom{x_{1.}}{y}\binom{x.. - x_{1.}}{x._{1} - y}$ is the number of ways to choose y red balls and $x._{1} - y$ white balls from $x..$ balls, we have (4.22).   □

Because (4.21) does not depend on the nuisance parameter, we can construct similar tests by evaluating the upper probability of a test statistic based on this conditional probability function. In the case of a one-sided test,

$$H_0 : \lambda = 0$$
$$H_1 : \lambda > 0,$$

the natural testing procedure is

$$X_{11} \geq c \implies \text{Reject } H_0,$$

and the *significance probability (p value)* is given by

$$p \text{ value} = \Pr(X_{11} \geq x_{11} \mid H_0) = \sum_{x=x_{11}}^{\min(x_{1\cdot},x_{\cdot 1})} p(x \mid x_{\cdot\cdot}, x_{1\cdot}, x_{\cdot 1}).$$

Calculating this value and comparing it with the given significance level $\alpha = 0.05$, for example, we can test the null hypothesis as

$$p \text{ value} \leq 0.05 \implies \text{Reject } H_0.$$

This testing procedure is called *Fisher's exact test.*

Perform Fisher's exact test for the data in Example 4.1.5.

Example 4.1.11 (Example 4.1.5, Continued). Using the values in Table 4.1, we have

$$p(x_{11} \mid 120, 20, 70) = \frac{\binom{20}{x_{11}}\binom{100}{70 - x_{11}}}{\binom{120}{70}}.$$

Based on this hypergeometric distribution, the p value of the observation $x_{11} = 14$ is calculated as

$$p \text{ value} = \sum_{x=14}^{20} p(x \mid 120, 20, 70) = 0.1818. \tag{4.23}$$

Therefore, with the significance level $\alpha = 0.05$, the p value is not significant, and H_0 cannot be rejected. We cannot conclude from Table 4.1 that the probability of smoking is higher for stomach cancer patients than it is for healthy people.

As seen in the above example, to test a hypothesis, we first must define the null hypothesis, the alternative hypothesis, the test statistics, and the significance level. After that, the only remaining task is to calculate the p value, i.e., the probability that the test statistic is equal to or more extreme than the observed value under the null hypothesis. This calculation can be performed by Markov bases (or Gröbner bases) in [11].

It may be difficult to calculate (4.23) by hand. We used the software R for the previous example. The source code and the output are as shown below.

```
> data <- matrix(c(14,6,56,44),2,2,byrow=T)
> data
     [,1] [,2]
[1,]   14    6
[2,]   56   44
> fisher.test(data,alternative="greater")

         Fisher's Exact Test for Count Data
data:    data
```

Table 4.4 Result of examinations of geometry and probability

Geometry\Probability	5	4	3	2	1—	Total
5	2	1	1	0	0	4
4	8	3	3	0	0	14
3	0	2	1	1	1	5
2	0	0	0	1	1	2
1—	0	0	0	0	1	1
Total	10	6	5	2	3	26

```
p-value = 0.1818
alternative hypothesis: true odds ratio is greater than 1
95 percent confidence interval:
 0.697064        Inf
sample estimates:
odds ratio
 1.824451
```

4.1.4 $I \times J$ Tables

As the last topic of Sect. 4.1, we extend 2×2 tables and consider general $I \times J$ tables. As with 2×2 tables, there are several sampling schemes, such as "I rows follow independent multinomial distributions", "the set of all entries follow one multinomial distribution", or "each entry follows an independent Poisson distribution". However, as with 2×2 tables, we can treat these cases in the same way when considering conditional probability functions. Here, we consider the case of one multinomial distribution as a whole.

Example 4.1.12 (Example of a Two-Way Contingency Table). Consider an example where 26 students take examinations in both geometry and probability. For both subjects, the students are placed into five categories according to their scores, as shown in Table 4.4.

Table 4.4 is an example of a 5×5 contingency table. In this example, it is natural to consider a multinomial distribution with the total number of students fixed at $n = 26$:

$$p(x) = \frac{n!}{\displaystyle\prod_{i=1}^{5}\prod_{j=1}^{5} x_{ij}!} \prod_{i=1}^{5}\prod_{j=1}^{5} p_{ij}^{x_{ij}}, \quad \sum_{i=1}^{5}\sum_{j=1}^{5} p_{ij} = 1, \quad \sum_{i=1}^{5}\sum_{j=1}^{5} x_{ij} = n, \tag{4.24}$$

where x_{ij} and p_{ij} are the frequency and the occurrence probability for the (i, j) cell (i.e., the geometry score is i and the probability score is j), respectively. The dimension of the saturated model is $IJ - 1 = 24$.

For this p_{ij}, a model of interest may be an independence model between the rows and the columns. In fact, Table 4.4 shows a positive correlation between the two scores. We want to judge the null hypothesis that "there is no actual correlation, and the observed positive correlation was obtained by chance". To construct a similar test, we consider a parameter transformation. Extending the relation (4.10) for the multinomial case of a 2×2 table, we consider the transform

$$
\begin{cases}
\psi_{1i} = \log \dfrac{p_{iJ}}{p_{IJ}}, & i = 1, \ldots, I-1, \\[2mm]
\psi_{2j} = \log \dfrac{p_{Ij}}{p_{IJ}}, & j = 1, \ldots, J-1, \\[2mm]
\lambda_{ij} = \log \dfrac{p_{ij} p_{IJ}}{p_{iJ} p_{Ij}}, & i = 1, \ldots, I-1, \; j = 1, \ldots, J-1.
\end{cases}
\tag{4.25}
$$

With $p_{..} = 1$, (4.25) is uniquely solved for positive p_{ij}, and we have the inverse transform

$$
\begin{cases}
p_{ij} = p_{IJ} e^{\psi_{1i} + \psi_{2j} + \lambda_{ij}}, & i = 1, \ldots, I-1, \; j = 1, \ldots, J-1, \\[2mm]
p_{iJ} = p_{IJ} e^{\psi_{1i}}, & i = 1, \ldots, I-1, \\[2mm]
p_{Ij} = p_{IJ} e^{\psi_{2j}}, & j = 1, \ldots, J-1, \\[2mm]
p_{IJ} = \left[1 + \displaystyle\sum_{i=1}^{I-1} e^{\psi_{1i}} + \sum_{j=1}^{J-1} e^{\psi_{2j}} + \sum_{i=1}^{I-1}\sum_{j=1}^{J-1} e^{\psi_{1i} + \psi_{2j} + \lambda_{ij}} \right]^{-1}.
\end{cases}
\tag{4.26}
$$

The dimension of the saturated model is

$$
(I-1) + (J-1) + (I-1)(J-1) = IJ - 1.
$$

Rewriting $p_{IJ} = e_0^{\psi}$, we have the expression

$$
\log p_{ij} = \psi_0 + \psi_{1i} + \psi_{2j} + \lambda_{ij},
\tag{4.27}
$$

which is known as a *log-linear model*. The log-linear model is one of the traditional models in statistical theory. The parameter of the log-linear model is relatively clear, and this is one of its merits. For (4.27), the conventional terminology is

- ψ_{1i} is a main effect of the level i of the factor 1 (a main effect of the score i of geometry),
- ψ_{2j} is a main effect of the level j of the factor 2 (a main effect of the score j of probability),
- λ_{ij} is an interaction effect of the level (i, j) of the factors $1, 2$ (an interaction effect of the level (i, j) of geometry and probability).

The hypothesis of the independence of rows and columns,

$$
H_0 : \; p_{ij} = p_{i \cdot} p_{\cdot j} \text{ for all } (i, j),
\tag{4.28}
$$

is written as

$$H_0 \ : \ \lambda_{ij} = 0 \text{ for all } (i, j). \tag{4.29}$$

Proposition 4.1.13. *Equations (4.29) and (4.28) are equivalent.*

Proof. Assume $p_{ij} = p_{i\cdot}p_{\cdot j}$ for all i, j. It is easy to check that substitution of this into λ_{ij} in (4.25) yields zero. Conversely, assume (4.29). From (4.26), we have

$$p_{i\cdot} = \sum_{j=1}^{J-1} p_{ij} + p_{iJ} = p_{IJ} e^{\psi_{1i}} \left(1 + \sum_{j=1}^{J-1} e^{\psi_{2j}} \right), \quad i = 1, \ldots, I-1,$$

$$p_{\cdot j} = \sum_{i=1}^{I-1} p_{ij} + p_{Ij} = p_{IJ} e^{\psi_{2j}} \left(1 + \sum_{i=1}^{I-1} e^{\psi_{1i}} \right), \quad j = 1, \ldots, J-1.$$

From these and (4.26), we have

$$p_{i\cdot}p_{\cdot j} = p_{IJ}^2 e^{\psi_{1i}+\psi_{2j}} \left(1 + \sum_{i=1}^{I-1} e^{\psi_{1i}} + \sum_{j=1}^{J-1} e^{\psi_{2j}} + \sum_{i=1}^{I-1}\sum_{j=1}^{J-1} e^{\psi_{1i}+\psi_{2j}} \right) = p_{ij}$$

for $i = 1, \ldots, I-1, j = 1, \ldots, J-1$. The cases of $i = I$ or $j = J$ can be proved in a similar way. $\qquad\square$

From Proposition 4.1.13, we see that the nuisance parameters for the null model (4.29) are $\{\psi_{1i}\}$ and $\{\psi_{2j}\}$.

Proposition 4.1.14. *The multinomial distribution (4.24) belongs to the exponential family. The sufficient statistic for the nuisance parameter under the null hypothesis (4.29) are the row sums and the columns sums $\{x_{i\cdot}\}$ and $\{x_{\cdot j}\}$.*

Proof. Note that

$$\prod_{i=1}^{I}\prod_{j=1}^{J} p_{ij}^{x_{ij}} = p_{IJ}^{n} \left[\prod_{i=1}^{I-1}\prod_{j=1}^{J-1} \left(\frac{p_{ij}}{p_{IJ}} \right)^{x_{ij}} \right] \left[\prod_{i=1}^{I-1} \left(\frac{p_{iJ}}{p_{IJ}} \right)^{x_{iJ}} \right] \left[\prod_{j=1}^{J-1} \left(\frac{p_{Ij}}{p_{IJ}} \right)^{x_{Ij}} \right].$$

By this relation and (4.26), we have an expression for the multinomial distribution of the $I \times J$ table as

$$p(x) = \frac{n!}{\prod\limits_{i=1}^{I}\prod\limits_{j=1}^{J} x_{ij}!} \prod_{i=1}^{I}\prod_{j=1}^{J} p_{ij}^{x_{ij}}$$

$$= \frac{n! p_{IJ}^{n}}{\prod\limits_{i=1}^{I}\prod\limits_{j=1}^{J} x_{ij}!} \exp\left(\sum_{i=1}^{I-1} x_{i\cdot}\psi_{1i} + \sum_{j=1}^{J-1} x_{\cdot j}\psi_{2j} + \sum_{i=1}^{I-1}\sum_{j=1}^{J-1} x_{ij}\lambda_{ij} \right).$$

This corresponds to the general form of the exponential family (4.2) with

$$h(x) = \frac{n!}{\prod\limits_{i=1}^{I}\prod\limits_{j=1}^{J} x_{ij}!},$$

$$c(\boldsymbol{\theta}) = -n\log p_{IJ} = n\log\left(1 + \sum_{i=1}^{I-1} e^{\psi_{1i}} + \sum_{j=1}^{J-1} e^{\psi_{2j}} + \sum_{i=1}^{I-1}\sum_{j=1}^{J-1} e^{\psi_{1i}+\psi_{2j}+\lambda_{ij}} \right).$$

We also see that the sufficient statistics for the nuisance parameters ψ_{1i} and ψ_{2j} are $x_{i\cdot}$ and $x_{\cdot j}$, respectively, under the null hypothesis (4.29). $\qquad\square$

We consider constructing a similar test for the conditional probability function, given the row and the column sums.

Proposition 4.1.15. *The conditional probability function, given the row sums and the column sums, for the multinomial distribution under the null hypothesis (4.29) is*

$$p(x \mid x_{i\cdot}, x_{\cdot j}, \lambda_{ij} = 0) = \frac{\left(\prod\limits_{i=1}^{I} x_{i\cdot}!\right)\left(\prod\limits_{j=1}^{J} x_{\cdot j}!\right)}{x_{\cdot\cdot}! \prod\limits_{i=1}^{I}\prod\limits_{j=1}^{J} x_{ij}!}. \tag{4.30}$$

Proof. As we saw in the proof of the factorization theorem (Theorem 4.1.3), for the probability function $p(x) = h(x)g(T(x),\boldsymbol{\theta})$ of a discrete random variable X, the conditional probability function of X given $T = t$ is obtained as the normalized $h(x)$, in such a way that summing $h(x)$ with respect to all x satisfying $T(x) = t$ equals 1. Therefore, from Lemma 4.1.14, we have the conditional probability function, given the row sums and the column sums, under the null hypothesis as

$$p(\boldsymbol{x} \mid x_{i\cdot}, x_{\cdot j}, \lambda_{ij} = 0) = \left(\prod_{i=1}^{I}\prod_{j=1}^{J}\frac{1}{x_{ij}!}\right) \bigg/ \left(\sum_{y}\prod_{i=1}^{I}\prod_{j=1}^{J}\frac{1}{y_{ij}!}\right),$$

where $\sum_{y}$ is the sum with respect to all $\boldsymbol{y} = \{y_{ij}\}$ satisfying $y_{i\cdot} = x_{i\cdot}$, $y_{\cdot j} = x_{\cdot j}$ for all i, j. Therefore we only need to show that

$$\left(\sum_{y}\prod_{i=1}^{I}\prod_{j=1}^{J}\frac{1}{y_{ij}!}\right) = \frac{x_{\cdot\cdot}!}{\left(\prod_{i=1}^{I}x_{i\cdot}!\right)\left(\prod_{j=1}^{J}x_{\cdot j}!\right)}. \tag{4.31}$$

Comparing the coefficient of $u_1^{x_{\cdot 1}}\cdots u_J^{x_{\cdot J}}$ in the relation

$$(u_1 + \cdots + u_J)^{x_{1\cdot}}(u_1 + \cdots + u_J)^{x_{2\cdot}} \times \cdots \times (u_1 + \cdots + u_J)^{x_{I\cdot}} = (u_1 + \cdots + u_J)^{x_{\cdot\cdot}},$$

we have

$$\sum_{y:y_{\cdot j}=x_{\cdot j}} \frac{x_{1\cdot}!}{y_{11}!\cdots y_{1J}!} \times \cdots \times \frac{x_{I\cdot}!}{y_{I1}!\cdots y_{IJ}!} = \frac{x_{\cdot\cdot}!}{x_{\cdot 1}!\cdots x_{\cdot J}!}.$$

This yields (4.31).   □

We will show another proof, based on combinatorics, in Example 4.1.20.

Note that the arguments above are similar to the case of 2×2 tables. In the case of 2×2 tables, we can directly compare them using similar tests (Fisher's exact tests) that are based on the hypergeometric distribution. However, because the conditional probability function is multidimensional in the case of general $I \times J$ tables, we have to consider the use of test statistics. An appropriate choice of test can be based on a consideration of the power of the test. As stated above, we consider that nuisance parameters exists in $\boldsymbol{\theta} = (\boldsymbol{\psi}, \boldsymbol{\lambda})$, and the alternative hypothesis is the saturated model, i.e.,

$$\begin{aligned} \mathrm{H}_0 &: \boldsymbol{\lambda} = \mathbf{0} \\ \mathrm{H}_1 &: \boldsymbol{\lambda} \neq \mathbf{0}. \end{aligned} \tag{4.32}$$

A test where the saturated model is the alternative hypothesis is called a *goodness-of-fit test*. Representative goodness-of-fit tests are the *chi-square goodness-of-fit test*

$$\chi^2(\boldsymbol{x}) = \sum_i\sum_j \frac{(x_{ij} - m_{ij})^2}{m_{ij}} \geq c_\alpha \Rightarrow \text{Reject } \mathrm{H}_0$$

and the (twice log) *likelihood ratio test*

$$G^2(\boldsymbol{x}) = 2\sum_i\sum_j x_{ij} \log \frac{x_{ij}}{m_{ij}} \geq c_\alpha \Rightarrow \text{Reject } \mathrm{H}_0,$$

where $m = (m_{ij})$ is a *fitted value* under the null hypothesis H_0 (or the maximum likelihood estimate of the expected frequency). Because $E(X_{ij}) = np_{ij}$ for the multinomial distribution, m is given by

$$m_{ij} = n\hat{p}_{ij}.$$

Here $\hat{p}_{ij}$ is the maximum likelihood estimate of p_{ij} under H_0, which is obtained by maximizing the log-likelihood

$$L = \text{Const} + \sum_{i,j} x_{ij} \log p_{ij}$$

under the constraint $\lambda = (0, \ldots, 0)$. The fitted value for the independence model of two-way contingency tables is given by

$$m_{ij} = \frac{x_{i.}x_{.j}}{x_{..}}.$$

This gives an estimate of the parameter as a function of the sufficient statistics under the model. This fact holds in general. In fact, the fitted value can be obtained as a nonnegative table that fits the model completely and for which the sufficient statistics have the same values as do the observations. Such a table is uniquely determined. The *iterative scaling procedure* is a method for calculating the fitted values.

Example 4.1.16 (Example 4.1.12, Continued). The fitted value under the null hypothesis of independence for Table 4.4 is calculated as follows.

	5	4	3	2	1−	Total
5	1.54	0.92	0.77	0.31	0.46	4
4	5.38	3.23	2.69	1.08	1.62	14
3	1.92	1.15	0.96	0.38	0.58	5
2	0.77	0.46	0.38	0.15	0.23	2
1−	0.38	0.23	0.19	0.08	0.12	1
Total	10	6	5	2	3	26

The goodness-of-fit test statistic is calculated as

$$\chi^2(x^o) = \sum_i \sum_j \frac{(x_{ij}^o - m_{ij})^2}{m_{ij}} = \frac{(2 - 1.54)^2}{1.54} + \cdots + \frac{(1 - 0.12)^2}{0.12} = 25.338,$$

where x^o indicates the observed frequencies.

Once the value of the test statistic is calculated, we only have to judge whether it is significantly large (i.e., the p value is less than or equal to the significance level α). For example, is the value 25.338 in Example 4.1.16 significantly large for rejecting H_0? This evaluation of the significance is based on the probability function of the test statistic under H_0. There are three strategies for calculating the p values, as follows.

(a)   Use the asymptotic distribution of the test statistic.
(b)   Exactly calculate the p value.
(c)   Estimate the p value by using the Monte Carlo method.

Though the aim of this chapter is to consider strategy (c), we will consider each strategy in order.

(a) Using the Asymptotic Distribution of the Test Statistics.

Although the distributions of test statistics are complicated in general, the asymptotic distributions of some test statistics are known. Here by the term asymptotic, we consider the limit with respect to the sample size as $n \to \infty$ under some regularity conditions based on the central limit theorem. In fact, the goodness-of-fit test statistics and the likelihood ratio test statistics have the same asymptotic distribution.

Theorem 4.1.17. *The goodness-of-fit test statistics and the (twice log) likelihood ratio test statistics for the test of (4.32) asymptotically follow the χ^2 distribution with $\dim\lambda$ degree of freedom under H_0.*

In the theorem, $\dim\lambda$ is the number of elements in λ that can vary freely, which also equals the difference between the dimensions of the manifolds of the statistical models H_0 and H_1. Refer to [10] or [24] for the proof of this theorem.

In Theorem 4.1.17, we consider the asymptotic distribution for $n \to \infty$ as $\left\{ \frac{x_{i\cdot}}{n} \right\}$ and $\left\{ \frac{x_{\cdot j}}{n} \right\}$ are fixed. Because of the simplicity of strategy (a), it is effective to rely on the asymptotic theory at the first stage of the analysis, even if we will also consider strategies (b) or (c). One of the disadvantages of strategy (a) is that there might not be a good fit with the asymptotic distribution. For example, it is doubtful that we can apply the asymptotic result of $n \to \infty$ to the data of Table 4.4, because the sample size is only $n = 26$. Besides, it is well known that the fit of the asymptotic distribution becomes poor for some types of data with relatively large sample sizes. One of these types is sparse data, and we note that Table 4.4 has many zero entries. Another problematic type of data is that for which the row sums and column sums are unbalanced. See [14] for an example for fitting the asymptotic distribution.

Example 4.1.18 (Example 4.1.16, Continued). Because Table 4.4 is a 5×5 table, the degrees of freedom is $\dim\lambda = (5 - 1)(5 - 1) = 16$. Therefore, to test at the significance level α, we compare the observed goodness-of-fit $\chi^2(x^o) = 25.338$ with the upper 100α percent point of the χ^2 distribution with 16 degrees of freedom.

If we set $\alpha = 0.05$, the upper 5 percent point of χ_{16}^2 is 26.30, and we cannot reject H_0.

The asymptotic tests are easy to conduct using R , as follows.

```
> data <- matrix(c(2,1,1,0,0,8,3,3,0,0,0,2,1,1,1,0,0,0,1,1,0,
                 0,0,0,1), byrow=T,ncol=5)
> data
     [,1] [,2] [,3] [,4] [,5]
[1,]    2    1    1    0    0
[2,]    8    3    3    0    0
[3,]    0    2    1    1    1
[4,]    0    0    0    1    1
[5,]    0    0    0    0    1
> res <- chisq.test(data)
Warning message:
in: chisq.test(data)
    Chi-squared approximation may be incorrect
> res

        Pearson's Chi-squared test

data:  data
X-squared = 25.3376, df = 16, p-value = 0.06409
> qchisq(0.95,16)
[1] 26.29623
> round(res$expected,2)
     [,1] [,2] [,3] [,4] [,5]
[1,] 1.54 0.92 0.77 0.31 0.46
[2,] 5.38 3.23 2.69 1.08 1.62
[3,] 1.92 1.15 0.96 0.38 0.58
[4,] 0.77 0.46 0.38 0.15 0.23
[5,] 0.38 0.23 0.19 0.08 0.12
```

(b) Exact Calculation of the p Value

As with Fisher's exact tests, it is very desirable to calculate the exact p value for the actual situation and for a finite sample size. The exact probability function of the test statistics is derived from the probability function of X under H_0, as follows. Recall that the probability function for X under the independent model of an $I \times J$ table is the multinomial hypergeometric distribution (4.30):

$$p(x \mid x_{i\cdot}, x_{\cdot j}, \lambda = 0) = \frac{\left(\prod_{i=1}^{I} x_{i\cdot}!\right)\left(\prod_{j=1}^{J} x_{\cdot j}!\right)}{x_{\cdot\cdot}! \prod_{i=1}^{I} \prod_{j=1}^{J} x_{ij}!} \quad (= h(x)). \tag{4.33}$$

The support of this probability function is the set of tables that have the same row sums and column sums as in the observed table x^o. We write it as

$$\mathscr{F} = \left\{ x \mid x_{i\cdot} = x_{i\cdot}^o, \ x_{\cdot j} = x_{\cdot j}^o, \ x_{ij} \in \{0, 1, 2, \ldots\} \right\}. \qquad (4.34)$$

The exact p value of the chi-square goodness-of-fit test statistic is

$$p = \Pr(\chi^2(x) \geq \chi^2(x^o) \mid H_0 \text{ is true}) = \sum_{x \in \mathscr{F}} g(x) h(x),$$

$$g(x) = \begin{cases} 1, & \text{if } \chi^2(x) \geq \chi^2(x^o), \\ 0, & \text{otherwise.} \end{cases}$$

It is very desirable to calculate the p value in this way. The disadvantage of this strategy is the possible lack of computational feasibility. In fact, as was estimated by [13], for example, the cardinality of $\mathscr{F}$ becomes huge as the sample size or the size of the tables increases, and an exact calculation becomes infeasible.

Example 4.1.19 (Example 4.1.18, Continued). We will calculate the exact p value for the goodness-of-fit test for Table 4.4. The multinomial hypergeometric distribution in this case is

$$h(x) = \frac{(4!14!5!2!1!) \, (10!6!5!2!3!)}{26!} \prod_{i=1}^{5} \prod_{j=1}^{5} \frac{1}{x_{ij}!}$$

and there are 229,174 elements in

$$\mathscr{F} = \left\{ x = (x_{ij}) \ \left| \ \begin{array}{ccccc|c} x_{11} & x_{12} & x_{13} & x_{14} & x_{15} & 4 \\ x_{21} & x_{22} & x_{23} & x_{24} & x_{25} & 14 \\ x_{31} & x_{32} & x_{33} & x_{34} & x_{35} & 5 \\ x_{41} & x_{42} & x_{43} & x_{44} & x_{45} & 2 \\ x_{51} & x_{52} & x_{53} & x_{54} & x_{55} & 1 \\ \hline 10 & 6 & 5 & 2 & 3 & 26 \end{array} \right. \right\}.$$

For each element x in $\mathscr{F}$, we calculate the value of $\chi^2(x)$, accumulate the value $h(x)$ for $\chi^2(x) \geq 25.338$, and finally obtain the exact p value 0.0609007. The conclusion is the same as that of strategy (a), i.e., we cannot reject H_0 at the significance level $\alpha = 0.05$.

(c) Estimate the p Value by Using the Monte Carlo Method

So far, we have seen two strategies for obtaining a p value: asymptotic evaluation and exact calculation. It is best to calculate the exact p value if this is possible. Even if the sample size is large and an exact calculation is infeasible, asymptotic evaluation can be effective if there is a good fit with the asymptotic distribution. But there is a problem when *the sample size is too large for an exact calculation, and the fit of the asymptotic distribution is poor due to sparseness or unbalanced row sums or column sums.* In these cases, the Monte Carlo method can be effective.

The Monte Carlo method estimates the p value as follows. The p value we want to estimate is given by $p = \sum_{x \in \mathscr{F}} g(x)h(x)$, where $h(x)$ is the conditional probability function under H_0. This value is estimated as $\hat{p} = \sum_{t=1}^{N} g(x_t)/N$ by the samples $x_1, \ldots, x_N$ from $h(x)$. This is an unbiased estimate of the p value. We can set N according to the performance of our computer. As an advantage of the Monte Carlo method, we can also estimate the variance of the estimate. For example, $\hat{p} \pm 1.96\sqrt{\hat{p}(1 - \hat{p})/N}$ is a conventional 95% confidence interval of p. Another advantage of the Monte Carlo method is that we can apply this method to arbitrary test statistics, in contrast to the exact calculation of strategy (b), which may not be feasible for some test statistics. For some test statistics, various efficient algorithms are known for calculating exact p values; one such is the network algorithm by Mehta and Patel [20] for the generalized Fisher's exact test of $I \times J$ tables. However, for complicated test statistics, the exact calculation is difficult. Such a dependence on the test statistics does not exist with the Monte Carlo method. The only problem is *how to generate samples from the null distribution.* The difficulty thus depends on the null distribution. First, we present the simple case of the multinomial hypergeometric distribution.

Example 4.1.20 (Generating Samples from the Multinomial Hypergeometric Distribution). We can generate samples from the multinomial hypergeometric distribution (4.30) by using the *urn model.* First, prepare $x_{..}$ balls. Write the numbers from 1 to I on the balls so that $x_{i.}$ balls are labeled $i, i = 1, \ldots, I$, and put them into an urn. Next, prepare J boxes. Pick balls from the urn one by one and at random (without seeing the labels), and put them into the boxes so that $x_{.j}$ balls are in the box labeled j, for $j = 1, \ldots, J$. After all the balls have been removed from the urn, let x_{ij} denote the number of the balls labeled i in the box labeled j. Then $x = \{x_{ij}\}$ is a sample from the multinomial hypergeometric distribution.

Proposition 4.1.21. x *obtained by the method described in Example 4.1.20 follows the multinomial hypergeometric distribution (4.30).*

Proof. The probability of occurrence of x is given by

$$p(x) = \frac{\text{Number of permutations of } x_{..} \text{ balls coinciding with } x}{\text{Number of permutations of } x_{..} \text{ balls}}$$

for $x_{..}$ balls where $x_{i.}$ balls are labeled i for $i = 1, \ldots, I$. The denominator of this probability is the number of permutations when we ignore the boxes, and it is given by

$$x_{..}! \bigg/ \left(\prod_{i=1}^{I} x_{i.}! \right).$$

The numerator of this probability is the product of the number of permutations of $x_{.j}$ balls in the box labeled j for $j = 1, \ldots, J$, given by

$$\prod_{j=1}^{J} \left(x_{.j}! \bigg/ \left(\prod_{i=1}^{I} x_{ij}! \right) \right).$$

The ratio of these values coincides with the multinomial hypergeometric distribution (4.30). □

In a similar manner, we can easily generate samples for some specific null distributions. However, such a direct sampling is difficult in general, especially for the distribution where the normalizing constant does not have a closed-form expression. For these cases, we consider *Markov chain Monte Carlo methods* (often abbreviated as *MCMC methods*).

In a Markov chain Monte Carlo method, we construct a *Markov chain with the stationary distribution as the null distribution*. Here, we consider a Markov chain over the finite conditional sample space $\mathscr{F}$. Let the elements of $\mathscr{F}$ be numbered as

$$\mathscr{F} = \{x_1, \ldots, x_s\}. \tag{4.35}$$

We write the null distribution as

$$\pi = (\pi_1, \ldots, \pi_s)$$

according to (4.35). By standard notation, we treat π as a row vector. We now write the *transition probability matrix* of the Markov chain $\{Z_t, \ t = 0, 1, 2, \ldots\}$ over $\mathscr{F}$ as $Q = (q_{ij})$, i.e., we define

$$q_{ij} = \Pr(Z_{t+1} = x_j \mid Z_t = x_i).$$

π is called a stationary distribution if it satisfies

$$\pi = \pi Q.$$

π is the eigenvector from the left of Q with the eigenvalue 1. The stationary distribution uniquely exists if the Markov chain is irreducible (i.e., connected in this case) and aperiodic. Therefore, we will consider connected and aperiodic Markov chains. Under these conditions, starting from an arbitrary state $Z_0 = x_i$, the distribution of Z_t for large t is close to the stationary distribution π. Therefore, if we can construct a Markov chain with the target stationary distribution π, then by running a Markov chain and discarding a large number t of initial steps (called *burn-in steps*), we can consider $Z_{t+1}, Z_{t+2}, \ldots$ to be samples from the stationary distribution π. Then, the problem becomes *how to construct a connected, aperiodic Markov chain with the stationary distribution as the null distribution over $\mathcal{F}$*. Among these conditions, the condition for the stationary distribution can be solved easily. Once we can construct an arbitrary connected chain over $\mathcal{F}$, we can easily modify the stationary distribution to the given null distribution π as follows.

Theorem 4.1.22 (Metropolis–Hastings Algorithm). *Let π be a probability distribution on $\mathcal{F}$. Let $R = (r_{ij})$ be the transition probability matrix of a connected, aperiodic, and symmetric Markov chain over $\mathcal{F}$. Define $Q = (q_{ij})$ by*

$$q_{ij} = r_{ij} \min\left(1, \frac{\pi_j}{\pi_i}\right), \quad i \neq j,$$

$$q_{ii} = 1 - \sum_{j \neq i} q_{ij}. \tag{4.36}$$

Then Q satisfies $\pi = \pi Q$.

This result is a special case of [15], and the symmetry assumption ($r_{ij} = r_{ji}$) can be removed relatively easily. In this chapter, we only consider symmetric R, and the simple statement of the above theorem is sufficient for our purpose.

Proof (Theorem 4.1.22). It suffices to show that the above Q is reversible in the following sense:

$$\pi_i q_{ij} = \pi_j q_{ji}. \tag{4.37}$$

In fact, under the reversibility

$$\pi_i = \pi_i \sum_{j=1}^{s} q_{ij} = \sum_{j=1}^{s} \pi_j q_{ji},$$

and we have $\pi = \pi Q$. Now, (4.37) clearly holds for $i = j$. For $i \neq j$, we have

$$\pi_i q_{ij} = \pi_i r_{ij} \min\left(1, \frac{\pi_j}{\pi_i}\right) = r_{ij} \min\left(\pi_i, \pi_j\right),$$

and therefore (4.37) holds if $r_{ij} = r_{ji}$. $\qquad\square$

An important advantage of the Markov chain Monte Carlo method is that it does not require the explicit evaluation of the normalizing constant of the stationary distribution π. We only need to know π up to a multiplicative constant. In fact, in (4.36), the stationary distribution π only appears in the form of ratios of its elements π_i / π_j, and the normalizing constant is canceled. With the Metropolis–Hastings algorithm, the remaining problem is to construct an arbitrary connected and aperiodic Markov chain over $\mathcal{F}$. This problem is solved by the Gröbner basis theory in Sect. 4.2.

4.2 Markov Basis

4.2.1 Markov Basis

In Sect. 4.1, we introduced the Markov chain Monte Carlo method for evaluating p values of various statistical models for the data summarized in contingency tables. In this approach, we consider constructing a connected Markov chain over a given sample space, which is defined from the sufficient statistics of the nuisance parameters. In general, this is a difficult problem. First, we will formalize the problem below.

Hereafter, we will treat a contingency table x as a column vector. For example, in the case of a $I \times J$ two-way contingency table, the frequency vector $x = (x_{11}, \ldots, x_{IJ})'$ is an $IJ \times 1$ column vector.[1] Let $t = Ax$ denote the sufficient statistic for the nuisance parameter, which will be fixed in the similar tests. Here A is an integer matrix. There are several ways to define the matrix A. For example, in the case of an independence model of 2×3 tables, where t is the row sums and column sums, A can be either

$$\begin{pmatrix} 1 & 1 & 1 & 0 & 0 & 0 \\ 0 & 0 & 0 & 1 & 1 & 1 \\ 1 & 0 & 0 & 1 & 0 & 0 \\ 0 & 1 & 0 & 0 & 1 & 0 \\ 0 & 0 & 1 & 0 & 0 & 1 \end{pmatrix} \tag{4.38}$$

or

$$\begin{pmatrix} 1 & 1 & 1 & 1 & 1 & 1 \\ 1 & 1 & 1 & 0 & 0 & 0 \\ 1 & 0 & 0 & 1 & 0 & 0 \\ 0 & 1 & 0 & 0 & 1 & 0 \end{pmatrix}.$$

Specifically, the vector space spanned by the rows of A is essential for defining A. In this section, we will not further consider the definition of A. For example, we will permit redundant (linearly dependent) rows in A, as in (4.38). We assume only the following.

Assumption 4.2.1. $(1, \ldots, 1)$ is in the row space of A.

By this assumption, the matrix A becomes a configuration matrix, as defined in Sect. 1.5. For a given configuration matrix A and a sufficient statistic t, we define the state space of the Markov chain by

$$\mathscr{F}_t = \{x \mid Ax = t, \text{ element of } x \in \{0, 1, 2, \ldots\}\}.$$

---

[1] In this chapter, we use $'$ to denote the transpose.

We call $\mathscr{F}_t$ a t-fiber. We have already used the symbol $\mathscr{F}$ for the state space in (4.34) and (4.35) of Sect. 4.1. We specified the sufficient statistic in those definitions. Note that, by Assumption 4.2.1, all the frequency vectors $x = (x_i)$ in the same t-fiber have the same sample size $\sum_i x_i$. We now consider the construction of a connected Markov chain over the t-fiber containing the observation x^o. For this purpose, we define

$$\mathscr{M}(A) = \mathrm{Ker}(A) \cap \mathbb{Z}^p$$
$$= \{z \mid Az = 0, \text{element of } z \in \{0, \pm 1, \pm 2, \ldots\}\}.$$

$\mathscr{M}(A)$ is the set of integer vectors for which the zero vector is a fixed sufficient statistic. Here p is the number of cells in the contingency table, which is also the number of columns of A. We call an element in $\mathscr{M}(A)$ a *move* for A. If A is clear from context, we simply write $\mathscr{M} = \mathscr{M}(A)$. In the notation of Sect. 1.5, we can also write $\mathscr{M}(A) = \mathrm{Ker}_{\mathbb{Z}} A$. For each move $z = \{z_i\}$, we define its positive part $z^+ = \{z_i^+\}$ and its negative part $z^- = \{z_i^-\}$ by

$$z_i^+ = \max(z_i, 0), \ z_i^- = \max(-z_i, 0).$$

Thus we have $z = z^+ - z^-$. We call $\sum_i z_i^+ = \sum_i z_i^-$ the degree of a move z. We construct a Markov chain from elements in $\mathscr{M}$. To consider the connectivity of the chain, we give the following definition.

Definition 4.2.2 (Mutually Accessibility). Let $\mathscr{B} \subset \mathscr{M}(A)$. $x, y(\neq x) \in \mathscr{F}_{Ax}$ is mutually accessible by $\mathscr{B} \subset \mathscr{M}$ if there exist $N > 0$, $z_j \in \mathscr{B}$, and $\varepsilon_j \in \{-1, 1\}$, $j = 1, \ldots, N$, satisfying the following two conditions.

$$y = x + \sum_{j=1}^{N} \varepsilon_j z_j, \tag{4.39}$$

$$x + \sum_{j=1}^{n} \varepsilon_j z_j \in \mathscr{F}_{Ax}, \ n = 1, \ldots, N. \tag{4.40}$$

The meaning of Definition 4.2.2 is explained as follows. Equation (4.39) shows that y can be reached from x by N steps by adding or subtracting elements in $\mathscr{B}$ to/from x. On the other hand, (4.40) shows that all the states from x to y are in the state space $\mathscr{F}_{Ax}$. Because $Az = 0$ for $z \in \mathscr{M}(A)$, it is obvious that $Ax = A(x \pm z)$ holds. Therefore the case $x \pm z \notin \mathscr{F}_{Ax}$ occurs if *some element becomes negative* in $x \pm z$. We say x and y are mutually accessible by $\mathscr{B}$ if we can choose a route from x to y by the elements in $\mathscr{B}$ without causing negative entries along the way. Obviously, the notion of mutual accessibility is symmetric and transitive. Allowing a zero move $0 \in \mathscr{M}(A)$ also yields reflexivity. Therefore, mutual accessibility by $\mathscr{B}$ is an equivalence relation and each $\mathscr{F}_t$ is partitioned into disjoint equivalence classes by moves of $\mathscr{B} \subset \mathscr{M}$. We call these equivalence classes $\mathscr{B}$-equivalence classes of $\mathscr{F}_t$. We now define a Markov basis.

Definition 4.2.3 (Markov Basis [11]). $\mathscr{B} \subset \mathscr{M}(A)$ is a Markov basis for A if $\mathscr{F}_t$ itself is a $\mathscr{B}$-equivalence class for arbitrary t.

$\mathscr{M}(A)$ itself is an obvious (infinite) Markov basis for A. We are interested in constructing a Markov basis with a simple structure and which a finite subset of $\mathscr{M}(A)$. How difficult this will be depends on the structure of A. In Sect. 4.2.2, we show two examples of a Markov basis, one with a simple structure and the other with a complicated structure.

Once a Markov basis has been obtained, it is easy to construct a connected and symmetric Markov chain over the t-fiber containing any observation x^o, i.e., $\mathscr{F}_{Ax^o}$. One simple method is the following. For each state x of the chain, randomly choose an element of $z \in \mathscr{B}$ and a sign $\varepsilon \in \{-1, +1\}$, move to $x + \varepsilon z$ if $x + \varepsilon z \in \mathscr{F}_t$, and remain at x if $x + \varepsilon z \notin \mathscr{F}_t$; we now have a connected and symmetric Markov chain. Applying the Metropolis–Hastings algorithm of Theorem 4.1.22 to this simple procedure, we have the following algorithm to evaluate p values.

Algorithm 4.2.4.

Input: Observation x^o, Markov basis $\mathscr{B}$, Number of steps N, configuration matrix A, null distribution $f(\cdot)$, test statistics $T(\cdot)$, (we set $t = Ax^o$)

Output: Estimate of the p value

Variables: obs, count, sig, x, x_{next}

Step 1: obs $= T(x^o)$, $x = x^o$, count $= 0$, sig $= 0$

Step 2: Choose $z \in \mathscr{B}$ randomly. Choose $\varepsilon \in \{-1, +1\}$ with probability $1/2$.

Step 3: If $x + \varepsilon z \notin \mathscr{F}_t$, then $x_{next} = x$, and go to Step 5. If $x + \varepsilon z \in \mathscr{F}_t$, then let u be a uniform random number between 0 and 1.

Step 4: If $u \leq \dfrac{f(x + \varepsilon z)}{f(x)}$, then $x_{next} = x + \varepsilon z$ and go to Step 5. If $u > \dfrac{f(x + \varepsilon z)}{f(x)}$, then $x_{next} = x$ and go to Step 5.

Step 5: If $T(x_{next}) \geq$ obs, then sig $=$ sig $+ 1$

Step 6: $x = x_{next}$, count $=$ count $+ 1$

Step 7: If count $< N$, then go to Step 2.

Step 8: The estimate of the p value is $\dfrac{\text{sig}}{N}$.

In the above algorithm, Step 2 corresponds to the transition probability matrix $R = \{r_{ij}\}$ in Theorem 4.1.22. In Step 2, the statement "Choose $z \in \mathscr{B}$ randomly" does not mean that we have to choose $z \in \mathscr{B}$ according to the uniform distribution over $\mathscr{B}$. In fact, we can use any distribution for which all the elements in $\mathscr{B}$ are in the support. For the condition $r_{ij} = r_{ji}$, we choose the sign ε such that the alternatives have equal probability $1/2$. We note an important point: In Step 5, we evaluate the value of the test statistic even if $x_{next} = x$ in Steps 3 and 4. It is necessary to include such a "reject" step in order to attain a stationary distribution (and to obtain the unbiased estimate of the p value).

Algorithm 4.2.4 is very simple, and various improvements are possible. For example, grouping several steps of Algorithm 4.2.4 into a single step would speed up the convergence to the stationary distribution. This can be achieved as follows.

Algorithm 4.2.5.
Modify Steps 2, 3 and 4 in Algorithm 4.2.4, as follows.
Step 2 : Choose $z \in \mathcal{B}$ randomly.
Step 3 : Let $I = \{n \mid x + nz \in \mathcal{F}_t\}$.
Step 4 : Choose x_{next} from $\{x + nz \mid n \in I\}$ according to the probability

$$p_n = \frac{f(x + nz)}{\displaystyle\sum_{n \in I} f(x + nz)}.$$

Note that in both Algorithms 4.2.4 and 4.2.5, the null distribution $f(\cdot)$ appears in the form of a ratio. Hence, we do not need to compute a normalizing constant for $f(\cdot)$, as was discussed in Sect. 4.1. Because the computation of normalizing constants is often difficult, this is an important advantage of the Markov chain Monte Carlo method.

4.2.2   Examples of Markov Bases

We now present two examples of Markov bases. The first example is a Markov basis for the independence model of the two-way contingency tables of Sect. 4.1.4. We consider $I \times J$ tables. The frequency vector is $x = (x_{11}, \ldots, x_{IJ})'$, $x_{ij} \in \{0, 1, 2, \ldots\}$, and the sufficient statistic is the row sums and column sums. Let A be the corresponding configuration matrix. The Markov basis for A is as follows.

Theorem 4.2.6 (Markov Basis for the Independence Model of an $I \times J$ Table).
A Markov basis for the independence model of an $I \times J$ table is constructed as the set of $\binom{I}{2}\binom{J}{2}$ *moves* $\{z_{i_1 i_2 j_1 j_2} \mid 1 \le i_1 < i_2 \le I,\ 1 \le j_1 < j_2 \le J\}$ *defined by*

$$z_{i_1 i_2 j_1 j_2} = \{z_{ij}\}, \quad z_{ij} = \begin{cases} +1, & (i, j) = (i_1, j_1), (i_2, j_2), \\ -1, & (i, j) = (i_1, j_2), (i_2, j_1), \\ 0, & otherwise. \end{cases} \tag{4.41}$$

Proof. Let $\mathcal{B}$ be the set of moves defined by (4.41). We offer a proof by contradiction. Suppose that $\mathcal{B}$ is not a Markov basis. Then there exists a fiber

$x \in \mathcal{F}_t$ and two elements $x, y \in \mathcal{F}_t$ such that we cannot move from x to y by the moves of $\mathcal{B}$. Let

$$\mathcal{N}_x = \{y \in \mathcal{F}_t \mid \text{we cannot move from } x \text{ to } y \text{ by moves of } \mathcal{B}\}.$$

Then $\mathcal{N}_x$ is not empty by assumption, and $x - y$ is a nonzero move for any $y \in \mathcal{N}_x$. For $z = (z_{ij}) \in \mathcal{M}$, let $|z| = \sum_{i=1}^{I} \sum_{j=1}^{J} |z_{ij}|$ denote its 1-norm. Define

$$y^* = \underset{y \in \mathcal{N}_x}{\arg\min} |x - y|. \tag{4.42}$$

y^* is one of the closest elements of $\mathcal{F}_t$ that cannot be reached from x by $\mathcal{B}$. We have

$$|x - y^*| = \min_{y \in \mathcal{N}_x} |x - y|.$$

Now let $w = x - y^*$, and consider the signs of the elements of w. Because w contains a positive element, let $w_{i_1 j_1} > 0$. Then, because w is a move, there exist $j_2 \neq j_1$ with $w_{i_1 j_2} < 0$ and $i_2 \neq i_1$ with $w_{i_2 j_1} < 0$. Hence for $y^* = (y_{ij}^*)$, we have $y_{i_1 j_2}^* > 0$ and $y_{i_2 j_1}^* > 0$. Then

$$y^* + z_{i_1 i_2 j_1 j_2} \in \mathcal{F}_t$$

holds. Because $y^* + z_{i_1 i_2 j_1 j_2}$ can be reached from y^* by $\mathcal{B}$, we have $y^* + z_{i_1 i_2 j_1 j_2} \in \mathcal{N}_x$. Now we check the value of $|x - (y^* + z_{i_1 i_2 j_1 j_2})|$.

- If $w_{i_2 j_2} > 0$, then $|x - (y^* + z_{i_1 i_2 j_1 j_2})| = |x - y^*| - 4$ holds.
- If $w_{i_2 j_2} \leq 0$, then $|x - (y^* + z_{i_1 i_2 j_1 j_2})| = |x - y^*| - 2$ holds.

Therefore for both cases, we have $|x - (y^* + z_{i_1 i_2 j_1 j_2})| < |x - y^*|$. However, this contradicts the minimality of y^* in (4.42). $\square$

In the case of the independence model for $I \times J$ tables, the minimum degree of the moves is 2. Hence Theorem 4.2.6 states that the set of moves with the minimum degree constitutes a Markov basis. In [1], a move with the minimum degree is called a basic move, and in [12], a move of degree 2 is called a primitive move. In addition, it is shown in [12] that the set of the primitive moves constitutes a Markov basis if the null model belongs to the class of so-called decomposable models. The independence model for two-way contingency tables is one of the decomposable models.

The above is an example where the set of the basic moves becomes a Markov basis. However, it is known that, in general, a Markov basis has a complicated structure. As an example of such a case, we now consider a $I \times J \times K$ three-way contingency table. For the frequency vector $x = (x_{111}, \ldots, x_{IJK})'$, $x_{ijk} \in \{0, 1, 2, \ldots\}$, the sufficient statistic is defined as the set of two-dimensional marginal totals $\{x_{ij\cdot}\}$, $\{x_{i\cdot k}\}$, $\{x_{\cdot jk}\}$. A corresponding configuration matrix A can be given, for example, by

$$A = \begin{pmatrix} \mathbf{1}'_I \otimes E_J \otimes E_K \\ E_I \otimes \mathbf{1}'_J \otimes E_K \\ E_I \otimes E_J \otimes \mathbf{1}'_K \end{pmatrix},$$

where E_n denotes an $n \times n$ identity matrix, $\mathbf{1}_n = (1, \ldots, 1)'$ denotes the n-dimensional vector consisting of 1s, and $\otimes$ is the Kronecker product. This configuration arises in the testing problem of the null hypothesis that there is no three-factor interaction:

$$H_0 : (\alpha\beta\gamma)_{ijk} = 0, \quad \text{for all } i, j, k,$$

in the log-linear model

$$\log p_{ijk} = \mu + \alpha_i + \beta_j + \gamma_k + (\alpha\beta)_{ij} + (\alpha\gamma)_{ik} + (\beta\gamma)_{jk} + (\alpha\beta\gamma)_{ijk}. \quad (4.43)$$

(Here we assume that the dimension of the parameter coincides with the dimension of $\{p_{ijk}\}$ with appropriate constraints.) The nuisance parameter for this null model is

$$\mu, \ \alpha_i, \ \beta_j, \ \gamma_k, \ (\alpha\beta)_{ij}, \ (\alpha\gamma)_{ik}, \ (\beta\gamma)_{jk}$$

and the sufficient statistics for them are the two-dimensional marginal totals $x_{ij\cdot}, \ x_{i\cdot k}, \ x_{\cdot jk}$. The conditional probability function for given two-dimensional marginal totals under the null hypothesis is given by

$$p(x_{ijk} \mid \{x_{ij\cdot}\}, \{x_{i\cdot k}\}, \{x_{\cdot jk}\}, H_0) = C^{-1} \prod_{i=1}^{I} \prod_{j=1}^{J} \prod_{k=1}^{K} \frac{1}{x_{ijk}!}$$

$$C = \sum_{y \in \mathcal{F}_{Ax}} \left(\prod_{i=1}^{I} \prod_{j=1}^{J} \prod_{k=1}^{K} \frac{1}{y_{ijk}!} \right).$$

In this case, the normalizing constant C cannot be written in an explicit form, and hence a direct sampling is difficult. On the other hand, the hypothesis that there is no three-factor interaction (4.43) is considered first in the analysis of three-way contingency tables. Therefore, it is valuable for applications if we obtain a Markov basis for this problem. In this problem, the basic move (the degree is 4) is defined by

$$z_{ijk} = \begin{cases} +1, & (i, j, k) = (i_1, j_1, k_1), (i_1, j_2, k_2), (i_2, j_1, k_2), (i_2, j_2, k_1), \\ -1, & (i, j, k) = (i_1, j_1, k_2), (i_1, j_2, k_1), (i_2, j_1, k_1), (i_2, j_2, k_2), \\ 0, & \text{otherwise} \end{cases} \quad (4.44)$$

for $i_1 \neq i_2, j_1 \neq j_2, k_1 \neq k_2$. This is a natural extension of the basic move of two-way contingency tables, $\begin{smallmatrix} +1 & -1 \\ -1 & +1 \end{smallmatrix}$, to the three-way tables. Interestingly, if $I, J, K \geq 3$, the Markov chain constructed from the basic move is not connected except for some special fibers.

Example 4.2.7 (Markov Basis for the Model for a $3 \times 3 \times 3$ Contingency Table Where the Null Hypothesis Is That There Is No Three-Factor Interaction). This is an example of the $I = J = K = 3$ case where the set of basic moves is not a Markov basis. We write the $3 \times 3 \times 3$ table $\boldsymbol{x} = \{x_{ijk}\}$ as three 3×3 tables for $i = 1, 2, 3$ as

$$
\begin{array}{|ccc|}
x_{111} & x_{112} & x_{113} \\
x_{121} & x_{122} & x_{123} \\
x_{131} & x_{132} & x_{133}
\end{array}
\begin{array}{|ccc|}
x_{211} & x_{212} & x_{213} \\
x_{221} & x_{222} & x_{223} \\
x_{231} & x_{232} & x_{233}
\end{array}
\begin{array}{|ccc|}
x_{311} & x_{312} & x_{313} \\
x_{321} & x_{322} & x_{323} \\
x_{331} & x_{332} & x_{333}
\end{array}.
$$

Consider the case that the fixed two-dimensional marginal totals $\{x_{ij\cdot}\}$, $\{x_{i\cdot k}\}$, $\{x_{\cdot jk}\}$ are common and given as

$$
\begin{array}{|ccc|}
2 & 1 & 1 \\
1 & 2 & 1 \\
1 & 1 & 2
\end{array}.
$$

We display this fiber as follows.

$$
\begin{array}{|ccc|c}
x_{111} & x_{112} & x_{113} & 2 \\
x_{121} & x_{122} & x_{123} & 1 \\
x_{131} & x_{132} & x_{133} & 1 \\
\hline
2 & 1 & 1 & 4
\end{array}
\begin{array}{|ccc|c}
x_{211} & x_{212} & x_{213} & 1 \\
x_{221} & x_{222} & x_{223} & 2 \\
x_{231} & x_{232} & x_{233} & 1 \\
\hline
1 & 2 & 1 & 4
\end{array}
\begin{array}{|ccc|c}
x_{311} & x_{312} & x_{313} & 1 \\
x_{321} & x_{322} & x_{323} & 1 \\
x_{331} & x_{332} & x_{333} & 2 \\
\hline
1 & 1 & 2 & 4
\end{array}
\begin{array}{ccc|c}
2 & 1 & 1 & 4 \\
1 & 2 & 1 & 4 \\
1 & 1 & 2 & 4 \\
\hline
4 & 4 & 4 & 12
\end{array}
$$

(The right-most table shows the marginal table $\{x_{\cdot jk}\}$.) In this case, there are $\#\mathscr{F}_t = 18$ elements in the t-fiber as follows.

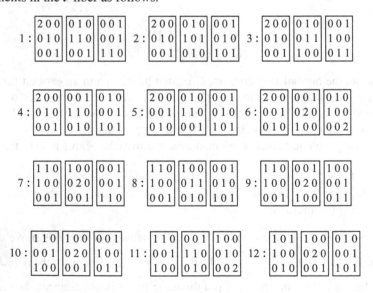

Fig. 4.1 Transition graph
obtained from the set of basic
moves

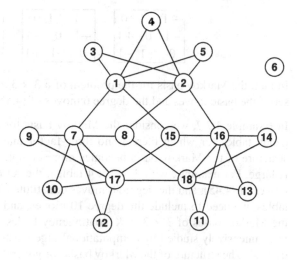

$$13: \begin{vmatrix} 1 & 0 & 1 \\ 1 & 0 & 0 \\ 0 & 1 & 0 \end{vmatrix} \begin{vmatrix} 0 & 1 & 0 \\ 0 & 1 & 1 \\ 1 & 0 & 0 \end{vmatrix} \begin{vmatrix} 1 & 0 & 0 \\ 0 & 1 & 0 \\ 0 & 0 & 2 \end{vmatrix} \quad 14: \begin{vmatrix} 1 & 0 & 1 \\ 0 & 1 & 0 \\ 1 & 0 & 0 \end{vmatrix} \begin{vmatrix} 1 & 0 & 0 \\ 0 & 1 & 1 \\ 0 & 1 & 0 \end{vmatrix} \begin{vmatrix} 0 & 1 & 0 \\ 1 & 0 & 0 \\ 0 & 0 & 2 \end{vmatrix} \quad 15: \begin{vmatrix} 1 & 0 & 1 \\ 0 & 1 & 0 \\ 1 & 0 & 0 \end{vmatrix} \begin{vmatrix} 0 & 1 & 0 \\ 1 & 1 & 0 \\ 0 & 0 & 1 \end{vmatrix} \begin{vmatrix} 1 & 0 & 0 \\ 0 & 0 & 1 \\ 0 & 1 & 1 \end{vmatrix}$$

$$16: \begin{vmatrix} 1 & 0 & 1 \\ 0 & 1 & 0 \\ 1 & 0 & 0 \end{vmatrix} \begin{vmatrix} 0 & 1 & 0 \\ 1 & 0 & 1 \\ 0 & 1 & 0 \end{vmatrix} \begin{vmatrix} 1 & 0 & 0 \\ 0 & 1 & 0 \\ 0 & 0 & 2 \end{vmatrix} \quad 17: \begin{vmatrix} 0 & 1 & 1 \\ 1 & 0 & 0 \\ 1 & 0 & 0 \end{vmatrix} \begin{vmatrix} 1 & 0 & 0 \\ 0 & 2 & 0 \\ 0 & 0 & 1 \end{vmatrix} \begin{vmatrix} 1 & 0 & 0 \\ 0 & 0 & 1 \\ 0 & 1 & 1 \end{vmatrix} \quad 18: \begin{vmatrix} 0 & 1 & 1 \\ 1 & 0 & 0 \\ 1 & 0 & 0 \end{vmatrix} \begin{vmatrix} 1 & 0 & 0 \\ 0 & 1 & 1 \\ 0 & 1 & 0 \end{vmatrix} \begin{vmatrix} 1 & 0 & 0 \\ 0 & 1 & 0 \\ 0 & 0 & 2 \end{vmatrix}$$

For these 18 elements, the transition graph for the set of basic moves of degree
4 is shown in Fig. 4.1. As we see in Fig. 4.1, state 6 is isolated, and there are two
equivalence classes in $\mathscr{F}_t$. It is easily checked that adding (or subtracting) any basic
move to (or from) state 6 will produce a negative entry.

Although there are only 18 elements in this fiber, a similar problem also occurs
for larger datasets. For example, consider the t-fiber defined by the two-dimensional
marginal totals

$$x_{ij\cdot} = x_{i\cdot k} = x_{\cdot jk} = n, \quad i, j, k = 1, 2, 3,$$

where n is any integer. The cardinality of this t-fiber increases rapidly as n increases.
However, for any n, an element such as

$$\begin{array}{|ccc|} n & 0 & 0 \\ 0 & n & 0 \\ 0 & 0 & n \end{array} \quad \begin{array}{|ccc|} 0 & n & 0 \\ 0 & 0 & n \\ n & 0 & 0 \end{array} \quad \begin{array}{|ccc|} 0 & 0 & n \\ n & 0 & 0 \\ 0 & n & 0 \end{array}$$

is an isolated equivalence class with a single element.

These two examples show that the Markov basis for this problem must contain
moves of a larger degree, such as

$$\begin{vmatrix} +1 & -1 & 0 \\ -1 & 0 & +1 \\ 0 & +1 & -1 \end{vmatrix} \quad \begin{vmatrix} -1 & +1 & 0 \\ +1 & 0 & -1 \\ 0 & -1 & +1 \end{vmatrix} \quad \begin{vmatrix} 0 & 0 & 0 \\ 0 & 0 & 0 \\ 0 & 0 & 0 \end{vmatrix}. \tag{4.45}$$

In fact, the Markov basis for the problem of a $3 \times 3 \times 3$ table is constructed as the set of the basic moves and the degree 6 moves of (4.45).

In Example 4.2.7, we consider the Markov basis for models of three-way contingency tables for which there is no three-factor interaction. In this problem, the structure of the Markov basis becomes more complicated if the size of the tables is large. For example, with $3 \times 3 \times 4$ tables, the set of basic moves, the degree 6 moves of (4.45), and the degree 8 moves constitute a Markov basis. For $3 \times 3 \times 5$ tables, we need to include the degree 10 moves, and so on. See [1] for details on the Markov bases of $3 \times 3 \times K$ contingency tables. The $4 \times 4 \times 4$ problem has been intensively studied by computational algebraists (in about 2003; see [17] for details). The structure of the Markov basis for general $I \times J \times K$ tables has not yet been determined.

4.2.3 Markov Bases and Ideals

An algorithm for calculating Markov bases for a given A was first presented in [11]. We summarize it in this section.

Let $u = \{u_1, \ldots, u_p\}$ denote a set of indeterminates, and consider the polynomial ring $k[u] = k[u_1, \ldots, u_p]$ over a field k. Here, p corresponds to the number of cells in the contingency tables. For the case of $I \times J$ tables, we have $p = IJ$. The table x corresponds to the monomial $u^x = u_1^{x_1} \cdots u_p^{x_p}$. For example, for 3×3 tables, if we write the indeterminates $u = \{u_1, \ldots, u_9\}$ as $u = \{u_{11}, \ldots, u_{33}\}$, the table

$$x = \begin{vmatrix} 2 & 0 & 1 \\ 1 & 1 & 0 \\ 0 & 1 & 2 \end{vmatrix}$$

corresponds to the monomial

$$u^x = u_{11}^2 u_{13} u_{21} u_{22} u_{32} u_{33}^2.$$

We let an integer array z correspond to the binomial $u^{z^+} - u^{z^-}$, in which both the positive and negative parts are included. For the example of a 3×3 table, the integer array

$$z = \begin{vmatrix} 2 & -1 & 0 \\ 1 & 0 & 1 \\ -1 & -2 & 1 \end{vmatrix} \tag{4.46}$$

corresponds to the binomial

$$u^z = u^{z^+} - u^{z^-} = u_{11}^2 u_{21} u_{23} u_{33} - u_{12} u_{31} u_{32}^2.$$

Note that the example of z in (4.46) is not a move for a configuration of A when using the independence model. In fact, any integer array can correspond to a binomial. Here we consider the set of the binomials corresponding to the elements in $\mathcal{M}(A)$. The binomial ideal in $k[u]$ generated by this set of binomials,

$$I_A = \langle \{ u^{z^+} - u^{z^-}, z \in \mathcal{M}(A) \} \rangle,$$

is the *toric ideal* of the configuration A defined in Sect. 1.5.

The main theorem of [11] states that a Markov basis is characterized as a set of generators of the toric ideal I_A.

Theorem 4.2.8 (Theorem 3.1 of [11]). $\mathcal{B} = \{z_1, \dots, z_L\} \subset \mathcal{M}(A)$ *is a Markov basis for* A *if and only if* $\{u^{z_i^+} - u^{z_i^-}, i = 1, \dots, L\}$ *generates* I_A.

Proof. Write $F = \{ u^{z_i^+} - u^{z_i^-}, i = 1, \dots, L \}$.
(Sufficiency) We assume that $\mathcal{B} = \{z_1, \dots, z_L\}$ is a Markov basis for A and show $f \in \langle F \rangle$ for arbitrary $f \in I_A$. Because I_A is a toric ideal, it is sufficient to assume that f is a binomial. Write $f = u^x - u^y$. Because $\mathcal{B}$ is a Markov basis, $x, y (\in \mathscr{F}_{Ax})$ are mutually accessible by $\mathcal{B}$. Therefore, we have

$$y = x + \sum_{j=1}^{s} \varepsilon_j z_{i_j}, \quad x + \sum_{j=1}^{s} \varepsilon_j z_{i_j} \in \mathscr{F}_{Ax}, \ 1 \le s \le S.$$

We now use induction on S. Let $\min(x, y)$ denote the vector of the element-wise minimum of x, y. First, in the case of $S = 1$, because $y = x + z_{i_1}$ or $y = x - z_{i_1}$ holds, $f = u^x - u^y$ is written as $f = -u^{\min(x,y)}(u^{z_{i_1}^+} - u^{z_{i_1}^-})$ or $f = u^{\min(x,y)}(u^{z_{i_1}^+} - u^{z_{i_1}^-})$, i.e., $f \in \langle F \rangle$ holds. Next, suppose $S > 1$ and assume that the theorem holds up to $S - 1$. If we write

$$y = x + \sum_{j=1}^{S} \varepsilon_j z_{i_j} = \underbrace{x + \sum_{j=1}^{S-1} \varepsilon_j z_{i_j}}_{x'} + \varepsilon_S z_{i_S} = x' + \varepsilon_S z_{i_S},$$

we have $u^x - u^{x'}, u^{x'} - u^y \in \langle F \rangle$ from the assumption, which yields

$$f = u^x - u^y = (u^x - u^{x'}) + (u^{x'} - u^y) \in \langle F \rangle.$$

(Necessity) We assume $I_A = \langle F \rangle$ and show arbitrary $x, y (\neq x) \in \mathscr{F}_{Ax}$ are mutually accessible by $\mathscr{B}$. Write a Gröbner basis of I_A as $\{g_1, \ldots, g_M\}$. Because $u^x - u^y \in I_A = \langle g_1, \ldots, g_M \rangle$, the binomial $u^x - u^y$ can be written as

$$u^x - u^y = \sum_j c_j u^{v_j} g_{i_j},$$

where i_j is one of $1, \ldots, M$ for each j. Note that, in general, $c_j u^{v_j}$ is a monomial in $k[u]$ and $c_j \in k$. However, the right-hand side represents division by a Gröbner basis. Recall that each step of a division algorithm is a procedure of eliminating an initial term by multiplying monomials with the coefficient $+1$ or -1. Therefore, we can assume that c_j is $+1$ or -1 regardless of the field k, providing that we allow overlapping on the right-hand side. From $g_1, \ldots, g_M \in I_A = \langle F \rangle$ and the fact that each element of a Gröbner basis is obtained by computing S-polynomials from each pair of elements of the set of generators, we have

$$g_j = \sum_\ell d_\ell u^{w_\ell} (u^{z_{i_\ell}^+} - u^{z_{i_\ell}^-}),$$

where d_ℓ is $+1$ or -1. From these considerations, the binomial $u^x - u^y$ can be written as

$$u^x - u^y = \sum_{j=1}^{S} \varepsilon_j u^{h_j} (u^{z_{i_j}^+} - u^{z_{i_j}^-}), \tag{4.47}$$

where each ε_j is $+1$ or -1 if we allow overlapping. We again use induction on the S of (4.47). For the case of $S = 1$, (4.47) shows that we can move from x to y by z_{i_1}, i.e., we can move from x to y by adding $-z_{i_1}$ if $\varepsilon_1 = 1$, and by adding z_{i_1} if $\varepsilon_1 = -1$. Next, assume that $S > 1$ and that (4.47) holds up to $S - 1$. Because (4.47) is an identity, some of the $2S$ terms derived from expansion of the right-hand side equal u^x. In other words, $u^x = \varepsilon_j u^{h_j} u^{z_{i_j}^+}$ or $u^x = -\varepsilon_j u^{h_j} u^{z_{i_j}^-}$ holds for some j. Without loss of generality, we can write $u^x = u^{h_S} u^{z_{i_S}^-}, \varepsilon_S = -1$. In this case, $x - z_{i_S}^-$ is a nonnegative vector. Therefore, $x + z_{i_S}$ is also a nonnegative vector, and we have $x + z_{i_S} = h_S + z_{i_S}^- + z_{i_S} = h_S + z_{i_S}^+$. After subtracting $\varepsilon_S u^{h_S} (u^{z_{i_S}^+} - u^{z_{i_S}^-})$ from each side of (4.47), we have

$$\text{right hand} = \sum_{j=1}^{S-1} \varepsilon_j u^{h_j} (u^{z_{ij}^+} - u^{z_{ij}^-})$$

$$
\begin{aligned}
\text{left hand} &= u^x - u^y - \varepsilon_S u^{h_S} (u^{z_{iS}^+} - u^{z_{iS}^-}) \\
&= u^{h_S} u^{z_{iS}^+} - u^y \\
&= u^{h_S + z_{iS}^+} - u^y \\
&= u^{x + z_{iS}} - u^y,
\end{aligned}
$$

i.e., from the assumption of the induction, we see that $x + z_{iS}$ and y are mutually accessible. From the result of the case $S = 1$, we see that x and $x + z_{iS}$ are also mutually accessible. Therefore, we have proved that x and y are mutually accessible. □

In the above proof, an important point is that the statements are proved by induction on S. In the proof of sufficiency, S represents the number of steps required, as in Definition 4.2.2. However, in the proof of necessity, S in (4.47) is the number of terms when $u^x - u^y$ is expressed as a combination of the elements of a set of generators, and does not equal the number of steps from x to y. In other words, (4.47) is not obtained as a simple rewriting of the transition from x to y. In this sense, Theorem 4.2.8 shows a nontrivial result.

To calculate a Markov basis for a given A, we can use the elimination theory of Theorem 1.4.1. For this purpose, treat the configuration matrix A as a $d \times p$ matrix, and prepare indeterminates $v = \{v_1, \dots, v_d\}$ for the sufficient statistic t. The relation $t = Ax$ can be expressed by the homomorphism

$$
\begin{aligned}
\phi_A &: k[u] \to k[v] \\
u_j &\mapsto v_1^{a_{1j}} v_2^{a_{2j}} \cdots v_d^{a_{dj}}.
\end{aligned}
$$

We then have the following.

Proposition 4.2.9. *The toric ideal I_A for the configuration A is given as*

$$I_A = \{f \in k[u] \mid \phi_A(f) = 0\}.$$

This proposition is stated as Lemma 1.5.9 in Chap. 1. We show an example of ϕ_A.

Example 4.2.10. For the independence model of a 3×3 table,

$$
z = \begin{array}{|ccc|}
\hline
2 & -2 & 0 \\
-1 & 0 & 1 \\
-1 & 2 & -1 \\
\hline
\end{array}
$$

is a move for the configuration matrix A and corresponds to the binomial $u_{11}^2 u_{23} u_{32}^2 - u_{12}^2 u_{21} u_{31} u_{33}$. On the other hand, if we let $v = (r_1, r_2, r_3, c_1, c_2, c_3)$, then ϕ_A is written as

$$\phi_A(u_{ij}) = r_i c_j, \quad 1 \le i, j \le 3.$$

For the binomial corresponding to z, we can check that

$$
\begin{aligned}
&\phi_A(u_{11}^2 u_{23} u_{32}^2 - u_{12}^2 u_{21} u_{31} u_{33}) \\
&= \phi_A(u_{11}^2)\phi_A(u_{23})\phi_A(u_{32}^2) - \phi_A(u_{12}^2)\phi_A(u_{21})\phi_A(u_{31})\phi_A(u_{33}) \\
&= (r_1 c_1)^2 (r_2 c_3)(r_3 c_2)^2 - (r_1 c_2)^2 (r_2 c_1)(r_3 c_1)(r_3 c_3) \\
&= 0,
\end{aligned}
$$

and $u_{11}^2 u_{23} u_{32}^2 - u_{12}^2 u_{21} u_{31} u_{33} \in I_A$ holds.

Using the elimination theory, we can give an algorithm for computing the Gröbner basis of I_A, as follows.

Corollary 4.2.11 (Theorem 3.2 of [11]). *Define an ideal of $k[u, v]$ as*

$$I_A^* = \langle -\phi_A(u_i) + u_i, \ i = 1, \ldots, p \rangle \subset k[u, v].$$

Then we have $I_A = I_A^* \cap k[u]$. Therefore, for the reduced Gröbner basis G^* of I_A^* for any term order satisfying $\{v_1, \ldots, v_d\} \succ \{u_1, \ldots, u_p\}$, $G^* \cap k[u]$ is a reduced Gröbner basis of I_A.

We have already stated Corollary 4.2.11 as Lemma 1.5.11 in Chap. 1. By this corollary, we can obtain a generator of I_A as its Gröbner basis, and we have a Markov basis as the set of moves corresponding to the Gröbner basis.

4.3 Design of Experiments and Markov Basis

In this section, as an example of an application of Gröbner bases, we consider the design of experiments. The design of experiments is an important field in applied statistics, and since the beginning of the field of algebraic statistics, it has also been connected to computational algebra. The first statistical paper in which Gröbner basis theory was used [22], considered the problem of the design of experiments. The term "computational algebraic statistics" first appeared in the textbook [23]. If we regard the work [11] on Markov bases as one of the founding studies of computational algebraic statistics, then we should also consider [22] to be a founding study. For more information on early works in this field, see the survey paper [5].

In this section, we consider the design of experiments as one of the applications of the Markov chain Monte Carlo method, i.e., we will use the Markov chain Monte Carlo method to estimate a p value for experimental data. As is shown in [7], we can use formalization that is similar to the case of contingency tables provided that the experimental data consists of nonnegative integers.

Run	A	B	C	D	E	F	G	x

Table 4.5 Design and number of defects x for the wave-soldering experiment [9]

Run	A	B	C	D	E	F	G	x
1	+1	+1	+1	+1	+1	+1	+1	69
2	+1	+1	+1	−1	−1	−1	−1	31
3	+1	+1	−1	+1	+1	−1	−1	55
4	+1	+1	−1	−1	−1	+1	+1	149
5	+1	−1	+1	+1	−1	+1	−1	46
6	+1	−1	+1	−1	+1	−1	+1	43
7	+1	−1	−1	+1	−1	−1	+1	118
8	+1	−1	−1	−1	+1	+1	−1	30
9	−1	+1	+1	+1	−1	−1	+1	43
10	−1	+1	+1	−1	+1	+1	−1	45
11	−1	+1	−1	+1	−1	+1	−1	71
12	−1	+1	−1	−1	+1	−1	+1	380
13	−1	−1	+1	+1	+1	−1	−1	37
14	−1	−1	+1	−1	−1	+1	+1	36
15	−1	−1	−1	+1	+1	+1	+1	212
16	−1	−1	−1	−1	−1	−1	−1	52

4.3.1 Two-Level Designs

In this section, we consider two-level factorial designs. We start with an example.

Example 4.3.1 (Wave-Soldering Experiment). We will consider the data from [9], which consists of the number of defects arising in a wave-soldering process for attaching components to an electric circuit card. In Chap. 7 of [9], the following seven factors for wave-soldering process are considered.

A: Prebake condition
B: Flux density
C: Conveyer speed
D: Preheat condition
E: Cooling time
F: Ultrasonic solder agitator
G: Solder temperature

These seven factors are each coded into two levels. For example, the two levels of factor B are {high density, low density}; for factor C, {high speed, low speed}; and similarly for the other factors. For simplicity, we denote the two levels as {−1, +1}. There are $2^7 = 128$ combinations of levels, and in the experiment presented in [9], the number of defects were observed in 16 of these combinations. Although three observations (for three circuit cards) were reported for each combination of levels, we use the rounded average of the three observations, as in [7]. Table 4.5 shows the summarized data.

One of the aims of this type of experiments is to seek the optimal combination of the levels. For the wave-soldering data, we want to decide which levels for each factor will minimize the soldering defects. However, the solution to this problem is not obvious, for two reasons.

(i) Out of $2^7 = 128$ cases, there are only 16 observations in Table 4.5. Therefore, we have to guess the number of defects for the combinations of levels not observed.
(ii) It is natural to consider that the observations contain errors. Therefore the number of defects in Table 4.5 should be treated as one realization of some random variables.

Concerning the second reason, we may say the following about Table 4.5.

- The combination of levels for run 8 seems good because the number of defects (30) is minimized.
- The number of defects between run 8 and run 2 differs by only one. However, there are four factors, B, C, E, and F, that differ in level between these two runs. Which levels are desirable for these factors? Or, is the influence of some or all of these factors negligible?
- For factor C, the total number of defects for the $+1$ level (308) is much smaller than that for the -1 level $(1, 067)$. It seems, therefore, that the influence of factor C is not negligible.
- On the other hand, for the factor D, there are only small differences between the total number of defects for $+1$ (651) and -1 (766). We note, however, that factor D was -1 in the two best runs, 2 and 8. Is this significant?

To answer these questions, we need to fit a statistical model, such as we considered in Sect. 4.1.

In Sect. 4.3.2, we will first ignore reason (i) above and consider a design that includes all the combinations of levels; this is called a *full factorial designs*. Following this, in Sect. 4.3.3, we will consider *fractional factorial designs*, such as we see in Table 4.5.

4.3.2 Analysis of Full Factorial Designs

In full factorial designs of p two-level factors, there are $k = 2^p$ runs. The *design matrix D* is a $k \times p$ matrix in which the (i, j)th element of D is the level of the jth factor in run i. We write the jth column of D as $\boldsymbol{d}_j$.

Example 4.3.2 (Wave-Soldering Experiment as a Four-Factor Full Factorial Design). We will use the wave-soldering experiment data again, this time as an example of a full factorial design. Because we are considering a full factorial design, for simplicity, we will ignore factors E, F, and G, and only consider the four factors A, B, C, and D. Therefore, the design matrix is as follows.

$$D = \begin{pmatrix} +1 & +1 & +1 & +1 \\ +1 & +1 & +1 & -1 \\ +1 & +1 & -1 & +1 \\ +1 & +1 & -1 & -1 \\ +1 & -1 & +1 & +1 \\ +1 & -1 & +1 & -1 \\ +1 & -1 & -1 & +1 \\ +1 & -1 & -1 & -1 \\ -1 & +1 & +1 & +1 \\ -1 & +1 & +1 & -1 \\ -1 & +1 & -1 & +1 \\ -1 & +1 & -1 & -1 \\ -1 & -1 & +1 & +1 \\ -1 & -1 & +1 & -1 \\ -1 & -1 & -1 & +1 \\ -1 & -1 & -1 & -1 \end{pmatrix} = \begin{pmatrix} d_1 & d_2 & d_3 & d_4 \end{pmatrix}$$

The observations of a full factorial design of p two-level factors can be treated as 2^p p-way contingency tables. For example, with $p = 4$, the observations are treated as the realization of the 2^p-dimensional random vector

$$X = (X_{1111}, X_{1112}, X_{1121}, X_{1122}, \ldots, X_{2222})'.$$

Note that we write the factor levels as $1, 2$, not $-1, 1$, according to the notation used with contingency tables. We can now consider the conditional tests for evaluating the fitting of statistical models, as we saw in Sect. 4.1. We present it below in detail.

First, it is natural to assume independent Poisson distributions for this type of data. This is written as

$$X_{abcd} \sim \mathrm{Po}(\mu_{abcd}), \ a, b, c, d = 1, 2, \ \text{independent},$$

where $\mu_{abcd} = E(X_{abcd})$ is the parameter of the expected value. For this parameter, we consider the log-linear model as

$$\begin{aligned} \log \mu_{abcd} = {}& \kappa + \alpha_a + \beta_b + \gamma_c + \delta_d + (\alpha\beta)_{ab} + (\alpha\gamma)_{ac} + (\alpha\delta)_{ad} \\ & + (\beta\gamma)_{bc} + (\beta\delta)_{bd} + (\gamma\delta)_{cd} + (\alpha\beta\gamma)_{abc} + (\alpha\beta\delta)_{abd} \\ & + (\alpha\gamma\delta)_{acd} + (\beta\gamma\delta)_{bcd} + (\alpha\beta\gamma\delta)_{abcd}. \end{aligned} \quad (4.48)$$

This is a one-to-one transform from the saturated model with appropriate constraints, which we define as

$$\begin{aligned} & \sum_a \alpha_a = \cdots = \sum_d \delta_d = 0 \\ & \sum_a (\alpha\beta)_{ab} = \sum_b (\alpha\beta)_{ab} = \cdots = \sum_c (\gamma\delta)_{cd} = \sum_d (\gamma\delta)_{cd} = 0 \\ & \sum_a (\alpha\beta\gamma)_{abc} = \sum_b (\alpha\beta\gamma)_{abc} = \cdots = \sum_d (\beta\gamma\delta)_{bcd} = 0 \\ & \sum_a (\alpha\beta\gamma\delta)_{abcd} = \cdots = \sum_d (\alpha\beta\gamma\delta)_{abcd} = 0. \end{aligned}$$

Then, because each factor has two levels, each parameter has one degree of freedom. For example, $(\alpha\beta)_{ab}$, the two-factor interaction effect from the combination of levels (a, b) for the factors A and B, can be written as

$$(\alpha\beta)_{11} = (\alpha\beta)_{22} = \psi_{AB}, \quad (\alpha\beta)_{12} = (\alpha\beta)_{21} = -\psi_{AB}$$

with the constraint

$$\sum_a (\alpha\beta)_{ab} = \sum_b (\alpha\beta)_{ab} = 0.$$

Similar expressions are also possible for the main effects, the three-factor interaction effects, and the four-factor interaction effect. Write the constant term κ as ψ_0. Denote $\odot$ an element-wise product of two vectors. We also write $d_{ab} = d_a \odot d_b$, $d_{abc} = d_a \odot d_b \odot d_c$, and $d_{1234} = d_1 \odot d_2 \odot d_3 \odot d_4$ for the columns of the design matrix $D = (d_1, \ldots, d_4)$. Then (4.48) can be written as follows.

$$\log \mu = M\psi \tag{4.49}$$

$$\log \mu = (\log \mu_{1111}, \log \mu_{1112}, \ldots, \log \mu_{2221}, \log \mu_{2222})'$$

$$\psi = (\psi_0, \psi_A, \psi_B, \psi_{AB}, \psi_C, \psi_{AC}, \psi_{BC}, \psi_{ABC},$$
$$\psi_D, \psi_{AD}, \psi_{BD}, \psi_{CD}, \psi_{ABD}, \psi_{ACD}, \psi_{BCD}, \psi_{ABCD})'$$

$$M = (1, d_1, d_2, d_{12}, d_3, d_{13}, d_{23}, d_{123}, d_4, d_{14}, d_{24}, d_{34}, d_{124}, d_{134}, d_{234}, d_{1234}),$$

where $1 = (1, \ldots, 1)'$ is the $k \times 1$ column vector consisting of 1s. The matrix M is called a *covariate matrix*. In this example, M is written as

$$
M = \begin{pmatrix}
1 & 1 & 1 & 1 & 1 & 1 & 1 & 1 & 1 & 1 & 1 & 1 & 1 & 1 & 1 & 1 \\
1 & 1 & 1 & 1 & 1 & 1 & 1 & 1 & -1 & -1 & -1 & -1 & -1 & -1 & -1 & -1 \\
1 & 1 & 1 & 1 & -1 & -1 & -1 & -1 & 1 & 1 & 1 & 1 & -1 & -1 & -1 & -1 \\
1 & 1 & 1 & 1 & -1 & -1 & -1 & -1 & -1 & -1 & -1 & -1 & 1 & 1 & 1 & 1 \\
1 & 1 & -1 & -1 & 1 & 1 & -1 & -1 & 1 & 1 & -1 & -1 & 1 & 1 & -1 & -1 \\
1 & 1 & -1 & -1 & 1 & 1 & -1 & -1 & -1 & -1 & 1 & 1 & -1 & -1 & 1 & 1 \\
1 & 1 & -1 & -1 & -1 & -1 & 1 & 1 & 1 & 1 & -1 & -1 & -1 & -1 & 1 & 1 \\
1 & 1 & -1 & -1 & -1 & -1 & 1 & 1 & -1 & -1 & 1 & 1 & 1 & 1 & -1 & -1 \\
1 & -1 & 1 & -1 & 1 & -1 & 1 & -1 & 1 & -1 & 1 & -1 & 1 & -1 & 1 & -1 \\
1 & -1 & 1 & -1 & 1 & -1 & 1 & -1 & -1 & 1 & -1 & 1 & -1 & 1 & -1 & 1 \\
1 & -1 & 1 & -1 & -1 & 1 & -1 & 1 & 1 & -1 & 1 & -1 & -1 & 1 & -1 & 1 \\
1 & -1 & 1 & -1 & -1 & 1 & -1 & 1 & -1 & 1 & -1 & 1 & 1 & -1 & 1 & -1 \\
1 & -1 & -1 & 1 & 1 & -1 & -1 & 1 & 1 & -1 & -1 & 1 & 1 & -1 & -1 & 1 \\
1 & -1 & -1 & 1 & 1 & -1 & -1 & 1 & -1 & 1 & 1 & -1 & -1 & 1 & 1 & -1 \\
1 & -1 & -1 & 1 & -1 & 1 & 1 & -1 & 1 & -1 & -1 & 1 & -1 & 1 & 1 & -1 \\
1 & -1 & -1 & 1 & -1 & 1 & 1 & -1 & 1 & -1 & -1 & -1 & 1 & 1 & -1 & 1
\end{pmatrix}, \tag{4.50}
$$

which is a *Hadamard matrix* of order 16.

Proposition 4.3.3. *$M'x$ is the sufficient statistic for the parameter ψ.*

Proof. Substituting (4.49) into the probability function of X, we have

$$
\prod_{a=1}^{2}\prod_{b=1}^{2}\prod_{c=1}^{2}\prod_{d=1}^{2} \frac{\mu_{abcd}^{x_{abcd}} e^{-\mu_{abcd}}}{x_{abcd}!}
$$

$$
= \left(\prod_{a=1}^{2}\prod_{b=1}^{2}\prod_{c=1}^{2}\prod_{d=1}^{2} \frac{1}{x_{abcd}!}\right) \exp\left(\sum_{a=1}^{2}\sum_{b=1}^{2}\sum_{c=1}^{2}\sum_{d=1}^{2}(x_{abcd}\log\mu_{abcd} - \mu_{abcd})\right)
$$

$$
= \left(\prod_{a=1}^{2}\prod_{b=1}^{2}\prod_{c=1}^{2}\prod_{d=1}^{2} \frac{1}{x_{abcd}!}\right) \exp\left(\psi'M'x - \sum_{a=1}^{2}\sum_{b=1}^{2}\sum_{c=1}^{2}\sum_{d=1}^{2}\mu_{abcd}\right).
$$

Then, by the factorization theorem, $M'x$ is the sufficient statistic for ψ. □

Now we define a null model (H_0). H_0 is a submodel of the saturated model, which we specify by setting some parameters to zero, on the left-hand side of the log-linear model (4.49). We will usually consider the following class of submodels.

Definition 4.3.4 (Hierarchical Models). A model is called a hierarchical model if, for each interaction term in the model, all the lower-order interaction terms included in the term are also contained in the model.

For example, the hierarchical models including ψ_{ABC} also include the lower-order terms

$$\psi_A, \psi_B, \psi_C, \psi_{AB}, \psi_{AC}, \psi_{BC}.$$

It is natural to restrict our consideration to the hierarchical models in view of the interpretation of the parameters. For example, when we consider the two-factor interaction effect ψ_{AB} of the factors A and B, we should assume that each main effect ψ_A, ψ_B exists; otherwise, the interpretation of ψ_{AB} becomes difficult. Conversely, we may say that a two-factor interaction effect is defined as an effect that cannot be explained taking separately each of the main factor effects.

Once we restrict our consideration to the class of hierarchical models, we can specify each hierarchical model concisely by using its generating set, i.e., the set of maximal interaction terms it contains. For example, "the model ABC/ABD" means the model consisting of the three-factor interaction effect for the factors A, B, and C; the three-factor interaction effect for the factors A, B, and D; and each of the main and two-factor interaction effects included in these. To evaluate the fit of the model, we use a goodness-of-fit test for the null hypothesis

$$H_0 : \psi_{CD} = \psi_{ACD} = \psi_{BCD} = \psi_{ABCD} = 0. \tag{4.51}$$

The degrees of freedom of the model and the degrees of freedom of the test statistic are calculated as in Sect. 4.1. For example, the degree of freedom of the model ABC/ABD is $(16 - 1) - 4 = 11$, and the degree of freedom of the test statistic is 4.

Example 4.3.5 (Calculation by R).   We used R to perform the test of (4.51) for
the wave-soldering experiment data with four factors. In R, a function glm of the
generalized linear model is available; as a default, it gives the value of the likelihood
ratio. We use this function as follows.

```
> data4 <- read.table("data4.txt", header=T)
> data4
    A  B  C  D   x
1   1  1  1  1  69
2   1  1  1 -1  31
3   1  1 -1  1  55
4   1  1 -1 -1 149
5   1 -1  1  1  46
6   1 -1  1 -1  43
7   1 -1 -1  1 118
8   1 -1 -1 -1  30
9  -1  1  1  1  43
10 -1  1  1 -1  45
11 -1  1 -1  1  71
12 -1  1 -1 -1 380
13 -1 -1  1  1  37
14 -1 -1  1 -1  36
15 -1 -1 -1  1 212
16 -1 -1 -1 -1  52
> data4.glm <- glm(x~A+B+C+D+A*B+A*C+A*D+B*C+B*D+A*B*C+A*B*D,
                   data4,family="poisson")
> summary(data4.glm)

Call:
glm(formula = x ~ A + B + C + D + A * B + A * C + A * D + B *
    C + B * D + A * B * C + A * B * D, family =
    "poisson", data = data4)

Deviance Residuals:
      1        2        3        4        5        6        7        8
  4.015   -4.037   -3.298    2.476   -2.080    2.746    1.503   -2.463
      9       10       11       12       13       14       15       16
  4.821   -3.131   -2.617    1.279   -2.446    3.450    1.196   -2.133


Coefficients:
              Estimate Std. Error z value Pr(>|z|)
(Intercept)   4.182653   0.034131 122.546  < 2e-16 ***
A            -0.072021   0.034131  -2.110  0.03485 *
B             0.142381   0.034131   4.172 3.03e-05 ***
C            -0.517643   0.031613 -16.374  < 2e-16 ***
D             0.020120   0.030598   0.658  0.51083
A:B          -0.001776   0.034131  -0.052  0.95850
A:C           0.212262   0.031613   6.714 1.89e-11 ***
A:D           0.089063   0.030598   2.911  0.00361 **
B:C          -0.069127   0.031613  -2.187  0.02877 *
B:D          -0.442261   0.030598 -14.454  < 2e-16 ***
A:B:C         0.018033   0.031613   0.570  0.56838
A:B:D         0.146741   0.030598   4.796 1.62e-06 ***
```

```
---
Signif. codes:   0 '***' 0.001 '**' 0.01 '*' 0.05 '.' 0.1 ' ' 1

(Dispersion parameter for poisson family taken to be 1)

    Null deviance: 1021.25  on 15  degrees of freedom
Residual deviance:  135.02  on  4  degrees of freedom
AIC: 255.19

Number of Fisher Scoring iterations: 5
>
> fitted(data4.glm)
         1          2          3          4          5          6
  40.78947   59.21053   83.21053  120.78947   61.58650   27.41350
         7          8          9         10         11         12
 102.41350   45.58650   18.61224   69.38776   95.38776  355.61224
        13         14         15         16
  53.93769   19.06231  195.06231   68.93769
>
> 1-pchisq(135.02, 4)
[1] 0
>
> qchisq(0.95, 4)
[1] 9.487729
```

From this output, we see that the value of the likelihood ratio test statistic (135.02) is much larger than 9.487729, the upper 5% point of the asymptotic χ^2 distribution with four degrees of freedom. In fact, the asymptotic p value is practically 0. From these results, we can conclude that the fit of the model is poor, and thus at least some of the four terms of (4.51) are important.

In the above example, we considered the null model (4.51) just as an illustration, and it turned out that it was a meaningless model. However, in actual data analysis, we should select models more carefully. In this example, we should first consider the hypothesis

$$H_0 : \psi_{ABCD} = 0.$$

If this null hypothesis is not rejected, then we next consider the following hypotheses.

$$H_0 : \psi_{ABC} = \psi_{ABCD} = 0$$

$$H_0 : \psi_{ABD} = \psi_{ABCD} = 0$$

$$H_0 : \psi_{ACD} = \psi_{ABCD} = 0$$

$$H_0 : \psi_{BCD} = \psi_{ABCD} = 0$$

From the results of these four, we select the best models, i.e., models that are not rejected and for which the p value is large. We then consider whether we can set additional terms equal to zero. Such a procedure is an example of *model selection*. When selecting a model, the significance level for each test is set to be relatively large, such as $\alpha = 0.20$, compared to the usual testing hypothesis.

Example 4.3.6 (Model Selection Using R). With the same data as used in Example 4.3.5, we consider selecting a model by setting the higher-order terms to zero. (Such a procedure is called a stepwise method.) The following results are based on the likelihood ratio statistic, and its asymptotic p values are calculated by using R. First, the hypothesis

$$H_0 : \psi_{ABCD} = 0$$

is not significant since the likelihood ratio is 0.026 with one degree of freedom; $p = 0.87$. Then we assume $\psi_{ABCD} = 0$ and obtain the model $ABC/ABD/ACD/BCD$. From the four models, which set each of the three-factor interaction terms to be zero, the model

$$H_0 : \psi_{ABC} = \psi_{ABCD} = 0$$

is the best of these, since the likelihood ratio is 0.204 with two degrees of freedom; $p = 0.90$. Therefore we may assume $\psi_{ABC} = 0$ and obtain the model $ABD/ACD/BCD$. From the three models, which set each of the remaining three-factor interaction terms to be zero, the model

$$H_0 : \psi_{ABC} = \psi_{ACD} = \psi_{ABCD} = 0$$

is the best of these, since the likelihood ratio is 0.485 with three degrees of freedom; $p = 0.92$. Therefore we may assume $\psi_{ACD} = 0$ and obtain the model $AC/ABD/BCD$. As the next step, we consider three models by setting each of ψ_{ABD}, ψ_{BCD}, and ψ_{AC} to be zero. The best model is

$$H_0 : \psi_{ABC} = \psi_{ACD} = \psi_{ABD} = \psi_{ABCD} = 0.$$

(The likelihood ratio is 8.446 with four degrees of freedom; $p = 0.077$.) However, we should note here the relatively small p value. As the next step, we consider four models by setting each of $\psi_{AB}, \psi_{AC}, \psi_{AD}$, and ψ_{BCD} to be zero. The best model is

$$H_0 : \psi_{AB} = \psi_{ABC} = \psi_{ACD} = \psi_{ABD} = \psi_{ABCD} = 0,$$

and we have the model $AC/AD/BCD$. (The likelihood ratio is 8.449 with five degrees of freedom; $p = 0.13$.) The next step yields poorly fitting models. In fact, the model with the largest p value is

$$H_0 : \psi_{AB} = \psi_{AD} = \psi_{ABC} = \psi_{ACD} = \psi_{ABD} = \psi_{ABCD} = 0,$$

for which the likelihood ratio is 21.56 with six degrees of freedom; $p = 0.0014$. Obviously it is unreasonable to assume $\psi_{AD} = 0$. As a result, considering the drastic increase of the likelihood ratio values when assuming $\psi_{ABD} = 0$ for the model $AC/ABD/BCD$, it is reasonable to select the model $AC/ABD/BCD$, with an alternative candidate, $AC/AD/BCD$, as a more simple model.

In Examples 4.3.5 and 4.3.6, we evaluated p values by using asymptotic distributions. We should make a similar consideration as in Sect. 4.1.

- It is easiest to rely on asymptotic theory. However, this may result in a poor fit for some types of data.
- We want to calculate the exact p value if possible. However, the problem of computational feasibility occurs for larger datasets.
- If the exact calculation is difficult, the Markov chain Monte Carlo method is useful.

For the observed data x^o, the conditional probability function of X given the sufficient statistics is

$$p(x \mid M'x = M'x^o) = C(M'x^o) \prod_{i=1}^{k} \frac{1}{x_i!},$$

where $C(M'x^o)$ is the normalizing constant defined by

$$C(M'x^o)^{-1} = \sum_{x \in \mathscr{F}_{M'x^o}} \left(\prod_{i=1}^{k} \frac{1}{x_i!} \right),$$

$$\mathscr{F}_{M'x^o} = \{x \mid M'x = M'x^o, \ x \in \{0, 1, 2, \ldots\}^k\}.$$

In other words, we *treat the transpose of the covariate matrix M as the configuration matrix*, and calculate a Markov basis for M' that connects the fiber $\mathscr{F}_{M'x^o}$.

4.3.3 Analysis of Fractional Factorial Designs

Obviously, if we ignore the cost of experiments, the full factorial design is the best design. However, the number of runs becomes huge if the number of factors is large, and it is difficult to conduct the experiments in actual situations. In particular, when each run is expensive, it is important to reduce the number of runs. One efficient strategy to reduce the number of runs is to restrict the experiment to only those runs for which the combinations of levels satisfy an *aliasing relation* (or a *defining relation*). This is called a *regular fractional factorial design*. In a regular fractional factorial design, the number of the runs is a $1/2^q$ fraction of the full factorial design, where q is the number of independent aliasing relations, as explained below. The regular fractional factorial design is known to be very efficient in the sense that it has various desirable statistical properties. The only drawback is that the number of the runs is restricted to the form 2^{p-q}. In this section, we consider conditional tests for the data obtained from regular fractional factorial designs.

Example 4.3.7 (Defining Relation of the Wave-Soldering Experiment). In Table 4.5, there are 16 runs, which is $1/8$ of the full factorial design of 7 two-level factors. These 16 combinations of the levels are chosen by the relation

$$ABDE = ACDF = BCDG = 1. \tag{4.52}$$

For example, the design of Table 4.5 is constructed by choosing

- the levels of the factor E as the product of the levels of A, B, D,
- the levels of the factor F as the product of the levels of A, C, D,
- the levels of the factor G as the product of the levels of B, C, D,

in addition to the full factorial design for the four factors A, B, C, and D. A relation such as (4.52) is called a *defining relation* or an *aliasing relation*.

In general, there are $k = 2^{p-q}$ runs in a regular fractional factorial design of p two-level factors with q independent aliasing relations. Such designs are called 2^{p-q} fractional factorial designs. The design of Table 4.5 is a 2^{7-3} fractional factorial design.

For fractional factorial designs, we define a $k \times p$ design matrix D for which the components are $\{+1, -1\}$. Recall that d_j is the jth column of D. The design matrix for Table 4.5 is given as

$$D = \begin{pmatrix}
+1 & +1 & +1 & +1 & +1 & +1 & +1 \\
+1 & +1 & +1 & -1 & -1 & -1 & -1 \\
+1 & +1 & -1 & +1 & +1 & -1 & -1 \\
+1 & +1 & -1 & -1 & -1 & +1 & +1 \\
+1 & -1 & +1 & +1 & -1 & +1 & -1 \\
+1 & -1 & +1 & -1 & +1 & -1 & +1 \\
+1 & -1 & -1 & +1 & -1 & -1 & +1 \\
+1 & -1 & -1 & -1 & +1 & +1 & -1 \\
-1 & +1 & +1 & +1 & -1 & -1 & +1 \\
-1 & +1 & +1 & -1 & +1 & +1 & -1 \\
& & & \vdots & & & \\
-1 & -1 & -1 & +1 & +1 & +1 & +1 \\
-1 & -1 & -1 & -1 & -1 & -1 & -1
\end{pmatrix} = \begin{pmatrix} d_1 & \cdots & d_7 \end{pmatrix}.$$

As for full factorial designs, we can define the model matrix M from the design matrix D and express the log-linear model as (4.49). Here, we should ask whether *all the parameters are estimable* in the model. The estimability of a parameter is judged from the defining relation (4.52).

Example 4.3.8 (Estimability of the Parameters for the 2^{7-3} Fractional Factorial Design). Carefully consider the defining relation for the 2^{7-3} fractional factorial design, in Table 4.5. Multiplying the terms of (4.52) under the condition that the values are -1 or $+1$, we have

$$ABDE = ABFG = ACDF = ACEG = BCDG = BCEF = DEFG = 1.$$
$$(4.53)$$

For the $8(= 2^q)$ monomials in (4.53), their index vectors constitute a vector space over GF(2). Such a characterization always holds in general. In the context of the design of experiments, the eight monomials in (4.53) (other than 1) are called *defining contrasts*, and the minimum number of the word lengths of the defining contrasts is called the *resolution*. The resolution is traditionally expressed by a Roman numeral. Therefore, this example is a design of resolution IV. Now we consider the estimability of the parameters of (4.53).

First, multiplying (4.53) by A, we have

$$A = BDE = BFG = CDF = CEG = ABCDG = ABCEF = ADEFG.$$
$$(4.54)$$

This relation shows that the main effect of the factor A and the three-factor interaction effect of the factors B, D, E are not simultaneously estimable. We say that these two effects are *confounded* with each other. In fact, in the design matrix D, $d_2 \odot d_4 \odot d_5$, the element-wise product of the three columns d_2, d_4, d_5, coincides with d_1. By similar considerations for (4.54), we see that at most one of the following can be included in the model: the main effect of A, the three-factor interaction of B, D, E, the three-factor interaction of B, F, G, the three-factor interaction of C, D, F, the three-factor interaction of C, E, G, the five-factor interaction of A, B, C, D, G, the five-factor interaction of A, B, C, E, F, and the five-factor interaction of A, D, E, F, G.

Similarly, multiplying (4.53) by AB, we have

$$AB = DE = FG = BCDF = BCEG = ACDG = ACEF = ABDEFG.$$
$$(4.55)$$

From this relation, we see that two-factor interactions of A, B, D, E, and F, G are confounded with each other, and therefore at most one of them can be included in the model.

Considering these confounding relations, we should construct a covariate matrix so that all the parameters in the model are estimable. For example, the parameters of the model $AC/BD/E/F/G$, i.e., the model with seven main effects and the 2 two-factor interactions of AC and BD, are simultaneously estimable because AC and BD are not confounded. The covariate matrix for this model is given by

$$M = \begin{pmatrix} +1 & +1 & +1 & +1 & +1 & +1 & +1 & +1 & +1 & +1 \\ +1 & +1 & +1 & +1 & -1 & -1 & -1 & -1 & +1 & -1 \\ +1 & +1 & +1 & -1 & +1 & +1 & -1 & -1 & -1 & +1 \\ +1 & +1 & +1 & -1 & -1 & -1 & +1 & +1 & -1 & -1 \\ +1 & +1 & -1 & +1 & +1 & -1 & +1 & -1 & +1 & -1 \\ +1 & +1 & -1 & +1 & -1 & +1 & -1 & +1 & +1 & +1 \\ +1 & +1 & -1 & -1 & +1 & -1 & -1 & +1 & -1 & -1 \\ +1 & +1 & -1 & -1 & -1 & +1 & +1 & -1 & -1 & +1 \\ +1 & -1 & +1 & +1 & +1 & -1 & -1 & +1 & -1 & +1 \\ +1 & -1 & +1 & +1 & -1 & +1 & +1 & -1 & -1 & -1 \\ & & & & \vdots & & & & & \\ +1 & -1 & -1 & -1 & +1 & +1 & +1 & +1 & +1 & -1 \\ +1 & -1 & -1 & -1 & -1 & -1 & -1 & -1 & +1 & +1 \end{pmatrix} = \begin{pmatrix} +1 \\ \vdots & D & d_1 \odot d_3 & d_2 \odot d_4 \\ +1 \end{pmatrix}.$$

On the other hand, the parameters of the model

$$AB/DE/C/F/G$$

are not simultaneously estimable, because AB and DE are confounded.

Remark 4.3.9. Pistone and Wynn [22] shows the correspondence between these confounded relations and ideal membership problems in polynomial rings. See [22] or [5] for details.

Remark 4.3.10. Following the standard procedures for the design of experiments, we should choose fractional factorial designs with high resolution. This strategy is based on the concept that interaction effects with lower degree are more important than those with higher degree. For designs of resolution IV, for example, because the main effect is confounded with the three-factor interaction effect, *we can estimate each of the main effects if we ignore the three-factor interaction effects.* It is similar for designs of resolution V: because the two-factor interaction effect is confounded with the three-factor interaction effect, *we can estimate each of the main effects and the two-factor interaction effect if we ignore the three-factor interaction effects.* One interpretation of the resolution is that it is a criterion for evaluating designs as if all the factors were equally important.

There are $k = 16$ parameters in the saturated model of the 2^{7-3} design, and the covariate matrix for the saturated model is the Hadamard matrix (4.50). However, the interpretation of the saturated model is not unique. In fact, both

$$ABC/AD/BD/CD/AG/EF \quad \text{and}$$
$$ABCD$$

are examples of hierarchical models with estimable parameters, and can thus be considered to be interpretations of the saturated model.

Once we have constructed the covariate matrix, our testing procedure is similar to that for the full factorial design. We treat M' as the configuration matrix, obtain a Markov basis for M', and generate samples from the fiber $\mathscr{F}_{M'x^o}$ for the observation x^o. Then we can use the Markov chain Monte Carlo method to evaluate the fitting of the null model specified by M.

4.4 Research Topics

We conclude by presenting some research topics related to the material in this chapter. The following problems are very difficult and have not yet been completely solved.

4.4.1 Topics with Markov Bases for Models without Three-Factor Interactions for Three-Way Contingency Tables

As the last topic of Sect. 4.2.2, we considered Markov bases for models without three-factor interactions for three-way contingency tables. The structure of the Markov bases for $3 \times 3 \times K$ tables is shown in [1], and we summarize this in Table 4.6. One of the interesting results in Table 4.6 is the that the minimal Markov basis can be constructed by the four types of the moves with degrees 4, 6, 8, and 10 for $3 \times 3 \times K$ tables for any $K \geq 5$. In fact, such an upper bound on the degrees in the minimal Markov basis exists for general $I \times J \times K$ problems. In other words, there exists a positive integer m such that all the elements in the minimal Markov basis for the model with no three-factor interactions for the $I \times J \times K$ tables are included in the $I \times J \times m$ tables (Corollary 2 of [25]). Santos and Sturmfels [25] calls this m the Markov complexity and evaluates the upper bound of m. However, it seems that the upper bound of m given in [25] is not so tight. For example, for $3 \times 3 \times K$ tables, we see in Table 4.6 that the Markov complexity is $m = 5$, while the upper bound of m, as given by Santos and Sturmfels [25], is 9. Of course, if we

Table 4.6 Minimal Markov bases of for the model with no three-factor interactions for $3 \times 3 \times K$ contingency tables

$3 \times 3 \times 3$	Basic move of degree 4	Move of degree 6
$3 \times 3 \times 4$	Basic move of degree 4	Move of degree 6
	Move of degree 8	
$3 \times 3 \times K$	Basic move of degree 4	Move of degree 6
$(K \geq 5)$	Move of degree 8	Move of degree 10

know the upper bound of the Markov complexity, it is sufficient to calculate Gröbner bases up to the degree. Therefore, the contribution of [25] is substantial.

Problem 4.4.1. Solve the Markov complexity m for $I = 3, J = 4$ tables, and classify the elements in the minimal Markov basis for $3 \times 4 \times m$ tables.

This problem is approached in [2] using an elementary method similar to [1], and it is concluded that "it is sufficient to consider up to $3 \times 4 \times 8$ tables and the unique minimal Markov basis is constructed as 20 kinds of moves with the degree up to 16." However, the proof for these results is not given in [2], and it is possible that this is not correct.[2]

Another related topic is the uniqueness of the minimal Markov bases. Here, we call a Markov basis minimal if there is no proper subset that is also a Markov basis. For the results on the minimality of Markov bases, see [26] or [5]. The minimal Markov basis always exists, but it is not always unique.[3]

For the Markov basis for the model with no three-factor interactions for three-way contingency tables, it has been shown that a unique minimal Markov basis exists for $3 \times 3 \times K$ and $4 \times 4 \times 4$ tables, and is conjectured in [2] to exist for $3 \times 4 \times K$ tables. However, the uniqueness of the minimal Markov basis for general cases has not yet been determined.

Problem 4.4.2. Investigate the uniqueness of the minimal Markov basis for the model with no three-factor interactions for three-way contingency tables. Prove that the unique minimal Markov basis always exists, or give a counterexample.

4.4.2 Topics Related to the Efficient Algorithm for a Markov Basis

In Sect. 4.2.3, we derived a Markov basis as a reduced Gröbner basis of the ideal I_A. However, as stated in Theorem 4.2.8, a Markov basis is a set of generators of the configuration A and is not necessarily a Gröbner basis. Therefore, it will be very useful if we can find a way to more efficiently calculate Markov bases (i.e., some way that does not use the elimination theory of Corollary 4.2.11). In particular, when the ideal I_A has a *symmetry structure*, there is a possibility that we can develop methods for computing Markov bases efficiently by using this symmetry. For example, the elementary method used in [1] directly derives a minimal set of generators, not a Gröbner basis, by considering the symmetry of the contingency tables; in particular, it considers the symmetry about the permutation of levels and

---

[2]In fact, another conjecture given in [2], "the unique minimal Markov basis for $4 \times 4 \times 4$ table is constructed as 14 kinds of moves with the degree up to 14", includes one mistake, and it was corrected to be "15 kinds of moves with the degree up to 14" by Hemmecke and Malkin [16].

[3]Here, we define minimality by ignoring the indeterminacy of the signs of the elements of a Markov basis.

the permutation of axes with the same levels. For example, with the Markov basis for $3 \times 3 \times 3$ tables, if we know that the move corresponding to the binomial

$$u_{111}u_{122}u_{212}u_{221} - u_{112}u_{121}u_{211}u_{222}$$

is needed for a minimal Markov basis, it is obvious from symmetry of Markov bases that the moves corresponding to the binomials

$$u_{i_1 j_1 k_1} u_{i_1 j_2 k_2} u_{i_2 j_1 k_2} u_{i_2 j_2 k_1} - u_{i_1 j_1 k_2} u_{i_1 j_2 k_1} u_{i_2 j_1 k_1} u_{i_2 j_2 k_2}$$

for $i_1 \neq i_2$, $j_1 \neq j_2$, $k_1 \neq k_2$ are also included in the minimal Markov basis.

In [3, 4], the symmetry of the contingency tables, i.e., the invariant structure of the contingency tables under permutations of levels or permutation of axes with the same levels, is formalized as a group action, and the concept of an "invariant Markov basis" is defined as the invariant set of moves for this group action. Similarly, we can consider the invariant set or the minimal invariance set of generators of the ideals. In general, a minimal invariant set of generators is larger than a minimal set of generators.

Problem 4.4.3. Develop efficient algorithms to compute the minimal invariant generating sets.

In the above discussion, we emphasized the minimality of Markov bases.[4] Though the minimality is an important concept in considering the structure of a Markov basis, minimal Markov bases are not always desirable due to the speed of convergence of Markov chains. In fact, higher-degree moves are expressed as the sum of lower-degree moves, which corresponds to several steps of the chain occurring at once. Obviously, a Markov chain with such higher-degree moves has rapid convergence.

Problem 4.4.4. For a given Markov basis, evaluate the speed of converging to the stationary distribution. Determine the characteristics of a Markov basis that quickly convergences to the stationary distribution, and then develop a method for deriving such a basis.

4.4.3 Topics on Modeling Experimental Data

In Sect. 4.3, we considered models that reflected an aliasing relation and evaluated the fit of such a model for designs with two-level factors. The essential point was to define the covariate matrix M so that the columns of M become the coefficient vectors of the sufficient statistics for the parameters. On the other hand, Aoki and

---

[4]See [26] or [8] for the minimality of the Markov bases.

Takemura [6] proposed a method for defining M for the designs of three-level factors with multiple degrees of freedom for each factor effect.

Problem 4.4.5. Extend the theory in Sect. 4.3 to the designs of factors with different numbers of levels.

Once we have defined the covariate matrix M, we can follow the arguments for analyzing contingency tables in a similar way, by treating M' as the configuration matrix. The configuration matrix M is constructed from the design matrix D in view of the models to be tested. On the other hand, there are various results on the structure of the fractional factorial designs (such as various types of optimality) in the theory of the design of experiments. We refer to [21] as a relatively recent textbook. The relations between these results and the structure of the ideal $I_{M'}$ are largely unknown at present.

Problem 4.4.6. Discuss the relationship between the structure of the design D and the ideal $I_{M'}$.

References

1. S. Aoki, A. Takemura, Minimal basis for a connected Markov chain over $3 \times 3 \times K$ contingency tables with fixed two-dimensional marginals. Aust. N. Z. J. Stat. **45**, 229–249 (2003)
2. S. Aoki, A. Takemura, The list of indispensable moves of unique minimal Markov basis for $3 \times 4 \times K$ and $4 \times 4 \times 4$ contingency tables with fixed two-dimensional marginals. Technical Report METR 03-38, Department of Mathematical Engineering and Information Physics, The University of Tokyo (2003)
3. S. Aoki, A. Takemura, Minimal invariant Markov basis for sampling contingency tables with fixed marginals. Ann. Inst. Stat. Math. **60**, 229–256 (2008)
4. S. Aoki, A. Takemura, The largest group of invariance for Markov bases and toric ideals. J. Symb. Comput. **43**, 342–358 (2008)
5. S. Aoki, A. Takemura, A short survey on design of experiments and Gröbner bases. Bull. Jpn. Soc. Symb. Algebr. Comput. **16**(2), 15–22 (2009) (in Japanese)
6. S. Aoki, A. Takemura, Markov basis for design of experiments with three-level factors, in *Algebraic and Geometric Methods in Statistics* (dedicated to Professor Giovanni Pistone on the occasion of his sixty-fifth birthday), ed. by P. Gibilisco, E. Riccomagno, M.P. Rogantin, H.P. Wynn (Cambridge University Press, Cambridge, 2009), pp. 225–238
7. S. Aoki, A. Takemura, Markov chain Monte Carlo tests for designed experiments. J. Stat. Plan. Inference **140**, 817–830 (2010)
8. S. Aoki, A. Takemura, R. Yoshida, Indispensable monomials of toric ideals and Markov bases. J. Symb. Comput. **43**, 490–507 (2008)
9. L.W. Condra, *Reliability Improvement with Design of Experiments* (Dekker, New York, 1993)
10. J. Cornfield, A statistical problem arising from retrospective studies, in *Proceedings of 3rd Berkeley Symposium on Mathematical Statistics and Probability*, University of California Press, vol. 4 (1956), pp. 135–148
11. P. Diaconis, B. Sturmfels, Algebraic algorithms for sampling from conditional distributions. Ann. Stat. **26**, 363–397 (1998)
12. A. Dobra, Markov bases for decomposable graphical models. Bernoulli **9**, 1093–1108 (2003)
13. M.H. Gail, N. Mantel, Counting the number of $r \times c$ contingency tables with fixed marginals. J. Am. Stat. Assoc. **72**, 859–862 (1977)

14. S.J. Haberman, A warning on the use of chi-squared statistics with frequency tables with small expected cell counts. J. Am. Stat. Assoc. **83**, 555–560 (1988)
15. W.K. Hastings, Monte Carlo sampling methods using Markov chains and their applications. Biometrika **57**, 97–109 (1970)
16. R. Hemmecke, P. Malkin, Computing generating sets of toric ideals. arXiv: math.CO/0508359 (2005)
17. R. Hemmecke, P.N. Malkin, Computing generating sets of lattice ideals and Markov bases of lattices. J. Symb. Comput. **44**, 1463–1476 (2009)
18. E.L. Lehmann, *Testing Statistical Hypotheses*, 2nd edn. (Wiley, New York, 1986)
19. E.L. Lehmann, G. Casella, *Theory of Point Estimation*, 2nd edn. (Springer, Berlin, 2001)
20. C.R. Mehta, N.R. Patel, A network algorithm for performing Fisher's exact test in $r \times c$ contingency tables. J. Am. Stat. Assoc. **78**, 427–434 (1983)
21. R. Mukerjee, C.F.J. Wu, *A Modern Theory of Factorial Designs*. Springer Series in Statistics (Springer, New York, 2006)
22. G. Pistone, H.P. Wynn, Generalised confounding with Gröbner bases. Biometrika **83**, 653–666 (1996)
23. G. Pistone, E. Riccomagno, H.P. Wynn, *Algebraic Statistics, Computational Commutative Algebra in Statistics* (Chapman & Hall, London, 2000)
24. R.L. Plackett, *The Analysis of Categorical Data*, 2nd edn. (Griffin, London, 1981)
25. F. Santos, B. Sturmfels, Higher Lawrence configurations. J. Comb. Theory Ser. A **103**, 151–164 (2003)
26. A. Takemura, S. Aoki, Some characterizations of minimal Markov basis for sampling from discrete conditional distributions. Ann. Inst. Stat. Math. **56**, 1–17 (2004)

Chapter 5
Convex Polytopes and Gröbner Bases

Hidefumi Ohsugi

Abstract Gröbner bases of toric ideals have applications in many research areas. Among them, one of the most important topics is the correspondence to triangulations of convex polytopes. It is very interesting that, not only do Gröbner bases give triangulations, but also "good" Gröbner bases give "good" triangulations (unimodular triangulations). On the other hand, in order to use polytopes to study Gröbner bases of ideals of polynomial rings, we need the theory of Gröbner fans and state polytopes. The purpose of this chapter is to explain these topics in detail. First, we will explain convex polytopes, weight vectors, and monomial orders, all of which play a basic role in the rest of this chapter. Second, we will study the Gröbner fans of principal ideals, homogeneous ideals, and toric ideals; this will be useful when we analyze changes of Gröbner bases. Third, we will discuss the correspondence between the initial ideals of toric ideals and triangulations of convex polytopes, and the related ring-theoretic properties. Finally, we will consider the examples of configuration matrices that arise from finite graphs or contingency tables, and we will use them to verify the theory stated above. If you would like to pursue this topic beyond what is included in this chapter, we suggest the books [2, 7].

5.1 Convex Polytopes

When we say "polytopes," you may imagine three-dimensional ones, such as cubes and tetrahedrons. In Chap. 5, however, we will discuss "polytopes" that are not necessarily three-dimensional. In Sect. 5.1, we present the minimum requirements

H. Ohsugi (✉)
Department of Mathematics, College of Science, Rikkyo University,
Toshima-ku, Tokyo 171-8501, Japan
e-mail: ohsugi@rikkyo.ac.jp

T. Hibi (ed.), *Gröbner Bases: Statistics and Software Systems*,
DOI 10.1007/978-4-431-54574-3_5, © Springer Japan 2013

about convex polytopes that will be used in Chap. 5. If you are interested in the details of this discussion, please refer to Ziegler's famous textbook [12] on convex polytopes.

In Chap. 5, we will let $\mathbb{Z}_{\geq 0}$ be the set of nonnegative integers and let $\mathbb{Q}_{\geq 0}$ be the set of nonnegative rational numbers.

5.1.1 Convex Polytopes and Cones

A subset $P \subset \mathbb{Q}^d$ is said to be *convex* if, for each $\alpha, \beta \in P$, the line

$$\{t\alpha + (1-t)\beta \ : \ t \in \mathbb{Q}, \ 0 \leq t \leq 1\}$$

which connects two points is contained in P. First, we define several convex sets (convex polytopes, polyhedral convex cones, and polyhedrons) which will later play important roles. For a finite subset $X = \{\alpha_1, \ldots, \alpha_n\}$ of $\mathbb{Q}^d$, let

$$\mathrm{CONV}(X) := \left\{ \sum_{i=1}^{n} r_i \alpha_i \ : \ 0 \leq r_i \in \mathbb{Q}, \ \sum_{i=1}^{n} r_i = 1 \right\} \subset \mathbb{Q}^d,$$

which we will call the *convex hull* of X. A nonempty subset P of $\mathbb{Q}^d$ is called a *convex polytope* if there exists a finite subset $X \subset \mathbb{Q}^d$ such that $P = \mathrm{CONV}(X)$. On the other hand, the set

$$\mathbb{Q}_{\geq 0} X := \left\{ \sum_{i=1}^{n} r_i \alpha_i \ : \ 0 \leq r_i \in \mathbb{Q} \right\} \subset \mathbb{Q}^d$$

is called a *polyhedral convex cone* generated by X. A nonempty subset $C \subset \mathbb{Q}^d$ is called a *cone* if, for any finite subset X of C, we have $\mathbb{Q}_{\geq 0} X \subset C$. Every polyhedral convex cone generated by a finite set is a cone. In Ziegler's textbook [12] on convex polytopes, the following proposition [12, Theorem 1.3] is introduced.

Proposition 5.1.1 (Main Theorem for Cones). *Let $C \subset \mathbb{Q}^d$ be a cone. Then, there exists a finite set $X \subset \mathbb{Q}^d$ such that $C = \mathbb{Q}_{\geq 0} X$ if and only if C is the intersection of finitely many linear closed half-spaces.*

Example 5.1.2. For a finite set $X = \{[1,1]^\top, [1,4]^\top, [4,1]^\top\} \subset \mathbb{Q}^2$, the convex sets are $\mathrm{CONV}(X)$ and $\mathbb{Q}_{\geq 0} X$, as shown in Fig. 5.1.

A (convex) *polyhedron* is the intersection of finitely many (not necessarily linear) closed half-spaces of $\mathbb{Q}^d$. In other words, a polyhedron is the set of solutions to a system of linear inequalities (with rational coefficients). By Proposition 5.1.1, a polyhedral cone is a polyhedron. Given two polyhedrons P, $Q \subset \mathbb{Q}^d$, the

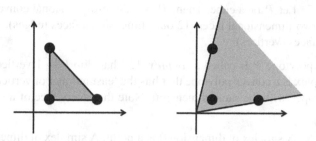

Fig. 5.1 CONV(X) and $\mathbb{Q}_{\geq 0} X$

Minkowski sum of P and Q is defined by $P + Q = \{\mathbf{p} + \mathbf{q} \in \mathbb{Q}^d : \mathbf{p} \in P, \mathbf{q} \in Q\} \subset \mathbb{Q}^d$. For example, if P and Q are convex polytopes, then $P + Q$ is also a convex polytope.

Proposition 5.1.3. *Let* $X_1, X_2 \subset \mathbb{Q}^d$ *be finite sets and let* $Y = \{\mathbf{x}_1 + \mathbf{x}_2 : \mathbf{x}_i \in X_i\}$. *Then,* CONV($Y$) = CONV($X_1$) + CONV($X_2$) *holds.*

We can prove Proposition 5.1.3 by using the definition of the convex hull; try this as an exercise. The following proposition follows from Proposition 5.1.1. (See [12, Theorem 1.2].)

Proposition 5.1.4 (Main Theorem for Polyhedrons). *A set* $P \subset \mathbb{Q}^d$ *is a polyhedron if and only if there exists a finite set* $X, Y \subset \mathbb{Q}^d$ *such that* $P = $ CONV(X) + $\mathbb{Q}_{\geq 0} Y$. *(In fact, any* $\mathbb{Q}_{\geq 0} Y$ *satisfying this condition is unique.)*

Example 5.1.5. For a polyhedron $P = \{[x, y] \in \mathbb{Q}^2 : x \geq 0, y \geq 0, x + 3y \geq 4, 3x + y \geq 4\}$, we have $P = $ CONV(X) + $\mathbb{Q}_{\geq 0} Y$ for $X = \{[0, 4], [1, 1], [4, 0]\}$, $Y = \{[0, 1], [1, 0]\}$. (Drawing the figure of this example is left as an exercise.)

Finally, we introduce the main theorem [12, Theorem 1.1] for convex polytopes.

Proposition 5.1.6 (Main Theorem for Convex Polytopes). *A set* $P \subset \mathbb{Q}^d$ *is a convex polytope if and only if it is a bounded polyhedron.*

5.1.2 Faces of Convex Polytopes

For a polyhedron $P \subset \mathbb{Q}^d$ and a vector $\mathbf{w} \in \mathbb{Q}^d$, the set

$$\text{FACE}_{\mathbf{w}}(P) := \{\mathbf{u} \in P : \mathbf{w} \cdot \mathbf{u} \geq \mathbf{w} \cdot \mathbf{v} \text{ for all } \mathbf{v} \in P\}$$

is called a *face* of P (with respect to $\mathbf{w}$). In particular, by considering $\mathbf{w} = \mathbf{0}$, P itself is a face of P. In addition, we regard the empty set $\emptyset$ as a face of P. A point α in a polyhedron P is called a *vertex* of P if $\{\alpha\}$ is a face of P. The *dimension* of a convex polytope $P \subset \mathbb{Q}^d$ is the dimension of the subspace of $\mathbb{Q}^d$ spanned by $\{\mathbf{x} - \alpha : \mathbf{x} \in P\} \subset \mathbb{Q}^d$, where $\alpha \in P$ is any fixed point. We denote the dimension of P by dim P.

Example 5.1.7. Let P be a cube. Then, P is a three-dimensional convex polytope which has 6 two-dimensional faces, 12 one-dimensional faces (edges), and 8 zero-dimensional faces (vertices).

A convex polytope P is called a *simplex* if P has dim $P + 1$ vertices. In other words, a simplex is a convex polytope that has the least number of vertices of all the convex polytopes with the same dimension. Note that every face of a simplex is a simplex.

Example 5.1.8. A simplex of dimension 0 is a point. A simplex of dimension 1 is a line segment. A simplex of dimension 2 is a triangle. A simplex of dimension 3 is a tetrahedron.

We now introduce several basic propositions about vertices and faces.

Proposition 5.1.9. *For a set* $X = \{\alpha_1, \ldots, \alpha_n\} \subset \mathbb{Q}^d$ *and* $\mathbf{w} \in \mathbb{Q}^d$, *let*

$$\lambda = \max(\mathbf{w} \cdot \alpha_i \ : \ 1 \le i \le n),$$

$$X_{\mathbf{w}} = \{\alpha_i \in X \ : \ \mathbf{w} \cdot \alpha_i = \lambda\}.$$

Then, $\mathrm{FACE}_{\mathbf{w}}(P) = \mathrm{CONV}(X_{\mathbf{w}})$ *holds for the convex hull* $P = \mathrm{CONV}(X)$. *In particular,* $\mathrm{FACE}_{\mathbf{w}}(P)$ *is a convex polytope, and* P *has only finitely many faces.*

Proof. First, we show $\lambda = \max(\mathbf{w} \cdot \alpha \ : \ \alpha \in P)$. Let $\alpha \in P$. Then, we have

$$\alpha = \sum_{i=1}^{n} r_i \alpha_i \ \ (0 \le r_i \in \mathbb{Q}, \ \sum_{i=1}^{n} r_i = 1).$$

If we take the inner product with the vector $\mathbf{w}$, then

$$\mathbf{w} \cdot \alpha = \sum_{i=1}^{n} r_i (\mathbf{w} \cdot \alpha_i) \le \sum_{i=1}^{n} r_i \lambda = \lambda$$

holds. Hence, $\lambda = \max(\mathbf{w} \cdot \alpha \ : \ \alpha \in P)$.

By changing indices if necessary, we may assume that $X_{\mathbf{w}} = \{\alpha_1, \ldots, \alpha_\ell\}$. Let $\beta \in \mathrm{CONV}(X_{\mathbf{w}})$. Then,

$$\beta = \sum_{i=1}^{\ell} s_i \alpha_i \ \ (0 \le s_i \in \mathbb{Q}, \ \sum_{i=1}^{\ell} s_i = 1).$$

If we take the inner product with the vector $\mathbf{w}$, then

$$\mathbf{w} \cdot \beta = \sum_{i=1}^{\ell} s_i (\mathbf{w} \cdot \alpha_i) = \sum_{i=1}^{\ell} s_i \lambda = \lambda$$

holds. Hence, β belongs to $\mathrm{FACE}_\mathbf{w}(P)$. Thus, we have $\mathrm{CONV}(X_\mathbf{w}) \subset \mathrm{FACE}_\mathbf{w}(P)$. Conversely, let $\gamma \in \mathrm{FACE}_\mathbf{w}(P)$. Then, $\mathbf{w} \cdot \gamma = \lambda$. Moreover, since $\gamma \in P$, we have $\gamma = \sum_{i=1}^n t_i \alpha_i$ $(0 \le t_i \in \mathbb{Q}, \sum_{i=1}^n t_i = 1)$. If we take the inner product with the vector $\mathbf{w}$, then $\lambda = \mathbf{w} \cdot \gamma = \sum_{i=1}^n t_i (\mathbf{w} \cdot \alpha_i)$ holds. However, since

$$\mathbf{w} \cdot \alpha_i \begin{cases} = \lambda & (i = 1, 2, \ldots, \ell) \\ < \lambda & (i = \ell + 1, \ldots, n), \end{cases}$$

we have $t_i = 0$ for all $i = \ell + 1, \ldots, n$. Thus, $\gamma = \sum_{i=1}^\ell t_i \alpha_i \in \mathrm{CONV}(X_\mathbf{w})$. Hence, $\mathrm{FACE}_\mathbf{w}(P) \subset \mathrm{CONV}(X_\mathbf{w})$ holds. Therefore, $\mathrm{FACE}_\mathbf{w}(P) = \mathrm{CONV}(X_\mathbf{w})$, and $\mathrm{FACE}_\mathbf{w}(P)$ is a convex polytope. Moreover, since the finite set X has only finitely many subsets, P has only finitely many faces.   □

Proposition 5.1.10. *Let V be the vertex set of a convex polytope P. Then, V is a finite set, and we have $P = \mathrm{CONV}(V)$.*

Proof. First, we choose a finite set X which is minimal among the sets satisfying $P = \mathrm{CONV}(X)$. We will show that X is equal to V. Since $V \subset X$ follows immediately from Proposition 5.1.9, it is enough to show that $X \subset V$. However, since a strict proof of this is very complex, here, we will just present a sketch.

For $\alpha \in X$, let $Y = X \setminus \{\alpha\}$ and $P' = \mathrm{CONV}(Y)$. By an appropriate choice of the set X, we have $P' \subsetneq P$ and $\alpha \in P \setminus P'$. Let β be one of the points in P' which is the closest to α. Let $\mathbf{w} = \alpha - \beta$. It then follows that, for any $\gamma \in Y$, we have $\mathbf{w} \cdot \alpha > \mathbf{w} \cdot \gamma$. Hence, $\mathrm{FACE}_\mathbf{w}(P) = \{\alpha\}$. Thus, $X \subset V$, as desired.   □

Proposition 5.1.11. *Let $F = \mathrm{FACE}_\mathbf{w}(P)$ be a face of a convex polytope P, and let $F' = \mathrm{FACE}_{\mathbf{w}'}(F)$ be a face of convex polytope F. Then, F' is a face of P. More precisely, for a sufficiently small $\varepsilon > 0$, we have $F' = \mathrm{FACE}_{\mathbf{w}+\varepsilon\mathbf{w}'}(P)$.*

Proof. Suppose that a finite set X satisfies $P = \mathrm{CONV}(X)$. Let $\lambda = \max(\mathbf{w} \cdot \alpha : \alpha \in X)$, $X_\mathbf{w} = \{\alpha \in X : \mathbf{w} \cdot \alpha = \lambda\}$, $\lambda' = \max(\mathbf{w}' \cdot \alpha : \alpha \in X_\mathbf{w})$, and $X_{\mathbf{w},\mathbf{w}'} = \{\alpha \in X_\mathbf{w} : \mathbf{w}' \cdot \alpha = \lambda'\}$. Let ε be a real number satisfying the condition

$$0 < \varepsilon < \frac{\min(\lambda - \mathbf{w} \cdot \beta : \beta \in X \setminus X_\mathbf{w})}{2 \max(|\lambda' - \mathbf{w}' \cdot \beta| : \beta \in X \setminus X_\mathbf{w})}.$$

(If the denominator is 0, then we let $0 < \varepsilon < \infty$, and we do not need to consider Case 3.) By Proposition 5.1.9, we have $F = \mathrm{CONV}(X_\mathbf{w})$ and $F' = \mathrm{CONV}(X_{\mathbf{w},\mathbf{w}'})$. Moreover, for any $\alpha \in X_{\mathbf{w},\mathbf{w}'}$, it follows that $(\mathbf{w} + \varepsilon\mathbf{w}') \cdot \alpha = \lambda + \varepsilon\lambda'$. Hence, it is enough to show that, for $\alpha \in X_{\mathbf{w},\mathbf{w}'}$ and $\beta \in X \setminus X_{\mathbf{w},\mathbf{w}'}$, we have $(\mathbf{w} + \varepsilon\mathbf{w}') \cdot \alpha > (\mathbf{w} + \varepsilon\mathbf{w}') \cdot \beta$.

Case 1. $\beta \in X_\mathbf{w} \setminus X_{\mathbf{w},\mathbf{w}'}$:

$$(\mathbf{w} + \varepsilon\mathbf{w}') \cdot (\alpha - \beta) = \mathbf{w} \cdot (\alpha - \beta) + \varepsilon\mathbf{w}' \cdot (\alpha - \beta)$$
$$= 0 + \varepsilon(\lambda' - \mathbf{w}' \cdot \beta)$$
$$> 0.$$

Case 2. $\beta \notin X_{\mathbf{w}}$ and $\mathbf{w}' \cdot (\alpha - \beta) \geq 0$:

$$(\mathbf{w} + \varepsilon \mathbf{w}') \cdot (\alpha - \beta) = \mathbf{w} \cdot (\alpha - \beta) + \varepsilon \mathbf{w}' \cdot (\alpha - \beta)$$
$$\geq \mathbf{w} \cdot (\alpha - \beta)$$
$$> 0.$$

Case 3. $\beta \notin X_{\mathbf{w}}$ and $\mathbf{w}' \cdot (\alpha - \beta) < 0$:

$$(\mathbf{w} + \varepsilon \mathbf{w}') \cdot (\alpha - \beta) = \mathbf{w} \cdot (\alpha - \beta) + \varepsilon \mathbf{w}' \cdot (\alpha - \beta)$$
$$> \mathbf{w} \cdot (\alpha - \beta) + \frac{\mathbf{w} \cdot (\alpha - \beta)}{2 |\mathbf{w}' \cdot (\alpha - \beta)|} \mathbf{w}' \cdot (\alpha - \beta)$$
$$= \frac{1}{2} \mathbf{w} \cdot (\alpha - \beta)$$
$$> 0.$$

Thus, for $\alpha \in X_{\mathbf{w}, \mathbf{w}'}$ and $\beta \in X \setminus X_{\mathbf{w}, \mathbf{w}'}$, we have $(\mathbf{w} + \varepsilon \mathbf{w}') \cdot (\alpha - \beta) > 0$, and hence $(\mathbf{w} + \varepsilon \mathbf{w}') \cdot \alpha > (\mathbf{w} + \varepsilon \mathbf{w}') \cdot \beta$ holds. $\qquad \square$

The following proposition holds for the Minkowski sum on faces of polytopes.

Proposition 5.1.12. *For convex polytopes* $P_1, \ldots, P_s \subset \mathbb{Q}^d$, *let* $P = \sum_{i=1}^{s} P_i$. *Then, for* $\mathbf{w} \in \mathbb{Q}^d$, *we have* $\mathrm{FACE}_{\mathbf{w}}(P) = \sum_{i=1}^{s} \mathrm{FACE}_{\mathbf{w}}(P_i)$.

Proposition 5.1.13. *For convex polytopes* $P_1, \ldots, P_s \subset \mathbb{Q}^d$, *let* $P = \sum_{i=1}^{s} P_i$. *If* $\mathrm{FACE}_{\mathbf{w}}(P)$ *is a vertex of* P, *then the expression*

$$\mathrm{FACE}_{\mathbf{w}}(P) = \mathbf{p}_1 + \cdots + \mathbf{p}_s \quad (\mathbf{p}_i \in P_i)$$

is unique. (More precisely, $\mathrm{FACE}_{\mathbf{w}}(P_i) = \mathbf{p}_i$ *for all* $1 \leq i \leq s$.)

Proposition 5.1.14. *For convex polytopes* $P_1, \ldots, P_s \subset \mathbb{Q}^d$, *let* $P = \sum_{i=1}^{s} P_i$. *Then, for* $\mathbf{w}, \mathbf{w}' \in \mathbb{Q}^d$, *the following conditions are equivalent:*

(i) $\mathrm{FACE}_{\mathbf{w}}(P) = \mathrm{FACE}_{\mathbf{w}'}(P)$;
(ii) *For all* $1 \leq i \leq s$, $\mathrm{FACE}_{\mathbf{w}}(P_i) = \mathrm{FACE}_{\mathbf{w}'}(P_i)$.

Proof. By Proposition 5.1.12, (ii) $\Longrightarrow$ (i) is trivial. Hence, we will show (i) $\Longrightarrow$ (ii). Suppose that condition (i) holds, and that, for some $1 \leq i \leq s$, $\mathrm{FACE}_{\mathbf{w}}(P_i) \neq \mathrm{FACE}_{\mathbf{w}'}(P_i)$. By replacing the role of $\mathbf{w}$ and $\mathbf{w}'$ if necessary, we may assume that there exists a vertex $\mathbf{v}$ of $\mathrm{FACE}_{\mathbf{w}}(P_i)$ such that $\mathbf{v} \notin \mathrm{FACE}_{\mathbf{w}'}(P_i)$. Then, there exists $\mathbf{u}$ such that $\mathrm{FACE}_{\mathbf{u}}(\mathrm{FACE}_{\mathbf{w}}(P_i)) = \{\mathbf{v}\}$. Let $\mathbf{u}'$ be a vector such that $\mathrm{FACE}_{\mathbf{u}'}(\mathrm{FACE}_{\mathbf{u}}(\mathrm{FACE}_{\mathbf{w}}(P)))$ is a vertex. Then, by condition (i), we have

$$\mathrm{FACE}_{\mathbf{u}'}(\mathrm{FACE}_{\mathbf{u}}(\mathrm{FACE}_{\mathbf{w}}(P))) = \mathrm{FACE}_{\mathbf{u}'}(\mathrm{FACE}_{\mathbf{u}}(\mathrm{FACE}_{\mathbf{w}'}(P))).$$

Thus, by Proposition 5.1.13, it follows that

$$\text{FACE}_{\mathbf{u}'}(\text{FACE}_{\mathbf{u}}(\text{FACE}_{\mathbf{w}}(P_i))) = \text{FACE}_{\mathbf{u}'}(\text{FACE}_{\mathbf{u}}(\text{FACE}_{\mathbf{w}'}(P_i))).$$

Thus, the left-hand side equals $\{\mathbf{v}\}$, and the right-hand side does not contain $\mathbf{v}$. This is a contradiction.   □

Let ∂P denote the boundary of a convex polytope P, and let $P \setminus \partial P$ denote the interior of P. Note that here by "interior" we mean the so-called "relative interior." For example, for a triangle P in three-dimensional space, ∂P is the union of its three edges. (If you are interested in the details, then please refer to Lemmas 2.8, 2.9, and the surrounding text in [12].) In general, the following holds.

Proposition 5.1.15. *Let* $\{\alpha_1, \ldots, \alpha_n\}$ *be the vertex set of a convex polytope* $P \subset \mathbb{Q}^d$. *Then,* $\alpha \in \mathbb{Q}^d$ *belongs to the interior* $P \setminus \partial P$ *of* P *if and only if* α *is expressed as*

$$\alpha = \sum_{i=1}^{n} r_i \alpha_i \quad (0 < r_i \in \mathbb{Q}, \sum_{i=1}^{n} r_i = 1).$$

If the dimension of a face F of a convex polytope $P \subset \mathbb{Q}^d$ equals $\dim P - 1$, then F is called a *facet* of P. The boundary of a convex polytope is described by its facets, as follows.

Proposition 5.1.16. *For a convex polytope* $P \subset \mathbb{Q}^d$, *let* Δ *be the set of all faces of* P *(which is different from* P*), and let* Δ' *be the set of all facets of* P. *Then, we have*

$$\partial P = \bigcup_{F \in \Delta} F = \bigcup_{F' \in \Delta'} F'.$$

A convex polytope P is said to be *integral* if all the vertices of P are integer vectors. For a configuration matrix $A = [\mathbf{a}_1, \ldots, \mathbf{a}_n]$, we call $\text{CONV}(A) := \text{CONV}(\{\mathbf{a}_1, \ldots, \mathbf{a}_n\})$ the convex hull of A. Then, $\text{CONV}(A)$ is an integral convex polytope. For an integral convex polytope $P = \text{CONV}(A)$, the dimension of P is equal to $\dim P = \text{rank } A - 1$.

Example 5.1.17. For the configuration matrix

$$A = \begin{bmatrix} 0 & 1 & 0 & 1 & 1 \\ 0 & 0 & 1 & 1 & -1 \\ 1 & 1 & 1 & 1 & 1 \end{bmatrix},$$

the convex hull $\text{CONV}(A)$ of A is the quadrangle shown in Fig. 5.2. Since rank $A = 3$, the dimension of $\text{CONV}(A)$ is equal to 2.

Fig. 5.2 CONV(A) in
Example 5.1.17

5.1.3 Polyhedral Complices and Fans

A finite set Δ of polyhedrons of $\mathbb{Q}^d$ is called a *complex* if the following conditions
are satisfied:

 (i) If F' is a face of $F \in \Delta$, then $F' \in \Delta$.
(ii) If $F, F' \in \Delta$, then $F \cap F'$ is a face of F and a face of F'.

In particular, for a complex Δ,

* If every polyhedron in Δ is a polyhedral convex cone, then Δ is called a *fan*.
* If every polyhedron in Δ is a convex polytope, then Δ is called a *polyhedral
  complex*.
* If every polyhedron in Δ is a simplex, then Δ is called a *simplicial complex*.

Given a polyhedron $P \subset \mathbb{Q}^d$ and its face F,

$$\mathcal{N}_P(F) := \{\mathbf{w} \in \mathbb{Q}^d \ : \ \mathrm{FACE}_\mathbf{w}(P) = F\}$$

is called a *normal cone* of F in P. With respect to the dimension, we have
$\dim \mathcal{N}_P(F) = d - \dim F$. Let

$$\mathcal{N}(P) := \left\{ \overline{\mathcal{N}_P(F)} \ : \ F \text{ is a face of } P \right\}$$

be the set of the closure of the normal cones of a polyhedron P. Since, for faces F
and F' of a polyhedron P, we have

$$F' \text{ is a face of } F \iff \overline{\mathcal{N}_P(F)} \text{ is a face of } \overline{\mathcal{N}_P(F')},$$

it follows that $\mathcal{N}(P)$ is a fan. This fan is called a *normal fan* of P.

5.2   Initial Ideals

In this section, we introduce some basic facts, such as those on weight vectors,
which will be needed later. In Chap. 1, we presented several monomial orders (e.g.,
the lexicographic order). It is known that every monomial order can be represented
by a weight vector. We assume that *none of the ideals that appear in this section are
zero ideals*.

5.2.1 Initial Ideals

First, we introduce two useful propositions about initial ideals.

Proposition 5.2.1. *Fix a monomial order $<$. If ideals $I, J \subset K[\mathbf{x}]$ satisfy $I \subsetneq J$, then we have $\mathrm{in}_<(I) \subsetneq \mathrm{in}_<(J)$.*

Proof. Suppose that ideals $I, J \subset K[\mathbf{x}]$ satisfy $I \subsetneq J$. Then, we have $\mathrm{in}_<(I) \subset \mathrm{in}_<(J)$. We now assume that $\mathrm{in}_<(I) = \mathrm{in}_<(J)$ holds. Let $\mathscr{G} \subset I$ be a Gröbner basis of I with respect to $<$. Then, $\mathrm{in}_<(I) = \mathrm{in}_<(J)$ is generated by $\{\mathrm{in}_<(g) \; : \; g \in \mathscr{G}\}$. Moreover, since $I \subset J$ holds, we have $\mathscr{G} \subset J$, and hence $\mathscr{G}$ is a Gröbner basis of J with respect to $<$. Thus, by Theorem 1.1.21, $\mathscr{G}$ is a set of generators of both I and J. Therefore we have $I = J$, but this contradicts the assumption $I \subsetneq J$. Thus, $\mathrm{in}_<(I) \subsetneq \mathrm{in}_<(J)$. $\qquad\square$

Proposition 5.2.2. *If monomial orders $<$, $<'$, and an ideal $I \subset K[\mathbf{x}]$ satisfy $\mathrm{in}_<(I) \subset \mathrm{in}_{<'}(I)$, then we have $\mathrm{in}_<(I) = \mathrm{in}_{<'}(I)$.*

Proof. Suppose that $\mathrm{in}_<(I) \subset \mathrm{in}_{<'}(I)$ holds. Then, the set of all standard monomials with respect to $\mathrm{in}_{<'}(I)$ is included in the set of all standard monomials with respect to $\mathrm{in}_<(I)$. By Theorem 1.6.2, since each of two sets is a basis of $K[\mathbf{x}]/I$, the two sets coincide. Thus, we have $\mathrm{in}_<(I) = \mathrm{in}_{<'}(I)$. $\qquad\square$

5.2.2 Weight Vectors and Monomial Orders

Fix a nonnegative vector $\mathbf{w} = [w_1, \ldots, w_n] \in \mathbb{Q}^n$. For any polynomial

$$(0 \neq) \; f = \sum_{i=1}^{m} c_i \mathbf{x}^{\mathbf{a}_i} \in K[\mathbf{x}] \qquad (0 \neq c_i \in K),$$

the *initial form* $\mathrm{in}_{\mathbf{w}}(f)$ of f is the sum of all terms $c_i \, \mathbf{x}^{\mathbf{a}_i}$ of f such that the inner product $\mathbf{w} \cdot \mathbf{a}_i$ is maximal. For any ideal I of $K[\mathbf{x}]$, the *initial form ideal* $\mathrm{in}_{\mathbf{w}}(I)$ is defined by

$$\mathrm{in}_{\mathbf{w}}(I) := \langle \mathrm{in}_{\mathbf{w}}(f) \; : \; 0 \neq f \in I \rangle.$$

Note that, in general, the initial form ideal is not necessarily a monomial ideal. On the other hand, as introduced in Example 1.1.13 of Chap. 1, if we define an order $<_{\mathbf{w}}$ by

$$\mathbf{x}^{\mathbf{a}} >_{\mathbf{w}} \mathbf{x}^{\mathbf{b}} \Leftrightarrow \text{ either } \mathbf{w} \cdot \mathbf{a} > \mathbf{w} \cdot \mathbf{b} \text{ or } \mathbf{w} \cdot \mathbf{a} = \mathbf{w} \cdot \mathbf{b} \text{ and } \mathbf{x}^{\mathbf{a}} > \mathbf{x}^{\mathbf{b}}$$

for a nonnegative vector $\mathbf{w} \in \mathbb{Q}^n$ and a monomial order $<$ on $K[\mathbf{x}]$, then $<_{\mathbf{w}}$ is a monomial order on $K[\mathbf{x}]$.

Proposition 5.2.3. *For an ideal $I \subset K[\mathbf{x}]$, a nonnegative vector $\mathbf{w} \in \mathbb{Q}^n$, and a monomial order $<$, we have $\mathrm{in}_<(\mathrm{in}_\mathbf{w}(I)) = \mathrm{in}_{<_\mathbf{w}}(I)$. In particular, if $\mathrm{in}_\mathbf{w}(I)$ is a monomial ideal, then $\mathrm{in}_\mathbf{w}(I) = \mathrm{in}_{<_\mathbf{w}}(I)$ holds.*

Proof. By the definition of a monomial order $<_\mathbf{w}$, for any nonzero polynomial $f \in I$, we have $\mathrm{in}_<(\mathrm{in}_\mathbf{w}(f)) = \mathrm{in}_{<_\mathbf{w}}(f)$. Hence, monomial ideals $\mathrm{in}_<(\mathrm{in}_\mathbf{w}(I))$ and $\mathrm{in}_{<_\mathbf{w}}(I)$ are generated by the same set of monomials. $\square$

Corollary 5.2.4. *For a nonnegative vector $\mathbf{w} \in \mathbb{Q}^n$, let $\mathcal{G}$ be a Gröbner basis of an ideal $I \subset K[\mathbf{x}]$ with respect to a monomial order $<_\mathbf{w}$. Then, $\{\mathrm{in}_\mathbf{w}(g) : g \in \mathcal{G}\}$ is a Gröbner basis of $\mathrm{in}_\mathbf{w}(I)$ with respect to $<$.*

Proposition 5.2.5. *For an ideal $I \subset K[\mathbf{x}]$ and a nonnegative vector $\mathbf{w}, \mathbf{w}' \in \mathbb{Q}^n$, we have $\mathrm{in}_{\mathbf{w}'}(\mathrm{in}_\mathbf{w}(I)) = \mathrm{in}_{\mathbf{w}+\varepsilon\mathbf{w}'}(I)$ for a sufficiently small $\varepsilon > 0$.*

Proof. First, we prepare a tie-breaking monomial order $<$. Let $<'$ denote the monomial order $<_{\mathbf{w}'}$, and let $\mathcal{G}$ be the reduced Gröbner basis of I with respect to the monomial order $<'_\mathbf{w}$. By Corollary 5.2.4, $\{\mathrm{in}_{\mathbf{w}'}(\mathrm{in}_\mathbf{w}(g)) : g \in \mathcal{G}\}$ is a Gröbner basis of $\mathrm{in}_{\mathbf{w}'}(\mathrm{in}_\mathbf{w}(I))$ with respect to $<$. By the same argument given in the proof of Proposition 5.1.11, for a sufficiently small $\varepsilon > 0$, $\mathrm{in}_{\mathbf{w}'}(\mathrm{in}_\mathbf{w}(g)) = \mathrm{in}_{\mathbf{w}+\varepsilon\mathbf{w}'}(g)$ holds for all $g \in \mathcal{G}$. Thus, we have $\mathrm{in}_{\mathbf{w}'}(\mathrm{in}_\mathbf{w}(I)) \subset \mathrm{in}_{\mathbf{w}+\varepsilon\mathbf{w}'}(I)$. We assume $\mathrm{in}_{\mathbf{w}'}(\mathrm{in}_\mathbf{w}(I)) \subsetneq \mathrm{in}_{\mathbf{w}+\varepsilon\mathbf{w}'}(I)$, and deduce a contradiction. Then, by Proposition 5.2.1, the initial ideals of these ideals with respect to $<$ satisfy $\mathrm{in}_<(\mathrm{in}_{\mathbf{w}'}(\mathrm{in}_\mathbf{w}(I))) \subsetneq \mathrm{in}_<(\mathrm{in}_{\mathbf{w}+\varepsilon\mathbf{w}'}(I))$. By Proposition 5.2.3, we have

$$\mathrm{in}_<(\mathrm{in}_{\mathbf{w}'}(\mathrm{in}_\mathbf{w}(I))) = \mathrm{in}_{<'}(\mathrm{in}_\mathbf{w}(I)) = \mathrm{in}_{<'_\mathbf{w}}(I),$$

$$\mathrm{in}_<(\mathrm{in}_{\mathbf{w}+\varepsilon\mathbf{w}'}(I)) = \mathrm{in}_{<_{\mathbf{w}+\varepsilon\mathbf{w}'}}(I),$$

and hence $\mathrm{in}_{<'_\mathbf{w}}(I) \subsetneq \mathrm{in}_{<_{\mathbf{w}+\varepsilon\mathbf{w}'}}(I)$. This contradicts Proposition 5.2.2. Thus, we have $\mathrm{in}_{\mathbf{w}'}(\mathrm{in}_\mathbf{w}(I)) = \mathrm{in}_{\mathbf{w}+\varepsilon\mathbf{w}'}(I)$. $\square$

If we fix an ideal, it is known that every monomial order can be represented by a weight vector. In order to prove this important fact, we use Farkas' Lemma. There are several propositions which are equivalent to Farkas' Lemma. Here, we adopt Farkas' Lemma II [12, Proposition 1.8]. (Please see [12] for more details.)

Lemma 5.2.6 (Farkas' Lemma II). *For a $p \times q$ matrix A and $\mathbf{z} \in \mathbb{Q}^p$, one and only one of the following two conditions holds.*

(i) *There exists a column vector $\mathbf{x} \in \mathbb{Q}_{\geq 0}^q$ such that $A\mathbf{x} = \mathbf{z}$.*
(ii) *There exists a row vector $\mathbf{c} \in \mathbb{Q}^p$ such that $\mathbf{c}A \in \mathbb{Q}_{\geq 0}^q$ and $\mathbf{c} \cdot \mathbf{z} < 0$.*

Proposition 5.2.7. *Let $<$ be a monomial order on $K[\mathbf{x}]$, and let I be an ideal of $K[\mathbf{x}]$. Then, there exists a nonnegative integer vector $\mathbf{w} \in \mathbb{Z}_{\geq 0}^n$ such that $\mathrm{in}_<(I) = \mathrm{in}_\mathbf{w}(I)$.*

Proof. Let $\{g_1, \ldots, g_s\}$ be a Gröbner basis of I with respect to $<$. For each g_k, let $\mathbf{x}^{\mathbf{a}_0^{(k)}}, \mathbf{x}^{\mathbf{a}_1^{(k)}}, \ldots, \mathbf{x}^{\mathbf{a}_{i_k}^{(k)}}$ be monomials appearing in g_k, and let $\mathrm{in}_<(g_k) = \mathbf{x}^{\mathbf{a}_0^{(k)}}$. We now define a subset $\mathscr{C}$ of $\mathbb{Q}_{\geq 0}^n$ by

$$\mathscr{C} = \{\mathbf{w} \in \mathbb{Q}_{\geq 0}^n \ : \ \mathbf{w} \cdot (\mathbf{a}_0^{(k)} - \mathbf{a}_\ell^{(k)}) > 0 \text{ for } 1 \leq k \leq s, \ 1 \leq \ell \leq i_k\}.$$

In order to prove $\mathscr{C} \neq \emptyset$, we use Farkas' Lemma II. Let B be the matrix whose row vectors are $\{\mathbf{a}_0^{(k)} - \mathbf{a}_\ell^{(k)} \ : \ 1 \leq k \leq s, \ 1 \leq \ell \leq i_k\}$. If $\mathscr{C} = \emptyset$ holds, then there exists no nonnegative row vector $\mathbf{w} \in \mathbb{Q}_{\geq 0}^n$ such that $B\mathbf{w}$ is a positive vector. Then, for the matrix $A = [\, -B \mid E\,]$ (E is an identity matrix) and $\mathbf{z} = [-1, \ldots, -1]^T$, there exist no nonnegative column vectors $\mathbf{w}$ and $\mathbf{v}$ such that

$$A \begin{bmatrix} \mathbf{w} \\ \mathbf{v} \end{bmatrix} = -B\mathbf{w} + \mathbf{v} = \mathbf{z}.$$

(Note that at least one component of the left-hand side is nonnegative.) Hence, by Farkas' Lemma II, there exists a row vector $\mathbf{c}$ such that $\mathbf{c}A = [-\mathbf{c}B \mid \mathbf{c}]$ is a nonnegative vector and $\mathbf{c} \cdot \mathbf{z} < 0$. Since $\mathbf{c}$ is a nonnegative vector, $\mathbf{c} \cdot \mathbf{z} < 0$ implies that the sum of all components is positive, and hence, in particular, $\mathbf{c}$ is not a zero vector. Thus, there exists a set $\Lambda = \{\lambda_\ell^{(k)} \in \mathbb{Z}_{\geq 0} \ : \ 1 \leq k \leq s, \ 1 \leq \ell \leq i_k\} \neq \{0\}$ such that

$$-\sum_{k=1}^{s} \sum_{\ell=1}^{i_k} \lambda_\ell^{(k)} (\mathbf{a}_0^{(k)} - \mathbf{a}_\ell^{(k)}) \in \mathbb{Z}_{\geq 0}^n.$$

This means that $\prod_{k=1}^{s} \prod_{\ell=1}^{i_k} (\mathbf{x}^{\mathbf{a}_\ell^{(k)}})^{\lambda_\ell^{(k)}}$ is divided by $\prod_{k=1}^{s} \prod_{\ell=1}^{i_k} (\mathbf{x}^{\mathbf{a}_0^{(k)}})^{\lambda_\ell^{(k)}}$. By Lemma 1.1.15, we have

$$\prod_{k=1}^{s} \prod_{\ell=1}^{i_k} (\mathbf{x}^{\mathbf{a}_\ell^{(k)}})^{\lambda_\ell^{(k)}} \geq \prod_{k=1}^{s} \prod_{\ell=1}^{i_k} (\mathbf{x}^{\mathbf{a}_0^{(k)}})^{\lambda_\ell^{(k)}}.$$

On the other hand, since $\mathrm{in}_<(g_k) = \mathbf{x}^{\mathbf{a}_0^{(k)}}$, it follows that $\mathbf{x}^{\mathbf{a}_\ell^{(k)}} < \mathbf{x}^{\mathbf{a}_0^{(k)}}$ for all $1 \leq k \leq s$, $1 \leq \ell \leq i_k$. When these are all multiplied together, a property of monomial orders implies

$$\prod_{k=1}^{s} \prod_{\ell=1}^{i_k} (\mathbf{x}^{\mathbf{a}_\ell^{(k)}})^{\lambda_\ell^{(k)}} < \prod_{k=1}^{s} \prod_{\ell=1}^{i_k} (\mathbf{x}^{\mathbf{a}_0^{(k)}})^{\lambda_\ell^{(k)}}.$$

(Note that since $\Lambda \neq \{0\}$, equality cannot hold.) This is a contradiction. Thus, we have $\mathscr{C} \neq \emptyset$.

We now choose a vector $\mathbf{w} \in \mathscr{C} \cap \mathbb{Z}^n (\subset \mathbb{Z}_{\geq 0}^n)$. Then, for all $1 \leq k \leq s$, we have $\mathrm{in}_\mathbf{w}(g_k) = \mathbf{x}^{\mathbf{a}_0^{(k)}} = \mathrm{in}_<(g_k)$. Since the initial ideal $\mathrm{in}_<(I)$ is generated by these monomials, it follows that $\mathrm{in}_<(I) \subset \mathrm{in}_\mathbf{w}(I)$. Then, the initial ideals of these ideals satisfy

$$\text{in}_<(I) = \text{in}_<(\text{in}_<(I)) \subset \text{in}_<(\text{in}_\mathbf{w}(I)) = \text{in}_{<_\mathbf{w}}(I).$$

By Proposition 5.2.2, $\text{in}_<(I) = \text{in}_{<_\mathbf{w}}(I)$. Thus, by Proposition 5.2.1, we have $\text{in}_<(I) = \text{in}_\mathbf{w}(I)$. □

5.2.3 Universal Gröbner Bases

For an ideal $I \subset K[\mathbf{x}]$, a finite set is called a *universal Gröbner basis* of I if it is a Gröbner basis of I with respect to any monomial order. By the following theorem, a universal Gröbner basis always exists.

Theorem 5.2.8. *Let $I \subset K[\mathbf{x}]$ be an ideal. Then, there exists only finitely many initial ideals for I.*

Proof. Suppose that the set $\Sigma_0 = \{\text{in}_<(I) \ : \ < \text{ is a monomial order }\}$ of all initial ideals of I is an infinite set. We choose a nonzero polynomial $f_1 \in I$. Then, since f_1 has only finitely many monomials, there exists a monomial m_1 appearing in f_1 such that $\Sigma_1 = \{M \in \Sigma_0 \ : \ m_1 \in M\}$ is an infinite set. Then, there exists a monomial order $<$ such that $m_1 \in \text{in}_<(I) \in \Sigma_1$. If $\langle m_1 \rangle = \text{in}_<(I)$, then by Proposition 5.2.2, we have $\Sigma_1 = \{\text{in}_<(I)\}$. This contradicts the assumption that Σ_1 is infinite. Hence, $\langle m_1 \rangle \subsetneq \text{in}_<(I)$. Thus, there exists a nonzero polynomial $f_2 \in I$ such that no monomial in f_2 belongs to $\langle m_1 \rangle$. Since f_2 has only finitely many monomials, there exists a monomial m_2 in f_2 such that $\Sigma_2 = \{M \in \Sigma_1 \ : \ m_2 \in M\}$ is an infinite set. Then, by Proposition 5.2.2 and by using a similar argument as before, it follows that there exists a monomial order $<$ such that $\langle m_1, m_2 \rangle \subsetneq \text{in}_<(I) \in \Sigma_2$. Thus, there exists a nonzero polynomial $f_3 \in I$ such that no monomial in f_3 belongs to $\langle m_1, m_2 \rangle$. By repeating such arguments, we have an infinite ascending chain of monomial ideals

$$\langle m_1 \rangle \subsetneq \langle m_1, m_2 \rangle \subsetneq \langle m_1, m_2, m_3 \rangle \subsetneq \cdots .$$

Let J be a monomial ideal of $K[\mathbf{x}]$ generated by $\{m_k \ : \ 0 < k \in \mathbb{Z}\}$. By Lemma 1.1.7, J is generated by a finite set $\{m_{\lambda_1}, \ldots, m_{\lambda_s}\}$. Let $\lambda = \max(\lambda_1, \ldots, \lambda_s)$. Since $J = \langle m_1, m_2, \ldots, m_k \rangle$ for all $k \geq \lambda$, this contradicts the above infinite ascending chain. □

Corollary 5.2.9. *For any ideal $I \subset K[\mathbf{x}]$, there exists a universal Gröbner basis of I.*

Proof. Let $\mathcal{G}_<$ be the reduced Gröbner basis of I with respect to a monomial order $<$. Then, by Theorem 5.2.8, the union

$$\bigcup_{<:\text{monomial order}} \mathcal{G}_<$$

is a finite set. Moreover, since this set contains the reduced Gröbner basis with respect to an arbitrary monomial order, it is a Gröbner basis of I with respect to an arbitrary monomial order.   □

In Sturmfels lecture [9], the Gröbner basis which is the union of all the reduced Gröbner bases and appears in the above proof is called *the* universal Gröbner basis. Therefore, in this book, we will call a universal Gröbner basis which is the union of all the reduced Gröbner bases of an ideal I, *the universal Gröbner basis* of I.

5.3   Gröbner Fans and State Polytopes

In this section, we introduce the Gröbner fan GF(I) and the state polytope State(I), which characterize the possible initial ideals of a given ideal I.

5.3.1   Gröbner Fans of Principal Ideals

As a principal ideal, we consider the ideal $I = \langle f \rangle \subset K[x_1, x_2]$ generated by

$$f = x_1^6 + x_1^5 x_2 + 3x_1^2 x_2^2 + x_1 x_2^3 + x_1 + x_2^2.$$

Since $\mathrm{in}_<(I) = \langle \mathrm{in}_<(f) \rangle$ holds for any monomial order $<$, $\{f\}$ is a universal Gröbner basis of I. Which monomial ideals may be the initial ideal of I? In this case, although six monomials appear in f, the number of initial ideals of I is not six. In fact, by Lemma 1.1.15, for any monomial order $<$, we have $x_1 x_2^3 > x_1$, $x_1 x_2^3 > x_2^2$, and hence neither x_1 nor x_2^2 is equal to $\mathrm{in}_<(f)$. In addition, any monomial order $<$ satisfies exactly one of $x_1 > x_2$ and $x_2 > x_1$. If $x_1 > x_2$, then $x_1^5 x_2 > x_1^4 x_2^2 > x_1^2 x_2^2$. If $x_2 > x_1$, then $x_1 x_2^3 > x_1^2 x_2^2$. Thus, $x_1^2 x_2^2$ cannot be equal to $\mathrm{in}_<(f)$. On the other hand, each of other three monomials is equal to $\mathrm{in}_<(f)$ for some $<$. In fact,

$$\begin{aligned}
\mathbf{w}_1 = [1, 0] &\Longrightarrow \mathrm{in}_{\mathbf{w}_1}(f) = x_1^6 \\
\mathbf{w}_2 = [0, 1] &\Longrightarrow \mathrm{in}_{\mathbf{w}_2}(f) = x_1 x_2^3 \\
\mathbf{w}_3 = [2, 3] &\Longrightarrow \mathrm{in}_{\mathbf{w}_3}(f) = x_1^5 x_2
\end{aligned}$$

hold. Is it possible to more easily reach these conclusions? In order to do so, we observe the set of exponent vectors of monomials appearing in f:

$$\{[6, 0], [5, 1], [2, 2], [1, 3], [1, 0], [0, 2]\} \subset \mathbb{Z}^2.$$

The convex hull of this set is the pentagon shown in Fig. 5.3. This polytope is called the Newton polytope of f, and denoted by New(f). Then, the normal fan

Fig. 5.3 Newton polytope of f

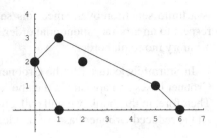

Fig. 5.4 Normal fan

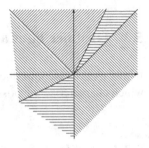

of New(f) is shown in Fig. 5.4. The intersection of this normal fan with the first quadrant consists of three regions:

$$\{[x, y] \in \mathbb{Q}^2 \; : \; 0 \le y \le x\},$$
$$\{[x, y] \in \mathbb{Q}^2 \; : \; 0 \le x \le y \le 2x\},$$
$$\{[x, y] \in \mathbb{Q}^2 \; : \; 0 \le 2x \le y\}.$$

This is called the Gröbner fan of the ideal I.

For a general principal ideal $I = \langle f \rangle$, it is defined as follows. For any polynomial

$$(0 \ne) \; f = \sum_{i=1}^{m} c_i \mathbf{x}^{\mathbf{a}_i} \in K[\mathbf{x}] \qquad (0 \ne c_i \in K),$$

the *Newton polytope* of f is defined by

$$\text{New}(f) = \text{CONV}(\{\mathbf{a}_1, \ldots, \mathbf{a}_m\}).$$

Then, the intersection of the normal fan $\mathcal{N}(\text{New}(f))$ of the Newton polytope New(f) and $\mathbb{Q}_{\ge 0}^n$ is called the *Gröbner fan* of I, and denoted by GF(I). By the same argument as in the above example, we have the following fact.

Proposition 5.3.1. *Let $I = \langle f \rangle \subset K[\mathbf{x}]$ be a principal ideal. Then, for positive vectors $\mathbf{w}, \mathbf{w}' \in \mathbb{Q}^n$, $\text{in}_{\mathbf{w}}(I) = \text{in}_{\mathbf{w}'}(I)$ if and only if $\mathbf{w}$ and $\mathbf{w}'$ belong to the interior of the same polyhedral cone in GF(I).*

We now present two useful propositions about Newton polytopes.

Proposition 5.3.2. *For nonzero polynomials* $f, g \in K[\mathbf{x}]$, *we have* $\mathrm{New}(f \cdot g) = \mathrm{New}(f) + \mathrm{New}(g)$.

Proof. Let $f = \sum_{i=1}^{m} c_i \mathbf{x}^{\mathbf{a}_i}$ $(0 \neq c_i \in K)$ and $g = \sum_{j=1}^{\ell} d_j \mathbf{x}^{\mathbf{b}_j}$ $(0 \neq d_j \in K)$. Then, since

$$f \cdot g = \sum_{i=1}^{m} \sum_{j=1}^{\ell} c_i d_j \mathbf{x}^{\mathbf{a}_i + \mathbf{b}_j} \in K[\mathbf{x}],$$

we have

$$\mathrm{New}(f \cdot g) = \mathrm{CONV}(\{\mathbf{a}_i + \mathbf{b}_j \; : \; i = 1, 2, \ldots, m, j = 1, 2, \ldots, \ell\}).$$

Thus, by Proposition 5.1.3, the assertion holds. $\qquad\square$

Proposition 5.3.3. *For a nonzero polynomial* $f \in K[\mathbf{x}]$ *and a nonnegative vector* $\mathbf{w} \in \mathbb{Q}^n$, *we have* $\mathrm{FACE}_{\mathbf{w}}(\mathrm{New}(f)) = \mathrm{New}(\mathrm{in}_{\mathbf{w}}(f))$.

Proof. For a polynomial $f = \sum_{i=1}^{m} c_i \mathbf{x}^{\mathbf{a}_i}$ $(0 \neq c_i \in K)$, let $X = \{\mathbf{a}_1, \ldots, \mathbf{a}_m\}$, $\lambda = \max(\mathbf{w} \cdot \mathbf{a}_i \; : \; 1 \leq i \leq m)$, and $X_{\mathbf{w}} = \{\mathbf{a}_i \in X \; : \; \mathbf{w} \cdot \mathbf{a}_i = \lambda\}$. Then, by the definition of the initial form, we have $\mathrm{in}_{\mathbf{w}}(f) = \sum_{\mathbf{a}_i \in X_{\mathbf{w}}} c_i \mathbf{x}^{\mathbf{a}_i}$. Thus, by Proposition 5.1.9, we have

$$\mathrm{FACE}_{\mathbf{w}}(\mathrm{New}(f)) = \mathrm{FACE}_{\mathbf{w}}(\mathrm{CONV}(X)) = \mathrm{CONV}(X_{\mathbf{w}}) = \mathrm{New}(\mathrm{in}_{\mathbf{w}}(f)),$$

as desired. $\qquad\square$

5.3.2 Gröbner Fans and State Polytopes of Homogeneous Ideals

Recall that an ideal is called a *homogeneous ideal* if it is generated by homogeneous polynomials (Sect. 1.6.3). Originally, as introduced in [9, Chap. 2], the Gröbner fan was defined for an arbitrary ideal. However, since arguments may become complicated if we do not assume that the ideals are homogeneous, in this section we will study only homogeneous ideals (with respect to an ordinary grading).

When we defined the initial form $\mathrm{in}_{\mathbf{w}}(f)$ of a polynomial f, we assumed that the weight vector $\mathbf{w}$ was *nonnegative*. That is because, if $\mathbf{w}$ has a negative component, then the order $<_{\mathbf{w}}$ defined by a monomial order $<$ does not satisfy one of the conditions of the definition of monomial orders, i.e., "1 is the smallest monomial." However, we can define "the sum of the terms whose inner product is maximal" even if the vector has a negative component. If we assume an ideal is homogeneous, then we have the following useful property.

Proposition 5.3.4. *Let $I \subset K[\mathbf{x}]$ be a homogeneous ideal. Then, for any vector $\mathbf{w} \in \mathbb{Q}^n$, there exists a nonnegative vector $\mathbf{w}' \in \mathbb{Q}_{\geq 0}^n$ such that $\mathrm{in}_\mathbf{w}(I) = \mathrm{in}_{\mathbf{w}'}(I)$.*

Proof. For any vector $\mathbf{w} \in \mathbb{Q}^n$, $\mathbf{w}' = \mathbf{w} + (\lambda, \ldots, \lambda)$ is nonnegative for a sufficiently large $\lambda > 0$. We will show that $\mathrm{in}_\mathbf{w}(I) = \mathrm{in}_{\mathbf{w}'}(I)$.

By Lemma 1.6.10, any polynomial $0 \neq f \in I$ is represented by homogeneous polynomials $g_i \in I$ as $f = g_1 + \cdots + g_r$. Then, there exist $\{k_1, \ldots, k_t\} \subset \{1, \ldots, r\}$ such that $\mathrm{in}_\mathbf{w}(f) = \mathrm{in}_\mathbf{w}(g_{k_1}) + \cdots + \mathrm{in}_\mathbf{w}(g_{k_t})$. Since each $g_{k_j} \in I$ is homogeneous, it follows that $\mathrm{in}_\mathbf{w}(g_{k_j}) = \mathrm{in}_{\mathbf{w}'}(g_{k_j}) \in \mathrm{in}_{\mathbf{w}'}(I)$. Hence, we have $\mathrm{in}_\mathbf{w}(f) \in \mathrm{in}_{\mathbf{w}'}(I)$. Thus, $\mathrm{in}_\mathbf{w}(I) \subset \mathrm{in}_{\mathbf{w}'}(I)$ holds. The proof of the converse inclusion, $\mathrm{in}_\mathbf{w}(I) \supset \mathrm{in}_{\mathbf{w}'}(I)$, is similar. $\qquad\square$

For a homogeneous ideal $I \subset K[\mathbf{x}]$ and a vector $\mathbf{w} \in \mathbb{Q}^n$, we define

$$C[\mathbf{w}] := \{\mathbf{w}' \in \mathbb{Q}^n \; : \; \mathrm{in}_{\mathbf{w}'}(I) = \mathrm{in}_\mathbf{w}(I)\}$$

and call it a *Gröbner cone*. We now show that this is, in fact, a cone.

Proposition 5.3.5. *For a nonnegative vector $\mathbf{w} \in \mathbb{Q}_{\geq 0}^n$ and a monomial order $<$, let $\mathscr{G}$ be the reduced Gröbner basis of I with respect to a monomial order $<_\mathbf{w}$. Then, we have*

$$C[\mathbf{w}] = \{\mathbf{w}' \in \mathbb{Q}^n \; : \; \mathrm{in}_{\mathbf{w}'}(g) = \mathrm{in}_\mathbf{w}(g) \text{ for all } g \in \mathscr{G}\}.$$

Proof. ($\supset$) Suppose that the vector $\mathbf{w}'$ belongs to the set of the right-hand side. Then, by Corollary 5.2.4, we have

$$\mathrm{in}_\mathbf{w}(I) = \langle \mathrm{in}_\mathbf{w}(g) \; : \; g \in \mathscr{G} \rangle = \langle \mathrm{in}_{\mathbf{w}'}(g) \; : \; g \in \mathscr{G} \rangle \subset \mathrm{in}_{\mathbf{w}'}(I).$$

By Proposition 5.2.3, their initial ideals satisfy

$$\mathrm{in}_{<_\mathbf{w}}(I) = \mathrm{in}_<(\mathrm{in}_\mathbf{w}(I)) \subset \mathrm{in}_<(\mathrm{in}_{\mathbf{w}'}(I)) = \mathrm{in}_{<_{\mathbf{w}'}}(I).$$

Then, by Proposition 5.2.2, $\mathrm{in}_{<_\mathbf{w}}(I) = \mathrm{in}_{<_{\mathbf{w}'}}(I)$. By Proposition 5.2.1, it follows that $\mathrm{in}_\mathbf{w}(I) = \mathrm{in}_{\mathbf{w}'}(I)$. Thus, we have $\mathbf{w}' \in C[\mathbf{w}]$.

($\subset$) Suppose that $\mathbf{w}'$ belongs to $C[\mathbf{w}]$. By Corollary 5.2.4, $\{\mathrm{in}_\mathbf{w}(g) \; : \; g \in \mathscr{G}\}$ is a Gröbner basis of $\mathrm{in}_\mathbf{w}(I)$ ($= \mathrm{in}_{\mathbf{w}'}(I)$) with respect to $<$. Assume that there exists $g \in \mathscr{G}$ such that $\mathrm{in}_\mathbf{w}(g) \neq \mathrm{in}_{\mathbf{w}'}(g)$. Since $\mathrm{in}_{\mathbf{w}'}(g) \in \mathrm{in}_{\mathbf{w}'}(I)$ holds, there exists $g' \in \mathscr{G}$ such that $\mathrm{in}_<(\mathrm{in}_\mathbf{w}(g')) = \mathrm{in}_{<_\mathbf{w}}(g')$ divides $\mathrm{in}_<(\mathrm{in}_{\mathbf{w}'}(g)) = \mathrm{in}_{<_{\mathbf{w}'}}(g)$. If $g \neq g'$, then it contradicts the hypothesis that $\mathscr{G}$ is the reduced Gröbner basis of I with respect to $<_\mathbf{w}$. Hence, we have $g = g'$. Thus, $\mathrm{in}_{<_\mathbf{w}}(g)$ divides $\mathrm{in}_{<_{\mathbf{w}'}}(g)$. By a property of monomial orders, it follows that $\mathrm{in}_{<_{\mathbf{w}'}}(g) = \mathrm{in}_{<_\mathbf{w}}(g)$.

By the hypothesis, $h = \mathrm{in}_\mathbf{w}(g) - \mathrm{in}_{\mathbf{w}'}(g)$ is a nonzero polynomial belonging to $\mathrm{in}_\mathbf{w}(I)$ ($= \mathrm{in}_{\mathbf{w}'}(I)$). However, since g belongs to the reduced Gröbner basis, any monomial other than the initial monomial $\mathrm{in}_{<_\mathbf{w}}(g)$ does not belong to $\mathrm{in}_{<_\mathbf{w}}(I)$. Thus, any monomial appearing in h does not belong to $\mathrm{in}_{<_\mathbf{w}}(I)$. This contradicts $\mathrm{in}_<(h) \in \mathrm{in}_<(\mathrm{in}_\mathbf{w}(I)) = \mathrm{in}_{<_\mathbf{w}}(I)$. $\qquad\square$

We can express Proposition 5.3.5 in terms of polytopes, as follows.

Corollary 5.3.6. *Work with the same assumption as in Proposition 5.3.5. Let $Q \subset \mathbb{Q}^n$ be a convex polytope defined by*

$$Q = \sum_{g \in \mathscr{G}} \mathrm{New}(g) = \mathrm{New}\left(\prod_{g \in \mathscr{G}} g\right).$$

Then, we have $C[\mathbf{w}] = \mathscr{N}_Q(\mathrm{FACE}_\mathbf{w}(Q))$.

Proof. Applying in order Propositions 5.3.5, 5.3.3, and 5.1.14, we have

$$\mathbf{w}' \in C[\mathbf{w}]$$

$$\Longleftrightarrow \mathrm{in}_\mathbf{w}(g) = \mathrm{in}_{\mathbf{w}'}(g) \text{ for all } g \in \mathscr{G}$$

$$\Longleftrightarrow \mathrm{New}(\mathrm{in}_\mathbf{w}(g)) = \mathrm{New}(\mathrm{in}_{\mathbf{w}'}(g)) \text{ for all } g \in \mathscr{G}$$

$$\Longleftrightarrow \mathrm{FACE}_\mathbf{w}(\mathrm{New}(g)) = \mathrm{FACE}_{\mathbf{w}'}(\mathrm{New}(g)) \text{ for all } g \in \mathscr{G}$$

$$\Longleftrightarrow \mathrm{FACE}_\mathbf{w}\left(\sum_{g \in \mathscr{G}} \mathrm{New}(g)\right) = \mathrm{FACE}_{\mathbf{w}'}\left(\sum_{g \in \mathscr{G}} \mathrm{New}(g)\right)$$

$$\Longleftrightarrow \mathbf{w}' \in \mathscr{N}_Q(\mathrm{FACE}_\mathbf{w}(Q)).$$

$\square$

By Corollary 5.3.6, since $C[\mathbf{w}]$ is the normal cone of a convex polytope, its closure $\overline{C[\mathbf{w}]}$ is a convex polyhedral cone. For a homogeneous ideal I, a set

$$\mathrm{GF}(I) := \left\{\overline{C[\mathbf{w}]} : \mathbf{w} \in \mathbb{Q}^d\right\}$$

of convex polyhedral cones is called the *Gröbner fan* of I. We now show that this is a fan.

Proposition 5.3.7. *For a homogeneous ideal I, $\mathrm{GF}(I)$ is a fan.*

Proof. First, we prove the following fact:

If $\mathbf{w}' \in \overline{C[\mathbf{w}]}$, then $\overline{C[\mathbf{w}']}$ is a face of $\overline{C[\mathbf{w}]}$.

Suppose that $\mathbf{w}'$ belongs to $\overline{C[\mathbf{w}]}$. Since $C[\mathbf{w}]$ is a cone, for a sufficiently small $\varepsilon > 0$,

$$\varepsilon \mathbf{w} + (1 - \varepsilon)\mathbf{w}' = \mathbf{w}' + \varepsilon(\mathbf{w} - \mathbf{w}')$$

belongs to $C[\mathbf{w}]$. Let $\mathbf{v} = \mathbf{w} - \mathbf{w}'$. Then, by Proposition 5.2.5, we have

$$\mathrm{in}_\mathbf{w}(I) = \mathrm{in}_{\mathbf{w}'+\varepsilon \mathbf{v}}(I) = \mathrm{in}_\mathbf{v}(\mathrm{in}_{\mathbf{w}'}(I)).$$

Let $<'$ denote the monomial order $<_v$. Then, the initial ideal of $in_w(I)$ with respect to the monomial order $<'$ satisfies

$$in_{<'_w}(I) = in_{<'_{w'+\varepsilon v}}(I) = in_{<'_{w'}}(I).$$

Let $\mathscr{G}$ be the reduced Gröbner basis of I with respect to $<'_w$. Then, since the initial ideals coincide, $\mathscr{G}$ is the reduced Gröbner basis of I with respect to both $<'_{w'+\varepsilon v}$ and $<'_{w'}$. (Note that the reduced Gröbner basis is unique for each monomial order, and a finite set of binomials of an ideal is the reduced Gröbner basis if and only if the initial monomials of elements form a minimal set of generators of the initial ideal and the non-initial monomial of each element does not belong to the initial ideal. Thus, the reduced Gröbner basis is completely determined by the initial ideal.) Let $Q \subset \mathbb{Q}^n$ be a convex polytope defined by $Q = \sum_{g \in \mathscr{G}} New(g)$. Then, by Corollary 5.3.6, we have

$$C[\mathbf{w}] = C[\mathbf{w}' + \varepsilon \mathbf{v}] = \mathscr{N}_Q(\mathrm{FACE}_{\mathbf{w}'+\varepsilon \mathbf{v}}(Q))$$
$$C[\mathbf{w}'] = \mathscr{N}_Q(\mathrm{FACE}_{\mathbf{w}'}(Q)).$$

Moreover, by Proposition 5.1.11, since $\mathrm{FACE}_{\mathbf{w}'+\varepsilon \mathbf{v}}(Q) = \mathrm{FACE}_v(\mathrm{FACE}_{\mathbf{w}'}(Q))$, it follows that $\mathrm{FACE}_{\mathbf{w}'+\varepsilon \mathbf{v}}(Q)$ is a face of $\mathrm{FACE}_{\mathbf{w}'}(Q)$. Hence, $\overline{C[\mathbf{w}']}$ is a face of $\overline{C[\mathbf{w}]}$. By using this fact, we will show that $GF(I)$ satisfies conditions (i) and (ii) in the definition of complices (fans).

Condition (i): For a face $F \neq \emptyset$ of the closure $\overline{C[\mathbf{w}]}$, let $\mathbf{w}' \in F \setminus \partial F$. Since $\mathbf{w}' \in C[\mathbf{w}]$, by the argument above, it follows that $C[\mathbf{w}']$ is a face of $C[\mathbf{w}]$. By Corollary 5.3.6, $\overline{C[\mathbf{w}]}$ belongs to the normal fan. Hence, the intersection of its face F and $\overline{C[\mathbf{w}']}$ is a face of F and a face of $\overline{C[\mathbf{w}']}$. However, since $\mathbf{w}'$ belongs to the interior of F and the interior of $\overline{C[\mathbf{w}']}$, by (the polyhedral convex cone version of) Proposition 5.1.16, $F = F \cap \overline{C[\mathbf{w}']} = \overline{C[\mathbf{w}']} \in GF(I)$ for a face $F \cap \overline{C[\mathbf{w}']}$ containing $\mathbf{w}'$.

Condition (ii): Suppose that the intersection $F = \overline{C[\mathbf{w}]} \cap \overline{C[\mathbf{w}']}$ of the closure $\overline{C[\mathbf{w}]}$ and $\overline{C[\mathbf{w}']}$ is not empty. Since F is the intersection of polyhedral cones, it is a polyhedral cone. For each $\mathbf{w}'' \in F$, $\overline{C[\mathbf{w}'']}$ is a face of $\overline{C[\mathbf{w}]}$ and a face of $\overline{C[\mathbf{w}']}$. Thus, in particular, we have $\overline{C[\mathbf{w}'']} \subset F$. It then follows that there exist $\mathbf{w}_1, \ldots, \mathbf{w}_s \in F$ such that $F = \overline{C[\mathbf{w}_1]} \cup \cdots \cup \overline{C[\mathbf{w}_s]}$. (Note that there exist only finitely many faces in a polyhedral cone $\overline{C[\mathbf{w}]}$.) We assume that s is minimal among such expressions. If $\mathbf{w}_i \in \overline{C[\mathbf{w}_j]}$ holds for some $1 \leq i \neq j \leq s$, then we have $\overline{C[\mathbf{w}_i]} \subset \overline{C[\mathbf{w}_j]}$, and hence $\overline{C[\mathbf{w}_i]}$ is redundant in the above expression for F. Thus, for any $1 \leq i \neq j \leq s$, we have $\mathbf{w}_i \notin \overline{C[\mathbf{w}_j]}$. Assume that $s \geq 2$. Since $\sum_{i=1}^{s} \frac{1}{s} \mathbf{w}_i$ does not belong to any $\overline{C[\mathbf{w}_i]}$, it does not belong to F. This contradicts that F is a polyhedral cone. Therefore, $s = 1$. Thus, $F = \overline{C[\mathbf{w}_1]}$ is a face of $\overline{C[\mathbf{w}]}$ and a face of $\overline{C[\mathbf{w}']}$. $\square$

We now present a concrete way to construct the state polytope $State(I)$ which was introduced in [9, Chap. 2]. Let $K[\mathbf{x}]_r$ denote the set of all homogeneous polynomials of degree r (and the zero polynomial). If $\mathscr{M} = \{\mathbf{x}^{\mathbf{a}_1}, \ldots, \mathbf{x}^{\mathbf{a}_m}\}$ is the set of all monomials in $K[\mathbf{x}]$ of degree r, then $K[\mathbf{x}]_r$ is an m-dimensional vector space over K with a basis $\mathcal{M}$. For a homogeneous ideal $I \subset K[\mathbf{x}]$, let $I_r = I \cap K[\mathbf{x}]_r$. Then, I_r is a subspace of $K[\mathbf{x}]_r$. For a homogeneous ideal I, an integer $r > 0$, and a monomial order $<$, we define

$$\mathbf{s}_{(I,r,<)} := \sum_{\mathbf{x}^{\mathbf{a}} \in \text{in}_<(I)_r} \mathbf{a} \in \mathbb{Z}^n,$$

$$\text{State}_r(I) := \text{CONV}(\{\mathbf{s}_{(I,r,<)} \; : \; < \text{ is a monomial order }\}).$$

Moreover, let D be the maximal degree of elements of a (the) universal Gröbner basis of a homogeneous ideal I, and let $\text{State}(I)$ be the Minkowski sum

$$\text{State}(I) := \sum_{r=1}^{D} \text{State}_r(I).$$

If a homogeneous ideal I is a principal ideal, then the state polytope equals the Newton polytope, which was introduced in the previous section.

Proposition 5.3.8. *Let I be a principal ideal generated by a homogeneous polynomial $f \in K[\mathbf{x}]$ of degree D. Then, we have $\text{State}(I) = \text{New}(f)$.*

Proof. Since I is a principal ideal, $\{f\}$ is a universal Gröbner basis. For a monomial order $<$, we have $\text{in}_<(I) = \langle \text{in}_<(f) \rangle$. Hence, for each $1 \leq r < D$, it follows that $\text{in}_<(I)_r = \{0\}$. Let $\text{in}_<(f) = \mathbf{x}^{\mathbf{a}}$. Then, we have $\mathbf{s}_{(I,D,<)} = \{\mathbf{a}\}$. Thus,

$$\begin{aligned}
\text{State}(I) &= \text{State}_D(I) \\
&= \text{CONV}(\{\mathbf{a} \; : \; < \text{ is a monomial order}, \; \text{in}_<(f) = \mathbf{x}^{\mathbf{a}}\}) \\
&= \text{CONV}(\{\mathbf{a} \; : \; \mathbf{w} \in \mathbb{Q}^n, \; \text{in}_{\mathbf{w}}(f) = \mathbf{x}^{\mathbf{a}}\}) \\
&= \text{CONV}(\{\mathbf{a} \; : \; \mathbf{w} \in \mathbb{Q}^n, \; \text{FACE}_{\mathbf{w}}(\text{New}(f)) = \{\mathbf{a}\}\}) \\
&= \text{New}(f)
\end{aligned}$$

holds. $\square$

In order to study general homogeneous ideals, we now present an important lemma.

Lemma 5.3.9. *For a homogeneous ideal $I \subset K[\mathbf{x}]$, an integer $r > 0$, and a vector $\mathbf{w} \in \mathbb{Q}^n$, we have $\text{FACE}_{\mathbf{w}}(\text{State}_r(I)) = \text{State}_r(\text{in}_{\mathbf{w}}(I))$.*

Proof. Suppose that a vector $\mathbf{w} \in \mathbb{Q}^n$ is sufficiently generic. This means that $\text{in}_{\mathbf{w}}(I)$ is a monomial ideal and that $\text{FACE}_{\mathbf{w}}(\text{State}_r(I))$ is a vertex.

Let $\{\mathbf{x}^{\mathbf{a}_1}, \ldots, \mathbf{x}^{\mathbf{a}_m}\}$ be the set of all monomials in $K[\mathbf{x}]$ of degree r. By changing indices if necessary, we may assume that $\{\mathbf{x}^{\mathbf{a}_1}, \ldots, \mathbf{x}^{\mathbf{a}_m}\} \cap \text{in}_{\mathbf{w}}(I)_r = \{\mathbf{x}^{\mathbf{a}_1}, \ldots, \mathbf{x}^{\mathbf{a}_\ell}\}$. Then, we have

$$\text{State}_r(\text{in}_{\mathbf{w}}(I)) = \{\mathbf{a}_1 + \cdots + \mathbf{a}_\ell\} = \{\mathbf{s}_{(I,r,\mathbf{w})}\}.$$

Let $\mathscr{G}$ be the reduced Gröbner basis of I with respect to a monomial order $\mathbf{w}$. By considering the standard form of each monomial in $\{\mathbf{x}^{\mathbf{a}_1}, \ldots, \mathbf{x}^{\mathbf{a}_\ell}\}$ with respect to $\mathscr{G}$, it follows that, for each $i = 1, 2, \ldots, \ell$, there exists $c_{ij} \in K$ such that

$$g_i = \mathbf{x}^{\mathbf{a}_i} - \sum_{j=\ell+1}^m c_{ij}\, \mathbf{x}^{\mathbf{a}_j} \in I_r,$$

where $\mathrm{in}_\mathbf{w}(g_i) = \mathbf{x}^{\mathbf{a}_i}$ and $\mathbf{w} \cdot \mathbf{a}_i > \mathbf{w} \cdot \mathbf{a}_j$ for all $1 \leq i \leq \ell < j \leq m$ satisfying $c_{ij} \neq 0$. Moreover, it then follows that any element of I_r is a linear combination of $g_1, \ldots, g_\ell$. Thus, $g_1, \ldots, g_\ell$ is a basis of I_r (as a K-vector space). In particular, we have $\ell = \dim_K(I_r)\ (\leq m)$.

The vertex $\mathrm{FACE}_\mathbf{w}(\mathrm{State}_r(I))$ coincides with $\mathbf{s}_{(I,r,<')}$ with respect to a monomial order $<'$. Let $\mathbf{x}^{\mathbf{a}_{j_1}}, \ldots, \mathbf{x}^{\mathbf{a}_{j_\ell}}$ denote the set of all monomials in $\mathrm{in}_{<'}(I)_r$. (We note that it has ℓ monomials since $\ell = \dim_K(I_r)$.) By an argument similar to the one used above, for each $k = 1, 2, \ldots, \ell$, there exist $b_{kj} \in K$ such that

$$h_k = \mathbf{x}^{\mathbf{a}_{j_k}} - \sum_{j \notin \{j_1, \ldots, j_\ell\}} b_{kj}\, \mathbf{x}^{\mathbf{a}_j} \in I_r,$$

where $h_1, \ldots, h_\ell$ form a basis of I_r (as a K-vector space). Let $C = [c_{ij}]$ and $B = [b_{ij}]$. Then, by considering a change of basis from $g_1, \ldots, g_\ell$ to $h_1, \ldots, h_\ell$, the matrix $M = \left[\, E_\ell \,\middle|\, {-}C \,\right]$ is transformed into a matrix for which the $j_1, \ldots, j_\ell$-th columns form the identity matrix, and the other columns form $-B$ by row elementary transformations. Thus, the $j_1, \ldots, j_\ell$-th columns of the matrix $M = [m_{ij}]$ form a nonsingular matrix, and hence there exists a permutation σ on $\{1, \ldots, \ell\}$ such that $\prod_{k=1}^\ell m_{\sigma(k)j_k} \neq 0$. If $m_{\sigma(k)j_k} \neq 0$, then each $\mathbf{x}^{\mathbf{a}_{j_k}}\ (1 \leq k \leq \ell)$ appears in $g_{\sigma(k)}$, and we have $\mathbf{w}\cdot\mathbf{a}_{\sigma(k)} \geq \mathbf{w}\cdot\mathbf{a}_{j_k}$. Thus, it follows that $\mathbf{w}\cdot(\mathbf{a}_1+\cdots+\mathbf{a}_\ell) \geq \mathbf{w}\cdot(\mathbf{a}_{j_1}+\cdots+\mathbf{a}_{j_\ell})$. Since the face with respect to $\mathbf{w}$ is the vertex $\mathbf{a}_{j_1} + \cdots + \mathbf{a}_{j_\ell}$, the assertion of the lemma, $\mathbf{s}_{(I,r,\mathbf{w})} = \mathbf{a}_1 + \cdots + \mathbf{a}_\ell = \mathbf{a}_{j_1} + \cdots + \mathbf{a}_{j_\ell} = \mathbf{s}_{(I,r,<')}$, holds.

Second, we prove the assertion for a general vector $\mathbf{w} \in \mathbb{Q}^n$ which is not necessarily generic. We consider another generic vector $\mathbf{w}' \in \mathbb{Q}^n$. By using the argument above for generic vectors, for sufficiently small $\varepsilon > 0$, we have

$$\mathrm{FACE}_{\mathbf{w}'}(\mathrm{FACE}_\mathbf{w}(\mathrm{State}_r(I))) = \mathrm{FACE}_{\mathbf{w}+\varepsilon\mathbf{w}'}(\mathrm{State}_r(I))$$

$$= \mathrm{State}_r(\mathrm{in}_{\mathbf{w}+\varepsilon\mathbf{w}'}(I))$$

$$= \mathrm{State}_r(\mathrm{in}_{\mathbf{w}'}(\mathrm{in}_\mathbf{w}(I)))$$

$$= \mathrm{FACE}_{\mathbf{w}'}(\mathrm{State}_r(\mathrm{in}_\mathbf{w}(I))).$$

Hence, $\mathrm{FACE}_\mathbf{w}(\mathrm{State}_r(I))$ and $\mathrm{State}_r(\mathrm{in}_\mathbf{w}(I))$ have the same vertices. Thus, they are the same convex polytopes. $\qquad\square$

Corollary 5.3.10. *Let I be a homogeneous ideal I, and let $<$ and $<'$ be monomial orders. If $\mathrm{in}_<(I)_r \neq \mathrm{in}_{<'}(I)_r$ holds, then we have $s_{(I,r,<)} \neq s_{(I,r,<')}$.*

Theorem 5.3.11. *For a homogeneous ideal $I \subset K[\mathbf{x}]$, the normal fan of the polytope $\mathrm{State}(I) \subset \mathbb{Q}^n$ equals the Gröbner fan. (A convex polytope is called the state polytope if its normal fan coincides with the Gröbner fan.)*

Proof. In order to prove that the Gröbner fan $\mathrm{GF}(I)$ equals the normal fan $\mathscr{N}(\mathrm{State}(I))$, it is enough to show that the maximal faces of the two fans coincide. This is equivalent to saying that, for any generic vectors $\mathbf{w}, \mathbf{w}' \in \mathbb{Q}^n$,

$$\mathrm{in}_{\mathbf{w}}(I) = \mathrm{in}_{\mathbf{w}'}(I) \iff \mathrm{FACE}_{\mathbf{w}}(\mathrm{State}(I)) = \mathrm{FACE}_{\mathbf{w}'}(\mathrm{State}(I))$$

holds.

Suppose that $\mathrm{in}_{\mathbf{w}}(I) = \mathrm{in}_{\mathbf{w}'}(I)$ holds. By Lemma 5.3.9, it then follows that

$$\mathrm{FACE}_{\mathbf{w}}(\mathrm{State}(I)) = \mathrm{FACE}_{\mathbf{w}}\left(\sum_{r=1}^{D} \mathrm{State}_r(I)\right)$$

$$= \sum_{r=1}^{D} \mathrm{FACE}_{\mathbf{w}}(\mathrm{State}_r(I))$$

$$= \sum_{r=1}^{D} \mathrm{State}_r(\mathrm{in}_{\mathbf{w}}(I)).$$

Hence, we have $\mathrm{FACE}_{\mathbf{w}}(\mathrm{State}(I)) = \mathrm{FACE}_{\mathbf{w}'}(\mathrm{State}(I))$.

On the other hand, assume that $\mathrm{FACE}_{\mathbf{w}}(\mathrm{State}(I)) = \mathrm{FACE}_{\mathbf{w}'}(\mathrm{State}(I))$ holds. Then, by the above formula, we have

$$\mathrm{FACE}_{\mathbf{w}}\left(\sum_{r=1}^{D} \mathrm{State}_r(I)\right) = \mathrm{FACE}_{\mathbf{w}'}\left(\sum_{r=1}^{D} \mathrm{State}_r(I)\right).$$

Since vectors $\mathbf{w}, \mathbf{w}' \in \mathbb{Q}^n$ are generic, both sides of this are vertices. Thus, by Proposition 5.1.13,

$$\mathrm{FACE}_{\mathbf{w}}(\mathrm{State}_r(I)) = \mathrm{FACE}_{\mathbf{w}'}(\mathrm{State}_r(I))$$

for all $r = 1, 2, \ldots, D$. By the argument in Proof of Lemma 5.3.9, this means that, for all $r = 1, 2, \ldots, D$, we have $s_{(I,r,\mathbf{w})} = s_{(I,r,\mathbf{w}')}$. Thus, by Corollary 5.3.10, $\mathrm{in}_{\mathbf{w}}(I)_r = \mathrm{in}_{\mathbf{w}'}(I)_r$ for all $r = 1, 2, \cdots, D$. Since the degree of each element of a minimal set of generators of initial ideals $\mathrm{in}_{\mathbf{w}}(I)$ and $\mathrm{in}_{\mathbf{w}'}(I)$ is less than or equal to D, we have $\mathrm{in}_{\mathbf{w}}(I) = \mathrm{in}_{\mathbf{w}'}(I)$. $\qquad\square$

5.4   State Polytopes of Toric Ideals

In the previous sections, we mainly considered general ideals. In this and later sections, we will study toric ideals. For toric ideals, there are interesting algorithms for computing such things as universal Gröbner bases or state polytopes.

5.4.1   Circuits and Graver Bases

Recall that **the** universal Gröbner basis is the union of all reduced Gröbner bases and a Gröbner basis with respect to any monomial order. For a configuration matrix A, let $\mathcal{U}_A$ denote the universal Gröbner basis of the toric ideal I_A. We now introduce two sets $\mathcal{C}_A$ and Gr_A which approximate $\mathcal{U}_A$.

For a nonzero polynomial $f \in K[\mathbf{x}]$, let $\mathrm{VAR}(f)$ denote the set of all variables appearing in f. An irreducible binomial f in the toric ideal I_A of a configuration matrix A is called a *circuit* if $\mathrm{VAR}(f)$ is minimal among the binomials in I_A. For a configuration matrix A, let $\mathcal{C}_A$ be the set of all circuits of I_A.

A binomial $\mathbf{x}^{\mathbf{u}} - \mathbf{x}^{\mathbf{v}}$ belonging to the toric ideal I_A of a configuration matrix A is said to be *primitive* if there exists no other binomial $\mathbf{x}^{\mathbf{u}'} - \mathbf{x}^{\mathbf{v}'}$ in I_A such that $\mathbf{x}^{\mathbf{u}'}$ divides $\mathbf{x}^{\mathbf{u}}$ and $\mathbf{x}^{\mathbf{v}'}$ divides $\mathbf{x}^{\mathbf{v}}$. For a configuration matrix A, we call Gr_A, the set of all primitive binomials of I_A, the *Graver basis* of I_A.

Proposition 5.4.1.   *For a configuration matrix A, we have $\mathcal{C}_A \subset \mathcal{U}_A \subset \mathrm{Gr}_A$.*

Proof. First, we show that $\mathcal{C}_A \subset \mathcal{U}_A$. Suppose that a binomial $f = \mathbf{x}^{\mathbf{u}} - \mathbf{x}^{\mathbf{v}}$ is a circuit. By changing indices if necessary, we may assume that $\mathrm{VAR}(\mathbf{x}^{\mathbf{u}}) = \{x_r, x_{r+1}, \ldots, x_s\}$ and $\mathrm{VAR}(\mathbf{x}^{\mathbf{v}}) = \{x_{s+1}, x_{s+2}, \ldots, x_n\}$, where $1 \le r \le s < n$. Let $\mathcal{G}$ be the reduced Gröbner basis of I_A with respect to a pure lexicographic order $<_{\mathrm{purelex}}$ induced by the ordering $x_1 > \cdots > x_n$. Since the universal Gröbner basis is the union of reduced Gröbner bases, it is enough to show that $f \in \mathcal{G}$. By the definition of a pure lexicographic order, $\mathrm{in}_{<_{\mathrm{purelex}}}(f) = \mathbf{x}^{\mathbf{u}} \in \mathrm{in}_{<_{\mathrm{purelex}}}(I_A)$. Hence, there exists $g = \mathbf{x}^{\mathbf{u}'} - \mathbf{x}^{\mathbf{v}'} \in \mathcal{G}$ such that $\mathrm{in}_{<_{\mathrm{purelex}}}(g) = \mathbf{x}^{\mathbf{u}'}$ divides $\mathbf{x}^{\mathbf{u}}$. Then, we have $\mathrm{VAR}(\mathbf{x}^{\mathbf{u}'}) \subset \{x_r, x_{r+1}, \ldots, x_s\}$. Since $\mathbf{x}^{\mathbf{u}'} >_{\mathrm{purelex}} \mathbf{x}^{\mathbf{v}'}$, it follows that $\mathrm{VAR}(\mathbf{x}^{\mathbf{v}'}) \subset \{x_r, x_{r+1}, \ldots, x_n\}$. Hence, $\mathrm{VAR}(g) \subset \mathrm{VAR}(f)$. Since f is a circuit, we have $\mathrm{VAR}(g) = \mathrm{VAR}(f)$. On the other hand, $\mathbf{w} = \mu'_n(\mathbf{u} - \mathbf{v}) - \mu_n(\mathbf{u}' - \mathbf{v}') = [0, \ldots, 0, \mu''_r, \ldots, \mu''_{n-1}, 0]$ belongs to $\mathrm{Ker}_{\mathbb{Z}}(A)$ since it is an integer combination of $\mathbf{u} - \mathbf{v} = [0, \ldots, 0, \mu_r, \ldots, \mu_n]$ and $\mathbf{u}' - \mathbf{v}' = [0, \ldots, 0, \mu'_r, \ldots, \mu'_n]$ belonging to $\mathrm{Ker}_{\mathbb{Z}}(A)$. If $\mathbf{w} \ne \mathbf{0}$, then we have $\mathrm{VAR}(h) \subsetneq \mathrm{VAR}(f)$, where $h \in I_A$ is the binomial corresponding to $\mathbf{w}$. This contradicts the hypothesis that f is a circuit. Hence, $\mathbf{w}$ is zero. Since the binomial f is irreducible, $\mathbf{u}' - \mathbf{v}'$ is an integral multiple of $\mathbf{u} - \mathbf{v}$. However, since $\mathbf{x}^{\mathbf{u}'}$ divides $\mathbf{x}^{\mathbf{u}}$, we have $f = g \in \mathcal{G}$.

Next, we will show that $\mathcal{U}_A \subset \mathrm{Gr}_A$. Suppose that a binomial $f = \mathbf{x}^{\mathbf{u}} - \mathbf{x}^{\mathbf{v}}$ belongs to the reduced Gröbner basis $\mathcal{G}$ of I_A with respect to a monomial order $<$, and that the initial monomial of f is $\mathbf{x}^{\mathbf{u}}$. We now assume that there exists a binomial

$g = \mathbf{x}^{\mathbf{u}'} - \mathbf{x}^{\mathbf{v}'} \in I_A$ with $g \neq f$ such that $\mathbf{x}^{\mathbf{u}'}$ divides $\mathbf{x}^{\mathbf{u}}$ and $\mathbf{x}^{\mathbf{v}'}$ divides $\mathbf{x}^{\mathbf{v}}$. If the initial term of g is $\mathbf{x}^{\mathbf{v}'}$, then it contradicts the hypothesis that $\mathscr{G}$ is reduced. Hence, the initial term of g is $\mathbf{x}^{\mathbf{u}'}$. Since the binomial f belongs to the reduced Gröbner basis, its initial monomial $\mathbf{x}^{\mathbf{u}}$ belongs to a minimal set of generators of $\mathrm{in}_<(I_A)$. Thus, we have $\mathbf{x}^{\mathbf{u}} = \mathbf{x}^{\mathbf{u}'}$. Then, $g - f = \mathbf{x}^{\mathbf{v}} - \mathbf{x}^{\mathbf{v}'}$ is a homogeneous binomial belonging to I_A. Since $\mathbf{x}^{\mathbf{v}'}$ divides $\mathbf{x}^{\mathbf{v}}$, we have $\mathbf{x}^{\mathbf{v}} = \mathbf{x}^{\mathbf{v}'}$. Thus, $f = g$, which is a contradiction. □

Since any Gröbner basis generates the ideal, we immediately have the following.

Corollary 5.4.2. *For a configuration matrix A, the toric ideal I_A is generated by the Graver basis Gr_A.*

5.4.2 Upper Bounds on the Degree

In this section, we introduce upper bounds on the degrees of circuits and elements of the Graver basis. Even though the bounds are only approximate, it follows that the Graver basis is a finite set. (This can be proved by using Lawrence liftings, which are presented in the next section.) In order to simplify the description of the upper bounds, in this section, we will assume that any configuration matrix $A \in \mathbb{Z}^{d \times n}$ satisfies $\mathrm{rank}(A) = d$. (Then, we have $d \leq n$.) This assumption is not unnatural. If $\mathrm{rank}(A) < d$, then, by deleting some rows of A, we can obtain a configuration matrix whose kernel is $\mathrm{Ker}_{\mathbb{Z}}(A)$ and which satisfies the hypothesis. We also assume that the columns of A are different from each other. First, since it is not useful to study a configuration matrix whose toric ideal is zero, we present the following lemma.

Lemma 5.4.3. *Suppose that a configuration matrix $A \in \mathbb{Z}^{d \times n}$ satisfies $\mathrm{rank}(A) = d$ (and hence $d \leq n$). Then, we have the following.*

(i) $2 \leq d < n \iff I_A \neq \{0\}$;

(ii) $2 \leq d = n - 1 \iff$ *There exists a binomial $f \neq 0$ such that $I_A = \langle f \rangle$.*

Proof. First, if $d = 1$, then we have $n = 1$ since A is a configuration matrix. Then, $I_A = \{0\}$, and hence (i) and (ii) hold when $d = 1$. We now assume that $d \geq 2$. Let $V = \{\mathbf{b} \in \mathbb{Q}^n : A\mathbf{b} = \mathbf{0}\}$. Then, since $\dim_{\mathbb{Q}} V = n - d$ and $\mathrm{Ker}_{\mathbb{Z}} A = V \cap \mathbb{Z}^n$, it follows that

$$d < n \iff \dim_{\mathbb{Q}} V > 0 \iff \mathrm{Ker}_{\mathbb{Z}} A \neq \{0\} \iff I_A \neq \{0\},$$

and that

$$d = n - 1 \iff \dim_{\mathbb{Q}} V = 1$$

$$\iff \text{There exists } \mathbf{0} \neq \mathbf{u} \in \mathbb{Z}^n \text{ such that } \mathrm{Ker}_{\mathbb{Z}} A = \{\alpha \mathbf{u} : \alpha \in \mathbb{Z}\}$$

$$\iff \text{There exist a binomial } f \neq 0 \text{ such that } I_A = \langle f \rangle.$$

Thus, we have (i) and (ii).   □

Next, we discuss an upper bound for the degree of circuits. For a configuration matrix $A = [\mathbf{a}_1, \ldots, \mathbf{a}_n] \in \mathbb{Z}^{d \times n}$, let

$$D(A) := \max(\, |\det[\mathbf{a}_{i_1}, \ldots, \mathbf{a}_{i_d}]| \, : \, 1 \le i_1 < \cdots < i_d \le n \,).$$

Then, we have the following.

Theorem 5.4.4. *For a configuration matrix $A \in \mathbb{Z}^{d \times n}$, let $f \in \mathscr{C}_A$. Then, VAR(f) consists of at most $d + 1$ elements, and we have $\deg(f) \le \frac{1}{2}(d + 1)D(A)$.*

Proof. By changing indices if necessary, we may assume that VAR(f) $= \{x_1, \ldots, x_r\}$.

First, in order to prove that VAR(f) consists of at most $d + 1$ elements, we assume that $r \ge d + 2$. Let $B = [\mathbf{a}_1, \ldots, \mathbf{a}_{r-1}] \in \mathbb{Z}^{d \times (r-1)}$. Since rank($B$) $\le d$, it follows that $(r - 1) - \text{rank}(B) \ge (d + 1) - d = 1$. Thus, $\text{Ker}_{\mathbb{Z}} B \ne \{\mathbf{0}\}$, and hence I_B possesses a binomial g. However, since g belongs to I_A and satisfies VAR(g) $\subset \{x_1, \ldots, x_{r-1}\} \subsetneq$ VAR(f), this contradicts that f is a circuit. Therefore, VAR(f) consists of at most $d + 1$ elements.

Next, let $B' = [\mathbf{a}_1, \ldots, \mathbf{a}_r] \in \mathbb{Z}^{d \times r}$. Since the rank of the configuration matrix A is d, adding column vectors of A to B', we may assume that the rank of $B'' = [\mathbf{a}_1, \ldots, \mathbf{a}_{d+1}] \in \mathbb{Z}^{d \times (d+1)}$ will be equal to d (changing indices if necessary). Then, by Cramer's rule, $\{\mathbf{u} \in \mathbb{Q}^n \, : \, B''\mathbf{u} = \mathbf{0}\}$ is a one-dimensional subspace spanned by the vector $\mathbf{w} \in \mathbb{Z}^n$ whose j-th component is $(-1)^j \det[\mathbf{a}_1, \ldots, \mathbf{a}_{j-1}, \mathbf{a}_{j+1}, \ldots, \mathbf{a}_{d+1}]$. Hence, if $f = \mathbf{x}^{\mathbf{u}} - \mathbf{x}^{\mathbf{v}}$, then there exists $m \in \mathbb{Q}$ such that $\mathbf{u} - \mathbf{v} = m\mathbf{w}$. Then, we have $\frac{1}{m}(\mathbf{u} - \mathbf{v}) = \mathbf{w} \in \mathbb{Z}^n$. Since f is irreducible, it follows that $\frac{1}{m} \in \mathbb{Z}$. Thus, the absolute value of the i-th component of $\mathbf{u} - \mathbf{v}$ is less than or equal to the absolute value of the i-th component of $\mathbf{w}$. On the other hand, since the absolute value of each component of $\mathbf{w}$ is less than or equal to $D(A)$, each component of $\mathbf{u}$ and $\mathbf{v}$ is less than or equal to $D(A)$. Moreover, since VAR(f) consists of at most $d + 1$ elements, the number of nonzero components of at least one of $\mathbf{u}$ and $\mathbf{v}$ is at most $\frac{1}{2}(d + 1)$. Since the degree of the binomial f is the sum of the components of $\mathbf{u}$ (and $\mathbf{v}$), we have $\deg(f) \le \frac{1}{2}(d + 1)D(A)$.   □

Example 5.4.5. For the configuration matrix

$$A = \begin{bmatrix} 0 & 1 & 0 & 1 \\ 0 & 0 & 1 & 1 \\ 1 & 1 & 1 & 1 \end{bmatrix},$$

the toric ideal I_A is a principal ideal generated by $f = x_1 x_4 - x_2 x_3$, and f is a (unique) circuit. Then, we have

$$\frac{1}{2}(d + 1)D(A) = \frac{1}{2} \times 4 \times 1 = 2 = \deg(f).$$

Lemma 5.4.6. *Let $A \in \mathbb{Z}^{d \times n}$ be a configuration matrix. Then, for any binomial $\mathbf{x}^{\mathbf{u}} - \mathbf{x}^{\mathbf{v}}$ of the toric ideal I_A, there exists a circuit $\mathbf{x}^{\mathbf{u}'} - \mathbf{x}^{\mathbf{v}'} \in \mathscr{C}_A$ such that*

$$\mathrm{VAR}(\mathbf{x}^{\mathbf{u}'}) \subset \mathrm{VAR}(\mathbf{x}^{\mathbf{u}}), \quad \mathrm{VAR}(\mathbf{x}^{\mathbf{v}'}) \subset \mathrm{VAR}(\mathbf{x}^{\mathbf{v}}).$$

Proof. If two monomials $\mathbf{x}^{\mathbf{u}}$ and $\mathbf{x}^{\mathbf{v}}$ of a binomial have a common factor, then the binomial obtained by removing it belongs to I_A. Hence, we may assume that $\mathrm{VAR}(\mathbf{x}^{\mathbf{u}}) \cap \mathrm{VAR}(\mathbf{x}^{\mathbf{v}}) = \emptyset$. Moreover, by Lemma 5.4.3 (i), we may assume that $2 \le d < n$. (In particular, we have $n \ge 3$.)

We will prove this by using induction on the number n of the columns of a configuration matrix A. First, if $n = 3$, then $d = 2$ by the hypothesis. Hence, $d = n - 1$ and by Lemma 5.4.3(ii), it is trivial that the assertion holds. Next, we assume that the assertion holds if the number of columns is at most $n - 1$, and we will show that the assertion holds for a configuration matrix $A \in \mathbb{Z}^{d \times n}$. Let $\mathrm{VAR}(\mathbf{x}^{\mathbf{u}} - \mathbf{x}^{\mathbf{v}}) = \{x_{i_1}, \ldots, x_{i_r}\}$. If $r < n$, then by deleting redundant rows from $B = [\mathbf{a}_{i_1}, \ldots, \mathbf{a}_{i_r}]$, it reduces to the case where the number of columns is at most $n - 1$. Thus, we may assume that $r = n$. Let $g = \mathbf{x}^{\mathbf{u}'} - \mathbf{x}^{\mathbf{v}'} \in \mathscr{C}_A$ be a circuit. (Since a circuit is irreducible, $\mathrm{VAR}(\mathbf{x}^{\mathbf{u}'}) \cap \mathrm{VAR}(\mathbf{x}^{\mathbf{v}'}) = \emptyset$ holds.) By changing indices of the variables if necessary, we may assume that $\mathbf{u} - \mathbf{v} = [\alpha_1, \ldots, \alpha_n] \in \mathbb{Z}^n$ and $\mathbf{u}' - \mathbf{v}' = [\beta_1, \ldots, \beta_n] \in \mathbb{Z}^n$ satisfy $\alpha_1 \beta_1 > 0$. We define a rational number $\delta > 0$ by

$$\delta = \min \left\{ \frac{\alpha_i}{\beta_i} : 1 \le i \le n, \ \alpha_i \beta_i > 0 \right\}.$$

For a suitable integer $m > 0$,

$$m\left((\mathbf{u} - \mathbf{v}) - \delta(\mathbf{u}' - \mathbf{v}')\right) = [\gamma_1, \ldots, \gamma_n]$$

belongs to $\mathrm{Ker}_{\mathbb{Z}} A$, and $\alpha_i \gamma_i \ge 0$ for all $1 \le i \le n$. If $[\gamma_1, \ldots, \gamma_n] = \mathbf{0}$, then g satisfies the assertion of this lemma. Hence, we may assume that $[\gamma_1, \ldots, \gamma_n] \ne \mathbf{0}$. Suppose that $\delta = \alpha_j / \beta_j$. Since $\gamma_j = 0$, the number of variables appearing in the binomial

$$h = \prod_{\gamma_i > 0} x_i^{\gamma_i} - \prod_{\gamma_j < 0} x_j^{-\gamma_j} \in I_A$$

is at most $n - 1$. By the hypothesis of induction, there exists $\mathbf{x}^{\mathbf{u}''} - \mathbf{x}^{\mathbf{v}''} \in \mathscr{C}_A$ such that

$$\mathrm{VAR}(\mathbf{x}^{\mathbf{u}''}) \subset \mathrm{VAR}\left(\prod_{\gamma_i > 0} x_i^{\gamma_i}\right), \quad \mathrm{VAR}(\mathbf{x}^{\mathbf{v}''}) \subset \mathrm{VAR}\left(\prod_{\gamma_j < 0} x_j^{-\gamma_j}\right).$$

Since $\alpha_i \gamma_i \ge 0$ holds for all $1 \le i \le n$, it follows that

$$\mathrm{VAR}(\mathbf{x}^{\mathbf{u}''}) \subset \mathrm{VAR}\left(\prod_{\gamma_i > 0} x_i^{\gamma_i}\right) \subset \mathrm{VAR}(\mathbf{x}^{\mathbf{u}})$$

$$\mathrm{VAR}(\mathbf{x}^{\mathbf{v}''}) \subset \mathrm{VAR}\left(\prod_{\gamma_j < 0} x_j^{-\gamma_j}\right) \subset \mathrm{VAR}(\mathbf{x}^{\mathbf{v}}).$$

□

Lemma 5.4.7. *Let $A \in \mathbb{Z}^{d \times n}$ be a configuration matrix. Then, for any irreducible binomial $f = \mathbf{x}^{\mathbf{u}} - \mathbf{x}^{\mathbf{v}}$ of the toric ideal I_A of A, there exist at most $n - d$ circuits*

$$\mathbf{x}^{\mathbf{u}_1} - \mathbf{x}^{\mathbf{v}_1}, \ldots, \mathbf{x}^{\mathbf{u}_N} - \mathbf{x}^{\mathbf{v}_N}$$

such that

$$\mathbf{x}^{m\mathbf{u}} = \prod_{i=1}^{N} \mathbf{x}^{m_i \mathbf{u}_i}, \quad \mathbf{x}^{m\mathbf{v}} = \prod_{i=1}^{N} \mathbf{x}^{m_i \mathbf{v}_i}$$

hold for some natural numbers $m, m_1, \ldots, m_N$.

Proof. By Lemma 5.4.6, there exists a circuit $\mathbf{x}^{\mathbf{u}_1} - \mathbf{x}^{\mathbf{v}_1} \in \mathscr{C}_A$ such that

$$\mathrm{VAR}(\mathbf{x}^{\mathbf{u}_1}) \subset \mathrm{VAR}(\mathbf{x}^{\mathbf{u}}), \quad \mathrm{VAR}(\mathbf{x}^{\mathbf{v}_1}) \subset \mathrm{VAR}(\mathbf{x}^{\mathbf{v}}).$$

Then, $\mathbf{u} - \mathbf{v} = [\alpha_1, \ldots, \alpha_n] \in \mathbb{Z}^n$ and $\mathbf{u}_1 - \mathbf{v}_1 = [\beta_1, \ldots, \beta_n] \in \mathbb{Z}^n$ satisfy $\alpha_i \beta_i \geq 0$ for all $1 \leq i \leq n$. We define $\delta > 0$ by

$$\delta = \min\left\{\frac{\alpha_i}{\beta_i} : 1 \leq i \leq n, \ \alpha_i \beta_i > 0\right\}.$$

For a suitable integer $m' > 0$,

$$m'\left((\mathbf{u} - \mathbf{v}) - \delta(\mathbf{u}_1 - \mathbf{v}_1)\right) = [\gamma_1, \ldots, \gamma_n]$$

belongs to $\mathrm{Ker}_{\mathbb{Z}} A$, and $\alpha_i \gamma_i \geq 0$ holds for all $1 \leq i \leq n$. If $[\gamma_1, \ldots, \gamma_n] = \mathbf{0}$, then the assertion of this lemma holds. Hence, we assume that $[\gamma_1, \ldots, \gamma_n] \neq \mathbf{0}$. If $\delta = \alpha_j / \beta_j$, then we have $\gamma_j = 0$. Thus, the number of variables appearing in the binomial

$$\prod_{\gamma_i > 0} x_i^{\gamma_i} - \prod_{\gamma_j < 0} x_j^{-\gamma_j} \in I_A$$

is less than the number of variables appearing in f. We replace f with this binomial and continue the same process. (The process terminates in a finite number of steps since the number of variables decreases.) For some natural numbers $m, m_1, \ldots, m_N$, we have

$$m(\mathbf{u} - \mathbf{v}) = m_1(\mathbf{u}_1 - \mathbf{v}_1) + \cdots + m_N(\mathbf{u}_N - \mathbf{v}_N).$$

Since the number of variables is reduced at each step, it follows that vectors $\mathbf{u}_1 - \mathbf{v}_1, \ldots, \mathbf{u}_N - \mathbf{v}_N$ belonging to the $n - d$-dimensional subspace $\{\mathbf{b} \in \mathbb{Q}^n \; : \; A\mathbf{b} = \mathbf{0}\}$ of $\mathbb{Q}^n$ are linearly independent over $\mathbb{Q}$. Thus, we have $N \leq n - d$, and it follows that at most $n - d$ circuits $\mathbf{x}^{\mathbf{u}_1} - \mathbf{x}^{\mathbf{v}_1}, \ldots, \mathbf{x}^{\mathbf{u}_N} - \mathbf{x}^{\mathbf{v}_N}$ satisfy

$$\mathbf{x}^{m\mathbf{u}} = \prod_{i=1}^{N} \mathbf{x}^{m_i \mathbf{u}_i}, \quad \mathbf{x}^{m\mathbf{v}} = \prod_{i=1}^{N} \mathbf{x}^{m_i \mathbf{v}_i}.$$

$\square$

Theorem 5.4.8. *Let $A \in \mathbb{Z}^{d \times n}$ be a configuration matrix, and let $f \in \mathrm{Gr}_A$. Then, we have*

$$\deg(f) \leq \frac{1}{2}(d + 1)(n - d)D(A).$$

Proof. By Theorem 5.4.4, we may assume that f is not a circuit. Applying Lemma 5.4.7 to a primitive binomial $f = \mathbf{x}^{\mathbf{u}} - \mathbf{x}^{\mathbf{v}}$, working with the same notation as in Lemma 5.4.7, we have

$$\mathbf{u} = \sum_{i=1}^{N} \frac{m_i}{m} \mathbf{u}_i, \quad \mathbf{v} = \sum_{i=1}^{N} \frac{m_i}{m} \mathbf{v}_i.$$

If $\frac{m_i}{m} \geq 1$ for some i, then $\mathbf{x}^{\mathbf{u}_i}$ divides $\mathbf{x}^{\mathbf{u}}$, and $\mathbf{x}^{\mathbf{v}_i}$ divides $\mathbf{x}^{\mathbf{v}}$. This contradicts that $f \in \mathrm{Gr}_A \setminus \mathscr{C}_A$. Hence, $\frac{m_i}{m} < 1$ for all i. Thus, by Theorem 5.4.4, we have

$$\deg(f) \leq \sum_{i=1}^{N} \frac{m_i}{m} \deg(\mathbf{x}^{\mathbf{u}_i} - \mathbf{x}^{\mathbf{v}_i})$$

$$< \sum_{i=1}^{N} \deg(\mathbf{x}^{\mathbf{u}_i} - \mathbf{x}^{\mathbf{v}_i})$$

$$\leq \sum_{i=1}^{N} \frac{1}{2}(d + 1)D(A)$$

$$= \frac{1}{2}N(d + 1)D(A)$$

$$\leq \frac{1}{2}(d + 1)(n - d)D(A),$$

as desired.

$\square$

Since f appearing in Example 5.4.5 is a circuit, f is, in particular, primitive. Note that for the degree inequality of Theorem 5.4.8, this f satisfies equality. By Theorem 5.4.8, we immediately have the following.

Corollary 5.4.9. *For a configuration matrix A, the Graver basis Gr_A is a finite set.*

Corollary 5.4.10. *For a configuration matrix A, the Graver basis Gr_A is a universal Gröbner basis.*

5.4.3   Lawrence Liftings

In this section, we introduce Lawrence liftings, which provide an algorithm for computing Graver bases. The *Lawrence lifting* of a configuration matrix $A \in \mathbb{Z}^{d \times n}$ is the configuration matrix

$$\Lambda(A) := \begin{bmatrix} A & O \\ E_n & E_n \end{bmatrix} \in \mathbb{Z}^{(d+n) \times 2n},$$

where E_n is the $n \times n$ identity matrix and O is the $d \times n$ zero matrix. Then, we have

$$\mathrm{Ker}_{\mathbb{Z}}(\Lambda(A)) = \left\{ \begin{bmatrix} \mathbf{b} \\ -\mathbf{b} \end{bmatrix} : \mathbf{b} \in \mathrm{Ker}_{\mathbb{Z}}(A) \right\}.$$

Hence, between two toric ideals

$$I_A \subset K[\mathbf{x}] = K[x_1, \dots, x_n]$$

$$I_{\Lambda(A)} \subset K[\mathbf{x}, \mathbf{y}] = K[x_1, \dots, x_n, y_1, \dots, y_n]$$

we have the following:

$$\mathbf{x}^{\mathbf{u}} \mathbf{y}^{\mathbf{v}} - \mathbf{x}^{\mathbf{v}} \mathbf{y}^{\mathbf{u}} \in I_{\Lambda(A)} \quad \Longleftrightarrow \quad \mathbf{x}^{\mathbf{u}} - \mathbf{x}^{\mathbf{v}} \in I_A$$

$$\mathbf{x}^{\mathbf{u}} \mathbf{y}^{\mathbf{v}} - \mathbf{x}^{\mathbf{v}} \mathbf{y}^{\mathbf{u}} \in \mathrm{Gr}_{\Lambda(A)} \quad \Longleftrightarrow \quad \mathbf{x}^{\mathbf{u}} - \mathbf{x}^{\mathbf{v}} \in \mathrm{Gr}_A$$

$$I_{\Lambda(A)} = \langle \mathbf{x}^{\mathbf{u}} \mathbf{y}^{\mathbf{v}} - \mathbf{x}^{\mathbf{v}} \mathbf{y}^{\mathbf{u}} : \mathbf{x}^{\mathbf{u}} - \mathbf{x}^{\mathbf{v}} \in I_A \rangle.$$

Theorem 5.4.11. *Let A be a configuration matrix. Then, for the toric ideal of the Lawrence lifting $\Lambda(A)$ of A, the following sets coincide (up to scalar multiples).*

(i) *The Graver basis $\mathrm{Gr}_{\Lambda(A)}$.*

(ii) *The universal Gröbner basis $\mathcal{U}_{A(A)}$.*

(iii) *The reduced Gröbner basis $\mathcal{G}_<$ of $I_{A(A)}$ with respect to a monomial order $<$.*

(iv) *A minimal set $\mathcal{F}$ of binomial generators of $I_{A(A)}$.*

Proof. First, we will show that, for any minimal set $\mathcal{F}$ of binomial generators of $I_{A(A)}$, $\mathrm{Gr}_{A(A)} = \mathcal{F}$. Suppose that a binomial $f = \mathbf{x}^\mathbf{u}\mathbf{y}^\mathbf{v} - \mathbf{x}^\mathbf{v}\mathbf{y}^\mathbf{u} \in I_{A(A)}$ belongs to the Graver basis $\mathrm{Gr}_{A(A)}$. Let $\mathcal{F}$ be a minimal set of binomial generators of $I_{A(A)}$. We can write

$$f = h_1 f_1 + \cdots + h_s f_s \quad (h_i \in K[\mathbf{x}, \mathbf{y}], \ f_i \in \mathcal{F}).$$

Then, (allowing for multiplication by a scalar, if necessary,) there exists $f_i = \mathbf{x}^{\mathbf{u}'}\mathbf{y}^{\mathbf{v}'} - \mathbf{x}^{\mathbf{v}'}\mathbf{y}^{\mathbf{u}'} \in \mathcal{F}$ such that $\mathbf{x}^{\mathbf{u}'}\mathbf{y}^{\mathbf{v}'}$ divides $\mathbf{x}^\mathbf{u}\mathbf{y}^\mathbf{v}$. However, by changing the roles of x and y, it follows that $\mathbf{x}^{\mathbf{v}'}\mathbf{y}^{\mathbf{u}'}$ divides $\mathbf{x}^\mathbf{v}\mathbf{y}^\mathbf{u}$. Since f is primitive by assumption, it follows that $f = f_i \in \mathcal{F}$. Thus, we have $\mathrm{Gr}_{A(A)} \subset \mathcal{F}$. On the other hand, by Corollary 5.4.2, since $\mathrm{Gr}_{A(A)}$ is a set of generators and $\mathcal{F}$ is a minimal set of generators, it follows that $\mathrm{Gr}_{A(A)} = \mathcal{F}$.

Thus, in particular, a minimal set of binomial generators of $I_{A(A)}$ is unique. Hence, $\mathcal{F} \subset \mathcal{G}_<$. By Proposition 5.4.1, up to scalar multiples, we have $\mathcal{F} \subset \mathcal{G}_< \subset \mathcal{U}_{A(A)} \subset \mathrm{Gr}_{A(A)}$. Thus, by $\mathrm{Gr}_{A(A)} = \mathcal{F}$, it follows that the four sets coincide. $\square$

By this theorem, we can, in general, compute the Graver basis Gr_A of the toric ideal I_A of a configuration matrix A. That is, if a set of generators (a Gröbner basis) of $I_{A(A)}$ is computed, then Gr_A is obtained by substituting 1 for each y_i. For example, we can compute a set of generators of the toric ideal $I_{A(A)}$ by Lemma 1.5.11, that is,

$$I_{A(A)} = \langle x_1 - \mathbf{t}^{\mathbf{c}_1} t_{n+1}, \ldots, x_n - \mathbf{t}^{\mathbf{c}_n} t_{2n}, y_1 - t_{n+1}, \ldots, y_n - t_{2n} \rangle \cap K[\mathbf{x}, \mathbf{y}]$$

$$= \langle x_1 - \mathbf{t}^{\mathbf{c}_1} y_1, \ldots, x_n - \mathbf{t}^{\mathbf{c}_n} y_n, y_1 - t_{n+1}, \ldots, y_n - t_{2n} \rangle \cap K[\mathbf{x}, \mathbf{y}]$$

$$= \langle x_1 - \mathbf{t}^{\mathbf{c}_1} y_1, \ldots, x_n - \mathbf{t}^{\mathbf{c}_n} y_n \rangle \cap K[\mathbf{x}, \mathbf{y}]$$

(with the same notation $\mathbf{t}^{\mathbf{a}_i} = \dfrac{\mathbf{t}^{\mathbf{c}_i}}{\mathbf{t}^\mathbf{b}}$ as in Lemma 1.5.11).

5.4.4 Computations of State Polytopes

For a toric ideal, we can compute the Graver basis Gr_A which contains the universal Gröbner basis $\mathcal{U}_A$. We now provide an algorithm that uses this to compute the state polytopes of toric ideals.

A vector $\mathbf{b} \in \mathbb{Z}^n$ is called a *Gröbner degree* if there exists a binomial $\mathbf{x}^\mathbf{u} - \mathbf{x}^\mathbf{v} \in \mathcal{U}_A$ such that $A\mathbf{u} = A\mathbf{v} = \mathbf{b}$. If a vector $\mathbf{b} \in \mathbb{Z}^n$ is a Gröbner degree, then a polytope

$$\mathrm{Fiber}(\mathbf{b}) := \mathrm{CONV}(\{\mathbf{u} \in \mathbb{Z}^n_{\geq 0} : A\mathbf{u} = \mathbf{b}\})$$

is called a *Gröbner fiber* .

Theorem 5.4.12. *For a configuration matrix A,*

$$P = \sum_{\mathbf{b}\,:\,Gröbner\ degree} \mathrm{Fiber}(\mathbf{b})$$

is the state polytope of I_A, in the sense that, for vectors $\mathbf{w}, \mathbf{w}' \in \mathbb{Q}^n$, $\mathrm{in}_{\mathbf{w}}(I_A) = \mathrm{in}_{\mathbf{w}'}(I_A)$ if and only if $\mathrm{FACE}_{-\mathbf{w}}(P) = \mathrm{FACE}_{-\mathbf{w}'}(P)$.

Proof. By a property of universal Gröbner bases, for any monomial order $<$, $\mathrm{in}_<(I_A)$ is generated by

$$\bigcup_{\mathbf{b}\,:\,Gröbner\ degree} \{\mathbf{x}^{\mathbf{u}} \in \mathrm{in}_<(I_A) \;:\; A\mathbf{u} = \mathbf{b}\}.$$

In addition, for each Gröbner degree $\mathbf{b}$, $\{\mathbf{x}^{\mathbf{u}} \notin \mathrm{in}_<(I_A) \;:\; A\mathbf{u} = \mathbf{b}\}$ consists of one element $\mathbf{x}^{\mathbf{v}}$ such that $\mathbf{x}^{\mathbf{v}} < \mathbf{x}^{\mathbf{u}}$ holds for all $\mathbf{x}^{\mathbf{u}} \in \mathrm{in}_<(I_A)$ $(A\mathbf{u} = \mathbf{b})$. Thus, for two generic vectors $\mathbf{w}, \mathbf{w}' \in \mathbb{Q}^n$,

$$\mathrm{in}_{\mathbf{w}}(I_A) = \mathrm{in}_{\mathbf{w}'}(I_A)$$

$$\Longleftrightarrow \{\mathbf{x}^{\mathbf{u}} \in \mathrm{in}_{\mathbf{w}}(I_A) \;:\; A\mathbf{u} = \mathbf{b}\} = \{\mathbf{x}^{\mathbf{u}} \in \mathrm{in}_{\mathbf{w}'}(I_A) \;:\; A\mathbf{u} = \mathbf{b}\}$$

for all Gröbner degrees $\mathbf{b}$

$$\Longleftrightarrow \{\mathbf{x}^{\mathbf{u}} \notin \mathrm{in}_{\mathbf{w}}(I_A) \;:\; A\mathbf{u} = \mathbf{b}\} = \{\mathbf{x}^{\mathbf{u}} \notin \mathrm{in}_{\mathbf{w}'}(I_A) \;:\; A\mathbf{u} = \mathbf{b}\}$$

for all Gröbner degrees $\mathbf{b}$

$$\Longleftrightarrow \mathrm{FACE}_{-\mathbf{w}}(\mathrm{Fiber}(\mathbf{b})) = \mathrm{FACE}_{-\mathbf{w}'}(\mathrm{Fiber}(\mathbf{b}))$$

for all Gröbner degrees $\mathbf{b}$

$$\Longleftrightarrow \mathrm{FACE}_{-\mathbf{w}}(P) = \mathrm{FACE}_{-\mathbf{w}'}(P)$$

holds. (The last "$\Longleftrightarrow$" follows from Proposition 5.1.13.) $\square$

5.5   Triangulations of Convex Polytopes and Gröbner Bases

The purpose of this section is to introduce the nice correspondence between Gröbner bases of toric ideals and triangulations of convex polytopes.

5.5.1   Unimodular Triangulations

In the remaining sections of this chapter, we will often regard a configuration matrix $A = [\mathbf{a}_1, \ldots, \mathbf{a}_n]$ as a set $A = \{\mathbf{a}_1, \ldots, \mathbf{a}_n\}$. Let Δ be a collection of simplices whose vertices belong to a configuration matrix A. Then, Δ is called a *covering* of A if

Fig. 5.5 Two triangulations
in Example 5.5.1

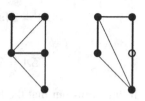

$$\text{CONV}(A) = \bigcup_{F \in \Delta} F$$

holds. In addition, if a covering Δ of a configuration matrix A is a simplicial complex, then it is called a *triangulation* of A. Note that we do not call it a "triangulation of CONV(A)." This is because the vertices of each simplex in a triangulation must belong to A. It is possible that two different configuration matrices A and B satisfy CONV(A) = CONV(B), but a triangulation of A is not necessarily a triangulation of B.

For a configuration matrix $A = [\mathbf{a}_1, \dots, \mathbf{a}_n] \in \mathbb{Z}^{d \times n}$, let

$$\mathbb{Z}A = \left\{ \sum_{i=1}^{n} z_i \mathbf{a}_i \; : \; z_i \in \mathbb{Z} \right\} \subset \mathbb{Z}^d.$$

Let $B \subset \{\mathbf{a}_1, \dots, \mathbf{a}_n\}$ be the vertex set of a maximal simplex $\sigma \in \Delta$ in a covering (triangulation) Δ of A.

Definition I. The *normalized volume* of σ is defined by $\text{VOL}(\sigma) := [\mathbb{Z}A : \mathbb{Z}B]$, that is, the index of a subgroup $\mathbb{Z}B$ in a group $\mathbb{Z}A$.

For readers who are not familiar with groups, we provide another definition, using minors, which was introduced in the textbook by Thomas [11].

Definition II. Suppose that the rank of a configuration matrix $A \in \mathbb{Z}^{d \times n}$ is equal to d. Let δ be the greatest common divisor of all $d \times d$ minors of A. Then, the *normalized volume* of σ is defined by $\text{VOL}(\sigma) := |\det(B)|/\delta$.

Using the Hermite normal form, it can be proved that the two definitions are equivalent. Using the same notation as above, we have $|\det(B)| = [\mathbb{Z}^d : \mathbb{Z}B]$ and $\delta = [\mathbb{Z}^d : \mathbb{Z}A]$.

A covering (triangulation) Δ of A is said to be *unimodular* if the normalized volume of any maximal simplex in Δ is equal to 1.

Example 5.5.1. Let A be the configuration matrix in Example 5.1.17. Then, we have $\mathbb{Z}A = \mathbb{Z}^3$. In Fig. 5.5, the figure on the left-hand side (three triangles and their edges and vertices) and the figure on the right-hand side (two triangles and their edges and vertices) are both triangulations of A. The normalized volume of each maximal simplex in these triangulations is computed by

$$\mathbb{Z}A = \mathbb{Z}\{\mathbf{a}_1, \mathbf{a}_2, \mathbf{a}_4\} = \mathbb{Z}\{\mathbf{a}_1, \mathbf{a}_2, \mathbf{a}_5\} = \mathbb{Z}\{\mathbf{a}_1, \mathbf{a}_3, \mathbf{a}_4\}$$

$$\mathbb{Z}A = \mathbb{Z}\{\mathbf{a}_1, \mathbf{a}_3, \mathbf{a}_5\}, \quad [\mathbb{Z}A : \mathbb{Z}\{\mathbf{a}_3, \mathbf{a}_4, \mathbf{a}_5\}] = 2.$$

It thus turns out that the figure on the left-hand side is unimodular and the one on the right is not. (As an exercise, try computing the minors.)

Example 5.5.2. A configuration matrix

$$A = \begin{bmatrix} 0 & 2 & 3 \\ 1 & 1 & 1 \end{bmatrix}$$

satisfies $\mathbb{Z}A = \mathbb{Z}^2$. It is easy to see that A has no unimodular covering (triangulation).

5.5.2 Regular Triangulations

For a configuration matrix $A = [\mathbf{a}_1, \ldots, \mathbf{a}_n] \in \mathbb{Z}^{d \times n}$ and a vector $\mathbf{w} = [w_1, \ldots, w_n] \in \mathbb{Q}^n$, let $\Delta_{\mathbf{w}}$ be the set of all convex polytopes $\mathrm{CONV}(\{\mathbf{a}_{i_1}, \ldots, \mathbf{a}_{i_r}\})$ satisfying the following condition:

$$\text{There exists } \mathbf{c} \in \mathbb{Q}^d \text{ such that } \begin{cases} \mathbf{a}_j \cdot \mathbf{c} = w_j & j \in \{i_1, \ldots, i_r\}, \\ \mathbf{a}_j \cdot \mathbf{c} < w_j & j \notin \{i_1, \ldots, i_r\}. \end{cases}$$

It is known that, if $\mathbf{w}$ is sufficiently generic, then $\Delta_{\mathbf{w}}$ is a triangulation of A. (For example, if $\mathbf{w} = \mathbf{0}$, then we have $\mathrm{CONV}(A) \in \Delta_{\mathbf{w}}$. Although the word "generic" is ambiguous, this means that we exclude such an exceptional vector.) A triangulation Δ of a configuration matrix A is said to be *regular* if there exists $\mathbf{w} \in \mathbb{Q}^d$ such that $\Delta = \Delta_{\mathbf{w}}$. A regular triangulation $\Delta_{\mathbf{w}}$ corresponds to the set of lower faces of the convex hull of the configuration obtained by lifting the configuration matrix A into the next dimension with the height vector $\mathbf{w}$. By multiplying $\mathbf{w}$ by a suitable integer, we may assume that $\mathbf{w} = [w_1, \ldots, w_n] \in \mathbb{Z}^n$. We define the configuration matrix

$$\hat{A} = \begin{bmatrix} \mathbf{a}_1 & \cdots & \mathbf{a}_n \\ w_1 & \cdots & w_n \end{bmatrix} \in \mathbb{Z}^{(d+1) \times n}.$$

Then, $\Delta_{\mathbf{w}}$ is the projection of the set of all faces of $\mathrm{CONV}(\hat{A})$ with respect to normal vectors whose last component is negative. Note that this normal vector is equal to $[\mathbf{c}, -1]$, where $\mathbf{c}$ is the vector in the definition of regular triangulations. (Please refer to [12, Definition 5.3] and the surrounding text for details.)

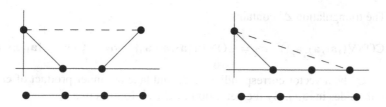

Fig. 5.6 Regular triangulations

Fig. 5.7 Regular and
nonregular triangulations

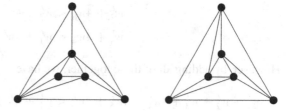

Example 5.5.3. For the configuration matrix

$$A = \begin{bmatrix} 0 & 1 & 2 & 3 \\ 1 & 1 & 1 & 1 \end{bmatrix}$$

in Example 1.5.12, the left- and right-hand side figures in Fig. 5.6 are regular triangulations of A with respect to $\mathbf{w} = [1, 0, 0, 1]$ and $\mathbf{w}' = [1, 0, 0, 0]$, respectively.

Example 5.5.4. For the configuration matrix

$$A = \begin{bmatrix} 4 & 0 & 0 & 2 & 1 & 1 \\ 0 & 4 & 0 & 1 & 2 & 1 \\ 0 & 0 & 4 & 1 & 1 & 2 \end{bmatrix}$$

in Exercise 1.5.13, we consider two triangulations Δ and Δ' in Fig. 5.7. Since the triangulation Δ for the figure on the left-hand side satisfies $\Delta = \Delta_{\mathbf{w}}$ for $\mathbf{w} = [3, 2, 1, 0, 0, 0]$, it is a regular triangulation. On the other hand, we will show that the triangulation Δ' for the figure on the right-hand side is a nonregular triangulation. Suppose that, for the triangulation Δ', there exists a vector $\mathbf{w}' = [w'_1, \ldots, w'_6] \in \mathbb{Q}^6$ such that $\Delta' = \Delta_{\mathbf{w}'}$. Note that, for column vectors $\mathbf{a}_1, \ldots, \mathbf{a}_6$ of a configuration matrix A, the equalities

$$\mathbf{a}_1 + 4\mathbf{a}_5 = \mathbf{a}_2 + 4\mathbf{a}_4 \qquad (5.1)$$

$$\mathbf{a}_2 + 4\mathbf{a}_6 = \mathbf{a}_3 + 4\mathbf{a}_5 \qquad (5.2)$$

$$\mathbf{a}_3 + 4\mathbf{a}_4 = \mathbf{a}_1 + 4\mathbf{a}_6 \qquad (5.3)$$

hold. The triangulation Δ' contains

$$\sigma_1 = \mathrm{CONV}(\{\mathbf{a}_1, \mathbf{a}_2, \mathbf{a}_5\}), \quad \sigma_2 = \mathrm{CONV}(\{\mathbf{a}_2, \mathbf{a}_3, \mathbf{a}_6\}), \quad \sigma_3 = \mathrm{CONV}(\{\mathbf{a}_1, \mathbf{a}_3, \mathbf{a}_4\}).$$

Let $\mathbf{c}_i \in \mathbb{Q}^3$ be a vector corresponding to σ_i and take the inner product of $\mathbf{c}_i$ with each of the sides in $(5.i)$. By the definition of $\Delta_{\mathbf{w}'}$, it follows that

$$w_1' + 4w_5' < w_2' + 4w_4'$$
$$w_2' + 4w_6' < w_3' + 4w_5'$$
$$w_3' + 4w_4' < w_1' + 4w_6'.$$

However, by adding all of these together, we have

$$w_1' + w_2' + w_3' + 4w_4' + 4w_5' + 4w_6' < w_1' + w_2' + w_3' + 4w_4' + 4w_5' + 4w_6',$$

which is a contradiction. Thus, Δ' is a nonregular triangulation.

5.5.3 Initial Complices

Recall the definition of radical ideals given in Sect. 3.3.3. For an ideal $I \subset K[\mathbf{x}]$, $\sqrt{I} := \{f \in K[\mathbf{x}] : f^m \in I$ for a natural number $m\}$ is called the *radical* of I. The radical $\sqrt{I}$ is an ideal which contains I. For a monomial $m = x_1^{a_1} \cdots x_n^{a_n}$, the *squarefree part* of m is defined by $\sqrt{m} = \prod_{a_i > 0} x_i$. In particular, if I is a monomial ideal, then we have the following.

Proposition 5.5.5. *Let $I \subset K[\mathbf{x}]$ be an ideal generated by monomials $m_1, \ldots, m_s$. Then, we have $\sqrt{I} = \langle \sqrt{m_1}, \ldots, \sqrt{m_s} \rangle$.*

Proof. First, we will show that $\sqrt{I} \supset \langle \sqrt{m_1}, \ldots, \sqrt{m_s} \rangle$. For each $m_i = x_1^{a_1} \cdots x_n^{a_n}$, let $a = \max(a_1, \ldots, a_n)$. Since $m_i \in I$ divides $\sqrt{m_i}^a$, we have $\sqrt{m_i}^a \in I$. Hence, $\sqrt{m_i}$ belongs to $\sqrt{I}$. Thus, the ideal generated by $\sqrt{m_1}, \ldots, \sqrt{m_s}$ is contained in $\sqrt{I}$.

Second, we will show that $\sqrt{I} \subset \langle \sqrt{m_1}, \ldots, \sqrt{m_s} \rangle$. Suppose that a nonzero polynomial $f \in \sqrt{I}$ does not belong to $\langle \sqrt{m_1}, \ldots, \sqrt{m_s} \rangle$. By deleting some terms if necessary, we may assume that any monomial in f does not belong to $\langle \sqrt{m_1}, \ldots, \sqrt{m_s} \rangle$. Since $f \in \sqrt{I}$, there exists a natural number m such that $f^m \in I$. Hence, there exists an expression $f^m = \sum_{i=1}^{s} h_i m_i$ ($h_i \in K[\mathbf{x}]$). Thus, any monomial in f^m is divided by some m_i. Fix a monomial order $<$. Then, $\mathrm{in}_<(f^m) = (\mathrm{in}_<(f))^m$ is divided by some m_i. Since $\mathrm{in}_<(f)$ is divided by $\sqrt{m_i}$, this is a contradiction. Therefore, $\sqrt{I} \subset \langle \sqrt{m_1}, \ldots, \sqrt{m_s} \rangle$ and hence $\sqrt{I} = \langle \sqrt{m_1}, \ldots, \sqrt{m_s} \rangle$. $\qquad\square$

A monomial m is said to be *squarefree* if m is equal to $\sqrt{m}$. By Proposition 5.5.5, for a monomial ideal I, $I = \sqrt{I}$ holds if and only if I has a minimal set of generators consisting of squarefree monomials.

For a configuration matrix $A = [\mathbf{a}_1, \ldots, \mathbf{a}_n] \in \mathbb{Z}^{d \times n}$ and a monomial order $<$,

$$\Delta(\mathrm{in}_<(I_A)) := \left\{ \mathrm{CONV}(B) \ : \ B \subset \{\mathbf{a}_1, \ldots, \mathbf{a}_n\}, \ \prod_{\mathbf{a}_i \in B} x_i \notin \sqrt{\mathrm{in}_<(I_A)} \right\}$$

is called the *initial complex*. By the following theorem, for any configuration matrix and any monomial order, it follows that the initial complex is a triangulation.

Theorem 5.5.6. *Let $A \in \mathbb{Z}^{d \times n}$ be a configuration matrix, and let $<$ be a monomial order. If $\mathbf{w} \in \mathbb{Q}^n$ satisfies $\mathrm{in}_<(I_A) = \mathrm{in}_\mathbf{w}(I_A)$, then we have $\Delta(\mathrm{in}_<(I_A)) = \Delta_\mathbf{w}$.*

Proof. Let $A = [\mathbf{a}_1, \ldots, \mathbf{a}_n]$ be a configuration matrix. Suppose that $\mathrm{in}_<(I_A) = \mathrm{in}_\mathbf{w}(I_A)$.

First, we will show that $\Delta(\mathrm{in}_<(I_A)) \supset \Delta_\mathbf{w}$. Suppose that $\mathrm{CONV}(B) \in \Delta_\mathbf{w}$ for $B = \{\mathbf{a}_{i_1}, \ldots, \mathbf{a}_{i_r}\}$. Then, by definition, there exists $\mathbf{c} \in \mathbb{Q}^d$ such that

$$\begin{cases} \mathbf{a}_j \cdot \mathbf{c} = w_j & j \in \{i_1, \ldots, i_r\}, \\ \mathbf{a}_j \cdot \mathbf{c} < w_j & j \notin \{i_1, \ldots, i_r\}. \end{cases}$$

We assume that $x_{i_1} \cdots x_{i_r} \in \sqrt{\mathrm{in}_<(I_A)}$. Then, there exists a natural number s such that $x_{i_1}^s \cdots x_{i_r}^s \in \mathrm{in}_<(I_A)$. Hence, there exists a monomial $\mathbf{x}^\mathbf{v} \notin \mathrm{in}_<(I_A)$ such that $f = x_{i_1}^s \cdots x_{i_r}^s - \mathbf{x}^\mathbf{v} \in I_A$. Let $\mathbf{v} = (v_1, \ldots, v_n)$. Since $f \in I_A$, it follows that $s\mathbf{a}_{i_1} + \cdots + s\mathbf{a}_{i_r} = v_1\mathbf{a}_1 + \cdots + v_n\mathbf{a}_n$. If $v_j = 0$ for all $j \notin \{i_1, \ldots, i_r\}$, then B is linearly dependent. This contradicts the fact that $\mathrm{CONV}(B)$ is a simplex. Thus, there exists $j \notin \{i_1, \ldots, i_r\}$ such that $v_j \neq 0$. Taking the inner products of both sides of $s\mathbf{a}_{i_1} + \cdots + s\mathbf{a}_{i_r} = v_1\mathbf{a}_1 + \cdots + v_n\mathbf{a}_n$ with the vector $\mathbf{c}$, by the hypothesis on $\mathbf{c}$, we have $sw_{i_1} + \cdots + sw_{i_r} = v_1\mathbf{a}_1 \cdot \mathbf{c} + \cdots + v_n\mathbf{a}_n \cdot \mathbf{c} < \mathbf{v} \cdot \mathbf{w}$. Therefore, $\mathrm{in}_\mathbf{w}(f) = \mathbf{x}^\mathbf{v} \in \mathrm{in}_<(I_A)$, which is a contradiction. Thus, $x_{i_1} \cdots x_{i_r} \notin \sqrt{\mathrm{in}_<(I_A)}$, and hence $\mathrm{CONV}(B) \in \Delta(\mathrm{in}_<(I_A))$.

Second, we will show that $\Delta(\mathrm{in}_<(I_A)) \subset \Delta_\mathbf{w}$. Suppose that $\mathrm{CONV}(B) \in \Delta(\mathrm{in}_<(I_A))$ for $B = \{\mathbf{a}_{i_1}, \ldots, \mathbf{a}_{i_r}\}$. Then, we have $x_{i_1} \cdots x_{i_r} \notin \sqrt{\mathrm{in}_<(I_A)}$. Suppose that $\mathrm{CONV}(B) \notin \Delta_\mathbf{w}$. Let

$$\alpha = \frac{1}{r} \sum_{k=1}^r \mathbf{a}_{i_k} \in \mathrm{CONV}(A).$$

Since $\Delta_\mathbf{w}$ is a triangulation of $\mathrm{CONV}(A)$, there exists $B' = \{\mathbf{a}_{j_1}, \ldots, \mathbf{a}_{j_s}\}$ such that $\mathrm{CONV}(B') \in \Delta_\mathbf{w}$ and

$$\alpha = \frac{1}{r} \sum_{k=1}^r \mathbf{a}_{i_k} = \sum_{\ell=1}^s \lambda_\ell \mathbf{a}_{j_\ell}.$$

for some $0 < \lambda_\ell \in \mathbb{Q}$ $(1 \le \ell \le s)$. Note that, since $\mathrm{CONV}(B) \notin \Delta_{\mathbf{w}}$, it follows that $B \neq B'$. By multiplying these two equations for α by a suitable positive integer, we have

$$\sum_{k=1}^{r} \gamma \mathbf{a}_{i_k} = \sum_{\ell=1}^{s} \delta_\ell \mathbf{a}_{j_\ell}$$

$(0 < \gamma, \delta_\ell \in \mathbb{Z})$. Taking the inner product of each side of this with a vector $\mathbf{c}$ corresponding to $\mathrm{CONV}(B')$, $B \neq B'$ implies that

$$\sum_{k=1}^{r} \gamma w_{i_k} > \sum_{k=1}^{r} \gamma \mathbf{a}_{i_k} \cdot \mathbf{c} = \sum_{\ell=1}^{s} \delta_\ell w_{j_\ell}.$$

Moreover, by the definition of toric ideals,

$$f = \prod_{k=1}^{r} x_{i_k}^{\gamma} - \prod_{\ell=1}^{s} x_{j_\ell}^{\delta_\ell}$$

satisfies $0 \neq f \in I_A$. By the above inequality, we have $\mathrm{in}_{\mathbf{w}}(f) = \prod_{k=1}^{r} x_{i_k}^{\gamma} \in \mathrm{in}_{\mathbf{w}}(I_A)$. This contradicts the first hypothesis $x_{i_1} \cdots x_{i_r} \notin \sqrt{\mathrm{in}_<(I_A)}$. $\qquad\square$

Using this theorem with respect to the prime decomposition introduced in Chap. 3, we have the following corollary.

Corollary 5.5.7. *Let $A \in \mathbb{Z}^{d \times n}$ be a configuration matrix. Suppose that $\mathrm{in}_{\mathbf{w}}(I_A)$ is a monomial ideal with respect to a vector $\mathbf{w} \in \mathbb{Q}^n$. Then, we have the following:*

$$\sqrt{\mathrm{in}_{\mathbf{w}}(I_A)} = \langle x_{i_1} \cdots x_{i_s} : \mathrm{CONV}(\mathbf{a}_{i_1}, \ldots, \mathbf{a}_{i_s}) \notin \Delta_{\mathbf{w}} \rangle$$

$$= \bigcap_{\sigma \in \Delta_{\mathbf{w}}} \langle x_i : \mathbf{a}_i \text{ is not a vertex of } \sigma \rangle.$$

Moreover, a "good" initial ideal corresponds to a "good" triangulation.

Theorem 5.5.8. *For a configuration matrix $A \in \mathbb{Z}^{d \times n}$ and a monomial order $<$, the regular triangulation $\Delta(\mathrm{in}_<(I_A))$ is unimodular if and only if $\sqrt{\mathrm{in}_<(I_A)} = \mathrm{in}_<(I_A)$.*

Proof. Suppose that a regular triangulation $\Delta(\mathrm{in}_<(I_A))$ is unimodular. We assume that a squarefree monomial $x_{i_1} \cdots x_{i_r}$ belongs to $\sqrt{\mathrm{in}_<(I_A)}$. We consider an element

$$\sum_{k=1}^{r} \frac{1}{r} \mathbf{a}_{i_k}$$

of $\mathrm{CONV}(A)$. Then, since $\Delta(\mathrm{in}_<(I_A))$ is a triangulation of A, there exists a maximal simplex $\mathrm{CONV}(B) \in \Delta(\mathrm{in}_<(I_A))$ with a vertex set $B = \{\mathbf{a}_{j_1}, \ldots, \mathbf{a}_{j_s}\}$ such that

$$\sum_{k=1}^{r} \frac{1}{r} \mathbf{a}_{i_k} = \sum_{\ell=1}^{s} \lambda_\ell \mathbf{a}_{j_\ell},$$

where $0 \leq \lambda_\ell \in \mathbb{Q}$. Since $\Delta(\mathrm{in}_<(I_A))$ is a unimodular triangulation, we have $\mathbb{Z}B = \mathbb{Z}A$. Thus,

$$\sum_{\ell=1}^{s} r\lambda_\ell \mathbf{a}_{j_\ell} = \sum_{k=1}^{r} \mathbf{a}_{i_k} \in \mathbb{Z}A = \mathbb{Z}B,$$

and hence it follows that $\sum_{\ell=1}^{s} r\lambda_\ell \mathbf{a}_{j_\ell} \in \mathbb{Z}B$. Since $\mathrm{CONV}(B)$ is a simplex, B is linearly independent. Thus, we have $r\lambda_\ell \in \mathbb{Z}$. Moreover, since $x_{i_1} \cdots x_{i_r} \in \sqrt{\mathrm{in}_<(I_A)}$, $\{\mathbf{a}_{i_1}, \ldots, \mathbf{a}_{i_r}\}$ is not a subset of B. Hence, we have

$$0 \neq x_{i_1} \cdots x_{i_r} - \prod_{\ell=1}^{s} x_{j_\ell}^{r\lambda_\ell} \in I_A.$$

However, since $\mathrm{CONV}(B) \in \Delta(\mathrm{in}_<(I_A))$, we have $\prod_{\ell=1}^{s} x_{j_\ell}^{r\lambda_\ell} \notin \mathrm{in}_<(I_A)$. Thus, $x_{i_1} \cdots x_{i_r} \in \mathrm{in}_<(I_A)$ holds. Therefore, we have $\sqrt{\mathrm{in}_<(I_A)} = \mathrm{in}_<(I_A)$.

Conversely, suppose that $\sqrt{\mathrm{in}_<(I_A)} = \mathrm{in}_<(I_A)$ with respect to a monomial order $<$. For a maximal simplex $\sigma \in \Delta(\mathrm{in}_<(I_A))$, let V_σ be the vertex set of σ. By Corollary 5.5.7, $\langle x_i : \mathbf{a}_i \notin V_\sigma \rangle$ appears in the prime decomposition of $\mathrm{in}_<(I_A)$. This implies that $\phi(\mathrm{in}_<(I_A)) = \langle x_i : \mathbf{a}_i \notin V_\sigma \rangle$ holds for the homomorphism $\phi : K[\mathbf{x}] \longrightarrow K[\{x_i : \mathbf{a}_i \notin V_\sigma\}]$ defined by

$$\phi(x_i) = \begin{cases} 1 & \text{if } \mathbf{a}_i \in V_\sigma, \\ x_i & \text{if } \mathbf{a}_i \notin V_\sigma. \end{cases}$$

Let $V = A \setminus V_\sigma$. Since $x_i \in \phi(\mathrm{in}_<(I_A))$ holds for each $\mathbf{a}_i \in V$, there exists $0 \neq f_i \in I_A$ such that

$$f_i = x_i \prod_{\mathbf{a}_j \in V_\sigma} x_j^{u_j^{(i)}} - \prod_{k=1}^{n} x_k^{v_k^{(i)}}$$

$$\mathrm{in}_<(f_i) = x_i \prod_{\mathbf{a}_j \in V_\sigma} x_j^{u_j^{(i)}}.$$

By the definition of toric ideals, it follows that

$$F_V = \prod_{\mathbf{a}_i \in V} \left(x_i \prod_{\mathbf{a}_j \in V_\sigma} x_j^{u_j^{(i)}} \right) - \prod_{\mathbf{a}_i \in V} \prod_{k=1}^{n} x_k^{v_k^{(i)}} \neq 0$$

belongs to I_A. Moreover, by the definition of monomial orders, the initial monomial of F_V is the first term. If $\prod_{\mathbf{a}_i \in V} x_i$ divides the second monomial of F_V, it follows that

$$\prod_{\mathbf{a}_i \in V} \prod_{\mathbf{a}_j \in V_\sigma} x_j^{u_j^{(i)}} - \frac{\prod_{\mathbf{a}_i \in V} \prod_{k=1}^n x_k^{v_k^{(i)}}}{\prod_{\mathbf{a}_i \in V} x_i} \neq 0$$

belongs to I_A. Moreover, by the definition of monomial orders, its initial monomial is the first monomial. Hence, $\prod_{\mathbf{a}_j \in V_\sigma} x_j$ belongs to $\sqrt{\mathrm{in}_<(I_A)}$. This contradicts that σ belongs to $\Delta(\mathrm{in}_<(I_A))$. Thus, the second monomial of F_V is not divided by $\prod_{\mathbf{a}_i \in V} x_i$, and hence there exists $\mathbf{a}_{i_1} \in V$ such that the second monomial of F_V is not divided by x_{i_1}. Next, let $V_1 = V \setminus \{\mathbf{a}_{i_1}\}$. Then, by considering the same thing for the element

$$F_{V_1} = \prod_{\mathbf{a}_i \in V_1} \left(x_i \prod_{\mathbf{a}_j \in V_\sigma} x_j^{u_j^{(i)}} \right) - \prod_{\mathbf{a}_i \in V_1} \prod_{k=1}^n x_k^{v_k^{(i)}} \neq 0$$

of I_A, it follows that there exists $\mathbf{a}_{i_2} \in V_1$ such that the second term of F_{V_1} is not divided by x_{i_2}. By repeating this process, we eventually obtain

$$F_{V_{N-1}} = x_{i_N} \prod_{\mathbf{a}_j \in V_\sigma} x_j^{u_j^{(i_N)}} - \prod_{k=1}^n x_k^{v_k^{(i_N)}} \neq 0,$$

where the second term of $F_{V_{N-1}}$ is not divided by any x_i ($\mathbf{a}_i \in V$).

Therefore, with respect to the pure lexicographic order ($x_{i_1} > x_{i_2} > \cdots > x_{i_N}$) on $K[\{x_i \ : \ \mathbf{a}_i \notin V_\sigma\}]$, the initial monomial of each $\phi(f_i)$ is x_i. Thus, by Lemma 1.3.1, $\{\phi(f_1), \ldots, \phi(f_N)\}$ is a Gröbner basis of the ideal $\langle \phi(f_1), \ldots, \phi(f_N) \rangle$ with respect to this pure lexicographic order. Since the reduced Gröbner basis is $\{x_i - 1 \ : \ \mathbf{a}_i \in V\}$, there exist $u_j, v_k \in \mathbb{Z}_{\geq 0}$ such that

$$x_i \prod_{\mathbf{a}_j \in V_\sigma} x_j^{u_j} - \prod_{\mathbf{a}_k \in V_\sigma} x_k^{v_k} \in I_A$$

for each i with $\mathbf{a}_i \in V$. By the definition of toric ideals, we have

$$\mathbf{a}_i + \sum_{\mathbf{a}_j \in V_\sigma} u_j \mathbf{a}_j = \sum_{\mathbf{a}_k \in V_\sigma} v_k \mathbf{a}_k.$$

Hence, by transposition,

$$\mathbf{a}_i = \sum_{\mathbf{a}_k \in V_\sigma} v_k \mathbf{a}_k - \sum_{\mathbf{a}_j \in V_\sigma} u_j \mathbf{a}_j \in \mathbb{Z} V_\sigma.$$

Thus, $\mathbb{Z} A = \mathbb{Z} V_\sigma$, and we have $\mathrm{VOL}(\sigma) = 1$. $\square$

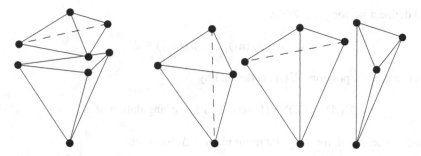

Fig. 5.8 Two triangulations

Example 5.5.9. For the configuration matrix

$$A = \begin{bmatrix} 0 & 1 & 1 & 0 & 1 \\ 0 & 1 & 0 & 1 & 1 \\ 0 & 0 & 1 & 1 & 1 \\ 1 & 1 & 1 & 1 & 1 \end{bmatrix}$$

in Example 1.5.5, the toric ideal is $I_A = \langle f \rangle$, where $f = x_1 x_5^2 - x_2 x_3 x_4$. In this case, for any monomial order, $\{f\}$ is a minimal Gröbner basis. Note that there exist two kinds of initial ideals. Suppose that monomial orders $<_1$ and $<_2$ satisfy $\mathrm{in}_{<_1}(f) = x_1 x_5^2$ and $\mathrm{in}_{<_2}(f) = x_2 x_3 x_4$, respectively. Then, we have $\mathrm{in}_{<_1}(I_A) = \langle x_1 x_5^2 \rangle$, $\sqrt{\mathrm{in}_{<_1}(I_A)} = \langle x_1 x_5 \rangle$, and $\mathrm{in}_{<_2}(I_A) = \langle x_2 x_3 x_4 \rangle = \sqrt{\mathrm{in}_{<_2}(I_A)}$. Hence, $\Delta(\mathrm{in}_{<_1}(I_A))$ is not unimodular, and its maximal simplices are

$$\mathrm{CONV}(A \setminus \{a_1\}), \mathrm{CONV}(A \setminus \{a_5\}).$$

On the other hand, $\Delta(\mathrm{in}_{<_2}(I_A))$ is unimodular and its maximal simplices are

$$\mathrm{CONV}(A \setminus \{a_2\}), \mathrm{CONV}(A \setminus \{a_3\}), \mathrm{CONV}(A \setminus \{a_4\}).$$

See Fig. 5.8.

5.5.4 Secondary Polytopes and State Polytopes

For a regular triangulation Δ of a configuration matrix $A = [a_1, \ldots, a_n] \in \mathbb{Z}^{d \times n}$, let $\Delta_i = \{\sigma \in \Delta : \sigma \text{ is a maximal simplex, and } a_i \text{ is a vertex of } \sigma\}$. In addition, let

$$\varphi_\Delta(a_i) = \sum_{\sigma \in \Delta_i} \mathrm{VOL}(\sigma) \quad \in \mathbb{Z},$$

and define a vector $\varphi_\Delta \in \mathbb{Z}^n$ by

$$\varphi_\Delta = (\varphi_\Delta(\mathbf{a}_1), \ldots, \varphi_\Delta(\mathbf{a}_n)) \in \mathbb{Z}^n.$$

Then, a convex polytope $\Sigma(A)$ is defined by

$$\Sigma(A) = \mathrm{CONV}(\{ -\varphi_\Delta \; : \; \Delta \text{ is a triangulation of } A\}).$$

On the other hand, for a regular triangulation Δ, we define

$$C_\Delta = \{\mathbf{w} \in \mathbb{Q}^n \; : \; \Delta_\mathbf{w} = \Delta\}.$$

Theorem 5.5.10. *Let $A \in \mathbb{Z}^{d \times n}$ be a configuration matrix of rank d. Then, the normal cone of $\Sigma(A)$ at each vertex $-\varphi_\Delta$ coincides with C_Δ. (In fact, $\dim \Sigma(A) = n - d$.)*

Proof. Let P denote the convex hull $\mathrm{CONV}(A)$ of a configuration matrix A. For a triangulation Δ of A and a vector $\mathbf{w} = [w_1, \ldots, w_n] \in \mathbb{Q}^n$, we define a piecewise linear function $g_{\mathbf{w},\Delta}$ on P by $g_{\mathbf{w},\Delta}(\mathbf{a}_i) = w_i$ for each simplex $\sigma \in \Delta$. More precisely, for a simplex σ with the vertex set $\{\mathbf{a}_{i_1}, \ldots, \mathbf{a}_{i_r}\}$, and for any point α in σ, there exists a unique expression

$$\alpha = \sum_{k=1}^r \lambda_k \mathbf{a}_{i_k} \quad (0 \le \lambda_k \in \mathbb{Q}, \; \sum_{k=1}^r \lambda_k = 1).$$

Then, we define $g_{\mathbf{w},\Delta}(\alpha) = \sum_{k=1}^r \lambda_k w_k$.

Let $\Delta = \Delta_\mathbf{w}$ be the regular triangulation. We will show that, for each regular triangulation Δ' of A, we have

$$g_{\mathbf{w},\Delta}(\mathbf{x}) \le g_{\mathbf{w},\Delta'}(\mathbf{x}) \text{ for all } \mathbf{x} \in P.$$

For any $\mathbf{x} \in P$, there exists a simplex $\sigma \in \Delta$ with its vertex set $\{\mathbf{a}_{i_1}, \ldots, \mathbf{a}_{i_r}\}$ such that

$$\mathbf{x} = \sum_{k=1}^r \lambda_k \mathbf{a}_{i_k} \quad (0 < \lambda_k \in \mathbb{Q}, \; \sum_{k=1}^r \lambda_k = 1).$$

If $\sigma \in \Delta'$, then we have $g_{\mathbf{w},\Delta}(\mathbf{x}) = g_{\mathbf{w},\Delta'}(\mathbf{x})$. Thus, we may assume that $\sigma \notin \Delta'$. Then, there exists a simplex $\sigma' \in \Delta'$ whose vertex set is $\{\mathbf{a}_{j_1}, \ldots, \mathbf{a}_{j_s}\}$ such that

$$\mathbf{x} = \sum_{k=1}^s \lambda'_k \mathbf{a}_{j_k} \quad (0 < \lambda'_k \in \mathbb{Q}, \; \sum_{k=1}^s \lambda'_k = 1).$$

Let $\mathbf{c}$ be a vector corresponding to the simplex $\sigma \in \Delta$. (Please refer to the definition of regular triangulations $\Delta_\mathbf{w}$.) We then have

$$g_{\mathbf{w},\Delta}(\mathbf{x}) = \sum_{k=1}^{r} \lambda_k w_{i_k} = \sum_{k=1}^{r} \lambda_k \mathbf{a}_{i_k} \cdot \mathbf{c} = \sum_{k=1}^{s} \lambda'_k \mathbf{a}_{j_k} \cdot \mathbf{c} < \sum_{k=1}^{s} \lambda'_k w_{j_k} = g_{\mathbf{w},\Delta'}(\mathbf{x}).$$

Hence, for any regular triangulation Δ' of A, we have

$$\int_{\mathbf{x} \in P} g_{\mathbf{w},\Delta}(\mathbf{x}) d\mathbf{x} \leq \int_{\mathbf{x} \in P} g_{\mathbf{w},\Delta'}(\mathbf{x}) d\mathbf{x},$$

and equality holds if and only if $\Delta = \Delta'$. For any regular triangulation Δ' of A, we have

$$\int_{\mathbf{x} \in P} g_{\mathbf{w},\Delta'}(\mathbf{x}) d\mathbf{x} = \sum_{\sigma \in M(\Delta')} \int_{\mathbf{x} \in \sigma} g_{\mathbf{w},\Delta'}(\mathbf{x}) d\mathbf{x}$$

$$= \sum_{\sigma \in M(\Delta')} \mathrm{VOL}(\sigma) \cdot g_{\mathbf{w},\Delta'}\left(\frac{1}{d} \sum_{\mathbf{a}_i \in V_\sigma} \mathbf{a}_i\right)$$

$$= \sum_{\sigma \in M(\Delta')} \mathrm{VOL}(\sigma) \cdot \frac{1}{d} \sum_{\mathbf{a}_i \in V_\sigma} g_{\mathbf{w},\Delta'}(\mathbf{a}_i)$$

$$= \frac{1}{d} \sum_{i=1}^{n} w_i \sum_{\sigma \in M(\Delta'), \, \mathbf{a}_i \in V_\sigma} \mathrm{VOL}(\sigma)$$

$$= \frac{1}{d} \mathbf{w} \cdot \varphi_{\Delta'},$$

where $M(\Delta')$ is the set of all maximal simplices of Δ', and V_σ is the set of all vertices of a simplex σ. It then follows that $\mathbf{w} \cdot \varphi_\Delta \leq \mathbf{w} \cdot \varphi_{\Delta'}$. Thus, $\mathbf{w} \cdot (-\varphi_\Delta) \geq \mathbf{w} \cdot (-\varphi_{\Delta'})$, and hence $-\varphi_\Delta$ is a vertex of a convex polytope $\Sigma(A)$ with respect to $\mathbf{w}$. $\square$

A convex polytope $\Sigma(A)$ is called the *secondary polytope* of A. By Theorem 5.5.6, we have the following theorem.

Theorem 5.5.11. *For a configuration matrix A, the secondary polytope $\Sigma(A)$ is a Minkowski summand of the state polytope $\mathrm{State}(I_A)$. That is, there exists an expression $\mathrm{State}(I_A) = \Sigma(A) + \cdots$.*

The normal fan $\mathcal{N}(\Sigma(A))$ of the secondary polytope $\Sigma(A)$ is called the *secondary fan*. The above theorem is stated in terms of fans, as follows:

Corollary 5.5.12. *For a configuration matrix A, the Gröbner fan is a refinement of the secondary fan.*

Note that, for the unimodular configurations which will be introduced in the next section (Sect. 5.6), the secondary polytope and the state polytope coincide, and the secondary fan and the Gröbner fan coincide.

5.6 Ring-Theoretic Properties and Triangulations

Although we will add definitions and further details below, we begin by stating that, for a configuration matrix A, the following conditions are well known:

(i) Any triangulation of A is unimodular;
(ii) Any reverse lexicographic triangulation of A is unimodular;
(iii) A has a unimodular regular triangulation;
(iv) A has a unimodular triangulation;
(v) A has a unimodular covering;
(vi) $K[A]$ is normal.

Then, (i) $\Rightarrow$ (ii) $\Rightarrow$ (iii) $\Rightarrow$ (iv) $\Rightarrow$ (v) $\Rightarrow$ (vi) holds, but the converse of each of them is false in general.

In this section, we introduce these ring-theoretic properties for configuration matrices related to triangulations and Gröbner bases.

5.6.1 Lexicographic Triangulations and Unimodular Configurations

First, although it is almost trivial, we study the configuration matrices which are the base cases for an inductive construction of lexicographic and reverse lexicographic triangulations.

Proposition 5.6.1. *Let A be a configuration matrix. Suppose that* $\mathrm{CONV}(A)$ *is a simplex and that A is the vertex set of* $\mathrm{CONV}(A)$. *Then, there exists only one triangulation of A, and it is the set of all faces of* $\mathrm{CONV}(A)$.

Next, we study lexicographic triangulations. A triangulation Δ of a configuration matrix $A \in \mathbb{Z}^{d \times n}$ is called a *lexicographic triangulation* if we have $\Delta = \Delta(\mathrm{in}_{<_{\mathrm{lex}}}(I_A))$ for a lexicographic order $<_{\mathrm{lex}}$ induced by an ordering $x_{i_1} > \cdots > x_{i_n}$ of variables. (It is also sometimes called a "placing triangulation" in the literature.) Since toric ideals are homogeneous ideals, there is no difference between lexicographic orders and pure lexicographic order for such ideals. It is known that every lexicographic triangulation can be computed recursively, as follows. Here, in the assertion of Proposition 5.6.2, "$\mathrm{CONV}(B)$ is visible from $\mathbf{a}_1$" means that, for any $\alpha \in \mathrm{CONV}(B)$, the line segment ℓ_α with end points α and $\mathbf{a}_1$ satisfies $\ell_\alpha \cap \mathrm{CONV}(A \setminus \{\mathbf{a}_1\}) = \{\alpha\}$.

Proposition 5.6.2. *For a configuration matrix* $A \in \mathbb{Z}^{d \times n}$, *let* $\Delta_{\text{lex}}(A)$ *be a lexicographic triangulation with respect to a lexicographic order* $<_{\text{lex}}$ *induced by an ordering* $x_1 > \cdots > x_n$ *of variables. If* $\mathbf{a}_1 \in \text{CONV}(A \setminus \{\mathbf{a}_1\})$, *then we have*

$$\Delta_{\text{lex}}(A) = \Delta_{\text{lex}}(A \setminus \{\mathbf{a}_1\}).$$

In addition, if $\mathbf{a}_1 \notin \text{CONV}(A \setminus \{\mathbf{a}_1\})$, *then we have*

$$\Delta_{\text{lex}}(A) = \Delta_{\text{lex}}(A \setminus \{\mathbf{a}_1\}) \cup \Delta$$

$$\Delta = \left\{ \text{CONV}(\{\mathbf{a}_1\} \cup B) : \begin{array}{l} B \subset A \setminus \{\mathbf{a}_1\} \\ \text{CONV}(B) \in \Delta_{\text{lex}}(A \setminus \{\mathbf{a}_1\}) \\ \text{CONV}(B) \text{ is visible from } \mathbf{a}_1 \end{array} \right\}.$$

Proof. By Corollary 1.4.2, we have $\text{in}_{<_{\text{lex}}}(I_A) \cap K[x_2, \ldots, x_n] = \text{in}_{<_{\text{lex}}}(I_{A \setminus \{\mathbf{a}_1\}})$. Hence, for any monomial m which is not divided by x_1, we have

$$m \notin \text{in}_{<_{\text{lex}}}(I_A) \iff m \notin \text{in}_{<_{\text{lex}}}(I_{A \setminus \{\mathbf{a}_1\}}).$$

Thus, for any squarefree monomial m' which is not divided by x_1, we have

$$m' \notin \sqrt{\text{in}_{<_{\text{lex}}}(I_A)} \iff m' \notin \sqrt{\text{in}_{<_{\text{lex}}}(I_{A \setminus \{\mathbf{a}_1\}})}.$$

Therefore, the set of all simplices in $\Delta_{\text{lex}}(A)$ whose vertex set does not contain $\mathbf{a}_1$ is equal to $\Delta_{\text{lex}}(A \setminus \{\mathbf{a}_1\})$.

On the other hand, suppose that $\Delta_{\text{lex}}(A)$ possesses a simplex σ whose vertex set is $\{\mathbf{a}_1\} \cup B$ (where $B \subset A \setminus \{\mathbf{a}_1\}$). Since $\text{CONV}(B) \in \Delta_{\text{lex}}(A)$, by the above fact, we have $\text{CONV}(B) \in \Delta_{\text{lex}}(A \setminus \{\mathbf{a}_1\})$. If $\mathbf{a}_1 \in \text{CONV}(A \setminus \{\mathbf{a}_1\})$, then there exists a natural number i_1 such that $f = x_1^{i_1} - x_2^{i_2} \cdots x_n^{i_n}$ belongs to I_A and such that $\text{in}_{<_{\text{lex}}}(f) = x_1^{i_1}$. Then, we have $x_1 \in \sqrt{\text{in}_{<_{\text{lex}}}(I_A)}$. This contradicts $\sigma \in \Delta_{\text{lex}}(A)$. Thus, $\mathbf{a}_1 \notin \text{CONV}(A \setminus \{\mathbf{a}_1\})$. Let $\alpha \in \text{CONV}(B)$, and let ℓ_α denote the line passing through α and $\mathbf{a}_1$. Suppose that there exists α' such that

$$\alpha \neq \alpha' \in \ell_\alpha \cap \text{CONV}(A \setminus \{\mathbf{a}_1\}).$$

Then, α' is expressed as

$$\alpha' = r_1 \mathbf{a}_1 + \sum_{\mathbf{a}_k \in B} r_k \mathbf{a}_k = \sum_{\ell=2}^{n} s_\ell \mathbf{a}_\ell$$

$$\left(0 < r_1 \in \mathbb{Q}, \ 0 \le r_k, s_\ell \in \mathbb{Q}, \ r_1 + \sum_{\mathbf{a}_k \in B} r_k = \sum_{\ell=2}^{n} s_\ell = 1 \right).$$

Hence, there exists a natural number p_1 such that

$$g = x_1^{p_1} \prod_{\mathbf{a}_k \in B} x_k^{p_k} - \prod_{\ell=2}^{n} x_\ell^{q_\ell}$$

belongs to I_A and such that $\mathrm{in}_{<_{\mathrm{lex}}}(g) = x_1^{p_1} \prod_{\mathbf{a}_k \in B} x_k^{p_k}$. Thus, we have $x_1 \prod_{\mathbf{a}_k \in B} x_k \in \sqrt{\mathrm{in}_{<_{\mathrm{lex}}}(I_A)}$. This contradicts that $\sigma \in \Delta_{\mathrm{lex}}(A)$. Therefore, $\mathrm{CONV}(B)$ is visible from $\mathbf{a}_1$, and hence σ belongs to Δ.

Conversely, let $\sigma \in \Delta$. That is, suppose that the vertex set of σ is $\{\mathbf{a}_1\} \cup B$ and that $\mathrm{CONV}(B) \in \Delta_{\mathrm{lex}}(A \setminus \{\mathbf{a}_1\})$ is visible from $\mathbf{a}_1$. We will now assume $\sigma \notin \Delta_{\mathrm{lex}}(A)$ and deduce a contradiction. First, by the definition of an initial complex, we have $x_1 \prod_{\mathbf{a}_i \in B} x_i \in \sqrt{\mathrm{in}_{<_{\mathrm{lex}}}(I_A)}$. Hence, there exists a natural number m such that $x_1^m \prod_{\mathbf{a}_i \in B} x_i^m \in \mathrm{in}_{<_{\mathrm{lex}}}(I_A)$. Thus, there exists a binomial

$$f = x_1^m \prod_{\mathbf{a}_i \in B} x_i^m - \prod_{j=1}^{n} x_j^{m_j}$$

belonging to I_A such that $\mathrm{in}_{<_{\mathrm{lex}}}(f) = x_1^m \prod_{\mathbf{a}_i \in B} x_i^m$. By the definition of the lexicographic order, we have $m \geq m_1$. Thus, removing $x_1^{m_1}$ from f, it follows that

$$f' = x_1^{m-m_1} \prod_{\mathbf{a}_i \in B} x_i^m - \prod_{j=2}^{n} x_j^{m_j}$$

is a binomial belonging to I_A and satisfies $\mathrm{in}_{<_{\mathrm{lex}}}(f') = x_1^{m-m_1} \prod_{\mathbf{a}_i \in B} x_i^m$. If $m - m_1 = 0$, then $\prod_{\mathbf{a}_i \in B} x_i \in \sqrt{\mathrm{in}_{<_{\mathrm{lex}}}(I_A)}$. This contradicts $\mathrm{CONV}(B) \in \Delta_{\mathrm{lex}}(A \setminus \{\mathbf{a}_1\}) \subset \Delta_{\mathrm{lex}}(A)$. Thus, $m - m_1 > 0$. Since $f' \in I_A$, it follows that

$$(m - m_1)\mathbf{a}_1 + \sum_{\mathbf{a}_i \in B} m\mathbf{a}_i = \sum_{j=2}^{n} m_j \mathbf{a}_j.$$

Let $r = \deg(f') (= \sum_{j=2}^{n} m_j)$. Then,

$$\frac{m - m_1}{r}\mathbf{a}_1 + \sum_{\mathbf{a}_i \in B} \frac{m}{r}\mathbf{a}_i = \sum_{j=2}^{n} \frac{m_j}{r}\mathbf{a}_j.$$

However, since the left-hand side belongs to the interior of σ and the right-hand side belongs to $\mathrm{CONV}(A \setminus \{\mathbf{a}_1\})$, this contradicts the hypothesis that $\mathrm{CONV}(B)$ is visible from $\mathbf{a}_1$. Thus, we have $\sigma \in \Delta_{\mathrm{lex}}(A)$. $\qquad\square$

A configuration matrix A is said to be *unimodular* if all triangulations of A are unimodular. Here, "all triangulations" means, of course, all regular triangulations and all nonregular triangulations. However, it will be turn out that it is enough to consider only the lexicographic triangulations.

Theorem 5.6.3. *For a configuration matrix $A \in \mathbb{Z}^{d \times n}$, the following conditions are equivalent.*

(i) *A is a unimodular configuration matrix.*
(ii) *Any regular triangulation of A is unimodular.*
(iii) *Any lexicographic triangulation of A is unimodular.*
(iv) *The normalized volume of any maximal simplex all of whose vertices belong to A is equal to 1.*
(v) *For an arbitrary $f \in \mathscr{C}_A$, any monomial appearing in f is squarefree.*

If $\mathrm{rank}(A) = d$, *then the following is also equivalent to the above.*

(vi) *All nonzero $d \times d$ minors of A have the same absolute value.*

Proof. By definition, it is trivial that (iv) $\implies$ (i) $\implies$ (ii) $\implies$ (iii). Let σ be a maximal simplex with the vertex set $\{\mathbf{a}_{i_1}, \ldots, \mathbf{a}_{i_r}\}$, for which the normalized volume is greater than or equal to 2. For any $j \notin \{i_1, \ldots, i_r\}$, let $<_{\mathrm{lex}}$ be a lexicographic order induced by the ordering $x_j > x_{i_1} > \cdots > x_{i_r}$ of variables. By Proposition 5.6.2, we have $\sigma \in \Delta_{\mathrm{lex}}(A)$, and hence $\Delta_{\mathrm{lex}}(A)$ is not unimodular. Thus, we have (iii) $\implies$ (iv).

Suppose that all monomials appearing in each circuit are squarefree. Assume that a binomial $g = \mathbf{x}^{\mathbf{u}} - \mathbf{x}^{\mathbf{v}} \in I_A$ is primitive. Then, by Lemma 5.4.6, there exists a circuit $f = \mathbf{x}^{\mathbf{u}'} - \mathbf{x}^{\mathbf{v}'} \in \mathscr{C}_A$ such that $\mathrm{VAR}(\mathbf{x}^{\mathbf{u}'}) \subset \mathrm{VAR}(\mathbf{x}^{\mathbf{u}})$, $\mathrm{VAR}(\mathbf{x}^{\mathbf{v}'}) \subset \mathrm{VAR}(\mathbf{x}^{\mathbf{v}})$. By assumption, $\mathbf{x}^{\mathbf{u}'}$ and $\mathbf{x}^{\mathbf{v}'}$ are squarefree monomials. Hence, $\mathbf{x}^{\mathbf{u}'}$ divides $\mathbf{x}^{\mathbf{u}}$, and $\mathbf{x}^{\mathbf{v}'}$ divides $\mathbf{x}^{\mathbf{v}}$. However, since g is primitive, it follows that $f = g$. By Proposition 5.4.1, we have $\mathscr{C}_A = \mathrm{Gr}_A$. Hence, $\mathscr{C}_A$ is a Gröbner basis with respect to any monomial order. Since all monomials appearing in each circuit are squarefree, $\sqrt{\mathrm{in}_<(I_A)} = \mathrm{in}_<(I_A)$ with respect to any monomial order $<$. By Theorem 5.5.8, this is equivalent to condition (ii). Thus, we have (v) $\implies$ (ii).

Suppose that there exists a circuit $f \in \mathscr{C}_A$ having a monomial $\mathbf{x}^{\mathbf{u}}$ which is not squarefree. By the argument in the proof of Proposition 5.4.1, there exists a lexicographic order $<_{\mathrm{lex}}$ such that $\mathbf{x}^{\mathbf{u}}$ appears in the minimal set of generators of $\mathrm{in}_{<_{\mathrm{lex}}}(I_A)$. Since the lexicographic triangulation $\Delta_{\mathrm{lex}}(A)$ is not unimodular, we have (iii) $\implies$ (v).

Therefore (i)–(v) are equivalent.

Assume $\mathrm{rank}(A) = d$. Then, by Definition II of the normalized volume, we have (iv) $\iff$ (vi). Therefore (i)–(vi) are equivalent. $\square$

We have shown the following fact in the above proof.

Corollary 5.6.4. *If a configuration matrix A is unimodular, then $\mathscr{C}_A = \mathscr{U}_A = \mathrm{Gr}_A$.*

5.6.2   Reverse Lexicographic Triangulations and Compressed Configurations

A triangulation Δ of a configuration matrix $A \in \mathbb{Z}^{d \times n}$ is called a *reverse lexicographic triangulation* if $\Delta = \Delta(\mathrm{in}_{<_{\mathrm{rev}}}(I_A))$ with respect to a reverse lexicographic order $<_{\mathrm{rev}}$ induced by the ordering $x_{i_1} > \cdots > x_{i_n}$ of variables. (This is sometimes called a "pulling triangulation" in the literature.) As in the case of lexicographic triangulations, every reverse lexicographic triangulation can be computed recursively, as follows.

Proposition 5.6.5. *For a configuration matrix $A \in \mathbb{Z}^{d \times n}$, let $\Delta_{\mathrm{rev}}(A)$ be a reverse lexicographic triangulation with respect to a reverse lexicographic order $<_{\mathrm{rev}}$ induced by the ordering $x_1 > \cdots > x_n$ of variables. Then, the set of all maximal simplices belonging to $\Delta_{\mathrm{rev}}(A)$ is*

$$\bigcup_{A' \in \mathrm{FCT}_n} \{ \mathrm{CONV}(\{\mathbf{a}_n\} \cup B) \; : \; \mathrm{CONV}(B) \text{ is a maximal simplex of } \Delta_{\mathrm{rev}}(A') \},$$

where

$$\mathrm{FCT}_n = \left\{ A' \subset A \; : \; \begin{array}{l} \text{There exists a facet } F \text{ of } \mathrm{CONV}(A) \\ \text{such that } \mathbf{a}_n \notin F, \; A' = F \cap A \end{array} \right\}.$$

Proof. Based on Proposition 5.6.1, we will prove this by induction on the number n of the columns of a configuration matrix A. First, we show that, if $\mathbf{a}_n \notin B \subset A$ and the vertex set of $\mathrm{CONV}(B)$ is B, then we have

$$\mathrm{CONV}(B) \in \Delta_{\mathrm{rev}}(A) \iff \mathrm{CONV}(\{\mathbf{a}_n\} \cup B) \in \Delta_{\mathrm{rev}}(A).$$

Since the triangulation $\Delta_{\mathrm{rev}}(A)$ is a complex, "$\impliedby$" is trivial. In order to prove "$\implies$", we assume that $B = \{\mathbf{a}_{i_1}, \ldots, \mathbf{a}_{i_r}\}$ satisfies both the above condition and the condition $\mathrm{CONV}(B) \in \Delta_{\mathrm{rev}}(A)$. Suppose that a squarefree monomial $x_{i_1} \cdots x_{i_r} x_n$ belongs to $\sqrt{\mathrm{in}_{<_{\mathrm{rev}}}(I_A)}$. Then, there exists a natural number m such that $x_{i_1}^m \cdots x_{i_r}^m x_n^m \in \mathrm{in}_{<_{\mathrm{rev}}}(I_A)$. Hence, there exists a monomial $\mathbf{x}^{\mathbf{b}}$ such that the initial monomial of $x_{i_1}^m \cdots x_{i_r}^m x_n^m - \mathbf{x}^{\mathbf{b}} \in I_A$ is $x_{i_1}^m \cdots x_{i_r}^m x_n^m$. However, by the definition of the reverse lexicographic order, x_n^m divides $\mathbf{x}^{\mathbf{b}}$. Thus, the initial monomial of $x_{i_1}^m \cdots x_{i_r}^m - \mathbf{x}^{\mathbf{b}}/x_n^m \in I_A$ is $x_{i_1}^m \cdots x_{i_r}^m$. This contradicts $\mathrm{CONV}(B) \in \Delta_{\mathrm{rev}}(A)$. Therefore, since $x_{i_1} \cdots x_{i_r} x_n \notin \sqrt{\mathrm{in}_{<_{\mathrm{rev}}}(I_A)}$, it follows that $\mathrm{CONV}(\{\mathbf{a}_n\} \cup B) \in \Delta_{\mathrm{rev}}(A)$.

By this fact, $\mathbf{a}_n$ is a vertex of any maximal simplex of $\Delta_{\mathrm{rev}}(A)$. Let $\{\mathbf{a}_{i_1}, \ldots, \mathbf{a}_{i_r}, \mathbf{a}_n\}$ be the vertex set of a maximal simplex $\sigma \in \Delta_{\mathrm{rev}}(A)$, and let $B = \{\mathbf{a}_{i_1}, \ldots, \mathbf{a}_{i_r}\}$ and $\sigma' = \mathrm{CONV}(B)$. Since $\sigma' \in \Delta_{\mathrm{rev}}(A)$, we have $x_{i_1} \cdots x_{i_r} \notin \sqrt{\mathrm{in}_{<_{\mathrm{rev}}}(I_A)}$. By a property of reverse lexicographic orders, it follows that

$$x_{i_1}^{u_1} \cdots x_{i_r}^{u_r} - \mathbf{x}^{\mathbf{b}} \in I_A \Rightarrow x_n \notin \text{VAR}(\mathbf{x}^{\mathbf{b}}).$$

Let $\alpha \in \sigma'$. Then, there exists an expression

$$\alpha = \sum_{k=1}^{r} \lambda_k \mathbf{a}_{i_k} \quad (0 \leq \lambda_k \in \mathbb{Q}, \sum_{k=1}^{r} \lambda_k = 1).$$

If α belongs to the interior of CONV(A), then, by Proposition 5.1.15, there exists an expression

$$\alpha = \sum_{\ell=1}^{n} \delta_\ell \mathbf{a}_\ell \quad (0 < \delta_\ell \in \mathbb{Q}, \sum_{\ell=1}^{n} \delta_\ell = 1).$$

However, by

$$\sum_{k=1}^{r} \lambda_k \mathbf{a}_{i_k} = \sum_{\ell=1}^{n} \delta_\ell \mathbf{a}_\ell,$$

there exists $x_{i_1}^{u_1} \cdots x_{i_r}^{u_r} - \mathbf{x}^{\mathbf{b}} \in I_A$ such that $x_n \in \text{VAR}(\mathbf{x}^{\mathbf{b}})$, which is a contradiction. Hence, σ' belongs to a facet F of CONV(A). If F contains $\mathbf{a}_n$, then the dimension of σ is equal to the dimension of F, and hence this contradicts that σ is a maximal simplex. By the hypothesis of induction, σ' belongs to $\Delta_{\text{rev}}(F \cap A)$. Thus, B satisfies the condition in the assertion of the proposition.

Conversely, suppose that B satisfies the assertion of the proposition. Let Δ_F be a triangulation of F obtained by restricting the triangulation $\Delta_{\text{rev}}(A)$ to F. Then, from the fact shown above, we have $\Delta_F \subset \Delta_{\text{rev}}(F \cap A)$. Since both of them are a triangulation of F, it follows that $\Delta_F = \Delta_{\text{rev}}(F \cap A)$. Thus, CONV($B$) $\in \Delta_{\text{rev}}(F \cap A) = \Delta_F \subset \Delta_{\text{rev}}(A)$, and hence CONV($\{\mathbf{a}_n\} \cup B$) $\in \Delta_{\text{rev}}(A)$. $\square$

A configuration matrix A is said to be *compressed* if, for any ordering $x_{i_1} > \cdots > x_{i_n}$ of n variables of $K[\mathbf{x}]$, the reverse lexicographic triangulation $\Delta(\text{in}_{<_{\text{rev}}}(I_A))$ of A with respect to the reverse lexicographic order $<_{\text{rev}}$ induced by this ordering is unimodular. The following theorem is due to Sullivant [10].

Theorem 5.6.6. *For a configuration matrix* $A = [\mathbf{a}_1, \ldots, \mathbf{a}_n] \in \mathbb{Z}^{d \times n}$, *let* $P = $ CONV(A). *Then, the following conditions are equivalent.*

  (i) *A is compressed.*
 (ii) *For any facet* $F = \text{FACE}_{\mathbf{w}}(P)$ *of* P, *$\{\mathbf{a}_1 \cdot \mathbf{w}, \ldots, \mathbf{a}_n \cdot \mathbf{w}\}$ consists of exactly two elements.*
(iii) *There exists a configuration matrix* $B \in \mathbb{Z}^{d \times n}$ *such that* $I_A = I_B$ *and* CONV(B) *is the intersection of a d-dimensional unit cube and an affine subspace.*

5.6.3   Normality of Toric Rings

In general, a configuration matrix A satisfies $\mathbb{Z}_{\geq 0}A \subset \mathbb{Z}A \cap \mathbb{Q}_{\geq 0}A$. The toric ring $K[A]$ is said to be *normal* if it satisfies $\mathbb{Z}_{\geq 0}A = \mathbb{Z}A \cap \mathbb{Q}_{\geq 0}A$. Originally, the formal definition of normal rings is "The integral domain $K[A]$ is called normal if it is integrally closed in its field of fractions." However, we employ an equivalent condition (see [9, Proposition 13.5]) as a definition since this one may be difficult to understand for readers not familiar with the theory of commutative rings.

With respect to the normality of the toric ring $K[A]$, the existence of unimodular triangulations and unimodular coverings of A plays an important role.

Theorem 5.6.7. *If a configuration matrix A has a unimodular covering, then the toric ring $K[A]$ is normal.*

Proof. Let $\alpha \in \mathbb{Z}A \cap \mathbb{Q}_{\geq 0}A$. Then, it is enough to show that $\alpha \in \mathbb{Z}_{\geq 0}A$. First, by $\alpha \in \mathbb{Q}_{\geq 0}A$, there exists an expression

$$\alpha = \sum_{i=1}^{n} r_i \mathbf{a}_i \quad (0 \leq r_i \in \mathbb{Q}).$$

Since $\mathbb{Z}_{\geq 0}A$ contains the zero vector, we may assume that α is not the zero vector. Let $r = \sum_{i=1}^{n} r_i \ (\neq 0)$. Then,

$$\frac{1}{r}\alpha = \sum_{i=1}^{n} \frac{r_i}{r} \mathbf{a}_i$$

belongs to $\mathrm{CONV}(A)$. By assumption, A has a unimodular covering. Hence, there exists a maximal simplex σ of normalized volume 1 for which the vertices belong to A and such that $\frac{1}{r}\alpha \in \sigma$. Let $V_\sigma = \{\mathbf{a}_{i_1}, \ldots, \mathbf{a}_{i_d}\}$ be the vertex set of σ. Then, there exists an expression

$$\frac{1}{r}\alpha = \sum_{k=1}^{d} s_k \mathbf{a}_{i_k},$$

where $0 \leq s_k \in \mathbb{Q}$ and $\sum_{k=1}^{d} s_k = 1$. Multiplying both sides by r, we have

$$\alpha = \sum_{k=1}^{d} r s_k \mathbf{a}_{i_k}.$$

On the other hand, since the normalized volume of σ is equal to 1, it follows that $\alpha \in \mathbb{Z}A = \mathbb{Z}V_\sigma$. Hence, there exists an expression

$$\alpha = \sum_{k=1}^{d} z_k \mathbf{a}_{i_k},$$

where $z_k \in \mathbb{Z}$. Since σ is a simplex, $0 \leq r s_k = z_k \in \mathbb{Z}$ for all k. Thus, $\alpha \in \mathbb{Z}_{\geq 0}A$. $\qquad\square$

Corollary 5.6.8. *Let A be a configuration matrix. If there exists a monomial order $<$ such that* $\mathrm{in}_<(I_A)$ *is generated by squarefree monomials, then the toric ring $K[A]$ is normal.*

Below, in Example 5.7.7, we will see that the converse of Corollary 5.6.8 does not hold in general. On the other hand, if any minimal set of generators has an element without squarefree monomials, then it is trivial that the hypothesis of Corollary 5.6.8 does not hold. In such a case, the following proposition says that the toric ring is not normal.

Proposition 5.6.9. *Let A be a configuration matrix. If there exists a minimal set of binomials generators of the toric ideal I_A of A which contains a binomial having no squarefree monomials, then $K[A]$ is not normal.*

Proof. Suppose that a binomial

$$f = x_i^2 \prod_{k=1}^{n} x_k^{u_k} - x_j^2 \prod_{k=1}^{n} x_k^{v_k} \in I_A$$

without squarefree monomials appears in a minimal set of binomial generators of I_A. Since f belongs to I_A, we have

$$2\mathbf{a}_i + \sum_{k=1}^{n} u_k \mathbf{a}_k = 2\mathbf{a}_j + \sum_{k=1}^{n} v_k \mathbf{a}_k.$$

By transforming this equation, we have

$$\mathbb{Z}A \ni \mathbf{a}_i - \mathbf{a}_j + \sum_{k=1}^{n} u_k \mathbf{a}_k = \sum_{k=1}^{n} \frac{u_k + v_k}{2} \mathbf{a}_k \in \mathbb{Q}_{\geq 0}A.$$

If $K[A]$ is normal, then this vector belongs to $\mathbb{Z}_{\geq 0}A = \mathbb{Z}A \cap \mathbb{Q}_{\geq 0}A$. Hence, there exist nonnegative integers $w_1, \ldots, w_n$ such that

$$\mathbf{a}_i - \mathbf{a}_j + \sum_{k=1}^{n} u_k \mathbf{a}_k = \sum_{\ell=1}^{n} w_\ell \mathbf{a}_\ell.$$

Thus, by

$$\mathbf{a}_i + \sum_{k=1}^{n} u_k \mathbf{a}_k = \mathbf{a}_j + \sum_{\ell=1}^{n} w_\ell \mathbf{a}_\ell,$$

it follows that $g = x_i \prod_{k=1}^{n} x_k^{u_k} - x_j \prod_{\ell=1}^{n} x_\ell^{w_\ell}$ belongs to I_A. Then, $h = x_i \prod_{\ell=1}^{n} x_\ell^{w_\ell} - x_j \prod_{k=1}^{n} x_k^{v_k}$ also belongs to I_A. Hence, f is expressed as $f = x_i g + x_j h$, where the degrees of g and h are both lower than that of f. This contradicts that f appears in a minimal set of generators. Thus, $K[A]$ is not normal. $\square$

Theorem 5.6.10. *For a configuration matrix A and its Lawrence lifting $\Lambda(A)$, the following conditions are equivalent.*

(i) *A is a unimodular configuration matrix.*
(ii) *$\Lambda(A)$ is a unimodular configuration matrix.*
(iii) *$K[\Lambda(A)]$ is normal.*

Proof. By Theorem 5.6.7, we have (ii) $\Longrightarrow$ (iii). As in the case of Graver bases, with regard to set of circuits, it follows that

$$\mathbf{x}^{\mathbf{u}}\mathbf{y}^{\mathbf{v}} - \mathbf{x}^{\mathbf{v}}\mathbf{y}^{\mathbf{u}} \in C_{\Lambda(A)} \quad \Longleftrightarrow \quad \mathbf{x}^{\mathbf{u}} - \mathbf{x}^{\mathbf{v}} \in C_A.$$

Thus, by Theorem 5.6.3, we have (i) $\Longleftrightarrow$ (ii).

In order to show (iii) $\Longrightarrow$ (i), we assume that A is not a unimodular configuration matrix. By the argument above, there exists a binomial f in $C_{\Lambda(A)}$ without squarefree monomials. By Proposition 5.4.1 and Theorem 5.4.11, f belongs to a minimal set of generators of $I_{\Lambda(A)}$. Hence, by Proposition 5.6.9, $K[\Lambda(A)]$ is not normal. $\qquad\square$

5.7 Examples of Configuration Matrices

In order to illustrate the above discussion, we show below some configuration matrices that arise from finite graphs and contingency tables.

5.7.1 Configuration Matrices of Finite Graphs

In this section, as examples of convex polytopes, we now consider edge polytopes arising from finite graphs. Since several properties of toric rings and toric ideals of edge polytopes are described in terms of graphs, this is a very useful example for better understanding the various concepts.

Let G be a finite graph on the vertex set $\{1, 2, \ldots, d\}$. We assume that G has no loops and no multiple edges. Let $E(G) = \{e_1, \ldots, e_n\}$ be the edge set of G. For each edge $e = \{i, j\} \in E(G)$ of G, let $\rho(e) := \mathbf{e}_i + \mathbf{e}_j \in \mathbb{Z}^d$. Let A_G be a configuration matrix whose column vectors are $\{\rho(e) : e \in E(G)\} \subset \mathbb{Z}^d$. Then, the convex hull $\mathrm{CONV}(A_G)$ of A_G is called the *edge polytope* of G.

Example 5.7.1. Let G be the graph with six vertices and ten edges shown in Fig. 5.9. Then, the corresponding configuration matrix is

Fig. 5.9 Wheel with six
vertices

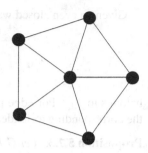

$$A_G = \begin{bmatrix} 1\ 1\ 1\ 1\ 1\ 0\ 0\ 0\ 0\ 0 \\ 1\ 0\ 0\ 0\ 0\ 1\ 0\ 0\ 0\ 1 \\ 0\ 1\ 0\ 0\ 0\ 1\ 1\ 0\ 0\ 0 \\ 0\ 0\ 1\ 0\ 0\ 0\ 1\ 1\ 0\ 0 \\ 0\ 0\ 0\ 1\ 0\ 0\ 0\ 1\ 1\ 0 \\ 0\ 0\ 0\ 0\ 1\ 0\ 0\ 0\ 1\ 1 \end{bmatrix} \in \mathbb{Z}^{6\times 10}.$$

We now present some graph-theory terminology. A sequence $\Gamma = (e_{j_1}, \ldots, e_{j_r})$ of edges of a finite graph G is called a *walk* of length r if Γ satisfies

$$e_{j_1} = \{i_1, i_2\}, e_{j_2} = \{i_2, i_3\}, \ldots, e_{j_r} = \{i_r, i_{r+1}\}.$$

In addition,

* If $i_1, \ldots, i_{r+1}$ are distinct vertices, then Γ is called a *path*.
* If $i_{r+1} = i_1$, then Γ is called a *closed walk* of length r. A closed walk of even length is called an *even closed walk* .
* If $i_{r+1} = i_1$ and $i_1, \ldots, i_r$ $(r \geq 3)$ are distinct, then Γ is called a *cycle* of length r. A cycle of odd length is called an *odd cycle*. A cycle of even length is called an *even cycle*.

A finite graph G is said to be *connected* if, for any two vertices i and j of G, there exists a walk from i to j. From now on, we always assume that G is a connected graph. If the vertex set V of a finite graph G is partitioned into $V = V_1 \cup V_2$, where $V_1 \cap V_2 = \emptyset$, and each edge of G joins a vertex in V_1 and a vertex in V_2, then G is called a *bipartite graph*. It is known that a finite graph G is a bipartite graph if and only if G has no odd cycles.

Proposition 5.7.2. *Let G be a finite connected graph. Then, we have*

$$\dim(\mathrm{CONV}(A_G)) = \begin{cases} d - 2 & G \text{ is a bipartite graph,} \\ d - 1 & \text{otherwise.} \end{cases}$$

Given an even closed walk $\Gamma = (e_{j_1}, \ldots, e_{j_{2r}})$, it is easy to see that

$$f_\Gamma = \prod_{k=1}^{r} x_{j_{2k-1}} - \prod_{k=1}^{r} x_{j_{2k}}$$

belongs to I_{A_G}. For edge polytopes, circuits and binomials of the Graver basis of the corresponding toric ideal can be characterized in terms of graphs.

Proposition 5.7.3. *Let G be a finite connected graph. Then, a binomial f belongs to C_{A_G} if and only if there exists an even closed walk Γ with $f = f_\Gamma$ satisfying one of the following.*

(i) *Γ is an even cycle.*
(ii) *Γ consists of two odd cycles having exactly one common vertex.*
(iii) *Γ consists of two odd cycles having no common vertex and a path which joins a vertex of one cycle to a vertex of the other cycle.*

Among the even closed walks appearing in Proposition 5.7.3, only even closed walks of type (iii) correspond to binomials with a nonsquarefree monomial. Thus, we have the following as a corollary.

Corollary 5.7.4. *Let G be a finite connected graph. Then, A_G is a unimodular configuration matrix if and only if any two odd cycles of G have a common vertex. In particular, if G is a bipartite graph, then A_G is a unimodular configuration matrix.*

For example, for the graph in Example 5.7.1, the corresponding configuration matrix is unimodular.

Proposition 5.7.5. *Let G be a finite connected graph. If a binomial $f \in I_{A_G}$ is primitive, then there exists an even closed walk Γ of G such that $f = f_\Gamma$ and satisfies one of the following.*

(i) *Γ is an even cycle.*
(ii) *Γ consists of two odd cycles having exactly one common vertex.*
(iii) *Γ consists of two odd cycles having no common vertex and a walk which joins a vertex of one cycle to a vertex of the other cycle.*

It is also known that the normality of edge polytopes is characterized by the following condition.

Theorem 5.7.6 ([3, 8]). *Let G be a finite connected graph. Then, the following conditions are equivalent.*

(i) *$K[A_G]$ is normal.*
(ii) *A_G has a unimodular covering.*
(iii) *For any two odd cycles C and C' of G without common vertices, there exists an edge of G which joins a vertex of C to a vertex of C'.*

Fig. 5.10 Graph for
Example 5.7.7

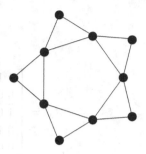

An edge polytope with certain valuable properties has been discovered, as shown below; no convex polytopes other than edge polytopes with the same property have yet been found.

Example 5.7.7 ([4]). Let G be the graph with 10 vertices and 15 edges shown in Fig. 5.10. Then, G has the following properties.

(i) For any monomial order $<$, we have $\sqrt{\text{in}_<(I_{A_G})} \neq \text{in}_<(I_{A_G})$.
(ii) A_G has unimodular triangulations (however, none of them is regular).
(iii) Any triangulation of A_G having the fewest maximal simplices is nonregular.

Among the above properties, (ii) and (iii) have been verified using the Puntos program developed by J.A. De Loera.

5.7.2 Configuration Matrices of Contingency Tables

In this section, we study configuration matrices which arise from the models with no n-way interactions that were introduced in Chap. 4. The configuration matrix arising from the model of an n-way $r_1 \times r_2 \times \cdots \times r_n$ contingency table ($r_1 \geq r_2 \geq \cdots \geq r_n \geq 2$) with no n-way interactions is the configuration matrix $A_{r_1 r_2 \cdots r_n}$, for which the columns are the set of all vectors

$$\mathbf{e}^{(1)}_{i_2 i_3 \cdots i_n} \oplus \mathbf{e}^{(2)}_{i_1 i_3 \cdots i_n} \oplus \cdots \oplus \mathbf{e}^{(n)}_{i_1 i_2 \cdots i_{n-1}},$$

where each i_k belongs to $\{1, 2, \ldots, r_k\}$ and $\mathbf{e}^{(k)}_{i_1 \cdots i_{k-1} i_{k+1} \cdots i_n}$ is a unit coordinate vector of $\mathbb{R}^{r_1 \cdots r_{k-1} r_{k+1} \cdots r_n}$. For example,

$$A_{32} = \begin{bmatrix} 1 & 1 & 1 & 0 & 0 & 0 \\ 0 & 0 & 0 & 1 & 1 & 1 \\ 1 & 0 & 0 & 1 & 0 & 0 \\ 0 & 1 & 0 & 0 & 1 & 0 \\ 0 & 0 & 1 & 0 & 0 & 1 \end{bmatrix}, \quad A_{222} = \left[\begin{array}{cccc|cccc} 1 & 1 & & & & & & \\ & & 1 & 1 & & & & \\ & & & & 1 & 1 & & \\ & & & & & & 1 & 1 \\ \hline 1 & & & 1 & & & & \\ & & 1 & & 1 & & & \\ & 1 & & & & 1 & & \\ & & & 1 & & & & 1 \\ \hline 1 & 1 & & & & & & \\ & & 1 & 1 & & & & \\ & & & & 1 & & & 1 \\ & & & & & 1 & 1 & \end{array} \right].$$

We now present two basic and important propositions.

Proposition 5.7.8. *A configuration matrix $A_{r_1 r_2 \cdots r_n 2}$ is isomorphic to the Lawrence lifting of a configuration matrix $A_{r_1 r_2 \cdots r_n}$.*

In general, for a configuration matrix A and its subconfiguration matrix B, $K[B]$ is called a *combinatorial pure subring* [6] of $K[A]$ if there exists a face F of CONV(A) such that $B = A \cap F$. For example, if $K[B] = K[A] \cap K[t_{i_1}, \ldots, t_{i_s}]$ holds, then $K[B]$ is a combinatorial pure subring of $K[A]$. If $K[B]$ is a combinatorial pure subring of $K[A]$, then it is known that major ring-theory properties, such as normality, are inherited.

Proposition 5.7.9. *If a configuration matrix $A_{r_1 r_2 \cdots r_n}$ and a configuration matrix $A_{s_1 s_2 \cdots s_n}$ satisfy $s_i \leq r_i$ for all $1 \leq i \leq n$, then $K[A_{s_1 s_2 \cdots s_n}]$ is a combinatorial pure subring of $K[A_{r_1 r_2 \cdots r_n}]$.*

Since the configuration matrix A_{333} is not unimodular, by Theorem 5.6.10, we have the following.

Proposition 5.7.10. *A configuration matrix $A_{r_1 r_2 \cdots r_n}$ is unimodular if and only if either $n = 2$ or $r_3 = 2$ holds.*

For compressed configurations, Sullivant [10] proved the following proposition.

Proposition 5.7.11. *A configuration matrix $A_{r_1 r_2 \cdots r_n}$ is compressed if and only if it satisfies one of the following.*

(i) $n = 2$,
(ii) $n \geq 3$ and $r_3 = 2$,
(iii) $n = 3$ and $r_2 = r_3 = 3$.

By using Proposition 5.7.9, we have the following proposition.

Proposition 5.7.12 ([5]). *If a configuration matrix $A_{r_1 r_2 \cdots r_n}$ satisfies one of the following, then $K[A_{r_1 r_2 \cdots r_n}]$ is not normal.*

(i) $n \geq 4$ and $r_3 \geq 3$,
(ii) $n = 3$ and $r_3 \geq 4$,
(iii) $n = 3$, $r_3 = 3$, $r_1 \geq 6$, and $r_2 \geq 4$.

By the above results, the remaining configurations whose properties are unknown are only A_{553}, A_{543}, and A_{433}. These examples became targets for developers of software. Finally, using the software programs 4ti2 and Normaliz2.5, it was verified that $K[A_{553}]$ is normal (see [1]). Therefore, by Proposition 5.7.9, it follows that $K[A_{543}]$ and $K[A_{433}]$ are normal. Summarizing these results, the classification is as follows.

Type of contingency tables	Ring-theoretic properties
$r_1 \times r_2$ or $r_1 \times r_2 \times 2 \times \cdots \times 2$	Unimodular
$r_1 \times 3 \times 3$	Compressed, but not unimodular
$4 \times 4 \times 3$	Normal
$5 \times 5 \times 3, 5 \times 4 \times 3$	but not compressed
otherwise, i.e., $n \geq 4$ and $r_3 \geq 3$ $n = 3$ and $r_3 \geq 4$ $n = 3$, $r_3 = 3$, $r_1 \geq 6$ and $r_2 \geq 4$	Not normal

Classification is almost complete; however, it is not yet known, for example, whether A_{553} has a unimodular triangulation.

References

1. W. Bruns, R. Hemmecke, B. Ichim, M. Köppe, C. Söger, Challenging computations of Hilbert bases of cones associated with algebraic statistics. Exp. Math. **20**, 25–33 (2011)
2. I.M. Gelfand, M.M. Kapranov, A.V. Zelevinski, *Discriminants, Resultants, and Multidimensional Determinants*. Mathematics: Theory & Applications (Birkhauser, Boston, 1994)
3. H. Ohsugi, T. Hibi, Normal polytopes arising from finite graphs. J. Algebra **207**, 409–426 (1998)
4. H. Ohsugi, T. Hibi, A normal (0,1)-polytope none of whose regular triangulations is unimodular. Discrete Comput. Geom. **21**, 201–204 (1999)
5. H. Ohsugi, T. Hibi, Toric ideals arising from contingency tables, in *Commutative Algebra and Combinatorics*, ed. by W. Bruns. Ramanujan Mathematical Society Lecture Notes Series, Number 4 (Ramanujan Mathematical Society, Mysore, 2007), pp. 91–115
6. H. Ohsugi, J. Herzog, T. Hibi, Combinatorial pure subrings. Osaka J. Math. **37**, 745–757 (2000)
7. M. Saito, B. Sturmfels, N. Takayama, *Gröbner Deformations of Hypergeometric Differential Equations*. Algorithms and Computation in Mathematics, vol. 6 (Springer, Berlin, 2000)
8. A. Simis, W.V. Vasconcelos, R.H. Villarreal, The integral closure of subrings associated to graphs. J. Algebra **199**, 281–289 (1998)
9. B. Sturmfels, *Gröbner Bases and Convex Polytopes* (American Mathematical Society, Providence, 1996)
10. S. Sullivant, Compressed polytopes and statistical disclosure limitation. Tohoku Math. J. **58**, 433–445 (2006)

11. R.R. Thomas, *Lectures in Geometric Combinatorics*. Student Mathematical Library, IAS/Park
    City Mathematical Subseries, vol. 33 (American Mathematical Society, Providence, 2006)
12. G.M. Ziegler, *Lectures on Polytopes* (Springer, New York, 1995)

Chapter 6
Gröbner Basis for Rings of Differential Operators and Applications

Nobuki Takayama

Abstract We introduce the theory and present some applications of Gröbner bases for the rings of differential operators with rational function coefficients R and for those with polynomial coefficients D.

The discussion with R, in the first half, is elementary. In the ring of polynomials, zero-dimensional ideals form the biggest class, and this is also true in R. However, in D, there is no zero-dimensional ideal, and holonomic ideals form the biggest class. Most algorithms for D use holonomic ideals.

As an application, we present an algorithm for finding local minimums of holonomic functions; it can be applied to the maximum-likelihood estimate.

The last part of this chapter considers A-hypergeometric systems; topics covered in other chapters will reappear in the study of A-hypergeometric systems. We have provided many of the proofs, but some technical proofs in the second half of this chapter have been omitted; these may be found in the references at the end of this chapter.

6.1 Gröbner Basis for the Ring of Differential Operators with Rational Function Coefficients R

A rational expression in $x_1, \ldots, x_n$ can be expressed as f/g where f, g are polynomials in $x_1, \ldots, x_n$ with complex number coefficients and $g \neq 0$. Since 1 is a polynomial, any polynomial can be regarded as a rational expression by setting $g = 1$. The field of rational expressions in n variables is denoted by $\mathbf{C}(x_1, \ldots, x_n)$. The operations of addition, subtraction, multiplication, and division can be performed in this field. Although in some discussions, the coefficient field $\mathbf{C}$

N. Takayama (✉)
Department of Mathematics, Graduate School of Science, Kobe University, Rokko,
Nadaku, Kobe 657-8501, Japan
e-mail: takayama@math.kobe-u.ac.jp

T. Hibi (ed.), *Gröbner Bases: Statistics and Software Systems*,
DOI 10.1007/978-4-431-54574-3_6, © Springer Japan 2013

may be replaced with the field of rational numbers or any field of characteristic 0, for simplicity, we will assume that the coefficient field is $\mathbf{C}$ throughout this chapter. Let $a(x) = a(x_1, \ldots, x_n)$ be a rational expression. We define the multiplication of differential operators ∂_i and $a(x)$ by

$$\partial_i a(x) = a(x)\partial_i + \frac{\partial a(x)}{\partial x_i}. \qquad (6.1)$$

We define that ∂_i and ∂_j commute. These rules, the associative law, and the distributive law can be extended to the sums and products of differential operators. The rational expressions in n variables, the differential operators $\partial_1, \partial_2, \ldots, \partial_n$, and these rules generate a ring, which is called the ring of differential operators with rational function coefficients and denoted by R_n:

$$R_n = \mathbf{C}(x_1, \ldots, x_n)\langle \partial_1, \ldots, \partial_n \rangle. \qquad (6.2)$$

When it is not necessary to specify the number of variables n, it can be omitted. We denote $\{0, 1, 2, \ldots\}$ by $\mathbf{N}_0$.

Any element of R can be expressed in the form $\sum_{\alpha \in E} a_\alpha(x)\partial^\alpha$, where, the ∂_i for each of the terms are collected to the right-most position. Here, ∂^α denotes the multi-index notation $\partial^\alpha = \partial_1^{\alpha_1} \cdots \partial_n^{\alpha_n}$, and E is a finite subset of $\mathbf{N}_0^n$. For the multi-index α, we define $|\alpha|$ by $\alpha_1 + \cdots + \alpha_n$.

In order to aid understanding, here is an example of a calculation in R:

$$\partial_1^2 x_1^2 x_2^2 = x_2^2 \partial_1(\partial_1 x_1^2) = x_2^2 \partial_1(x_1^2 \partial_1 + 2x_1) = x_2^2 \partial_1(x_1^2 \partial_1) + x_2^2 \partial_1(2x_1)$$
$$= x_2^2(x_1^2 \partial_1^2 + 2x_1 \partial_1) + x_2^2(2x_1 \partial_1 + 2) = x_1^2 x_2^2 \partial_1^2 + 4x_1 x_2^2 \partial_1 + 2x_2^2.$$

The ring R_n can be regarded as an infinite-dimensional vector space over the field $\mathbf{C}(x_1, \ldots, x_n)$. When we consider R_n as a vector space, we ignore the multiplication structure of R_n, and instead look only at the structures of addition and scalar multiplication by rational expressions from the left. For example, in the case of $n = 2$, the basis of the linear vector space of R_2 is the set $\{1, \partial_1, \partial_2, \partial_1^2, \partial_1 \partial_2, \partial_2^2, \ldots\}$. In some contexts, R_n is primarily regarded as a vector space.

An element of R acts on a function f as

$$a(x)\partial_x^\alpha \bullet f(x) = a(x)\frac{\partial^{|\alpha|} f}{\partial x_1^{\alpha_1} \cdots \partial x_n^{\alpha_n}}. \qquad (6.3)$$

In order to distinguish between multiplication in R and action to a function by an element of R, we denote the second operation by the symbol $\bullet$. However, we will omit $\bullet$ if no confusion would arise. It is known that

$$(pq) \bullet f = p \bullet (q \bullet f) \qquad (6.4)$$

holds for $p, q \in R$. In other words, when action to a function by two elements in R is performed, the same result is obtained if the multiplication first takes place between the two elements in R, followed by action to the function, or if one acts to the function and then the second acts to the result. The multiplication relation in R (6.1) is defined so that the above identity holds.

This is an intuitive introduction to action to functions by elements in R. To state this rigorously, we define a map $R \times \mathscr{F} \to \mathscr{F}$ by (6.3), where $\mathscr{F}$ is the additive group of the smooth functions. It satisfies the axioms of the action described in Sect. 6.7, but since we will focus on the computational aspects of R, the proof is omitted.

Theorem 6.1.1. *Set* $n = 1$. *When* $k > 0$, *we define* $[\alpha]_k = \alpha(\alpha - 1) \cdots (\alpha - k + 1)$. *When* $k = 0$, *we set* $[\alpha]_0 = 1$. *We denote* x_1 *by* x, *and* ∂_1 *by* ∂_x. *Then, we have*

$$\partial_x^\alpha a(x) = \sum_{k=0}^{\alpha} \frac{1}{k!} \frac{\partial^k a}{\partial x^k} [\alpha]_k \partial_x^{\alpha-k}. \tag{6.5}$$

Proof. We prove this by induction on α. When $\alpha = 1$, this is just the definition of multiplication. We denote by $a^{(k)}(x)$ the k-th derivative of $a(x)$. Since we have $\partial_x^\alpha a(x) = \partial_x(\partial_x^{\alpha-1} a(x))$, we obtain

$$\partial_x^\alpha a(x) = \partial_x \sum_{k=0}^{\alpha-1} \frac{1}{k!} a^{(k)} [\alpha - 1]_k \partial_x^{\alpha-1-k}$$

by the induction hypothesis. The expression on the right-hand side can be rewritten as

$$\sum_{k=0}^{\alpha-1} \frac{1}{k!} a^{(k+1)} [\alpha - 1]_k \partial_x^{\alpha-(1+k)} + \sum_{k=0}^{\alpha-1} \frac{1}{k!} a^{(k)} [\alpha - 1]_k \partial_x^{\alpha-k}.$$

Collecting the coefficients of $a^{(k)} \partial_x^{\alpha-k}$, we can see that the coefficient is equal to $\frac{1}{(k-1)!} [\alpha - 1]_{k-1} + \frac{1}{k!} [\alpha - 1]_k$. This can be simplified to $\frac{1}{k!} [\alpha - 1]_{k-1} (k + (\alpha - 1 - k + 1)) = \frac{1}{k!} [\alpha]_k$. Here, $0! = 1$. We note that the collection of the coefficients should be carefully performed on the border values of $k = 0$ and $k = \alpha - 1$. We suggest that the reader carefully considers the results at the borders.

In the above theorem, we note that the expression $\alpha(\alpha - 1) \cdots (\alpha - k + 1) \partial_x^{\alpha-k}$ can be written as $\frac{\partial^k \xi^\alpha}{\partial \xi^k}|_{\xi \to \partial_x}$. In other words, differentiating the polynomial ξ^α by ξ for k-times, and replacing ξ by ∂, we obtain the expression $[\alpha]_k \partial_x^{\alpha-k}$. This expression will be used in the theorem below.

The case for a general n can be proved analogously (see the introductory book by Oaku [26, Proposition 2.5, p. 55] and [29, Theorem 1.1.1, p. 3]).

Theorem 6.1.2 (Leibniz Formula). *Let p, q be polynomials in $x_1, \ldots, x_n, \partial_1, \ldots, \partial_n$, and the ∂ are collected to the right-most position in each term. We set $k! = k_1! \cdots k_n!$ for a multi-index k. We have*

$$p(x, \partial) q(x, \partial) = \sum_k \frac{1}{k!} \frac{\partial^{|k|} p(x, \xi)}{\partial \xi^k} \frac{\partial^{|k|} q(x, \xi)}{\partial x^k} |_{\xi \to \partial}. \qquad (6.6)$$

Here, the replacement of the commutative variable ξ by ∂ is done after collecting ξ to the right-most position in each term. We note that the right-hand side is a finite sum.

Until the end of the next proposition, we will assume that x is a single variable. We denote by θ_x the operator $x \partial_x$. The operator θ_x is called *the Euler operator*. The Leibniz formula is useful, and the following identities for the Euler operators are also useful for computations in R.

Proposition 6.1.3. *Let $b(\theta_x)$ be a polynomial in θ_x. Then, the following identities hold:*

1. $b(\theta_x) \bullet x^n = b(n) x^n$;
2. $x^k \partial_x^k = \theta_x (\theta_x - 1) \cdots (\theta_x - k + 1)$;
3. $\partial_x^k b(\theta_x) = b(\theta_x + k) \partial_x^k$;
4. $x^k b(\theta_x) = b(\theta_x - k) x^k$.

Let $\prec$ be a monomial order in the ring of polynomials $\mathbf{C}[\xi_1, \ldots, \xi_n]$ (see Sect. 1.1.4); we call this a *term order* (we do this to be consistent with the terminology of the book [29]). The term order in the ring of polynomials naturally induces an order $\prec$ in R_n. In other words, we define the order as

$$a(x) \partial^\alpha \prec b(x) \partial^\beta \Leftrightarrow \xi^\alpha \prec \xi^\beta. \qquad (6.7)$$

We note that elements in $\mathbf{C}(x_1, \ldots, x_n)$ are regarded as coefficients. Consider $f \in R$. We expand f so that the ∂ are collected to the right-most position in each term. Assume that the leading term of the expanded f by the order $\prec$ is $a_\alpha(x) \partial^\alpha$. We define the $\prec$-initial term of f by

$$\mathrm{in}_\prec(f) = a_\alpha(x) \xi^\alpha \in \mathbf{C}(x_1, \ldots, x_n)[\xi_1, \ldots, \xi_n]. \qquad (6.8)$$

Here, $\mathbf{C}(x_1, \ldots, x_n)[\xi_1, \ldots, \xi_n]$ is the ring of polynomials in $\xi_1, \ldots, \xi_n$ with rational function coefficients. Note that the coefficient of the initial term has been normalized to 1 in the previous chapters, but we do not do that here. Two elements which are not necessarily monomials are compared by their initial terms.

We now fix a term order for the sequence. The theory of Gröbner bases for R with a term order is analogous to that for the ring of polynomials. We will provide a sketch of this, and we suggest that the reader develop the proofs to the claims by referring to the analogous proofs for the ring of polynomials. We note that important constructions in D, which is the ring of differential operators with polynomial coefficients, require non-term orders, and they are no longer analogous to the case of the ring of polynomials. Some of these, for example, the integration algorithm, will be discussed further below.

We assume that $\alpha_i \leq \beta_i$, $1 \leq i \leq n$ hold for $a_\alpha(x)\xi^\alpha$, $a_\beta(x)\xi^\beta$. When they hold, we say that the term $a_\alpha(x)\xi^\alpha$ divides $a_\beta(x)\xi^\beta$. Note that the expressions $a_\alpha(x)$ and $a_\beta(x)$ are regarded as coefficients, and they are ignored when determining if one expression divides another. The following algorithm is an R analog of the division algorithm for the ring of polynomials, that was provided in Theorem 1.2.1.

Algorithm 6.1.4. NormalForm(f, G)

Input: $f \in R$, $G = \{g_1, \ldots, g_m\} \subset R$

Output: The normal form $r \in R$ (which is also called a remainder) and the multi-quotient $q_1, \ldots, q_m$ where the following relations hold: (a) $f = \sum_{i=1}^{m} q_i g_i + r$; (b) $f \succeq q_i g_i$; and (c) the monomial $\text{in}_\prec(g_i)$ does not divide any term of $r|_{\partial \to \xi}$.

1. $r \leftarrow 0$, $q_i \leftarrow 0$.
2. Call the procedure wNormalForm(f, G). Let $r', q_1', \ldots, q_m'$ be the outputs of the procedure.
3. $f \leftarrow r' - \text{in}_\prec(r')|_{\xi \to \partial}$, $r \leftarrow r + \text{in}_\prec(r')|_{\xi \to \partial}$, $q_i \leftarrow q_i + q_i'$. If $f = 0$ holds, then return $r, q_1, \ldots, q_m$, else go to 2.

When we make the replacement of $\xi \to \partial$, the variables ξ are collected on the right of each term.

Algorithm 6.1.5. (wNormalForm(f, G)) (weak normal form)

1. $r \leftarrow f$, $q_i \leftarrow 0$.
2. If there exists $\text{in}_\prec(g_i)$ which divides $\text{in}_\prec(r)$,
   rewrite $r \leftarrow r - c(x)\partial^\beta g_i$ and   rewrite $q_i \leftarrow q_i + c(x)\partial^\beta$.
   Here, $c(x)\partial^\beta$ is chosen so that $\text{in}_\prec(r) - c(x)\xi^\beta \text{in}_\prec(g_i) = 0$ holds.
   If no $\text{in}_\prec(g_i)$ exists which divides the initial term of r, then return $r, q_1, \ldots, q_m$.
3. Go to 2.

A non-empty subset I of R is called *a left ideal* of R when it satisfies the following two conditions:

1. For any $f, g \in I$, $f - g \in I$ holds;
2. For any $f \in I$ and any $r \in R$, $rf \in I$ holds.

If the condition $rf \in I$ in the second condition is replaced by the condition $fr \in I$, the set I is called *a right ideal*.

Example 6.1.6. We derive the normal form of $f = \partial_1 \partial_2^3$ by $g_1 = \partial_1 \partial_2 + 1$ and $g_2 = 2x_2 \partial_2^2 - \partial_1 + 3\partial_2 + 2x_1$ with the (graded) reverse lexicographic order (the initial terms are underlined).

$$\partial_1 \partial_2^3 - \partial_2^2 g_1 = -\partial_2^2$$

$$-\partial_2^2 + \frac{1}{2x_2} g_2 = \frac{1}{2x_2}(-\partial_1 + 3\partial_2 + 2x_1) =: f^*.$$

The normal form is f^* and the multi-quotient is $q_1 = \partial_2^2$, $q_2 = -\frac{1}{2x_2}$.

When $g = \text{NormalForm}(f, G)$ holds, we sometimes denote it by

$$f \longrightarrow^* g \quad \text{by } G.$$

The arrow indicates the division algorithm, which rewrites f as g. In this sense, division is often called *reduction*. Each step of the division $f - qg = r$ is denoted by

$$f \to r \quad \text{by } g.$$

Let I be a subset of R. The symbol $\text{in}_\prec(I)$ denotes the set $\{\text{in}_\prec(f) \mid f \in I\}$ or the ideal generated by the set. In most cases, the symbol denotes the ideal, but in some contexts, it denotes the set.

Definition 6.1.7. Let I be a left ideal in R. When $G = \{g_1, \ldots, g_m\}$ is a set of generators of I and satisfies the condition

$$\text{in}_\prec(I) = \langle \text{in}_\prec(g_1), \ldots, \text{in}_\prec(g_m) \rangle,$$

the set G is called a *Gröbner basis* of I with respect to the order $\prec$.

The existence of a Gröbner basis for a given left ideal can be proved by Dickson's lemma (Lemma 1.1.3).

Assume that $\text{in}_\prec(g_i) = a(x)\xi^u$, $\text{in}_\prec(g_j) = b(x)\xi^v$. Define the integer vector c by

$$c = (\max(u_1, v_1), \ldots, \max(u_n, v_n)),$$

and define the *S-polynomial* (*S*-differential operator) of g_i and g_j by

$$\text{sp}(g_i, g_j) = \partial^{c-u} g_i - \frac{a(x)}{b(x)} \partial^{c-v} g_j.$$

Note that $a(x)$ and $b(x)$ are coefficients. The S-polynomial is obtained by canceling the initial terms of g_i and g_j. We may modify the definition of the S-polynomial as

$$\text{sp}(g_i, g_j) = b(x)\partial^{c-u} g_i - a(x)\partial^{c-v} g_j.$$

With these definitions, the following theorem holds.

Theorem 6.1.8. *Let G be a set of generators of a left ideal I in R. If $\text{sp}(g_i, g_j) \longrightarrow^* 0$ by G holds for any pair g_i and g_j of elements of G, then the set G is a Gröbner basis of I.*

The proof is analogous to the case of the ring of polynomials (the Theorem 1.3.3 (Buchberger's criterion)) under our definition of the normal form and the S-polynomial. The Buchberger algorithm is analogous to the case of the ring of polynomials.

Example 6.1.9. We set $n = 2$, and we use the notation $x_1 = x, x_2 = y$. Let $\prec$ be the graded lexicographic order such that $\partial_x \succ \partial_y$. Set $f_1 = \partial_x^2 + y^2$, $f_2 = \partial_y^2 + x^2$. Here, the underlined terms are the $\prec$-initial terms. Consider the left ideal I of R generated by f_1, f_2. Let us derive a Gröbner basis for I. It follows from

$$\mathrm{sp}(f_1, f_2) = \partial_y^2 f_1 - \partial_x^2 f_2 = y^2 \partial_y^2 + 4y \partial_y + 2 - (x^2 \partial_x^2 + 4x \partial_x + 2)$$

that

$$\mathrm{sp}(f_1, f_2) = \underline{-x^2 \partial_x^2} + y^2 \partial_y^2 - 4x \partial_x + 4y \partial_y$$

$$\rightarrow \underline{y^2 \partial_y^2} - 4x \partial_x + 4y \partial_y + x^2 y^2 \text{ by } f_1$$

$$\rightarrow -4x \partial_x + 4y \partial_y \text{ by } f_2.$$

Set $f_3 = \underline{x \partial_x} - y \partial_y$. Then, we have

$$\mathrm{sp}(f_3, f_1) = \partial_x f_3 - x f_1 \rightarrow^* 0 \text{ by } \{f_1, f_2, f_3\}$$

and

$$\mathrm{sp}(f_3, f_2) = \partial_y^2 f_3 - x \partial_x f_2 \rightarrow^* 0 \text{ by } \{f_1, f_2, f_3\}.$$

Therefore, the set $\{f_1, f_2, f_3\}$ is a Gröbner basis. The reduced Gröbner basis is $\{f_2, f_3\}$.

We note that criterion 1 given in Lemma 1.3.1, which claims that the S-polynomial is reduced to 0 if the initial terms are relatively prime, holds only in the ring of polynomials and not in R. Here is a counterexample; if this lemma holds in R, the S polynomial of f_1 and f_2 is reduced to 0, and consequently $\{f_1, f_2\}$ is a Gröbner basis; this contradicts our previous calculation.

The following theorem can be shown analogously to that for the case of the ring of polynomials.

Theorem 6.1.10. *Let G be a Gröbner basis of a left ideal I in R.*

1. *The normal form r of $f \in R$ by G is unique (see Lemma 1.2.4 for the uniqueness of the remainder for the division algorithm).*
2. *The necessary and sufficient condition that f belongs to I is that the normal form of f by G is 0 (see Corollary 1.2.5, the ideal membership problem).*
3. *The standard monomials of G form a vector space basis of R/I over the field $\mathbf{C}(x_1, \ldots, x_n)$ (see Theorem 1.6.9, Macaulay's theorem).*

Here, we take a *standard monomial* to mean a monomial ∂^α which is obtained by replacing ξ by ∂ for a monomial ξ^α which does not belong to the ideal $\mathrm{in}_\prec(I)$.

When the quotient space R/I is a finite-dimensional vector space over $\mathbf{C}(x)$, the left ideal I is called a 0-*dimensional ideal*. Let G be a Gröbner basis of I with respect to $\prec$. The necessary and sufficient condition that I is a 0-dimensional ideal is that the number of standard monomials for $\mathrm{in}_{\prec}(G)$ is finite.

The following theorem holds as an analog to the elimination theorem 1.4.1 for the ring of polynomials; both the proof and the elimination algorithm are also analogous.

Theorem 6.1.11. *If I is a 0-dimensional ideal in R, then, for any i, we have*

$$I \cap \mathbf{C}(x_1, \ldots, x_n)\langle \partial_i \rangle \neq \{0\}.$$

The converse also holds.

The intersection $I \cap \mathbf{C}(x_1, \ldots, x_n)\langle \partial_i \rangle$ is a left ideal in the ring of differential operators in one variable with coefficients in $\mathbf{C}(x_1, \ldots, x_n)$, which is a principal ideal domain. It is generated by a single element. The generator can be regarded as a linear ordinary differential operator with respect to the variable x_i with parameters $x_1, \ldots, x_{i-1}, x_{i+1}, \ldots, x_n$.

6.2 Zero-Dimensional Ideals in R and Pfaffian Equations

Let G be a Gröbner basis of a 0-dimensional ideal I in R. Assume that the monomial ∂^β is a standard monomial with respect to the basis G. An expression of the form $c(x)\partial^\beta$, $0 \neq c(x) \in \mathbf{C}(x_1, \ldots, x_n)$ is called a (nonmonic) standard monomial. Let $S = \{s_1 = 1, s_2, \ldots, s_r\}$ be a set of linearly independent nonmonic standard monomials, and assume that $r = \sharp S = \dim_{\mathbf{C}(x_1, \ldots, x_n)} R/I$. In other words, the set S is a vector space basis of R/I. When a function $f(x)$ of the variables $x_1, \ldots, x_n$ is a solution of any operator of I, the function f is called *a solution* of I. In other words, when $\ell \bullet f = 0$ holds for any $\ell \in I$, f is a solution of I. When $\ell \bullet f = 0$, we say that f is *annihilated* by ℓ, and when $\ell \bullet f = 0$ holds for any $\ell \in I$, we say that f is annihilated by I. If the left ideal I is generated by $\ell_1, \ldots, \ell_p$, the condition $\ell \bullet f = 0$ for any $\ell \in I$ is equivalent to $\ell_i \bullet f = 0$, $i = 1, \ldots, p$. Let f be a solution of I. We set $Q = (s_j \bullet f \mid j = 1, \ldots, r)^T$. The normal form of $\partial_i s_j$ by G can be written as $\sum_k c^i_{jk} s_k$. Here, c^i_{jk} is an element of $\mathbf{C}(x_1, \ldots, x_n)$. Let P_i be an $r \times r$ square matrix of which the (j, k)-th entry is c^i_{jk}. Since we have $\ell \bullet f = 0$, $\ell \in I$, the following identities hold:

$$\frac{\partial Q}{\partial x_i} = P_i Q, \quad i = 1, \ldots, n. \tag{6.9}$$

This system is called a *Pfaffian system* or a *Pfaffian system of equations*. The zero set of the least common multiple of the denominator polynomials of c^i_{jk} is called the singular locus of the Pfaffian system.

Example 6.2.1. Set $n = 1$, and set $I = \langle \partial_x^2 - x \rangle$. For any term order, we have a Gröbner basis $G = \{\partial_x^2 - x\}$. We may assume that $S = \{1, \partial_x\}$. The normal form of $\partial_x 1 = \partial_x$ is ∂_x, and the normal form of $\partial_x \partial_x$ is $x = x \cdot 1$. Then, we have $P_1 = \begin{pmatrix} 0 & 1 \\ x & 0 \end{pmatrix}$. In the case of $n = 1$, I is a left ideal generated by an ordinary differential operator, and the transformation to a Pfaffian system is nothing but the well-known transformation of an ordinary differential equation of higher order to a system of first-order ordinary differential equations.

Example 6.2.2. Let us derive a Pfaffian system by using the Gröbner basis of Example 6.1.9. We have

$$\partial_x \to^* (y/x)\partial_y \text{ by } G$$

$$\partial_x \partial_y \to^* -xy + (1/x)\partial_y \text{ by } G$$

$$\partial_y \to^* \partial_y \text{ by } G$$

$$\partial_y^2 \to^* -x^2 \text{ by } G.$$

Then, setting $Q = (f, \partial_y \bullet f)^T$, we have

$$\frac{\partial Q}{\partial x} = \begin{pmatrix} 0 & y/x \\ -xy & 1/x \end{pmatrix} Q, \quad \frac{\partial Q}{\partial y} = \begin{pmatrix} 0 & 1 \\ -x^2 & 0 \end{pmatrix} Q.$$

Note that if we set $f = \cos(xy)$, the vector-valued function Q satisfies the Pfaffian system.

Theorem 6.2.3. *For the Pfaffian system (6.9), the relation*

$$\frac{\partial P_i}{\partial x_j} + P_i P_j = \frac{\partial P_j}{\partial x_i} + P_j P_i \tag{6.10}$$

holds for any i, j.

Proof. The proof of the general case uses complicated indices. In order to make the idea of the proof clear, we will first show the theorem for the case of $n = 2, r = 2$. We denote by $c_{k\ell}^i$ the (k, ℓ)-th element of P_i. It is a rational expression. It follows from the definition of P_i that the relations

$$\partial_1 \begin{pmatrix} s_1 \\ s_2 \end{pmatrix} = P_1 \begin{pmatrix} s_1 \\ s_2 \end{pmatrix} = \begin{pmatrix} c_{11}^1 & c_{12}^1 \\ c_{21}^1 & c_{22}^1 \end{pmatrix} \begin{pmatrix} s_1 \\ s_2 \end{pmatrix} \mod I$$

$$\partial_2 \begin{pmatrix} s_1 \\ s_2 \end{pmatrix} = P_2 \begin{pmatrix} s_1 \\ s_2 \end{pmatrix} = \begin{pmatrix} c_{11}^2 & c_{12}^2 \\ c_{21}^2 & c_{22}^2 \end{pmatrix} \begin{pmatrix} s_1 \\ s_2 \end{pmatrix} \mod I$$

hold in R_2^2. Here, the expression $\begin{pmatrix} a_1 \\ a_2 \end{pmatrix} = \begin{pmatrix} b_1 \\ b_2 \end{pmatrix}$ mod I indicates componentwise congruence, i.e., $a_1 - b_1 \in I$ and $a_2 - b_2 \in I$. Multiply both sides of the first identity by ∂_2, and multiply both side of the second identity by ∂_1. Since they are equal, the relation

$$\partial_2 \begin{pmatrix} c_{11}^1 & c_{12}^1 \\ c_{21}^1 & c_{22}^1 \end{pmatrix} \begin{pmatrix} s_1 \\ s_2 \end{pmatrix} = \partial_1 \begin{pmatrix} c_{11}^2 & c_{12}^2 \\ c_{21}^2 & c_{22}^2 \end{pmatrix} \begin{pmatrix} s_1 \\ s_2 \end{pmatrix} \text{ mod } I$$

holds in R_2^2. When c is a rational expression, it follows from the multiplication rule $\partial_i c = \frac{\partial c}{\partial x_i} + c \partial_i$ in R that we obtain the identity

$$\begin{pmatrix} \frac{\partial c_{11}^1}{\partial x_2} & \frac{\partial c_{12}^1}{\partial x_2} \\ \frac{\partial c_{21}^1}{\partial x_2} & \frac{\partial c_{22}^1}{\partial x_2} \end{pmatrix} \begin{pmatrix} s_1 \\ s_2 \end{pmatrix} + \begin{pmatrix} c_{11}^1 & c_{12}^1 \\ c_{21}^1 & c_{22}^1 \end{pmatrix} \partial_2 \begin{pmatrix} s_1 \\ s_2 \end{pmatrix}$$

$$= \begin{pmatrix} \frac{\partial c_{11}^2}{\partial x_1} & \frac{\partial c_{12}^2}{\partial x_1} \\ \frac{\partial c_{21}^2}{\partial x_1} & \frac{\partial c_{22}^2}{\partial x_1} \end{pmatrix} \begin{pmatrix} s_1 \\ s_2 \end{pmatrix} + \begin{pmatrix} c_{11}^2 & c_{12}^2 \\ c_{21}^2 & c_{22}^2 \end{pmatrix} \partial_1 \begin{pmatrix} s_1 \\ s_2 \end{pmatrix} \text{ mod } I.$$

A matrix presentation of this identity is

$$\frac{\partial P_1}{\partial x_2} \begin{pmatrix} s_1 \\ s_2 \end{pmatrix} + P_1 \partial_2 \begin{pmatrix} s_1 \\ s_2 \end{pmatrix} = \frac{\partial P_2}{\partial x_1} \begin{pmatrix} s_1 \\ s_2 \end{pmatrix} + P_2 \partial_1 \begin{pmatrix} s_1 \\ s_2 \end{pmatrix} \text{ mod } I.$$

By using $\partial_2 \begin{pmatrix} s_1 \\ s_2 \end{pmatrix} = P_2 \begin{pmatrix} s_1 \\ s_2 \end{pmatrix}$ mod I, $\partial_1 \begin{pmatrix} s_1 \\ s_2 \end{pmatrix} = P_1 \begin{pmatrix} s_1 \\ s_2 \end{pmatrix}$ mod I, we obtain

$$\frac{\partial P_1}{\partial x_2} \begin{pmatrix} s_1 \\ s_2 \end{pmatrix} + P_1 P_2 \begin{pmatrix} s_1 \\ s_2 \end{pmatrix} = \frac{\partial P_2}{\partial x_1} \begin{pmatrix} s_1 \\ s_2 \end{pmatrix} + P_2 P_1 \begin{pmatrix} s_1 \\ s_2 \end{pmatrix} \text{ mod } I.$$

If the relation

$$\frac{\partial P_1}{\partial x_2} + P_1 P_2 - \frac{\partial P_2}{\partial x_1} - P_2 P_1 = 0$$

does not hold, then there exists a linear dependence among s_1, s_2 in mod I. Since there is no dependence among standard monomials in mod I, then we obtain the above identity.

We now consider the case of general n and r. Multiplying both sides of $\partial_i s_k = \sum_\ell c_{k\ell}^i s_\ell$ mod I by ∂_j, we obtain

$$\partial_j \partial_i s_k = \sum_\ell \partial_j c_{k\ell}^i s_\ell = \sum_\ell \frac{\partial c_{k\ell}^i}{\partial x_j} s_\ell + \sum_\ell c_{k\ell}^i \partial_j s_\ell.$$

Rewriting $\partial_j s_\ell$, we obtain

$$\partial_j \partial_i s_k = \sum_\ell \frac{\partial c^i_{k\ell}}{\partial x_j} s_\ell + \sum_\ell c^i_{k\ell} \sum_m c^j_{\ell m} s_m \bmod I.$$

Exchanging the roles of i and j, we obtain

$$\partial_i \partial_j s_k = \sum_\ell \frac{\partial c^j_{k\ell}}{\partial x_i} s_\ell + \sum_\ell c^j_{k\ell} \sum_m c^i_{\ell m} s_m \bmod I.$$

Since the right-hand sides of both identities agree, we can obtain the conclusion by utilizing the linearly independent property of s_m in the vector space R/I.

6.3 Solutions of Pfaffian Equations

In this section, we prove the existence of solutions of a Pfaffian system. The theorem is proved by construction; thus the proof provides a method for constructing solutions. However, this method consists of infinite steps and these solutions are expressed as series of terms. We can obtain approximate solutions by performing a finite number of steps.

In order to avoid complicated indices, we consider the case of two variables; the general case is analogous. Let A and B each be an $r \times r$ matrix-valued formal power series. We will assume that they can be expressed as

$$A = \sum_{(p,q)\in \mathbf{N}_0^2} A_{pq} x^p y^q$$

$$B = \sum_{(p,q)\in \mathbf{N}_0^2} B_{pq} x^p y^q.$$

Here, A_{pq} and B_{pq} are $r \times r$ matrices for which the elements are complex numbers. The set running the indices p, q is omitted when it is clear in context. Two indices may be separated by a comma, such as in $A_{p+1,q}$. We note that a rational function (expression) f/g can be expanded into a power series if $g(0) \neq 0$.

In order to follow our proof, it is sufficient to consider that a formal power series is one that can be calculated as a polynomial with infinitely many terms. To understand our proof rigorously, please refer to textbooks on the theory of complex analysis [1, 15].

Theorem 6.3.1. *We assume that the integrability condition*

$$\frac{\partial A}{\partial y} + AB = \frac{\partial B}{\partial x} + BA$$

holds. Let C be an $r \times r$ constant matrix. Then, there exists a unique matrix-valued formal power series F in x, y satisfying

$$\frac{\partial F}{\partial x} = AF, \frac{\partial F}{\partial y} = BF, F(0,0) = C. \tag{6.11}$$

Moreover, if the series A and B converge at the origin, then the power series F converges at the origin.

Remark. Any column vector of the matrix F is a solution of the Pfaffian system.

Proof. Set $F = \sum F_{m,n} x^m y^n$. From the Pfaffian system, we have recurrence relations (difference equations) for $F_{m,n}$ where $F_{0,0} = C$ is the initial condition. Indices are separated by commas, as in $F_{m,n}$; the comma can be omitted if no confusion would arise. We now derive these recurrence relations. Note that the relations $x\partial_x x^m y^n = m x^m y^n$ and $y\partial_y x^m y^n = n x^m y^n$ hold. Multiplying both sides of $\partial_x F = AF$ by x, we have $x\partial_x F = xAF$. From the above relation, we have

$$x\partial_x F = \sum m F_{m,n} x^m y^n.$$

On the other hand, we have

$$xAF = x \left(\sum A_{p,q} x^p y^q \right) \left(\sum F_{r,s} x^r y^s \right)$$

$$= \sum_{m,n \in \mathbf{N}_0^2} x^{m+1} y^n \sum_{p+r=m,q+s=n} A_{p,q} F_{r,s}.$$

Comparing the coefficients of $x^{m+1} y^n$, we obtain the recurrence relation

$$(m+1) F_{m+1,n} = \sum_{p+r=m,q+s=n} A_{p,q} F_{r,s}. \tag{6.12}$$

From $\partial_y F = BF$, we analogously obtain the recurrence relation

$$(n+1) F_{m,n+1} = \sum_{p+r=m,q+s=n} B_{p,q} F_{r,s}. \tag{6.13}$$

Exercise. Derive the relation (6.13), specifying the computations in detail.

Note that the recurrence relations are overdetermined. For example, there are two ways to determine $F_{1,1}$: use $F_{0,0}$ to determine $F_{1,0}$ by using (6.12) and then determine $F_{1,1}$ by using (6.13), or use $F_{0,0}$ to determine $F_{0,1}$ by using (6.13) and then determine $F_{1,1}$ by using (6.12). To prove the existence of a solution, we must show that both ways give the same $F_{1,1}$. This can be shown by using the integrability condition.

Exercise. If the integrability condition does not hold, we may have two possible values of $F_{1,1}$, which implies that no solution exists. Give an example of a pair A, B for which this happens.

In order to prove the existence of a solution, we utilize a reduction of the Pfaffian system. The outline of our proof is as follows.

1. Set $F = (E - x^m y^n Q_{m,n})G$, and derive a Pfaffian system for G, where E is the identity matrix and $Q_{m,n}$ is a constant matrix to be determined later.
2. Prove that the Pfaffian system for G also satisfies the integrability condition. Hint 1: We denote by $'$ differentiation with respect to a variable. For square matrices P and Q, we have $(PQ)' = P'Q + PQ'$. Assume Q is the inverse matrix of P. Then, we have $(P^{-1})' = -P^{-1}P'P^{-1}$. Hint 2: Use the new variable $H = E - x^m y^n Q_{m,n}$, $(m,n) \neq (0,0)$ to avoid messy computations. Utilize the fact that $H_{xy} = H_{yx}$, which means that changing the order of differentiation does not change the output. Hint 3: Let the inverse matrix H be $\sum R_{mn} x^m y^n$, derive a recurrence to determine R_{mn}, and prove that there exists an inverse matrix H expressed as a formal power series. Moreover show that $R_{00} = E$ and $R_{ij} = 0$, $(0 < i + j < m + n)$.
3. Choose suitable constant matrices $Q_{m,n}$ such that the composite of the transformations of the form in item 1 translates the Pfaffian system into the trivial system $\partial_x F = 0, \partial_y F = 0$.

We now begin our proof.

1. Let $H = E - x^m y^n Q_{m,n}$ where E is the identity matrix and $Q_{m,n}$ is an undetermined constant matrix. Set $F = HG$, and substitute HG for F in the Pfaffian system. Then, we have $HG_x = (AH - H_x)G$, $HG_y = (BH - H_y)G$, where G_x and G_y are partial derivatives of G with respect to x and y, respectively, and H_x and H_y are partial derivatives of H with respect to x and y, respectively. Multiplying both sides by H^{-1}, we obtain

$$G_x = H^{-1}(AH - H_x)G, \quad G_y = H^{-1}(BH - H_y)G.$$

2. We leave as exercises the computation of the inverse of H and the proof that the Pfaffian for G satisfies the integrability condition.
3. Assume that the matrix A is expressed as $A = \sum_n A_{0n} y^n + x^{m-1} \sum_n A_{m-1,n} y^n + O(x^m)$. Since we have $H^{-1} = E + x^m y^n Q_{m,n} + O(x^{2m})$ and $H_x = -mx^{m-1}y^n Q_{m,n}$, we obtain the identity

$$H^{-1}(AH - H_x) = \sum_n A_{0n} y^n + x^{m-1} \sum_n A_{m-1,n} y^n - mx^{m-1} y^n Q_{m,n} + O(x^m).$$

Hence, if we choose $Q_{m,n}$ such that $mQ_{m,n} = A_{m-1,n}$, then the coefficient of $x^{m-1}y^n$ in $H^{-1}(AH - H_x)$, which will be called a new A, can be made to equal 0. We make an analogous computation for the y direction. Assume that the matrix B is expressed as $B = \sum_m B_{m0} x^m + y^{n-1} \sum_m B_{m,n-1} x^m + O(y^n)$. We have

$$H^{-1}(BH - H_y) = \sum_m B_{m0}x^m + y^{n-1} \sum_m B_{m,n-1}x^m - nx^m y^{n-1}Q_{m,n} + O(y^n).$$

We will call $H^{-1}(BH - H_y)$ a new B. If we choose $Q_{m,n}$ such that $nQ_{m,n} = B_{m,n-1}$ holds, then we can make the coefficient of $x^m y^{n-1}$ of the new B equal to 0.

Repeating the substitution $F = HG$ and the construction of H, we update the matrix A. After infinite repetitions, we finally obtain an A such that $A(y) = \sum_n A_{0n}y^n$ holds. Set $F = \exp(xA(y))G$, and derive a Pfaffian system for G. Replace G with F, and the new Pfaffian system can be written as $F_x = 0$, $F_y = BF$. We apply the integrability condition, and obtain $B_x = 0$, which means that B depends only on y. We apply an analogous transformation for $F_y = B(y)F$. After infinite repetitions of this type of transformation, we finally obtain the equation $F_y = B_{00}F$. Since these transformations have the form $H = E - y^n Q_{0n}$, the equation $F_x = 0$ retains this form under the transformation. In conclusion, we obtain the trivial Pfaffian system $F_x = 0$, $F_y = 0$ by the transformation $F = \exp(yB_{00})G$. The solution of the trivial Pfaffian system is a constant matrix. Multiplying the constant matrix by the composite of the transformation matrices H's, we obtain a solution of the original Pfaffian system.

It is suggested that the reader complete the proof sketched above. Completing the details after hearing a sketch allows for a deeper understanding than is obtained by simply reading a proof; this is especially true for theorems on power series.

Uniqueness follows from the recurrence relations.

Exercise. Give an example of a pair A and B which satisfies the integrability condition, and determine a power series solution.

We have proved the unique existence of the power series solution. We now prove the convergence of the matrix-valued power series solution. In order to prove convergence, we need to use the elementary theory of complex analysis (see, e.g., [1] or [15]). For a matrix $C = (c_{ij})$, we define the norm of C by $|C| = \max|c_{ij}|$. This notation can be confused with that for the determinant, but in this proof, we will use only the norm. We have the estimate $|C_1 C_2| \le r|C_1||C_2|$ for $r \times r$ matrices C_i.

We will prove that the series $\sum |F_{pq}||x|^p|y|^q$ converges when $|x|, |y| < \epsilon$ for sufficiently small ϵ. Here, by the convergence of $\sum |F_{pq}||x|^p|y|^q$, we mean that the partial sum $\sum_{p+q<N} |F_{pq}||x|^p|y|^q$ converges when $N \to \infty$.

Since A and B are matrix-valued convergent power series, there exist positive numbers C and α such that $|A_{pq}|, |B_{pq}| \le C\alpha^{p+q}$. We note that $\sum_{pq} C\alpha^{p+q}x^p y^q = \frac{C}{(1-\alpha x)(1-\alpha y)}$ holds. The constant matrices $F_{m,n}$ are determined by (6.12) and (6.13), and we have the inequality

$$(m + 1)|F_{m+1,n}| \le \sum_{p+r=m, q+s=n} r|A_{pq}||F_{rs}|.$$

(The index r in the expression does not mean the matrix size; it is used to avoid messy indices.) Consider the sequence $f_{m,n}$ determined by

$$(m+1) f_{m+1,n} = rC \sum_{p+r=m,q+s=n} \alpha^{p+q} \binom{p+q}{p} f_{rs}$$

$$(n+1) f_{m,n+1} = rC \sum_{p+r=m,q+s=n} \alpha^{p+q} \binom{p+q}{p} f_{rs}.$$

If we set $f_{0,0} = |F_{00}|$, we have the estimation $|F_{mn}| \le f_{mn}$. Now, we consider the system of differential equations

$$\frac{\partial f}{\partial x} = \frac{rC}{(1 - \alpha x - \alpha y)} f, \quad \frac{\partial f}{\partial y} = \frac{rC}{(1 - \alpha x - \alpha y)} f.$$

The series solution of this system is $\sum f_{mn} x^m y^n$. Note that the solution is a constant multiple of the function $(1 - \alpha x - \alpha y)^{-rC/\alpha}$. Expanding it as a Taylor series, we obtain an estimation of $|F_{mn}|$, and consequently, we can prove the convergence of our formal series solution. This method of constructing the series $f_{m,n}$, which bounds the norm of $F_{m,n}$ from above, is called the method of majorant series. We do not try to estimate $f_{m,n}$ directly, but rather estimate it by solving a simple system of differential equations. This method can be applied to several problems.

In the numerical analysis of F, we usually use the finite difference method to obtain approximate values, as follows. Let h_x and h_y be sufficiently small positive numbers. Note that we have the system $\partial_x \bullet F = AF, \partial_y \bullet F = BF$. Since the partial derivative $\partial_x \bullet F$ is approximated by $(F(x + h_x, y) - F(x, y))/h_x$ and the partial derivative $\partial_y \bullet F$ is approximated by $(F(x, y + h_y) - F(x, y))/h_y$, we have

$$F(x + h_x, y) = F(x, y) + h_x A(x, y) F(x, y),$$

$$F(x, y + h_y) = F(x, y) + h_y B(x, y) F(x, y).$$

Here, the symbol $=$ indicates that the left- and right-hand sides are approximately equal. The left-hand sides are the values of F at $(x + h_x, y)$ and $(x, y + h_y)$, respectively. On the other hand, the right-hand sides are expressed in terms of only the value of F at (x, y). In summary, the values of F at $(x + h_x, y)$ and $(x, y + h_y)$ are approximately determined by the value of F at (x, y). By repeating this procedure, we can use an initial value of F to determine approximate values of F at several points. This method of obtaining approximate values of F is called the *finite difference method*.

Theorem 6.3.2. *Assume that matrix valued functions A and B are holomorphic on the closed domain $E = [a_x, b_x] \times [a_y, b_y]$. Let the initial value of F at $(a_x, a_y) \in E$ be $F_{0,0}$. Take sufficiently small positive numbers h_x and h_y and determine a sequence $F_{m,n}$ by the recurrences*

$$F_{m+1,n} = F_{m,n} + h_x A(mh_x, nh_y)F_{m,n},$$

$$F_{m,n+1} = F_{m,n} + h_y B(mh_x, nh_y)F_{m,n},$$

where the second recurrence is applied, followed by the first one. Set $M + 1 = \lfloor (b_x - a_x)/h_x \rfloor$ and $N + 1 = \lfloor (b_y - a_y)/h_y \rfloor$. Suppose that the function $F(x, y)$ is a solution of the Pfaffian system satisfying $F(0,0) = F_{0,0}$. Then there exists a constant C which does not depend on h_x, h_y, such that the estimation

$$|F(mh_x, nh_y) - F_{m,n}| \leq C\max(h_x, h_y)$$

holds for any m and n satisfying $0 \leq m \leq M, 0 \leq n \leq N$.

This theorem says that when h_x and h_y converge to 0, then the approximate solution found by the finite difference method converges to a solution of the Pfaffian system.

Proof. The solution F is holomorphic on E; in particular, the derivatives $F_{xx} = \partial_x^2 \bullet F$ and $F_{yy} = \partial_y^2 \bullet F$, are continuous. Without loss of generality, we may assume $a_x = 0 \leq x \leq b_x$ and $b_x = 0 \leq y \leq b_y$. In other words, we assume that the domain E is defined by these inequalities, and the initial value is given at the point $(x, y) = (0, 0)$. We denote AF by $g_1(x, y, F)$ and BF by $g_2(x, y, F)$. Note that g_i is a vector-valued function. We define the norm $|v|$ of an r-dimensional vector v as $\sqrt{v_1^2 + \cdots + v_r^2}$. Since the matrix-valued functions A and B do not have singularities in E, there exists a constant L_i such that the inequality

$$|g_i(x, y, F) - g_i(x, y, G)| \leq L_i|F - G|$$

holds for any $(x, y) \in E$ and any vector F, G. This property is called the Lipschitz continuity of g_i. Set $x_m = mh_x$ and $y_n = nh_y$. When F is a solution, we denote by f_{mn} the value of F at $(x, y) = (x_m, y_n)$; i.e., $f_{mn} = F(x_m, y_n)$. We denote by F_{mn} the sequence determined by the finite difference method. The value of F_{mn} may depend on the choices made when applying the recurrence relations, as explained above. We fix the choices by determining F_{0n} from F_{00} by recurrence with respect to n, and then determining F_{mn} by recurrence with respect to m.

When we need to clearly distinguish between two indices, we will separate them by a comma, as in $F_{m+1,n}$. Consider the Taylor expansion of F with respect to x:

$$f_{m+1,n} = F(x_m + h_x, y_n) = F(x_m, y_n) + F_x(x_m, y_n)h_x + \frac{1}{2}F_{xx}(x_m + \theta h_x, y_n)h_x^2.$$

Here, θ satisfies $0 < \theta < 1$. Rewriting the right-hand side by using the Pfaffian system, we obtain

$$f_{m+1,n} = f_{mn} + g_1(x_m, y_n, f_{mn})h_x + \frac{1}{2}F_{xx}(x_m + \theta h_x, y_n)h_x^2.$$

On the other hand, the sequence satisfies

$$F_{m+1,n} = F_{m,n} + h_x g_1(x_m, y_n, F_{mn}).$$

From these two identities, we obtain a recurrence for the error

$$f_{m+1,n} - F_{m+1,n}$$

$$= f_{mn} - F_{mn} + (g_1(x_m, y_n, f_{mn}) - g_1(x_m, y_n, F_{mn})) h_x + \frac{1}{2} F_{xx}(x_m + \theta h_x, y_n) h_x^2.$$

We obtain the following estimate of the error from the existence of a constant M' such that $\left|\frac{1}{2} F_{xx}(x_m + \theta h_x, y_n)\right| \leq M'$ holds on E and from the Lipschitz continuity of g_1:

$$|f_{m+1,n} - F_{m+1,n}| \leq |f_{mn} - F_{mn}| + L_1 |f_{mn} - F_{mn}| h_x + M' h_x^2.$$

Recursively applying this estimate, we obtain

$$|f_{m+1,n} - F_{m+1,n}|$$

$$\leq (1 + L_1 h_x)^{m+1} |f_{0n} - F_{0n}| + (1 + (1 + L_1 h_x) + \cdots + (1 + L_1 h_x)^m) M' h_x^2$$

$$\leq (1 + L_1 h_x)^{m+1} \left(|f_{0n} - F_{0n}| + \frac{M' h_x}{L_1} \right).$$

Analogously, we have

$$|f_{m,n+1} - F_{m,n+1}| \leq (1 + L_2 h_y)^{n+1} \left(|f_{m0} - F_{m0}| + \frac{M'' h_y}{L_2} \right).$$

In particular, we have

$$|f_{0,n+1} - F_{0,n+1}| \leq \frac{M'' h_y}{L_2} (1 + L_2 h_y)^{n+1}.$$

From these two estimates, we have

$$|f_{m,n} - F_{m,n}| \leq (1 + L_1 h_x)^m \left(\frac{M'' h_y}{L_2} (1 + L_2 h_y)^n + \frac{M' h_x}{L_1} \right).$$

Note that when $(x, y) \in E$, we have $m \leq b_x / h_x$ and $n \leq b_y / h_y$. We then can obtain the estimate $(1 + L_1 h_x)^m \leq (1 + L_1 b_x/m)^m \leq e^{L_1 b_x}$ and $(1 + L_2 h_y)^n \leq e^{L_2 b_y}$. Summarizing these estimates, we finally obtain the conclusion

$$|f_{mn} - F_{mn}| \leq \left(e^{L_1 b_x} e^{L_2 b_y} \frac{M''}{L_2} + e^{L_1 b_x} \frac{M'}{L_1} \right) \max(h_x, h_y).$$

Tip 1 When the finite difference method gives overdetermined recurrence relations, in general, the values of F_{mn} depend on the order in which the relations are applied. Finding good system of recurrence relations which satisfy the compatibility condition (or the difference analog of the integrability condition) is currently a very active topic of research in the theory of integrable systems.

Below, we will apply Pfaffian systems to the holonomic gradient descent; for that, we will need a system of the form (6.14), which is called an *inhomogeneous Pfaffian system*. When $U_i \neq 0$, the zero function $F = 0$ is not a solution of the inhomogeneous system. The following theorem gives a condition for the existence of a solution of the inhomogeneous Pfaffian system.

Theorem 6.3.3. *Let U_i be a vector-valued function which is holomorphic at the origin. The matrix-valued holomorphic function P_i satisfies the integrability condition (6.10) around the origin. The necessary and sufficient condition that the system*

$$\frac{\partial F}{\partial x_i} = P_i F + U_i, \quad i = 1, \ldots, n \tag{6.14}$$

has a matrix-valued holomorphic solution F, such that $F(0) = E$, is that

$$P_i U_j + \frac{\partial U_i}{\partial x_j} = P_j U_i + \frac{\partial U_j}{\partial x_i} \tag{6.15}$$

holds for any pair of i and j.

Proof. Assume that there exists a matrix-valued solution. Differentiate (6.14) by x_j. Replace i by j in (6.14) and differentiate it by x_i. We obtain two identities. The left-hand sides of both identities are now equal, and so the right-hand sides are equal. Apply relation (6.10), and we obtain condition (6.15). Hence, this condition is necessary.

We now show that this condition is sufficient. Let Q be an invertible matrix-valued function (we assume that the inverse is also holomorphic at the origin), and let G be a new dependent vector. Substitute $F = QG$ in the system, and we obtain a system for G as $Q_i G + Q G_i = P_i Q G + U_i$. Here, we use the abbreviated notation $Q_i = \frac{\partial Q}{\partial x_i}$, $G_i = \frac{\partial G}{\partial x_i}$. The new system is expressed as

$$G_i = (Q^{-1} P_i Q - Q^{-1} Q_i) G + Q^{-1} U_i.$$

We can regard the function $(Q^{-1} P_i Q - Q^{-1} Q_i)$ as a new P_i and $Q^{-1} U_i$ as a new U_i. We want to show the following identity:

$$(Q^{-1} P_i Q - Q^{-1} Q_i) Q^{-1} U_j + \frac{\partial (Q^{-1} U_i)}{\partial x_j} = (Q^{-1} P_j Q - Q^{-1} Q_j) Q^{-1} U_i + \frac{\partial (Q^{-1} U_j)}{\partial x_i}. \tag{6.16}$$

Here, we use the abbreviation $U_{ij} = \frac{\partial U_i}{\partial x_j}$. (Note that $U_{ij} = U_{ji}$ does not necessarily hold.) The left-hand side of (6.16) is $(Q^{-1} P_i Q - Q^{-1} Q_i) Q^{-1} U_j - Q^{-1} Q_j Q^{-1} U_i + Q^{-1} U_{ij} = Q^{-1} P_i U_j - Q^{-1} Q_i Q^{-1} U_j - Q^{-1} Q_j Q^{-1} U_i + Q^{-1} U_{ij}$. The right-hand side of (6.16) is $Q^{-1} P_j U_i - Q^{-1} Q_j Q^{-1} U_i - Q^{-1} Q_i Q^{-1} U_j + Q^{-1} U_{ji}$. It follows from the condition (6.15) that the left- and right-hand sides agree. Thus, we see that the new system for G also satisfies condition (6.15). By construction of transformation matrices Q that is analogous to the homogeneous case (the proof of Theorem 6.3.1), we can finally translate the system into the form $\frac{\partial F}{\partial x_i} = U_i$. Here, we note that the identity $\frac{\partial U_i}{\partial x_j} = \frac{\partial U_j}{\partial x_i}$ holds for any i and j. The famous Poincaré Lemma claims that there exists a solution F for this system. In order to prove it, we use Stokes' theorem. More precisely, for a path C_x which connects the origin and x, we define F by the integral $\int_{C_x} \sum_{i=1}^n U_i dx_i$. By Stokes' theorem, the vector-valued function F does not depend on the choice of path C_x. The reader is asked to prove that the function F is holomorphic and satisfies the system. Another way of proving the existence of the solution of the simplified system is to derive the series expansion of F from the series expansion of U_i.

Tip 2 For simplicity, we assume $r = 1$. In this case, P_i is a scalar-valued function. It follows from the integrability condition (6.10) that the set $G = \{\underline{\partial_i} - P_i \mid i = 1, \ldots, n\}$ is a Gröbner basis in R with the graded lexicographic order. In fact, we have

$$\mathrm{sp}(\partial_i - P_i, \partial_j - P_j)$$
$$= \partial_j(\partial_i - P_i) - \partial_i(\partial_j - P_j)$$
$$= -\frac{\partial P_i}{\partial x_j} - P_i \partial_j + \frac{\partial P_j}{\partial x_i} - P_j \partial_i$$
$$= -P_i(\partial_j - P_j) + P_j(\partial_i - P_i).$$

If we introduce $e_i = \partial_i - P_i$, then we have

$$(\partial_j - P_j)e_i - (\partial_i - P_i)e_j = 0 \qquad (6.17)$$

from the above expressions for S polynomials. From the discussion in Sect. 3.5.3, we can see that these are generators of the syzygies of G. Let us replace e_i in (6.17) formally by $\bullet U_i$. Then, we obtain

$$\frac{\partial U_i}{\partial x_j} - P_j U_i = \frac{\partial U_j}{\partial x_i} - P_i U_j.$$

This is the condition (6.15) for the existence of a solution to the inhomogeneous system. A generalization of this observation is one of the foundations of the theory of D modules.

In order to create an algorithm for the construction of a series solution of the system (6.14), it is more useful to determine the coefficients of the series expansion by recurrence, rather than by following the construction of the proof. These recurrence relations form an overdetermined system, but the existence of a solution was ensured by our theorem, and the overdetermined system is consistent. We consider this method for the case of two variables. Let

$$F = \sum_{i,j=0}^{\infty} F_{ij} x^i y^j.$$

Here, F_{ij} is an r-dimensional vector of complex numbers. The $(1, 1)$ degree of $x^i y^j$ is defined by $\mathrm{ord}_{(1,1)}(x^i y^j) = i + j$. We also set the $(1, 1)$ degree of F_{ij} to $i + j$. We will determine F_{ij} degree by degree. For $k = 1, 2$, we assume that P_k and U_k are expanded as

$$P_k = \sum_{i,j=0}^{\infty} P_{ij}^k x^i y^j$$

$$U_k = \sum_{i,j=0}^{\infty} U_{ij}^k x^i y^j.$$

Since $\frac{\partial F}{\partial x} = \sum_{i,j=0}^{\infty} i F_{ij} x^{i-1} y^j$ holds, by comparing the coefficients of $x^i y^j$, we have

$$(i + 1) F_{i+1,j} = \sum_{s+u=i, t+v=j} P_{st}^1 F_{uv} + U_{ij}^1 \tag{6.18}$$

$$(j + 1) F_{i,j+1} = \sum_{s+u=i, t+v=j} P_{st}^2 F_{uv} + U_{ij}^2. \tag{6.19}$$

Set $m = i + j + 1$. The $(1, 1)$ degree of F_{uv} on the right-hand side is less than or equal to $m - 1$. On the other hand, the $(1, 1)$ degree of the $F_{i+1,j}$ on the left-hand side is m; the $(1, 1)$ degree of $F_{i,j+1}$ is also m. Hence, if we obtain the value of F_{00}, the coefficient F_{ij} is determined from those of less degree. The coefficient F_{m0} is determined only from (6.18), and that of F_{0m} only from (6.19); however, other coefficients for which the $(1, 1)$ degree is m may be determined by (6.18) or (6.19). Recurrence relations form an overdetermined system. By virtue of the existence theorem, the coefficients are uniquely determined without dependence on the order of recurrence.

In order to perform this algorithm on a computer, it is necessary to retain the coefficients F_{uv} that are computed. Thus, if we want to compute coefficients of high degree or to evaluate the series efficiently and precisely, further development may be necessary, depending on the situation.

Tip 3 When the matrix-valued functions P_i and U_i are not holomorphic, it is important in both theory and applications to describe the local solutions of the Pfaffian system. Several cases of this situation have been studied. When the singular locus crosses normally, and P_i has a first-order pole, a method for constructing solutions was given in [35]. The problem of an efficient method, however, does not seem to be well studied. Higher-order poles are currently being studied. Note that a search in the mathsci-net for the book [19] yields several active studies. In [18], Majima discusses attempts to construct solutions. This direction will be important, and the author expects future advances of the computational study of Pfaffian systems.

6.4 Holonomic Functions

The function defined by the series

$$\sum_{\beta \in \mathbf{N}_0^n} c_\beta (x-a)^\beta, \quad c_\beta \in \mathbf{C}, \ (x-a)^\beta = \prod_{i=1}^n (x_i - a_i)^{\beta_i} \tag{6.20}$$

which is absolutely convergent at $a \in \mathbf{C}^n$ is called a *holomorphic function* at $x = a$. We note, for those readers who are not familiar with complex analysis, that holomorphic functions have several useful properties. For instance, by regarding infinite series as polynomials with infinitely many terms, the sums, products, and differentials of holomorphic functions can be obtained by computing the sums, products, and differentials of the series; this method is called *formal computation* (see, e.g., [1, 15], and the references cited therein).

We are given a holomorphic function f at $x = a$ in $\mathbf{C}^n$. When there exists a 0-dimensional ideal I in R and the function f is a solution of I, (in other words, when $L \bullet f = 0$ holds for any $L \in I$) f is a *holonomic analytic function*. When f is a distribution and is a solution of a holonomic ideal defined in Sect. 6.8, the function is called a holonomic distribution. We will simply call a holonomic analytic function a *holonomic function*.

The following theorem says that a holonomic function, which is defined as a series around the point a, can be extended to a broader domain. The proof requires several theorems from complex analysis, and so we will not prove it here.

Theorem 6.4.1. *Let f be a holonomic function. There exists a polynomial p such that the function f can be analytically continued to the universal covering space of $\mathbf{C}^n \setminus V(p)$. The polynomial p is a divisor of the least common multiple of the denominator polynomials of the coefficient matrices P_i's of a Pfaffian system associated with the function f.*

The analytic set $V(p)$ is called the *singular locus* of the holonomic function f. The zero set of the least common multiple is called the singular locus of the Pfaffian system. In Example 6.5.4, presented later, the set $y = 0$ is the singular locus of the Pfaffian system, although the solution does not have singularities, and so $p = 1$. Thus, in general, the singular locus of the holonomic function and the singular locus of the Pfaffian system do not agree. The theory of D-modules will help us to understand this gap.

Tip 4 *Analytic continuation is one of the most important and exciting ideas in complex analysis.* **Exercise:** *Explain how analytic continuation along a circle around the origin changes the function $\sqrt{x}$ to $-\sqrt{x}$.*

A remarkable property of holonomic functions is that the integral of a holonomic function is again a holonomic function. This property will be presented in Theorem 6.10.14, but first we present some examples of holonomic functions.

Example 6.4.2. The polynomials and rational expressions (with complex coefficients) are holonomic functions. Let f be a rational expression. We denote by f_i the differential $\frac{\partial f}{\partial x_i}$. For $1 \leq i \leq n$, we have $\partial_i - f_i/f \in R$. These generate a 0-dimensional left ideal in R, and f is annihilated by these operators.

Example 6.4.3. Let h be a rational expression. Then, the function $\cos(h)$ is a holonomic function. Set $h_{ii} = \frac{\partial^2 h}{\partial x_i^2}$. The function $\cos(h)$ is annihilated by $\partial_i^2 - \frac{h_{ii}}{h_i}\partial_i + (h_i)^2 \in R$ for $1 \leq i \leq n$, and these operators generate a 0-dimensional left ideal.

Example 6.4.4. The function $\frac{1}{\sin x}$ is not a holonomic function because it has singularities at $x = 2\pi k$, $k \in \mathbf{Z}$. If the function $\frac{1}{\sin x}$ were a holonomic function, then by the above theorem, the singular points would be finite. Therefore, this function is not a holonomic function.

We note that the notion of holonomic functions was introduced by Zeilberger, who gave exciting applications of them to special function identities and combinatorial identities [36].

Tip 5 A goal of the study of holonomic functions is to be able to manipulate and understand holonomic functions as we do polynomials and trigonometric functions. Readers will be able to suggest new problems using holonomic functions by keeping this goal in mind.

6.5 Gradient Descent for Holonomic Functions

There are several applications for Gröbner bases in R. To illustrate, in this section we present an application of approximating a local minimum of a holonomic function g. This is a new method that was introduced in [21] to solve problems in statistics.

Following [22], we consider the recurrence relation

$$z^{(k+1)} = z^{(k)} + a_k d^{(k)} \quad k = 0, 1, 2, \ldots. \tag{6.21}$$

Here, $\{z^{(k)}\}$ is a sequence in $\mathbf{R}^n$ such that $g(z^{(k)})$ converges to a local minimum of the function g. The number $a_k \in \mathbf{R}_{>0}$ is called the *step length*, and the vector $d^{(k)}$ is called the *search direction*. In many optimization algorithms, the search direction has the following form

$$- H_k^{-1} \nabla g(z^{(k)}). \tag{6.22}$$

Here, H_k^{-1} is an $n \times n$ matrix. When the matrix H_k is the identity matrix, the method for finding a local minimum by this sequence is called the method of *gradient descent*, and when the matrix is the Hessian matrix of g, it is called *Newton's method*.

For a holonomic function g, the gradient ∇g and the Hessian H_k of g have the following "good" expressions in terms of a Gröbner basis. Let I be a 0-dimensional ideal in $R = R_n$ which annihilates the function g. Assume that the dimension of the vector space of R/I over the field of the rational expressions is r. We fix a term order and obtain a Gröbner basis B of the ideal I. Let $S = \{s_1, \ldots, s_r\}$ be a set of (nonmonic) standard monomials of the basis, which is a vector-space basis of R/I. Define a vector of functions G as $G = (s_1 \bullet g, \ldots, s_r \bullet g)^T$.

Lemma 6.5.1. *1. Let $\sum_{j=1}^r a_{ij} s_j$ be the normal form of $\partial_i = \partial/\partial z_i$ expressed by the Gröbner basis B. Here, $a_{ij} \in \mathbf{C}(z_1, \ldots, z_n)$. Let A be a matrix for which the (i, j)-th entry is a_{ij}. Then, the gradient of G can be expressed as*

$$\nabla g(z^{(k)}) = A(z^{(k)}) G(z^{(k)}).$$

2. Let $\sum_k u_{ijk} s_k$ be the normal form of $\partial_i \partial_j$ expressed by the Gröbner basis B. Here, $u_{ijk} \in \mathbf{C}(z_1, \ldots, z_n)$. Then, the Hessian of g is expressed as

$$\frac{\partial^2 g}{\partial z_i \partial z_j}(z^{(k)}) = (u_{ij1}(z^{(k)}), \ldots, u_{ijr}(z^{(k)})) G(z^{(k)}).$$

Note that a_{ij}, $(j = 1, \ldots, r)$ agrees with the first row of the matrix P_i of the Pfaffian equation (6.9).

Proof. Since the relations $\partial_i - \sum_j a_{ij} s_j \in I$ and $I \bullet g = 0$ hold, we have $\partial_i \bullet g = \sum_{s_j \in S} a_{ij} (s_j \bullet g)$. Thus, we obtain (1). Statement (2) can be shown analogously.

Let p' be the polynomial for the singular locus of a Pfaffian system associated with the holonomic function g, and let E be a simply connected domain in $\mathbf{R}^n \setminus V(p')$. We assume that the sequence $z^{(k)}$ stays in the domain E.

Algorithm 6.5.2 (Holonomic Gradient Descent).

1. Compute a Gröbner basis of the left ideal I in R. Find the set S of the standard monomials of the Gröbner basis.
2. Compute the matrix P_i in the Pfaffian equation (6.9) by using the normal form algorithm, the Gröbner basis, and S.

3. Compute the normal form of ∂_i by using the Gröbner basis, and obtain the matrix (a_{ij}) to express the gradient of g.
4. Take a point $z^{(0)}$ as a starting point. Numerically evaluate the vector-valued function G at the point $z = z^{(0)}$. Define the approximate value as $\bar{G}$. Set $k = 0$.
5. Numerically evaluate $(a_{ij}(z^{(k)}))\bar{G}$. The value is an approximation of the gradient $-\tilde{g} = \nabla g$ at $z^{(k)}$. If a stopping condition of the loop is satisfied (for instance $\tilde{g} = 0$), then stop the loop.
6. Let $z^{(k+1)} = z^{(k)} + a_k\tilde{g}$. (Move to the new evaluation point $z^{(k)} + a_k\tilde{g}$.)
7. Approximately evaluate G at $z = z^{(k+1)}$ by numerically solving the Pfaffian system (6.9). Let $\bar{G}$ be the approximate value. Increase k by 1. Go to the step 5.

In step 6, the number a_k is the step length; its size should be determined by standard recipes for the gradient descent method. The simplest step length is $a_k = \frac{\varepsilon}{|\tilde{g}|}$, where ε is a sufficiently small positive number. In order to numerically solve the Pfaffian system in the step 7, we use the finite difference and series expansion methods that were explained in Sect. 6.3. Methods that use the Pfaffian system to evaluate the vector-valued function G are called *holonomic gradient methods*.

Example 6.5.3. Here is an example for the case $n = 1$. We consider the function $g(x) = \exp(-x + 1) \int_0^\infty \exp(xt - t^3)dt$, which satisfies the differential equation $(3\partial_x^2 + 6\partial_x + (3 - x)) \bullet g = \exp(-x + 1)$. A method to derive this equation by a computer or in an algorithmic way will be explained in Example 6.10.12. Define $S = \{1, \partial_x\}$ as a set of standard monomials. Then, we have the inhomogeneous Pfaffian system

$$\frac{dG}{dx} = \begin{pmatrix} 0 & 1 \\ (-3 + x)/3 & -2 \end{pmatrix} G + \begin{pmatrix} 0 \\ \exp(-x + 1)/3 \end{pmatrix}.$$

The gradient is expressed as $\nabla g = G_2 = \begin{pmatrix} 0 & 1 \end{pmatrix} G$. Here, G_2 denotes the second element of the vector G. We evaluate $G(0) = (g(0), g'(0))^T$ using a numerical integration algorithm and obtain $\bar{G}(0) = (2.427, -1.20)^T$. In the closed interval $E = [0, 5]$, we apply the fourth-order Runge–Kutta method and the holonomic gradient descent method by putting $H_k = 1$, $h_k\tilde{g} = -0.1$. Then, we find an approximate local minimum $g(3.4) = 1.016$ at the point $x = 3.4$. The graph of the function $g(x)$ is illustrated in Fig. 6.1. The gradient descent method in one variable can be viewed as Euler's method for solving an ordinary differential equation.

In the example above, we used an inhomogeneous ordinary differential equation. We note that the gradient vector can also be obtained by using a Gröbner basis in the general case of inhomogeneous systems of linear differential equations. This is a generalization of Lemma 6.5.1. Suppose that the 0-dimensional ideal I is generated by $\ell_1, \ldots, \ell_m$. The target holonomic function g satisfies the inhomogeneous system of linear differential equations $\ell_i \bullet g = u_i$. Let B be a Gröbner basis, and let $S = \{s_1, \ldots, s_r\}$ be the set of the standard monomials. As in the lemma, we obtain

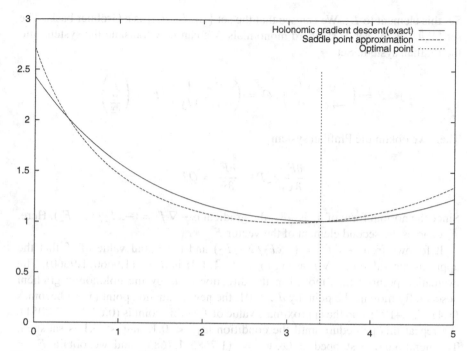

Fig. 6.1 Graph of $g(x)$ (*solid line*)

$\sum a_{ij} s_j$ by computing the normal form of ∂_i by using the Gröbner basis B. Then, there exist $c_k^i \in R$ satisfying $\partial_i - \sum a_{ij} s_j = \sum_k c_k^i \ell_k$. Applying this to the function, we obtain an expression of the gradient

$$\partial_i \bullet g = \sum a_{ij} s_j \bullet g + \sum_k c_k^i u_i$$

in terms of G and u_i.

Example 6.5.4. An example when $n = 2$. A holonomic function f satisfies the system of linear partial differential equations

$$L_1 \bullet f = L_2 \bullet f = 0, L_1 = y\partial_x - x\partial_y, L_2 = \partial_x\partial_y + 4xy$$

and takes approximate values $f = 0.7120, \partial_y \bullet f = -1.9660$ at $x = y = 1.4$. We will find a local minimum of f by using the holonomic gradient descent method, starting from this point. We translate this system into a Pfaffian system, then we compute a Gröbner basis in R with the graded lexicographic order, such that $\partial_x \succ \partial_y$. We then calculate the S-polynomial of L_1, L_2 as

$$\partial_y L_1 - yL_2 = y\partial_x\partial_y + \partial_x - x\partial_y^2 - y\partial_x\partial_y - 4xy^2 = \underline{-x\partial_y^2 + \partial_x - 4xy^2}.$$

Set this element to L_3. We can see that the set $\{L_1, L_2, L_3\}$ is a Gröbner basis. The set $\{1, \partial_y\}$ is the set of standard monomials. We can now translate the system into the Pfaffian system. Set

$$
P = \begin{pmatrix} 0 & x/y \\ -4xy & 0 \end{pmatrix}, Q = \begin{pmatrix} 0 & 1 \\ -4y^2 & 1/y \end{pmatrix}, F = \begin{pmatrix} f \\ \frac{\partial f}{\partial y} \end{pmatrix}.
$$

Then, we obtain the Pfaffian system

$$
\frac{\partial F}{\partial x} = PF, \quad \frac{\partial F}{\partial y} = QF.
$$

Since the normal form of ∂_x is $\frac{x}{y}\partial_y$, the gradient is $-\nabla f = (-xF_2/y, -F_2)$. Here, F_2 denotes the second element of the vector F.

It follows from $-\nabla f = (-xF_2/y, -F_2)$ and the initial value of f that the approximate value of $-\nabla f$ at $(x, y) = (1.4, 1.4)$ is $d = (1.9660, 1.9660)$. The evaluation point is thus moved in the direction of d by the holonomic gradient descent. By moving the point by $d \times 0.01$, the new evaluation point (x, y) becomes $(1.4197, 1.4197)$, and the approximate value of F at this point is $(0.6300, -2.2051)$. We repeat this procedure until the condition $|d_1| < 0.1, |d_2| < 0.1$ is satisfied. The iterations are stopped at $(x, y) = (1.7685, 1.7685)$, and we obtain $F = (-0.9996, -0.09865)$; thus, an approximate local minimum is located at -0.997.

In fact, the function f of this example is $-\cos(x^2 + y^2)$. A graph of this function on $[1.4, 3.4] \times [1.4, 3.4]$ is shown as Fig. 6.2. We note that in order to obtain a local minimum, the holonomic gradient descent method uses only differential equations of f and an initial value. This method is useful when f cannot be expressed in terms of special functions. Research problem: This function has a local minimum at $x = y = 0$. How do we find this point?

6.6 Gröbner Bases in the Ring of Differential Operators with Polynomial Coefficients D

The ring of differential operators with polynomial coefficients D has a finer structure than the ring with rational function coefficients R, and we thus need more complicated algorithms for constructing in D. Our goal is to present the theory of the D-module integration algorithm. We will assume that readers are familiar with quotient spaces by equivalence classes and abstract linear algebra. For quotient spaces, we refer the reader to Sect. 1.6.1 or introductory books on modern mathematics. More advanced algebra is necessary for the study of D than that for R, and we will cover this in the following sections.

We consider a noncommutative polynomial ring

$$
D_n = \mathbf{C}\langle x_1, \ldots, x_n, \partial_1, \ldots, \partial_n \rangle
$$

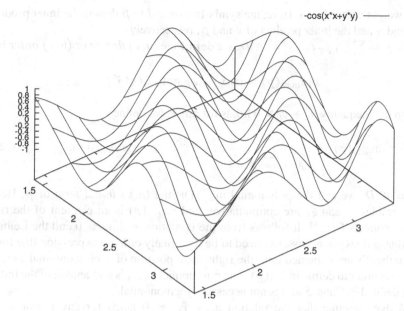

-cos(x*x+y*y) ———

Fig. 6.2 Graph of $-\cos(x^2 + y^2)$, $(x, y) \in [1.4, 3.4] \times [1.4, 3.4]$

which is generated by $x_1, \ldots, x_n, \partial_1, \ldots, \partial_n$ with the following relations and properties:

1. The associative and distributive laws hold;
2. $x_i x_j = x_j x_i, \partial_i \partial_j = \partial_j \partial_i, \partial_i x_j = x_j \partial_i$ $(1 \le i \ne j \le n)$;
3. $\partial_i x_i = x_i \partial_i + 1$ $(1 \le i \le n)$.

By virtue of these, in monomials, we can collect all the ∂_i's onto the right-hand side. For example, we have $\partial_1^2 x_1 = \partial_1(\partial_1 x_1) = \partial_1(x_1 \partial_1 + 1) = (\partial_1 x_1)\partial_1 + \partial_1 = (x_1 \partial_1 + 1)\partial_1 + \partial_1 = x_1 \partial_1^2 + 2\partial_1$. It is important to prove that these rules are consistent. Intuitively speaking, the rule $\partial_i x_i = x_i \partial_i + 1$ comes from the well-known Leibniz rule $\frac{\partial}{\partial x_i}(x_i f) = x_i \frac{\partial f}{\partial x_i} + \frac{\partial x_i}{\partial x_i} f$, where f is a function and ∂_i is regarded as the partial differential operator with respect to the variable x_i. Since the relation comes from the Leibniz rule, we can expect that these rules for D_n are consistent. In order to give a rigorous proof, we define D_n as a subring of $\mathrm{End}_{\mathbf{C}}(\mathbf{C}[x_1, \ldots, x_n])$ generated by x_i, ∂_i. See, e.g., [12, Chap. 1] for details.

The ring of differential operators D_n is called the *Weyl algebra*, and it is often denoted by A_n. Below, we will omit the subscript n of D_n if to do so will not cause confusion.

Let $u, v \in \mathbf{R}^n$ be n-dimensional real vectors. When the relation $u_i + v_i \ge 0$ holds for any i, the pair of vectors (u, v) is called the *weight vector* in the ring of differential operators D. We put

$$\mathrm{ord}_{(u,v)}(x^\alpha \partial^\beta) = u \cdot \alpha + v \cdot \beta$$

for a weight vector (u, v). Here, the symbols $u \cdot \alpha$ and $v \cdot \beta$ denote the inner product of u and α and the inner product of v and β, respectively.

For $p = \sum_{(\alpha,\beta) \in E} c_{\alpha,\beta} x^\alpha \partial^\beta \in D$, we define the (u, v) *degree* or (u, v) *order* by

$$\mathrm{ord}_{(u,v)}(p) = \max_{(\alpha,\beta) \in E} \mathrm{ord}_{(u,v)}(x^\alpha \partial^\beta).$$

When the inequality $u_i + v_i > 0$ holds for any i, we define

$$\mathrm{in}_{(u,v)}(p) = \sum_{(\alpha,\beta) \in E, \alpha \cdot u + \beta \cdot v = m} c_{\alpha,\beta} x^\alpha \xi^\beta, \quad m = \mathrm{ord}_{(u,v)}(p). \qquad (6.23)$$

For $p \in D$, we call the polynomial $\mathrm{in}_{(u,v)}(p)$ the (u, v) *initial term* of p. Here, the variables x_i and ξ_j are commutative, and $\mathrm{in}_{(u,v)}(p)$ is an element of the ring of polynomials $\mathbf{C}[x, \xi]$. It follows from the condition $u_i + v_i > 0$ and the Leibniz rule that p does not necessarily need to be a normally ordered expression (the form where the ∂'s are collected onto the right-most position of each monomial as in p above) in order to define $\mathrm{in}_{(u,v)}(p)$. The polynomial $\mathrm{in}_{(u,v)}$ is a D analog of the initial form defined in Chap. 5 and is not necessarily a monomial.

We next assume that the relation $u_i + v_i = 0$ holds for any i. For $p = \sum_{(\alpha,\beta) \in E} c_{\alpha,\beta} x^\alpha \partial^\beta \in D$, we define the initial term as

$$\mathrm{in}_{(u,v)}(p) = \sum_{(\alpha,\beta) \in E, \alpha \cdot u + \beta \cdot v = m} c_{\alpha,\beta} x^\alpha \partial^\beta, \quad m = \mathrm{ord}_{(u,v)}(p). \qquad (6.24)$$

Note that the operator ∂_i is not replaced by ξ_i, and the initial term $\mathrm{in}_{(u,v)}(p)$ is again an element of D. Note that all terms in the relation $\partial_i x_i = x_i \partial_i + 1$ have the (u, v) degree 0 under the condition $u_i + v_i = 0$.

The mixed case of $u_i + v_i = 0$ and $u_j + v_j > 0$ is not used in the book; for a consideration of this case, refer to [29].

In this chapter, an element of the form $c_{\alpha\beta} x^\alpha \partial^\beta$, $c_{\alpha\beta} \in \mathbf{C}$ is called a monomial in D. The constant $c_{\alpha\beta}$ is called the coefficient of the monomial. If the order satisfies the condition of Lemma 1.1.16, it is called a *well-order*. (Note that we ignore the coefficients of the monomials when comparing them with this order. For example, the monomials ∂_i and $2\partial_i$ have equal order.) A well-order $\prec$ among monomials of D that satisfies the following is called the *term order* in D: (1) $1 \prec x_i \partial_i$; and (2) if $x^\alpha \partial^\beta \succ x^a \partial^b$, then $x^{\alpha+s} \partial^{\beta+t} \succ x^{a+s} \partial^{b+t}$ holds for any $x^s \partial^t$. For example, the (pure) lexicographic order is a term order. The term order in this chapter is a D analog of the monomial order in the ring of polynomials that was defined in Sect. 1.1.4. We note that the term "monomial order" is used with a different meaning in [29], which is one of standard textbooks for Gröbner bases in D. In order to avoid confusion and to be consistent with [29], we call the above order the term order.

Let (u, v) be a weight vector, and let $\prec$ be a term order. We define an induced order $\prec_{(u,v)}$ by

$$x^\alpha \partial^\beta \prec_{(u,v)} x^{\alpha'} \partial^{\beta'} \tag{6.25}$$

$$\Leftrightarrow u \cdot \alpha + v \cdot \beta < u \cdot \alpha' + v \cdot \beta'$$

$$\text{or } (u \cdot \alpha + v \cdot \beta = u \cdot \alpha' + v \cdot \beta'$$

$$\text{and } x^\alpha \partial^\beta \prec x^{\alpha'} \partial^{\beta'}).$$

$$\tag{6.26}$$

This order refines the partial order defined by (u, v). We define the initial monomial (leading monomial) of p $\text{in}_{\prec_{(u,v)}}(p)$ by the largest monomial with respect to the order $\prec_{(u,v)}$ in the polynomial obtained by replacing ∂_i by ξ_i in p. Note that the coefficient of an initial monomial in R is a rational expression of x, and the coefficient of an initial monomial in D is an element of $\mathbf{C}$. In the following, when we say $p \prec_{(u,v)} q$ for $p, q \in D$, we mean that $\text{in}_{\prec_{(u,v)}}(p) \prec_{(u,v)} \text{in}_{\prec_{(u,v)}}(q)$.

Weight vectors in which either u_i or v_i takes a negative value play an important role in the theory of Gröbner bases for the ring of differential operators D. The order $\prec_{(u,v)}$ is not necessarily a well-order. When the inequality $u_i, v_i \geq 0$ holds for any i, the order $\prec_{(u,v)}$ is a term order and consequently is a well-order.

Example 6.6.1. We set $n = 2$ and $(u, v) = (0, 0, 1, 1)$. Let $\succ$ be the lexicographic order such that $\partial_1 \succ \partial_2 \succ x_1 \succ x_2$. When the f_i's are nonzero monomials in x_1, x_2, we have

$$f_1(x_1, x_2)\partial_1 \succ_{(u,v)} f_2(x_1, x_2)\partial_2 \succ_{(u,v)} f_3(x_1, x_2).$$

We set $n = 2$ and $(u, v) = (-1, -1, 1, 1)$. Let $\succ$ be the lexicographic order such that $\partial_1 \succ \partial_2 \succ x_1 \succ x_2$. Set $\theta_i = x_i \partial_i$. When the f_i's are nonzero monomials in two variables, we have

$$f_1(\theta_1, \theta_2)\partial_1 \succ_{(u,v)} f_2(\theta_1, \theta_2) \succ_{(u,v)} f_3(\theta_1, \theta_2)x_1.$$

We note that there exists an infinite descending sequence

$$1 \succ_{(u,v)} x_1 \succ_{(u,v)} x_1^2 \succ_{(u,v)} x_1^3 \succ_{(u,v)} \cdots$$

and so this order is not a well-order.

Example 6.6.2. Here are some examples of frequently used weight vectors (u, v):

1. $(\mathbf{0}, \mathbf{1})$ where $u = (0, 0, \ldots, 0)$, $v = (1, 1, \ldots, 1)$.
2. $(\mathbf{1}, \mathbf{1})$ where $u = (1, 1, \ldots, 1)$, $v = (1, 1, \ldots, 1)$.
3. $(-w, w)$ for $w \in \mathbf{R}^n$.

Let G be a subset of D. We denote by $\text{in}_\prec(G)$ the set $\{\text{in}_\prec(g) \mid g \in G\}$. The left ideal generated by $\text{in}_\prec(G)$ in $\mathbf{C}[x, \xi]$ is denoted by $\langle \text{in}_\prec(G) \rangle$ and is a monomial ideal. When G is a left ideal in D and no confusion arises, we may omit the symbols $\langle,\rangle$. We denote by $\mathrm{in}_{(u,v)}(G)$ the set $\{\mathrm{in}_{(u,v)}(g) \mid g \in G\}$. The ideal $\langle\mathrm{in}_{(u,v)}(G)\rangle$ lies in $\mathbf{C}[x,\xi]$ when $u_i + v_i > 0$, or in D when $u + v = 0$. When G is a left ideal and no confusion arises, we may omit the symbols $\langle,\rangle$.

Let I be a left ideal in D, and let $\prec$ be a term order in D or an order $\prec_{(u,v)}$ defined by the weight vector (u, v). A finite subset G of D is called a *Gröbner basis* of I with respect to $\prec$ when the set G generates I and the relation $\langle\mathrm{in}_\prec(G)\rangle = \mathrm{in}_\prec(I)$ holds.

When the order $\prec$ is a well-order, Gröbner bases can be obtained in a way analogous to the cases of the ring of polynomials and R. When we perform divisions and compute S polynomials, we multiply monomials to cancel initial terms, and these monomials must be multiplied from the left. Almost all the fundamental properties that were presented in Chap. 1 also hold here. It is left as an exercise to prove them rigorously in the case of D. If you need guidance, we refer you to [26, 29]. To clarify the meaning of the analogy, we give the definition of the S polynomial. When $\mathrm{in}_\prec(f) = ax^p\xi^q$, and $\mathrm{in}_\prec(g) = a'x^{p'}\xi^{q'}$, we define r and c by

$$r = (\max(p_1, p_1'), \ldots, \max(p_n, p_n')), c = (\max(q_1, q_1'), \ldots, \max(q_n, q_n')),$$

and define the S polynomial of f and g as

$$\mathrm{sp}(f, g) = x^{r-p}\partial^{c-q} f - \frac{a}{a'}x^{r-p'}\partial^{c-q'} g.$$

We have called this a "polynomial", but, strictly speaking, it is a differential operator.

In the integration algorithm explained below, we will use the order $\prec_{(u,v)}$ for a weight vector satisfying $u + v = 0$. This is not a well-order, as we have seen, and we cannot use the Gröbner basis method for term orders.

Example 6.6.3. Set $n = 1$ and $(u, v) = (-1, 1)$. We apply the normal form algorithm for x by $\underline{x} + x^2$. This algorithm does not stop since

$$x \to x - (x + x^2) = \underline{-x^2} \to -x^2 + x(x + x^2) = x^3 \to \cdots.$$

We thus present a method that uses a homogenized Weyl algebra

$$D_n^{(h)} = \mathbf{C}[h]\langle x_1, \ldots, x_n, \partial_1, \ldots, \partial_n\rangle$$

to obtain Gröbner bases for orders which are not term orders. In a homogenized Weyl algebra, multiplication is defined by the relation

$$\partial_i x_i = x_i \partial_i + h^2.$$

The new variable h commutes with x_i and ∂_i and is called the *homogenization variable*. The order $\prec_{(u,v)}$ in D naturally induces an order in $D^{(h)}$, as follows.

$$x^\alpha \partial^\beta h^\gamma \prec_{(u,v)} x^{\alpha'} \partial^{\beta'} h^{\gamma'} \tag{6.27}$$

$$\Leftrightarrow u \cdot \alpha + v \cdot \beta + \gamma < u \cdot \alpha' + v \cdot \beta' + \gamma'$$

$$\text{or } (u \cdot \alpha + v \cdot \beta + \gamma = u \cdot \alpha' + v \cdot \beta' + \gamma'$$

$$\text{and } x^\alpha \partial^\beta \prec_{(u,v)} x^{\alpha'} \partial^{\beta'}).$$

$$\tag{6.28}$$

Suppose that $p \in D$ is written as $p = \sum_{(\alpha,\beta)\in E} c_{\alpha\beta} x^\alpha \partial^\beta$. Set $m = \max_{(\alpha,\beta)\in E} |\alpha| + |\beta|$. We define the homogenization of p as $H(p) = \sum c_{\alpha\beta} x^\alpha \partial^\beta h^{m-|\alpha|-|\beta|}$. The element $H(p)$ is homogeneous in x, ∂, and h. It follows from the multiplication rule that the product of homogeneous polynomials (operators) is also homogeneous.

Consider monomials of the form $x^\alpha \partial^\beta h^\gamma$. When m is fixed, the number of monomials satisfying $|\alpha| + |\beta| + \gamma = m$ is finite. Then, if the input is homogeneous, the normal form algorithm stops for any weight vector (u, v) in D. Thus, if we restrict the input to only homogeneous elements, we can obtain Gröbner bases in $D^{(h)}$. (It is left as an exercise to prove this rigorously.)

We can obtain Gröbner bases in D with respect to an order which is not a well-order by utilizing the homogenized Weyl algebra.

Theorem 6.6.4. *Let F be a set of generators of a left ideal I in D. Let $G^{(h)}$ be a Gröbner basis in $D^{(h)}$ for the homogenized input $H(f)$, $f \in F$ with respect to the order $\prec_{(u,v)}$. Set $G = \{g|_{h=1} \mid g \in G^{(h)}\}$. Then the following properties hold.*

1. *Any element p of I has a standard representation in terms of G. In other words, there exists $c_i \in D$ such that*

$$p = \sum_{g_i \in G} c_i g_i, \quad p \succeq_{(u,v)} c_i g_i$$

   holds.
2. $\langle \text{in}_{\prec_{(u,v)}}(G) \rangle = \text{in}_{\prec_{(u,v)}}(I)$.
3. $\langle \text{in}_{(u,v)}(G) \rangle = \text{in}_{(u,v)}(I)$.

Proof. (1) Since the homogenized element $H(f)$, $f \in F$ belongs to the left ideal generated by $G^{(h)}$, the polynomial (operator) f belongs to the left ideal generated by G. Since the polynomial p is an element of the ideal generated by F, it is expressed as $p = \sum_{g_i \in G} d_i g_i$. For $g_i \in G$, the corresponding homogeneous element in $G^{(h)}$ is denoted by $H'(g_i)$. Set $\tilde{p} = \sum_{g_i \in G} h^{e_i} H(d_i) H'(g_i)$. Here, h^{e_i} is chosen so that all the terms in the sum have the same total degree. Hence, the polynomial $\tilde{p}$ belongs to the left ideal generated by $G^{(h)}$. Since the polynomial $\tilde{p}$ is homogeneous, it has the standard representation $\tilde{p} = \sum_{g_i \in G} \tilde{c}_i H'(g_i)$, $\tilde{p} \succeq_{(u,v)} \tilde{c}_i H'(g_i)$, and the total degrees of $\tilde{p}$ and $\tilde{c}_i H'(g_i)$ agree. Since the total degrees are equal, the order relations hold if we set $h = 1$. Therefore, we have $p \succeq_{(u,v)} \tilde{c}_i|_{h=1} g_i$.

Statement (2) follows from (1).

(3) We note that since $\prec_{(u,v)}$ is not a well-order, statement (3) is not trivial. Suppose that the equality does not hold. Then there exists a polynomial $p \in I$ such that $\text{in}_{(u,v)}(p) \notin \langle \text{in}_{(u,v)}(G) \rangle$. It follows from the definition of the order $\prec_{(u,v)}$ that $\text{in}_{\prec_{(u,v)}}(\text{in}_{(u,v)}(p)) = \text{in}_{\prec_{(u,v)}}(p)$ holds. Since $\prec$ is a well-order, there exists a polynomial p such that $\text{in}_{\prec}(\text{in}_{(u,v)}(p))$ is minimal with respect to the term order $\prec$ among p satisfying the "not-in" condition. We denote it by p. It follows from (2) that there exists $\text{in}_{\prec_{(u,v)}}(g_i)$, $g_i \in G$ which divides $\text{in}_{\prec_{(u,v)}}(p)$. Take a monomial $c \in D$ which annihilates the initial monomial of p, and consider $p' = p - c g_i \neq 0$. It is again an element of I. When $\text{ord}_{(u,v)}(p') < \text{ord}_{(u,v)}(p)$, the relations $\text{in}_{(u,v)}(p) = \text{in}_{(u,v)}(c g_i) = \text{in}_{(u,v)}(c) \, \text{in}_{(u,v)}(g_i) \in \langle \text{in}_{(u,v)}(G) \rangle$ hold. Then, the (u,v) degrees of p and p' agree. When it holds, we have $\text{in}_{(u,v)}(p') = \text{in}_{(u,v)}(p) - \text{in}_{(u,v)}(c g_i) \notin \langle \text{in}_{(u,v)}(G) \rangle$. This contradicts the minimality of $\text{in}_{\prec_{(u,v)}}(p)$.

The Buchberger criterion presented in Theorem 1.3.3 can be applied when the $\prec_{(u,v)}$ is a well-order. In order to apply it for non-well-orders, one way is to use it in $D^{(h)}$ and show that a set is a Gröbner basis by the theorem above. The ideal membership problem cannot be solved for non-well-orders with the normal form algorithm (for the membership problem, refer to Lemma 1.2.4). This is because the normal form algorithm (the division algorithm) does not necessarily stop. We need a tangent cone algorithm to solve the ideal membership problem. For more on this topic, refer to [6].

Finally, we generalize the notion of the left ideal in D and orders in D to a subset of the r direct product of D, which is denoted by D^r. When a subset L of D^r satisfies the following, it is called a *left submodule* of D^r: (1) $DL \subseteq L$; and (2) $p - q \in L$ for any $p, q \in L$. When $r = 1$, a left submodule of D is a left ideal of D.

Let e_i be an element of D^r such that the i the component of e_i is 1 and the other components are 0. The set $\{e_1, \ldots, e_r\}$ is a basis of D^r as a D-free module. Let (u, v) be a weight vector in D, let $w = (w_1, \ldots, w_r)$ be a vector of integers, and let $\prec$ be a term order in D. From these, we define an order $\prec_{(u,v,w)}$ in D^r as

$$x^\alpha \partial^\beta e_i \prec_{(u,v,w)} x^{\alpha'} \partial^{\beta'} e_j \tag{6.29}$$

$$\Leftrightarrow u \cdot \alpha + v \cdot \beta + w_i < u \cdot \alpha' + v \cdot \beta' + w_j$$

$$\text{or } (u \cdot \alpha + v \cdot \beta + w_i = u \cdot \alpha' + v \cdot \beta' + w_j \text{ and } i < j)$$

$$\text{or } (u \cdot \alpha + v \cdot \beta + w_i = u \cdot \alpha' + v \cdot \beta' + w_j \text{ and } i = j$$

$$\text{and } x^\alpha \partial^\beta \prec x^{\alpha'} \partial^{\beta'}).$$

$$\tag{6.30}$$

For (u, v) and w, we define the order (degree) by

$$\text{ord}_{(u,v,w)}(x^\alpha \partial^\beta e_i) = u \cdot \alpha + v \cdot \beta + w_i.$$

We call w a *shift vector*.

Example 6.6.5. We illustrate the definition by an example. We consider the case of $n = 1, r = 2, (u, v) = (0, 1)$, and $w = (0, 0)$.

1. We have $(\partial^2, 0) \succ_{(u,v,w)} (0, \partial)$, because the order $\text{ord}_{(u,v,w)}$ of the left-hand side is larger than that of the right-hand side.
2. We have $(\partial, 0) \prec_{(u,v,w)} (0, \partial)$, because the values of $\text{ord}_{(u,v,w)}$ of both sides agree. The left-hand side is ∂e_1, the right-hand side is ∂e_2, and $1 < 2$ holds.

When $(u, v) = (0, 0)$ and $w = 0$, the order $\prec_{(u,v,w)}$ is called the *POT order*. Note that in Sect. 3.5.3, the case of $i > j$ instead of $i < j$ is called the POT order. In other words, the order of indices is reversed in this chapter.

Tip 6 These orders are essential for several of the cohomology groups when they are computed by utilizing Gröbner bases in D [25, 27]. We use only the POT order in the integration algorithm in this chapter.

6.7   Filtrations and Weight Vectors

In the theory of D-modules, filtrations of modules are used as key structures. As will be seen in the following sections, several weight vectors are used as the foundation for algorithms for D-modules. We will begin by discussing the idea of modules and summarizing the fundamental facts for filtrations and weight vectors. We will use the ideas of modules and filtrations only for proving that the integration algorithm is correct (Theorems 6.10.8 and 6.10.11); those readers who do not need to understand the proof may skip all of this section except for the explanation of weight vectors.

First, we define modules. A left module M over D (a left D-module M) is an additive group for which an action D is defined. By "an action of D on M", we mean a map

$$D \times M \ni (p, m) \mapsto m' \in M$$

which satisfies the following conditions (m' is denoted by pm):

1. $1m = m, m \in M$;
2. $p(qm) = (pq)m, p, q \in D, m \in M$;
3. $(p + q)m = pm + qm, p, q \in D, m \in M$;
4. $p(m + m') = pm + pm', p \in D, m, m' \in M$.

Example 6.7.1. Let I be a left ideal in D. The additive group $M = D/I$ can be regarded as a left D module by a natural action of D. Let us explain what we mean by a natural action. Assume $m \in D$, and let $[m] = m + I$ be the equivalence class of m. The class $[m]$ is an element of M. For $p \in D$, we define the (natural) action p on $[m]$ by $p[m] = [pm] \in M$. When $[m'] = [m]$, then $p(m - m') \in I$ because $m - m' \in I$. Hence, we have $[pm'] = [pm]$. This implies that our definition of the action is well defined. It is left as an exercise to show the four conditions listed above.

We now define a filtration of D and left modules over D. For any integer i, we are given an additive subgroup $F_i D$ of D that satisfies the following four conditions:

1. $F_i D \subseteq F_{i+1} D$;
2. $\bigcup F_i D = D$;
3. $(F_i D)(F_j D) \subseteq F_{i+j} D$;
4. $1 \in F_0 D$.

The set $\{F_i D\}$ is called a *filtration* of D, and we denote it by (D, F). The symbol $F_i D$ might be confused with the product of F_i and D, but $F_i D$ is a symbol. In order to avoid confusion, we sometimes will write it as $F_i(D)$.

Let M be a left D-module. For any integer i, we are given an additive subgroup $F_i M$ of M that satisfies the following three conditions:

1. $F_i M \subseteq F_{i+1} M$;
2. $\bigcup F_i M = M$;
3. $(F_i D)(F_j M) \subseteq F_{i+j} M$.

The set $\{F_i M\}$ is called a *filtration* of M, and we denote it by (M, F). We may denote $F_i M$ by $F_i(M)$. If we have $F_i D = 0$ for any $i < 0$ and there exists $i_0 \le 0$ such that $F_i M = 0$ holds for any $i \le i_o$, then (D, F), (M, F) is called a *filtration bounded from below*. (We may relax the condition for $F_i D$ such that, for sufficiently small i, we have $F_i D = 0$. For simplicity, we will assume $F_i D = 0$ for $i < 0$.) When a filtration bounded from below satisfies the following two conditions, it is called a *good filtration*.

1. The subgroup $F_i M$ is finitely generated over $F_0 D$.
2. There exists a constant k_o such that $k_0 \ge 0$ holds, and for any nonnegative integers i and $k \ge k_0$, the relation $F_i D \, F_k M = F_{i+k} M$ holds.

The role of filtrations in the theory of D-modules is analogous to the role of weight vectors in algorithms with Gröbner bases. In fact, when a weight vector is given, we can define an associated filtration. Let (u, v) be an integer weight vector of D. For an integer m, we define the **C**-vector space $F_m D$ by

$$F_m D = \{p \in D \mid \mathrm{ord}_{(u,v)}(p) \le m\}.$$

Then, we have $(F_i D)(F_j D) \subseteq (F_{i+j} D)$, because we have $\mathrm{ord}_{(u,v)}(pq) = \mathrm{ord}_{(u,v)}(p) \, \mathrm{ord}_{(u,v)}(q)$ from the Leibniz formula.

Tip 7 *When $u \ge 0$ and $v \ge 0$ hold, we have $F_m D = 0$ for $m < 0$. Let $G_m = F_m D / F_{m-1} D$. Then we have $G_i G_j \subseteq G_{i+j}$, and D is equipped with the structure of a graded algebra. The ring D can be regarded as an infinite-dimensional vector space over **C**. This vector space can be studied as a collection of finite-dimensional vector spaces with the structure of the graded algebra. When we change the weight vector which defines G_m, we obtain a different structure of D as a graded algebra.*

Example 6.7.2. Let $w = (1, 0, \ldots, 0)$. The filtration defined by the weight vector $(-w, w)$ is called the Kashiwara–Malgrange filtration or the V-filtration with respect to x_1.

Tip 8 *For the same ring, the theory of spectral sequences can use several different structures as the graded algebra. The theory and algorithms for Gröbner fans, which lie in the space of the weight vectors, sometimes play roles that are analogous to those of spectral sequences [29, Theorem 1.4.12, Sect. 2.2].*

For an integer weight vector (u, v), let (D, F) be the filtration on D defined by (u, v). Let I be a left ideal of D. We consider the left D-module $M = D/I$. Let $F_m M = (F_m D)/((F_m D) \cap I)$.

Theorem 6.7.3. *1. $F_m M$ is a C-vector space.*
2. When $u \geq 0$ and $v \geq 0$ hold, the pair (D, F), (M, F) is a good filtration.

It is easy to prove these statements from the relevant definitions.

Example 6.7.4. Let $n = 2$. We consider a left ideal I generated by x_1, x_2. We define the filtration $F_m D$ by the weight vector $(u, v) = (1, 1, 1, 1)$. The set $\{x_1, x_2\}$ is a Gröbner basis with respect to the weight vector (u, v). Any monomial of ∂_1, ∂_2 is a standard monomial with respect to this Gröbner basis. It follows from Lemma 1.2.4 and the D-analog of Theorem 1.6.9 that we can show that the set $\{\partial_1^i \partial_2^j \mid v \cdot (i, j) = i + j \leq m\}$ is a basis of the C-vector space $M_m = (F_m D)/((F_m D) \cap I)$. More precisely, we have

$$M_0 = F_0/(F_0 \cap I) = \mathbf{C} \cdot 1$$
$$M_1 = F_1/(F_1 \cap I) = \mathbf{C} \cdot 1 + \mathbf{C} \cdot \partial_1 + \mathbf{C} \cdot \partial_2$$
$$M_2 = \ldots$$

(here, $F_m = F_m D$).

Let L be a left submodule of D^r, and let $M = D^r/L$ be the left D-module defined by L. Let (u, v) be a weight vector with nonnegative integer entries, and let w be an integer vector of length r, which we will call a shift vector. Define

$$(D^r)_k = \{p \in D^r \mid \mathrm{ord}_{(u,v,w)}(p) \leq k\}$$
$$F_k M = (D^r)_k/(L \cap (D^r)_k).$$

Let (D, F) be the filtration defined by the weight vector (u, v). Then we have $(F_m D)(F_k M) \subseteq F_{m+k} M$. This is a filtration of the D-module M. Thus, when a weight vector and a shift vector (u, v, w) are given, we can define a filtration on a left D-module of the form D^r/L.

When the D-module M admits a good filtration, M is finitely generated over $D = D_n$. In other words, there exist finite elements $m_i \in M$, $i = 1, \ldots, p$, and

M can be expressed as $M = \sum_{i=1}^{p} D_n m_i$. Let us show an example of a D-module which is not finitely generated. We can regard D_n as a left D_{n-1}-module. An action is defined by the multiplication of differential operators. Since the infinite number of elements $x_n^i \partial_n^j$, $i, j = 0, 1, 2, \ldots$ are generators of the left D_{n-1}-module D_n, it is not finitely generated. However, D_n is finitely generated when we regard it as a left D_n-module. In fact, D_n is generated by $1 \in D_n$. We will regard D_n as a left D_{n-1}-module in the integration algorithm.

6.8 Holonomic Systems

The Hilbert function for a homogeneous ideal was introduced in Sect. 1.6.3, and we can use it to measure the "size" of an ideal. We need the Hilbert function in order to define and study holonomic D-modules.

Let $S = \mathbb{C}[y_1, \ldots, y_m]$ be the ring of polynomials in m variables. Let J be an ideal of S which is not necessarily homogeneous. The initial ideal $\mathrm{in}_1(J)$ for the weight vector $\mathbf{1} = (1, \ldots, 1)$ is a homogeneous ideal. The Hilbert function of this ideal is denoted by $H(S/\mathrm{in}_1(J); i)$. In this chapter, we call the sum

$$h(S/J; k) = \sum_{i=0}^{k} H(S/\mathrm{in}_1(J); i), \tag{6.31}$$

the *Hilbert polynomial* of J (or S/J).

Theorem 6.8.1. *When k is sufficiently large, the function $h(S/J; k)$ is a polynomial in k.*

It follows from Theorem 1.6.15 that the number of standard monomials for which the degree is less than or equal to k is the value of the Hilbert polynomial at k. This number can be expressed in terms of a sum of binomial coefficients when k is sufficiently large. These coefficients are polynomials in k, and we have already proved Theorem 6.8.1. It is left as an exercise to provide a detailed proof (for this, we refer the reader to Sect. 5.2 and the surrounding text in [7]).

The degree of the Hilbert polynomial of J is called the *Krull dimension* of J, and we denote it by $\dim J$. When the Krull dimension is d, the Hilbert function can be written as $h(S/J; k) = \frac{p}{d!}k^d + O(k^{d-1})$. In the asymptotic form of the Hilbert polynomial, we denote p by degree (J). The 0-dimensional ideals are characterized as the ideals for which the Hilbert polynomials are constants; that is, the ideals for which the Krull dimension is 0 are called 0-dimensional ideals.

We now return to considering the ring of differential operators. When a left ideal I of D_n satisfies $\dim \mathrm{in}_{(0,1)}(I) = n$, the left idea I is called a *holonomic ideal*. The ideal $\mathrm{in}_{(0,1)}(I)$ is called a *characteristic ideal* of I. As we will see in the next theorem, holonomic ideals are the largest ideals or the ideals for which characteristic ideals are minimal with respect to their Krull dimensions.

Theorem 6.8.2. *If $I \neq D$, then we have*

$$\dim \mathrm{in}_{(0,1)}(I) \geq n.$$

This inequality follows from Theorems 6.8.5 and 6.8.6, presented below.
   We now define holonomic D-modules.

Definition 6.8.3. Let M be a finitely generated left D_n-module. Let (D_n, F) be the filtration on D_n associated with the weight vector $(1, 1)$; this is called the Bernstein filtration. Assume that there exists a good filtration $F_i M$ on M. Under this assumption, and for sufficiently large k, the vector space dimension $\dim_{\mathbf{C}} F_k M$ can be expressed as a polynomial. When $M = 0$ or the degree of the polynomial is n, the left D_n module M is called a *holonomic left D_n-module*. This definition does not depend on the choice of a good filtration [11, 12].

   Let I be a holonomic ideal. We set $M = D/I$. If $F_k M = (F_k D)/((F_k D) \cap I)$, then it is a good filtration on M. From Theorem 6.8.6, presented below, we have the following theorem.

Theorem 6.8.4. *A left ideal I is a holonomic ideal if and only if the left D-module $M = D/I$ is a holonomic left D-module.*

   Holonomic ideals and holonomic D-modules are related as in the theorem. When a left submodule L of the free module D^r is given, and D^r/L is a holonomic left D-module, in order to avoid confusion, we never say that L is a holonomic submodule, as in the case of left ideals.

Theorem 6.8.5 (Bernstein Inequality). *We fix a filtration of D by the weight vector $(1, 1) = (1, 1, \ldots, 1)$. Let $M \neq 0$ be a left D-module, and let (M, F) be a good filtration. We denote by $h(k)$ the dimension of $F_k M$ as a $\mathbf{C}$-vector space. When k is sufficiently large, the function $h(k)$ is a polynomial in k, and the degree of $h(k)$ is greater than or equal to n.*

Proof. The following clever proof is called Joseph's proof. We assume that $F_0 M \neq 0$, $F_i M = 0$, $(i < 0)$. We consider the $\mathbf{C}$-linear map

$$\rho : F_i D \ni P \mapsto (m \mapsto Pm) \in \mathrm{Hom}_{\mathbf{C}}(F_i M, F_{2i} M). \tag{6.32}$$

We use induction on i to show that the map ρ is injective. When $i = 0$, it is clearly injective because $F_i D = \mathbf{C}$. Proving that ρ is injective is equivalent to claiming that if $P \neq 0$, then $\rho(P) \neq 0$ (i.e., $\rho(P)$ is not a 0-linear map). Thus, alternatively, we may prove $P F_i M \neq 0$. Suppose that x_1 is contained in $\mathrm{in}_{(1,1)}(P)$. Since we have $[P, \partial_1] = P \partial_1 - \partial_1 P \in F_{i-1} D$, then by the induction hypothesis, there exists $m \in F_{i-1} M$ such that $(P \partial_1 - \partial_1 P)m \neq 0$ holds. Since we have $\partial_1 m \in F_i M$, $m \in F_{i-1} M \subset F_i M$, the relations $P \partial_1 m = 0$ and $Pm = 0$ contradict $[P, \partial_1]m \neq 0$; one of them is not 0. We can make analogous arguments for the other variables, and we thus conclude that ρ is injective.

Since the map ρ is injective, we have the following inequality for the dimensions:

$$\dim_\mathbf{C} F_i D \leq \dim_\mathbf{C} \operatorname{Hom}_\mathbf{C}(F_i M, F_{2i} M) = (\dim_\mathbf{C} F_i M)(\dim_\mathbf{C} F_{2i} M) = h(i)h(2i).$$

Therefore, we have

$$\binom{2n + i}{2n} \leq h(i)h(2i),$$

which implies that the degree of $h(i)$ with respect to i is greater than or equal to n.

A research problem: When M is a holonomic D-module and an operator $P \in D$ is given, determine an algorithm that finds all the $m \in M$ that are annihilated by P.

Theorem 6.8.6. *The Krull dimension of* $\operatorname{in}_{(0,1)}(I)$ *agrees with that of* $\operatorname{in}_{(1,1)}(I)$. *In particular, the left ideal I is a holonomic ideal if and only if the Krull dimension of* $\operatorname{in}_{(1,1)}(I)$ *is n.*

In order to prove this theorem, we need a proof of the existence of a Gröbner fan. Since the proof is very long, we omit it, but refer the reader to [29, Theorem 1.4.12, Sect. 2.2]. From this theorem, we can prove Theorem 6.8.2 (see [29, p. 65]). We note that, in Björk's [3] standard textbook on D-modules, there is a proof that uses spectral sequences.

We now consider the problem of determining if a given left ideal I is holonomic. By computing a Gröbner basis with the order $\prec_{(1,1)}$, we find a set of generators of $\operatorname{in}_{(1,1)}(I)$. The Krull dimension of this ideal can be obtained by constructing its Hilbert polynomial, and we can then determine if a given ideal is holonomic. Prior to the Buchberger algorithm, there was no general method for determining this.

Example 6.8.7. Below are some examples of holonomic ideals. The proof that they are holonomic is left as an exercise.

1. Let I be a left ideal generated by $\partial_1, \ldots, \partial_n$. The ideal I is a holonomic ideal of D_n.
2. Let I be a left ideal generated by $x_1, \ldots, x_n$. The ideal I is a holonomic ideal of D_n.
3. Let I be a left ideal in $D_2 = \mathbf{C}\langle x, y, \partial_x, \partial_y \rangle$ generated by the operators L_1, L_2 in Example 6.5.4. The ideal I is a holonomic ideal of D_2.

6.9 Relationship Between D and R

Theorem 6.9.1. *Let I be a holonomic ideal in D. Then the left ideal RI in R is a 0-dimensional ideal.*

Proof. Let (D, F) be the filtration on D by the weight vector $(\mathbf{1}, \mathbf{1})$. We consider the $\mathbf{C}$-linear map

$$\mu_k \ : \ F_k D \cap \mathbf{C}\langle x_1, \dots, x_n, \partial_i \rangle \ni p \mapsto [p] \in F_k D/((F_k D) \cap I). \qquad (6.33)$$

The dimension of the right-hand side as a vector space over $\mathbf{C}$ is $h(k) = \frac{m}{n!}k^n + O(k^{n-1})$, by the definition of the holonomic ideal. The dimension of the left-hand side as a vector space is $\binom{n+1+k}{n+1} = O(k^{n+1})$. When k becomes large, the dimension of the left-hand side becomes strictly larger than that of the right-hand side, and we thus have $\ker \mu_k \neq 0$. This implies $\mathbf{C}\langle x_1, \dots, x_n, \partial_i \rangle \cap I \neq 0$. For each i, we take from it a nonzero element $\sum_{j=0}^{m_i} a_j^i(x)\partial_i^j$. This is an ordinary differential operator of x_i with parameters $(x_1, \dots, x_{i-1}, x_{i+1}, \dots, x_n)$. This expression yields $\dim_{\mathbf{C}(x)} R/(RI) \leq \prod_{i=1}^n m_i < +\infty$, and thus we have the zero dimensionality of RI.

Theorem 6.9.2. *Let J be a zero-dimensional ideal of R. Then the left ideal $J \cap D$ of D is a holonomic ideal.*

Proof. We present only a sketch of the proof. Let $s_1, \dots, s_m$ be the standard monomials of a Gröbner basis of J in R. There exists a polynomial $p(x)$ such that

$$p\partial_i s_j = \sum p_{ij}^k s_k, \quad p_{ij}^k \text{ is a polynomial}$$

by mod $J \cap D$. By using this relation, we can define a structure for the left holonomic D-module in $\mathbf{C}[x, 1/p] \otimes_{\mathbf{C}[x]} D/(J \cap D)$. Since $D/(J \cap D)$ is a sub D-module of this holonomic D-module, we conclude that it is a holonomic D-module. For further details, see the appendix of [31]. See also [28].

A set of generators of a left ideal I of R is not necessarily a set of generators of $J \cap D$ as the ideal in D (see Example 6.10.13). The left ideal $J \cap D$ is called the *Weyl closure* of J. An algorithm for obtaining a set of generators of the Weyl closure was given by Tsai [34]. Some constructions are possible with only the computation of a Gröbner basis for zero-dimensional ideals in R, but there are constructions which require holonomic ideals as inputs. A typical example of these is the integration algorithm with a D-module, which will be discussed in the next section. The computation of the Weyl closure for $J \cap D$ requires extensive computer resources, and it is often a bottle neck for such constructions.

The computation of a Gröbner basis in R is sometimes slow because it requires the computation of rational functions. In such cases, it is often more efficient to obtain a Gröbner basis by performing the computations in D and using the following theorem.

Theorem 6.9.3. *Let $\prec$ be a block order in D satisfying $\partial \succ x$. The Gröbner basis G of I in D with respect to the order $\prec$ is a Gröbner basis of RI in R with respect to the order $\prec'$. Here the order $\prec'$ is defined by $\partial^\alpha \prec' \partial^\beta \Leftrightarrow \partial^\alpha \prec \partial^\beta$.*

Proof. Let p be an element of RI. Multiplying a polynomial f from the left, we assume that $fp = \sum_i c_i g_i$, $g_i \in G$, $c_i \in D$. Since $fp \in I$, we may assume that the expression above is a standard representation. In other words, we may assume that

$fp \succeq c_i g_i$. When we assume $\mathrm{in}_{\prec}(fp) = c_{\alpha\beta} x^{\alpha} \xi^{\beta}$, the leading term of p in R by the order $\prec'$ is a polynomial multiple of ξ^{β} because $\prec$ is a block order. Consider the leading terms of c_i and g_i with respect to $\prec$. Then, by a property of the block order, we have $\partial^{\beta} \succeq' c_i g_i$. Therefore the set G is a Gröbner basis of RI with respect to $\prec'$.

Example 6.9.4. Below is an example of using Risa/Asir with asir-contrib to compute a Gröbner basis in D for Example 6.1.9; for details, refer to Sect. 7.4 and [24].

```
F=[dx^2+y^2,dy^2+x^2];
dp_gr_print(1);
/* Computing a Groebner basis by the block order
   given in the theorem. */
G=nd_weyl_gr(F,[x,y,dx,dy],0,poly_r_omatrix(2));
/* Get the initial term by the matrix [[0,0,1,1],[0,0,0,-1]],
   which defines the order. */
G1=map(poly_in_w,G,[x,y,dx,dy],[0,0,1,1]);
G2=map(poly_in_w,G1,[x,y,dx,dy],[0,0,0,-1]);

/* Result */
[x*dx,dy^2,y*dy*dx,dx^2]
```

6.10   Integration Algorithm

We now introduce the integral of a left holonomic D-module. We begin with a theorem which leads us to the idea of an integral of a D-module. We consider a function $K(x, y)$ of the variables x, y, and we assume that K is smooth.

Theorem 6.10.1. *If there exist differential operators $\ell \in D_1 = \mathbf{C}\langle y, \partial_y \rangle$ and $\ell_1 \in D_2 = \mathbf{C}\langle x, y, \partial_x, \partial_y \rangle$ such that*

$$\left[\ell(y, \partial_y) + \partial_x \ell_1(x, y, \partial_x, \partial_y) \right] \bullet K = 0$$

holds, then we have

$$\ell \bullet \int_a^b K(x, y) dx + [\ell_1 \bullet K]_a^b = 0.$$

In particular, if $[\ell_1 \bullet K]_a^b = 0$, then we have a differential equation $\ell \bullet \int_a^b K(x, y) dx = 0$ that is satisfied by the definite integral of K. Here, $[\ell_1 \bullet K]_a^b$ denotes $(\ell_1 \bullet K)|_{x=b} - (\ell_1 \bullet K)|_{x=a}$.

Proof. We have $\int_a^b \left[\ell(y, \partial_y) + \partial_x \ell_1(x, y, \partial_x, \partial_y) \right] \bullet K dx = 0$. Therefore, by exchanging the integral and the differentiation and by applying the fundamental theorem of calculus, we have $\ell(y, \partial_y) \bullet \int_a^b K(x, y) dx + \int_a^b \partial_x \bullet (\ell_1 \bullet K) dx = 0$.

Suppose that K satisfies $[\ell_1 \bullet K]_a^b = 0$ for any operator $\ell_1 \in D_2$. Let I be the left ideal of D_2 that annihilates the function $K(x, y)$. Then, any element $\ell(y, \partial_y)$ of the left ideal $(I + \partial_x D_2) \cap D_1$ in D_1 can be written as $\ell_2 - \partial_x \ell_1$, $\ell_2 \in I$, and, by the above theorem, ℓ annihilates the definite integral of $K(x, y)$ with respect to y.

Example 6.10.2. We set $K(x, y) = \exp(-\frac{y}{2}x^2)$, $a = -\infty$, $b = +\infty$. We have $(\partial_y + x^2/2) \bullet K = 0$ and $(\partial_x + xy) \bullet K = 0$. Eliminating x^2, we obtain an operator $y(\partial_y + x^2/2) - x(\partial_x + xy)/2 = y\partial_y - x\partial_x/2$ which annihilates K. The right-hand side can be written as $y\partial_y - \partial_x x/2 + 1/2$. Since we have $[xK(x, y)/2]_a^b = 0$, the integral $\int_a^b K(x, y)dx$ satisfies $y\partial_y + 1/2$. By using the algorithm given in this section, we can prove that $(I + \partial_x D_2) \cap D_1$ is generated by $y\partial_y + 1/2$ when $I = \langle \partial_y + x^2/2, \partial_x + xy \rangle$.

The integral $\int_{-\infty}^{\infty} K(x, y)dx$ can be regarded as a *normalizing constant* of the (unnormalized) normal distribution K with a parameter y. For the K in the example, it is easy to see by solving the differential equation that the function defined by the integral is a constant multiple of the function $y^{-1/2}$. The term normalizing constant is used in statistics. We consider a nonnegative function $p(y, t)$ in t with a parameter y. For the integral $Z(y) = \int_{-\infty}^{\infty} p(y, t)dt$, the function $\frac{p(y,t)}{Z(y)}$ is nonnegative, and its integral with respect to t on $\mathbf{R}$ is 1. This function is a probability density function. The function $Z(y)$ normalizes the integral to 1 and is called a normalizing constant. When t is discrete, the corresponding sum is called the normalizing constant.

Let I be a left ideal of D_n. The left ideal

$$(I + \partial_n D_n) \cap D_{n-1}$$

in D_{n-1} is called the *integration ideal* with respect to the variable x_n of I. In the note after Theorem 6.10.1, we showed that operators of the integration ideal with respect to the variable x can be regarded as differential equations satisfied by the integral with the parameter y. Are there sufficiently many such operators in the integration ideal? The answer is yes, as we will see in the next theorem. The purpose of this section is to prove the following theorem and to provide an algorithm for the construction of the integration ideal.

Theorem 6.10.3. *If I is a holonomic ideal in D_n, then the integration ideal $(I + \partial_n D_n) \cap D_{n-1}$ is a holonomic ideal in $D_{n-1} = \mathbf{C}\langle x_1, \ldots, x_{n-1}, \partial_1, \ldots, \partial_{n-1} \rangle$.*

We begin with some preparatory claims before proving this theorem.

The following claim was shown as Theorem 6.6.4 in (3). It is the foundation of the integration algorithm, and we paraphrase it here.

Theorem 6.10.4. *Let G be a Gröbner basis of a left ideal I of D with respect to the order $\prec_{(u,v)}$. Then the set $\{\text{in}_{(u,v)}(g) \mid g \in G\}$ is a set of generators of the ideal $\text{in}_{(u,v)}(I)$.*

The b-function is of key importance for proving the properties of integration ideals and for the integration algorithm. Intuitively speaking, we will see that we can use the roots of b-functions to reduce questions on infinite-dimensional vector spaces to questions on finite-dimensional vector spaces.

Definition 6.10.5. Let $(-w, w)$ be a weight vector. The set

$$\text{in}_{(-w,w)}(I) \cap \mathbf{C}[w_1\theta_1 + \cdots + w_n\theta_n]$$

is the ideal in the ring of polynomials $\mathbf{C}[w_1\theta_1 + \cdots + w_n\theta_n]$ of one variable. This ideal is principal and is generated by the monic element $b(s)$, $s = w_1\theta_1 + \cdots + w_n\theta_n$. This polynomial $b(s)$ is called the b-*function* of I with respect to the weight vector $(-w, w)$.

Example 6.10.6. Let $n = 2$ and $(-w, w) = (0, -1, 0, 1)$. Set $L_1 = x_2 - x_1^2$ and $L_2 = 2x_1\partial_2 + \partial_1$. We consider the left ideal $I = \langle L_1, L_2 \rangle$. (Note: This ideal annihilates the delta function $\delta(x_2 - x_1^2)$.) Let $\prec$ be the (pure) lexicographic order satisfying $\partial_1 \succ x_1 \succ \partial_2 \succ x_2$. The S polynomial of L_1, L_2 by the order $\prec_{(-w,w)}$ is $2\partial_2 L_1 + x_1 L_2 = x_1\partial_1 + 2x_2\partial_2 + 2$. We denote it by L_3. The S polynomial of L_2 and L_3 is $\partial_1 L_2 - 2\partial_2 L_3 = -4x_2\partial_2^2 - 6\partial_2 + \partial_1^2$. The initial term $\text{in}_{(-w,w)}$ of this element is $-4x_2\partial_2^2 - 6\partial_2$. Multiplying x_2 from the left and rewriting the operator in terms of the Euler operator θ_2, we obtain $-2\theta_2(2\theta_2 + 1)$. Therefore, a monic polynomial which divides $s(s + 1/2)$ is the b-function. We can use the algorithm (presented below) for computing the b-function to prove that $b(s) = s(s + 1/2)$.

Theorem 6.10.7. *When the left ideal I is holonomic, there exists a nonzero b-function for any weight vector $(-w, w)$.*

Proving this theorem requires additional preparation; we omit it here but refer the interested reader to [29, Theorems 5.1.2, 5.1.3].

Once existence has been proved, the computation can be performed as follows. Compute a Gröbner basis G of I with the order $\prec_{(-w,w)}$. The set $G' = \text{in}_{(-w,w)}(G)$ is a Gröbner basis of $\text{in}_{(-w,w)}(I)$ with the order $\prec$. Set $p = w_1\theta_1 + \cdots + w_n\theta_n$. Let the c_i's be undetermined coefficients. Compute the normal form of $c_0 + c_1 p + c_2 p^2 + \cdots + c_{m-1}p^{m-1} + p^m$ by using the Gröbner basis G'. Setting the coefficients of the standard monomials appearing in the normal form to 0, we obtain a system of linear equations for the undetermined coefficients c_i. Take the minimal m such that the system has a nontrivial solution. Then, the polynomial $c_0 + c_1 s + \cdots + s^m$ is the b-function $b(s)$. For details, see Sect. 3.4.1 and [23].

If I is not holonomic, the b-function does not always exist. In order to determine if it exists, and, if it does, to obtain it, we compute the intersection of an ideal and a subring. See, e.g., [25, 27, Algorithm 4.6].

The left D_{n-1}-module $D_n/(I + \partial_n D_n)$ is called the *integration module* of the left D_n-module D_n/I with respect to the variable x_n.

Theorem 6.10.8. *If the left D_n-module D_n/I is holonomic, then the integration module $D_n/(I + \partial_n D_n)$ is a holonomic D_{n-1}-module.*

Since $D_{n-1}/(D_{n-1} \cap (I + \partial_n D_n))$ is a submodule of $D_n/(I + \partial_n D_n)$ (use the POT order, explained below, to prove this), then we have Theorem 6.10.3 from this theorem. We now prove this theorem.

Proof. Let $V_k D$ be the set of the elements of D_n for which the order by $(0, \ldots, 0, 1; 0, \ldots, 0, -1)$ is less than or equal to k (the set of elements for which $\mathrm{ord}_{(-w,w)}$, $w = (0, \ldots, 0, -1)$ is less than or equal to k). Below, we will denote D_n by D, except when we want to emphasize the number of variables n. From Theorem 6.10.7, there exists a b-function $b(s)$ which satisfies $b(-\partial_n x_n) \in V_{-1}D + I$. Multiplying x_n^k from the left, we have $x_n^k b(-x_n \partial_n - 1) = b(-(x_n \partial_n - k) - 1)x_n^k = b(-\partial_n x_n + k)x_n^k$ (see Proposition 6.1.3), and we have

$$b(k)x_n^k \in V_{k-1}D + I + \partial_n D$$

from $b(-\partial_n x_n + k) - b(k) \in \partial_n D$. It follows from this inclusion that if k_0 is the maximal integral root of $b(s) = 0$ and $k > k_0$, then x_n^k can be expressed in terms of an element in $D_{n-1}x_n^0 + \cdots + D_{n-1}x_n^{k_0}$ modulo $I + \partial_n D$. Therefore, the integration module $M' = D_n/(I + \partial_n D_n)$ is a quotient module of $D_{n-1}^{k_0+1}$ and, in particular, is a finitely generated left D_{n-1}-module. This construction is also a key step in the algorithm. When $k_0 < 0$ or there exists no integral solution, we have $1 \in V_{-1}D + I + \partial_n D = I + \partial_n D$, and then $D/(I + \partial_n D) = 0$. In the following, we will consider the case where k_0 is a nonnegative integer.

Let $M = D_n/I$. The map

$$\partial_n : M \ni m \longmapsto \partial_n m \in M$$

is a morphism of left D_{n-1}-modules. The integration module can also be expressed as $M' = M/\partial_n M$.

Let (D_n, F) be the filtration on D_n defined by the weight vector $(1, 1, \ldots, 1)$. Suppose that the map ∂_n is injective. The filtration $F_k(M) := (F_k D_n)/(I \cap (F_k D_n))$ is a good filtration. Since the D-module M is holonomic, the dimension is asymptotically $\dim_{\mathbb{C}} F_k(M) = \frac{m}{n!}k^n + O(k^{n-1})$. Let $F_k(M/\partial_n M) = F_k(M)/(\partial_n M \cap F_k(M))$. It is then a filtration of the D_{n-1}-module $M/\partial_n M$. Since it is finitely generated D_{n-1}-module, it is a good filtration. Since we have $\partial_n F_{k-1}(M) \subset F_k(M)$, we can evaluate the dimension from the injectivity of ∂_n as $\dim_{\mathbb{C}} F_k(M/\partial_n M) \leq \frac{m}{n!}k^n - \frac{m}{n!}(k-1)^n + O(k^{n-2})$. Simplifying the right-hand side, we obtain $\frac{m}{(n-1)!}k^{n-1} + O(k^{n-2})$. Therefore, we conclude that M' is a holonomic D_{n-1}-module.

Suppose that the map ∂_n is not injective. Let $N = \{p \in M \mid \text{There exists } k \text{ such that} \partial_n^k p = 0\}$. Then we can prove $N \subseteq \partial_n M$. We can also prove that N is a left D_n-module. These proofs will be presented below. Let $\bar{M} = M/N$. The D_n-module $\bar{M}$ is also holonomic. Since the map

$$\partial_n : \bar{M} \ni m \longmapsto \partial_n m \in \bar{M}$$

is injective and we have $M/\partial_n M \simeq \bar{M}/\partial_n \bar{M}$, we are done.

We now prove that N is a left D_n-module. Since variables other than x_n commute with ∂_n, we may show that when $m \in N$, we have $x_n m \in N$. Suppose $\partial_n^k m = 0$, $k > 0$. Then, we have $\partial_n^{k+1} x_n m = \partial_n(\partial_n^k x_n)m = \partial_n(x_n \partial_n^k + k\partial_n^{k-1})m = 0$. Hence, we have $x_n m \in N$.

We now prove that $N \subseteq \partial_n M$. It follows from the Leibniz formula that

$$\partial_n^k x_n^k = \sum_{i=0}^{k} \frac{1}{(k-i)!}(k(k-1)\cdots(i+1))^2 x_n^i \partial_n^i$$

holds in D. We set the coefficient to 1 when $i = k$. Moreover, we have $x_n^i \partial_n^i = \theta_n(\theta_n - 1)\cdots(\theta_n - i + 1)$, and then there exists a differential operator ℓ_i such that $x_n^i \partial_n^i = (\partial_n x_n - 1)\cdots(\partial_n x_n - i) = \partial_n \ell_i + (-1)^i i!$. Take $m \in N$, and suppose $\partial_n^k m = 0$. It follows from the formula above that

$$\partial_n^k x_n^k m = \sum_{i=0}^{k-1} \frac{1}{(k-i)!}(k(k-1)\cdots(i+1))^2 \partial_n \ell_i m + c_k m.$$

Here, we have $c_k = \sum_{i=0}^{k-1} \frac{1}{(k-i)!}(k(k-1)\cdots(i+1))^2(-1)^i i!$. Since we can show that c_k is not 0, we conclude $m \in \partial_n M$.

It is left as a research problem to determine an algorithm for constructing N in the above proof.

The *Fourier transformation*

$$\mathscr{F} : x_n \mapsto -\partial_n, \quad \mathscr{F} : \partial_n \mapsto x_n$$

is a ring isomorphism of D_n. Let M be a holonomic D-module. Since it is finitely generated, there exists a submodule L of D^{r_0} satisfying $M \simeq D^{r_0}/L$. Set $\mathscr{F}(L)$ as the Fourier transforms of the elements of L. We define $\mathscr{F}(M)$ by $D^{r_0}/\mathscr{F}(L)$. Since the D_{n-1}-module $M/x_n M$ is isomorphic to $\mathscr{F}^{-1}(\mathscr{F}(M)/\partial_n \mathscr{F}(M))$, we conclude, based on the theorem proved above, that $M/x_n M$ is a left holonomic D_{n-1}-module.

Let I be a left ideal of D. The left ideal $(I + x_n D) \cap D_{n-1}$, which lies in D_{n-1}, is called the *restriction ideal* of I to $x_n = 0$. The restriction ideal and the integration ideal are Fourier transforms of each other. Note that the Fourier transforms of the elements of $\partial_n D_n$ are $x_n D_n$. Let us explain it more precisely. When $f_1, \ldots, f_m$ are generators of I, we find $L_1, \ldots, L_{m'}$ such that $D_{n-1}^{k_0}/(D_{n-1}L_1 + \cdots + D_{n-1}L_{m'})$ is isomorphic to the restriction module $D_n/(\langle \mathscr{F}(f_1), \ldots, \mathscr{F}(f_m)\rangle + x_n D_n)$. Then, the integration module $D_n/(I + \partial_n D_n)$ is isomorphic to $D_{n-1}^{k_0}/(D_{n-1}\mathscr{F}^{-1}(L_1) + \cdots + D_{n-1}\mathscr{F}^{-1}(L_{m'}))$.

Since the procedures for computing integration modules and restriction modules are transformations of each other, we will describe the procedure for computing restriction modules in the case of two variables. We let $x = x_1$ and $y = x_2$. The case of n variables is analogous.

Algorithm 6.10.9 (Restriction Algorithm).

1. Compute a Gröbner basis of I with respect to $(-w, w) = (0, -1, 0, 1)$. We set
   the basis $\{g_1, \ldots, g_p\}$.
2. For i satisfying $0 \le i \le m_j$, write $\partial_y^i g_j$ in the form

$$\sum_s \ell_{js}^i(x, \partial_x) \partial_y^s + y(\cdots).$$

   We move all the terms containing y to $y(\cdots)$. The determination of m_j will be
   explained later. When $m_j < 0$ holds, we exclude g_j.
3. Set s_0 to the maximum value of s that appears in the above procedure. Set the
   free basis of $\mathbf{C}\langle x, \partial_x \rangle^{s_0+1}$ as $e_0 = 1, e_1 = \partial_y, e_2 = \partial_y^2, \ldots, e_{s_0} = \partial_y^{s_0}$.
4. We eliminate $e_{s_0}, \cdots, e_1$ from $\sum_s \ell_{js}^i(x, \partial_x) e_s$ by the POT order satisfying
   $e_{s_0} \succ \cdots \succ e_0$. The elements obtained by the elimination are generators of the
   restriction ideal.

The constant m_j is determined as follows.

1. Let $b(\theta_y)$ be the generator of $\text{in}_{(-w,w)}(I) \cap \mathbf{C}[\theta_y]$ where $\theta_y = y\partial_y$.
2. If $b(s) = 0$ has no nonnegative integral root, then the restriction ideal agrees
   with the whole ring and we stop.
3. Let r_0 be the maximal nonnegative integral root of $b(s) = 0$.
4. Put

$$m_j = r_0 - \text{ord}_{(0,-1,0,1)}(g_j).$$

Let L be the submodule of $D_1^{s_0+1}$ generated by $\sum_s \ell_{js}^i(x, \partial_x) e_s$. We will prove that
the left D_1-module $D_1^{s_0+1}/L$ is isomorphic to the restriction module $D_2/(I + yD_2)$
as a left D_1-module.

Example 6.10.10 (Continuation of Example 6.10.2). The operators $\hat{L}_1 = \partial_x + y^2/2$
and $\hat{L}_2 = y\partial_y - 2 + xy^2$ annihilate the function $y^2 \exp(-xy^2/2)$. We apply the
Fourier transformation $y \mapsto -\partial_y$, $\partial_y \mapsto y$ with respect to the variable y to these
operators. The results are L_1 and L_2, which can be written as $L_1 = \partial_x + (1/2)\partial_y^2$
and $L_2 = -\partial_y y - 2 + x\partial_y^2 = -y\partial_y - 3 + x\partial_y^2$. Let I be the left ideal generated by
L_1 and L_2. Taking the weight vector $(-w, w) = (0, -1, 0, 1)$, we will compute the
restriction ideal for $y = 0$. Let us compute a Gröbner basis with the order $\prec_{(-w,w)}$.

  We have $\text{sp}(L_1, L_2) = xL_1 - (1/2)L_2 = x\partial_x + (1/2)y\partial_y + 3/2 = x\partial_x + 3/2 + y(\partial_y/2)$. Call this element L_3. We can show that the set $\{L_1, L_2, L_3\}$ is a Gröbner
basis.

  The b-function is $s(s-1)$, and we have $(m_1, m_2, m_3) = (0, 0, 1)$. We remove the
terms of the form $y(\cdots)$ from

$$x\partial_x + 3/2 + y(\partial_y/2)$$

and

$$\partial_y(x\partial_x + 3/2 + y(\partial_y/2)) = (x\partial_x + 2)\partial_y + y(\partial_y^2/2).$$

Let L be the submodule of $D_1^2 = D_1 e_0 + D_1 e_1$ generated by $(x\partial_x + 3/2)e_0$ and $(x\partial_x + 2)e_1$. The module D^2/L is the restriction module. Eliminating e_1 by the POT order, then we obtain $x\partial_x + 3/2$. This operator annihilates the differentiation of the normalizing constant for the normal distribution with respect to x. In other words, it annihilates the function $\frac{\partial}{\partial x}\int_{-\infty}^{\infty} K(x,y)dy$, which is a constant multiple of $x^{-3/2}$.

Theorem 6.10.11. *Algorithm 6.10.9 outputs the restriction module and the restriction ideal.*

Proof. We set $n = 2$ and omit $(-w, w)$ in the symbol $\mathrm{ord}_{(-w,w)}$. We may assume $r_0 \geq 0$. Suppose $p \in I + x_n D$. Applying the first part of the proof of Theorem 6.10.8 to the case of the restriction module, we may assume $\mathrm{ord}(p) \leq r_0$. We show that the expression $p = \sum c_j g_j + x_n r, c_j, r \in D_n$ can be reduced to an expression of the same form satisfying $\mathrm{ord}(r) \leq r_0 + 1$. In other words, we can decrease the degree of r (the order $\mathrm{ord}_{(-w,w)}(r)$) so that it is less than or equal to $r_0 + 1$. We note that $b(\theta_n) = c_I + q$, $c_I \in I$, $\mathrm{ord}(q) \leq -1$. We decompose r as $r = r' + r''$, where r' is the sum of the $(-w, w)$ homogeneous terms which have the highest degree (order). Let the degree be k. We have $r'b(\theta_n) = b(\theta_n + k)r' = b(\partial_n x_n - 1 + k)r'$. Expanding b at $k - 1$ as a Taylor series, we have

$$b(\partial_n x_n + k - 1) - b(k - 1) = ax_n, \ \mathrm{ord}(a) \leq 1.$$

Therefore, we have

$$\begin{aligned}
b(k-1)r' &= b(\partial_n x_n + k - 1)r' - ax_n r' \\
&= r'b(\theta_n) - ax_n r' \\
&= r'b(\theta_n) - a(x_n r) + ax_n(r - r') \\
&= r'c_I + r'q - a(p - \sum c_j g_j) + ax_n(r - r').
\end{aligned}$$

In the last expression, the degree of the elements which do not belong to I is $\min(k - 1, \mathrm{ord}(ap))$. Hence, we can decrease the degree of r to $r_0 + 1$. Thus, we can have r satisfying $p - x_n r \in I$, $\mathrm{ord}(p - x_n r) \leq r_0$.

We show that $p - x_n r$ has a standard representation by mod $x_n D$. In other words, we show that there exist c_j's satisfying

$$p - x_n r = \sum c_j g_j \quad \mathrm{mod}\, x_n D$$

and

$$\mathrm{ord}(c_j g_j) \leq r_0, \ c_j \in D.$$

We divide the $p - x_n r$ by the g_j's. We choose the multi-quotients so that ord decreases by 1 or more after one division. Since our order is not a well-order, this division does not necessarily stop. However, if we repeat the division procedure until the degree of the remainder r is less than or equal to -1, then the remainder will be divided by x_n. We have thus shown the existence of the standard representation of the form above. Since c_j can be written as $c_j = \sum_{i=0}^{m_j} c_{ji}(x, \partial_x) \partial_n^i + x_n \tilde{c}_j$, we have completed the proof. (Note $c_{ji} \in D_{n-1}$. See also [26, Theorem 5.9, p. 154].)

We are ready to derive the differential equation in Example 6.5.3 algorithmically.

Example 6.10.12. The function $\exp(xy - y^3)$ is annihilated by $\hat{L}_1 = \partial_x - y$, $\hat{L}_2 = \partial_y - (x - 3y^2)$. Applying the Fourier transformation with respect to y, we obtain $L_1 = \partial_x + \partial_y$, $L_2 = y - (x - 3\partial_y^2)$. Compute a Gröbner basis with the weight vector $(0, -1, 0, 1)$. The output is $L_1, L_2, L_3 = 3\partial_x^2 - x + y$. The b-function is s. Therefore, the restriction module is $D_1/D_1(3\partial_x^2 - x)$, and the restriction ideal is generated by $(3\partial_x^2 - x)$. Define a function a by

$$a(x) = \int_0^{+\infty} \exp(xy - y^3) dy.$$

From L_3, the function $a(x)$ satisfies the differential equation

$$(3\partial_x^2 - x)a(x) - 1 = 0.$$

This is because $[\exp(xy - y^3)]_0^{+\infty} = -1$. As we have seen, the restriction/integration algorithm can be used to derive differential equations satisfied by definite integrals with parameters.

When an input for the restriction/integration algorithm is a set of generators of the zero-dimensional ideal of R, the algorithm does not always work, as seen in the following example.

Example 6.10.13. Let I be the left ideal in D_3 generated by $(x^3 - y^2z^2)^2\partial_x + 3x^2$, $(x^3 - y^2z^2)^2\partial_y - 2yz^2$, and $(x^3 - y^2z^2)^2\partial_z - 2y^2z$. The left ideal I annihilates the function $\exp(1/(x^3 - y^2z^2))$. The ideal RI is a 0-dimensional ideal in R. The degree of the Hilbert polynomial of I from the Bernstein filtration is 4, and so I is not a holonomic ideal. We cannot apply the integration algorithm to I with respect to the variable ∂_y because the b-function does not exist.

Let g be a holonomic function annihilated by a holonomic ideal $I \subset D = D_n$. Let a and b be numbers. We assume that the intersection of the singular locus of g and ($x_n = a$ or $x_n = b$) is an algebraic set for which the dimension is at most $n - 2$. Under this assumption, the restriction of g to $x_n = a, b$ is a holonomic function of $n - 1$ variables. The 0-dimensional ideals R_{n-1} generated by the restriction ideals $(I + (x_n - a)D_n) \cap D_{n-1}$ or $(I + (x_n - b)D_n) \cap D_{n-1}$ are strictly smaller than R_{n-1}. Under the above assumptions, we have the following theorem.

Theorem 6.10.14. *Assume that the integral*

$$\tilde{g}(x_1, \ldots, x_{n-1}) = \int_a^b g(x_1, \ldots, x_n) dx_n$$

has finite values in the neighborhood of a point $(x_1, \ldots, x_{n-1}) = c'$. *Then, the function* $\tilde{g}$ *is a holonomic function in* $n - 1$ *variables.*

Proof. This follows from the assumption that the function $\tilde{g}$ is a holomorphic function defined in the neighborhood of c'. It is sufficient to show that it is annihilated by a 0-dimensional ideal in R_{n-1}. Take a set of generators $\{\ell_i\}$ of $(I + \partial_n D_n) \cap D_{n-1}$. From Theorem 6.10.3, the set $\{\ell_i\}$ generates a holonomic ideal in D_{n-1}. Adding the generators and reordering the index, we assume that the first $n - 1$ generators ℓ_i, $i = 1, \ldots, n - 1$ are ordinary differential operators with respect to x_i (variables other than x_i appear as parameters). For ℓ_i, there exists an element r_i of D_n such that $\ell_i + \partial_i r_i \in I$. For this decomposition, we have $\ell_i \bullet \tilde{g} = [r_i \bullet g]_{x_n=a}^{x_n=b}$. Since derivatives, sums, and products of holonomic functions are holonomic (the proof is left as an exercise), we can show from the assumption that the function on the right-hand side $g_i = [r_i \bullet g]_{x_n=a}^{x_n=b}$ is a holonomic function in $n - 1$ variables. Thus, there exists an ordinary differential operator p_i with respect to x_i, and $p_i \bullet g_i = 0$ holds. Finally, we have $(p_i \ell_i) \bullet \tilde{g} = 0$. Here, the operator $p_i \ell_i$ is the ordinary differential operator with respect to the variable x_i; then the operators $p_i \ell_i$, $i = 1, \ldots, n - 1$ generate a 0-dimensional ideal in R_{n-1} that annihilates the function $\tilde{g}$.

Our way to introduce the theory of D-modules follows Chap. 5 of the textbook by Hotta [11]. We have presented algorithms for constructing several of the necessary objects. Various researches have revisited the theory of D-modules with a view to determining computational methods. One of the remarkable discoveries is the restriction and integration algorithm found by Oaku [25], and we recommend his introductory book on the subject [26]. In the theory of D-modules, the notions of restriction and the integration of modules play a central role, and they lead to algorithms for several objects. For these constructions, it is best to refer to the original papers, such as [27]. As an area of future research, the book [12] could be reexamined from an algorithmic point of view.

The main purpose of this chapter is to explain a new application of the algorithms for D-modules. We are now ready to explain the holonomic gradient descent method for definite integrals with parameters.

6.11   Finding a Local Minimum of a Function Defined by a Definite Integral

As an application of Gröbner bases in R, we explained above a method for finding a local minimum of a holonomic function. We have provided a method to derive a system of differential equations for a definite integral with parameters by using the integration algorithm. A combination of these two methods gives a general method for finding a local minimum of a function defined by a definite integral of a holonomic function.

Let us sketch our method with an example in which the integral can be expressed in terms of an elementary function.

Example 6.11.1. (Parameter estimation of the normal distribution by the maximal likelihood estimate)
We set

$$g(x, m, \beta) = e^{-\frac{\beta}{2}(x-m)^2}.$$

When real numbers X_i, $i = 1, \ldots, n$ are given, we want to find m and β which minimize the function

$$f(m, \beta) = \left(\prod g(X_i, m, \beta)\right)^{-1/n} \cdot \int_{-\infty}^{\infty} g(x, m, \beta)dx.$$

This problem is explained in all introductory textbooks in statistics, so we will just explain it briefly here. When we set the normalizing constant as $Z(m, \beta) = \int_{-\infty}^{\infty} g(x, m, \beta)dx$, the function $\frac{g}{Z}$ is a probability density function with respect to x with parameters m and β. Let m and β be unknown parameters of the distribution; we want to estimate them from the observed data $X_1, \ldots, X_n$. The *maximal likelihood estimate* takes a parameter vector which maximizes the likelihood function $\prod_{i=1}^{n} \frac{g(X_i)}{Z}$. The n-root of the reciprocal of the function is $f(m, \beta)$, and we want to find m and β which minimize f. Standard statistics books consider the logarithm of the likelihood function and minimize it by regarding it as a quadratic function of the parameters. We will solve it by the holonomic gradient descent method, which can be applied to a broad class of problems.

The function $g(x, m, \beta)$ satisfies the system of differential equations

$$\frac{\partial g}{\partial m} = \beta(x - m)g,$$

$$\frac{\partial g}{\partial \beta} = -\frac{(x - m)^2}{2}g,$$

$$\frac{\partial g}{\partial x} = -\beta(x - m)g.$$

We have

$$\left(\prod g(X_i, m, \beta)\right)^{-1/n} = \exp\left(\frac{\beta}{2n} \sum_{i=1}^{n}(X_i - m)^2\right).$$

We will call this function D. The function D satisfies

$$\frac{\partial D}{\partial m} = \left(\beta m - \frac{\beta}{n} \sum X_i \right) D,$$

$$\frac{\partial D}{\partial \beta} = \left(\frac{1}{2n} \sum (X_i - m)^2 \right) D.$$

Applying the integration algorithm with respect to the variable x for the system of differential equations for g, we can see that the normalizing constant $G(m, \beta) = \int_{-\infty}^{\infty} g(x, m, \beta) dx$ is annihilated by $2\beta \partial_\beta + 1$ and ∂_m. In other words, we have $(2\beta \partial_\beta + 1) \bullet G = 0$ and $\partial_m \bullet G = 0$. In fact, we have $G = \frac{\sqrt{2\pi}}{\sqrt{\beta}}$, which is a famous exercise in calculus.

The Pfaffian system satisfied by $G = \int_{-\infty}^{\infty} g(x, m, \beta) dx$ is

$$\partial_m G = 0, \beta \partial_\beta G = -\frac{1}{2} G.$$

Then, the Pfaffian system satisfied by DG is

$$\partial_m(DG) = (\partial_m D)G + D(\partial_m G) = \left(\beta m - \frac{\beta}{n} \sum X_i \right)(DG), \quad (6.34)$$

$$\partial_\beta(DG) = (\partial_\beta D)G + D(\partial_\beta G)$$

$$= \left(\frac{1}{2n} \sum (X_i - m)^2 - \frac{1}{2\beta} \right)(DG). \quad (6.35)$$

In the general algorithms for the holonomic gradient descent method, we first evaluate an approximate value of DG at a point and then find a local minimum by using the system of differential equations (6.34) and (6.35). In this problem, we can conclude that

$$\beta m - \frac{\beta}{n} \sum X_i = 0, \frac{1}{2n} \sum (X_i - m)^2 - \frac{1}{2\beta} = 0$$

is $\nabla(DG) = 0$, by examination. This condition determines the values of m and β. These are well-known expressions for the average (mean) and the variance of given data.

It is not necessary to use this method for the normal distribution, but we want to emphasize that this method works even when the integral cannot be expressed in terms of elementary functions.

The next example is nontrivial, and it illustrates that the holonomic gradient descent method can be applied to a broad class of normalizing constants.

Example 6.11.2. This example is a continuation of Examples 6.5.3 and 6.10.12. We consider the unnormalized distribution $g(x, t) = \exp(xt - t^3)$ on $t \in [0, +\infty)$.

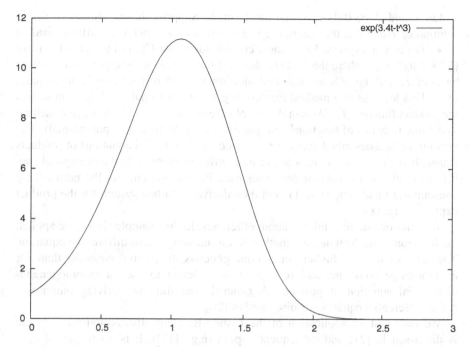

Fig. 6.3 Graph of $\exp(3.4t - t^3)$

Here, x is a parameter of the distribution. The normalizing constant is $a(x) = \int_0^{+\infty} g(x,t)dt$. As we have seen in Example 6.10.12 of the integration algorithm, the normalizing constant satisfies the differential equation $(3\partial_x^2 - x)a(x) - 1 = 0$. Assume we are given data T_i, $i = 1, \ldots, n$ and let

$$f(x) = \left(\prod g(x, T_i)\right)^{-1/n} \cdot \int_{-\infty}^{\infty} g(x,t)dt.$$

We want to find the value of x which locally minimizes f. The expression $(\prod g(x, T_i))^{-1/n}$ can be rewritten as

$$\exp\left(-x\frac{\sum_{i=1}^n T_i}{n} + \frac{\sum_{i=1}^n T_i^3}{n}\right).$$

We assume that the average of the data $\left(\sum_{i=1}^n T_i\right)/n$ and $\left(\sum_{i=1}^n T_i^3\right)/n$ are both equal to 1. The problem is to find a local minimum of $f(x) = \exp(-x + 1)a(x)$ and the corresponding x. We can derive a differential equation for $f(x)$ from that for $a(x)$: it is the differential equation of Example 6.5.3. We apply the holonomic gradient descent method to this equation, and find that the parameter value $x = 3.4$ gives an approximate local minimum. In fact, the graph of $\exp(3.4t - t^3)$ has a peak around $t = 1$ (see Fig. 6.3).

Let I and J be 0-dimensional ideals in R. Suppose that the function f is a solution of I and that the function g is a solution of J. Then the partial derivative $\partial^\alpha \bullet (fg)$ can be expressed as a linear combination over $\mathbf{C}(x)$ of terms of the form $(\partial^\beta \bullet f)(\partial^\gamma \bullet g)$, where the ∂^β's are standard monomials with respect to a Gröbner basis of I, and the ∂^γ's are standard monomials with respect to a Gröbner basis of J. This leads us to a method for deriving a 0-dimensional ideal that annihilates the product function fg. We can then apply the holonomic gradient descent method to definite integrals of functions with parameters of the form exp(polynomial). The general method described above for constructing differential equations of products is usually not efficient. We thus need a more efficient method. In the example above, it is more efficient if we first derive separate Pfaffian systems for the normalizing constant $a(x)$ and $\exp(-x + 1)$, and then derive a Pfaffian system for the product $\exp(-x + 1)a(x)$.

We offer one more comment about efficiency. In the example above, we applied the holonomic gradient descent method to an inhomogeneous differential equation. The rank of $a(x)$ is higher for a homogeneous differential equation than for an inhomogeneous one, and so it is more efficient to use an inhomogeneous differential equation, if possible. A general algorithm for deriving inhomogeneous differential equations is discussed in [20].

We note that an application of the Fisher–Bingham distribution on a sphere is discussed in [21] and subsequent papers (e.g., [17]). It is no longer of only theoretical interest. It is not possible to derive Pfaffian systems for the n-dimensional Fisher–Bingham distribution by computing Gröbner bases, because of the high computational complexity; thus the Pfaffian systems are computed "by hand" or by a mathematical insight. This is analogous to the derivation of a Markov basis "by hand" when applying the MCMC statistical test. The study of normalizing constants for systems of differential equations is an exciting research area, and it is analogous with the study of Markov bases in view of developments in combinatorics and commutative algebra. It is an important open problem to reexamine various statistical distributions in view of the holonomic gradient method and the holonomic gradient descent method, as we did for the Fisher–Bingham distributions. See [9, 16, 30, 33] for recent research achievements.

6.12 A-Hypergeometric Systems

Previous chapters of this book discussed toric ideals and their applications to statistics. Another interesting application of toric ideals is that of A-hypergeometric systems, which uses the ring of differential operators D. The A-hypergeometric system is a system of differential equations of the normalizing constant of the A-distribution [10]. There are a many attractive topics in A-hypergeometric systems, but we only present some introductory topics. An understanding of complex analysis is necessary for a thorough understanding of this section (see, e.g., [1, 15]), but we will present things so that the basic ideas can be understood without it.

Let A be a $d \times n$ matrix with integer entries. We denote by a_i the i-th column vector of A. We assume that the a_i's generate the lattice $\mathbf{Z}^d$ over $\mathbf{Z}$. In other words, we assume $\sum_{i=1}^{n} \mathbf{Z}a_i = \mathbf{Z}^d$. Let $\beta = (\beta_1, \ldots, \beta_d) \in \mathbf{C}^d$ be a parameter vector. We will use the following symbols to denote the set of nonnegative integers, negative integers, and nonpositive integers, respectively: $\mathbf{Z}_{\geq 0} = \mathbf{N}_0 = \{0, 1, 2, 3, \ldots\}$, $\mathbf{Z}_{<0} = \{-1, -2, -3, \ldots\}$, and $\mathbf{Z}_{\leq 0} = \{0, -1, -2, -3, \ldots\}$.

The main problem of the previous sections was to find differential equations satisfied by a definite integral with parameters. In this section, we consider the integral

$$\int_C \exp(\sum_{i=1}^{n} x_i t^{a_i}) t^{-\beta-1} dt. \tag{6.36}$$

Here, we define $t^{a_i} = \prod_{j=1}^{d} t_i^{a_{ji}}$, $t^{-\beta-1} = \prod_{j=1}^{d} t_j^{-\beta_j - 1}$ and $dt = dt_1 \cdots dt_d$.

A function of the form $t_j^{-\beta_j - 1}$ appears in this integral. When α is a complex number, we define z^α by $\exp(\alpha \log z)$. When α and z are positive real numbers, this is a power function. The complex function $\log z$ is multivalued, so its behavior is complicated. For example, the formula $(zw)^\alpha = z^\alpha w^\alpha$ does not hold in general, and a constant factor appears depending on the choice of the branch of $\log z$.

The simplest integral of the form (6.36) is $\int_0^{+\infty} \exp(-xt) t^{\alpha-1} dt$, where $d = n = 1$. When $x > 0$ and $\alpha > 0$, this integral can be expressed in terms of the Gamma function as $x^{-\alpha} \Gamma(\alpha)$ by the change of variable $xt = s$. The Gamma function is defined by the integral

$$\Gamma(\alpha) = \int_0^{+\infty} \exp(-s) s^{\alpha-1} ds.$$

This integral converges when $\operatorname{Re} \alpha > 0$, and it is analytically continued in the domain $\alpha \in \mathbf{C} \setminus \mathbf{Z}_{\leq 0}$. The Γ function may be the most special of the special functions. For instance, consider the n-dimensional sphere, which is one of the most fundamental objects in mathematics. Its volume is expressed in terms of π and a value of the Γ function. Elementary properties of the Γ function are explained in textbooks on calculus, and its properties as a complex function are explained in textbooks on complex analysis (e.g., [1]). Moreover, there are books about the Γ function that have been written by notable mathematicians, which implies that the subject is both exciting and deep. We now present some formulas for the Γ function, which are necessary to show the elementary properties of hypergeometric series.

$$\Gamma(\alpha + m) = \Gamma(\alpha)(\alpha)_m,$$

where $(\alpha)_m = \alpha(\alpha + 1) \cdots (\alpha + m - 1)$ is the Pochhammer symbol,

$$\Gamma(1 - z) = \frac{\pi}{\sin(\pi z) \Gamma(z)} \text{ (reflection formula)}.$$

The system of linear differential equations that is satisfied by the integral (6.36) is the A-hypergeometric system, defined in Definition 6.12.1 below. Here, the domain of the integration C is a rapidly decaying twisted cycle in general. This new and exciting research topic is discussed by Esterov and Takeuchi [4]. See also [29, p. 221] for more elementary cases. We note that the differential operators appearing in the system are elements of the integration ideal for $\exp(\sum_{i=1}^{n} x_i t^{a_i}) t^{-\beta-1}$.

Definition 6.12.1 ([5]). The following system of linear differential equations is called the *A-hypergeometric system* or the *GKZ hypergeometric system*.

$$(E_i - \beta_i) \bullet f = 0, \quad \text{where } E_i - \beta_i = \sum_{j=1}^{n} a_{ij} x_j \partial_j - \beta_i, \quad (i = 1, \dots, d)$$

$$\Box_u \bullet f = 0, \quad \text{where } \Box_u = \prod_{\{i \,|\, 1 \le i \le n, u_i > 0\}} \partial_i^{u_i} - \prod_{\{j \,|\, 1 \le j \le n, u_j < 0\}} \partial_j^{-u_j}$$

and $u \in \mathbf{Z}^n$ runs over u satisfying $Au = 0, u \ne 0$.

It follows from the definition of $\Box_u$ that the set $\Box_u$ generates the toric ideal I_A in the ring of polynomials $\mathbf{C}[\partial_1, \dots, \partial_n]$. The left ideal in D generated by $E_i - \beta_i$, $i = 1, \dots, d$, $\Box_u \in I_A$ is denoted by $H_A(\beta)$ and is called the A-hypergeometric ideal or the GKZ hypergeometric ideal.

The operator $x_j \partial_j$ will be written as θ_j. The operator θ_j is called the Euler operator, and E in the symbol E_i stands for Euler.

Example 6.12.2. When $A = \begin{pmatrix} 1\,1\,1\,1 \\ 0\,1\,0\,1 \\ 0\,0\,1\,1 \end{pmatrix}$, we have $a_1 = \begin{pmatrix} 1 \\ 0 \\ 0 \end{pmatrix}$, $a_2 = \begin{pmatrix} 1 \\ 1 \\ 0 \end{pmatrix}$, $a_3 = \begin{pmatrix} 1 \\ 0 \\ 1 \end{pmatrix}$, and $a_4 = \begin{pmatrix} 1 \\ 1 \\ 1 \end{pmatrix}$. The integral (6.36) can be written as

$$\int_C \exp(x_1 t_1 + x_2 t_1 t_2 + x_3 t_1 t_3 + x_4 t_1 t_2 t_3) t_1^{-\beta_1 - 1} t_2^{-\beta_2 - 1} t_3^{-\beta_3 - 1} \, dt_1 dt_2 dt_3.$$

Example 6.12.3. The A-hypergeometric ideal $H_A(\beta)$ for Example 6.12.2 is generated by

$$\theta_1 + \theta_2 + \theta_3 + \theta_4 - \beta_1$$
$$\theta_2 + \theta_4 - \beta_2$$
$$\theta_3 + \theta_4 - \beta_3$$
$$\partial_1 \partial_4 - \partial_2 \partial_3.$$

In most articles on A-hypergeometric systems, there are few discussions on integral representations at the initial point, and they begin with the left ideal $H_A(\beta)$. We introduce integral representations in this section in order to motivate the introduction of A-hypergeometric systems and to understand their properties intuitively. We will discuss the construction of series solutions to $H_A(\beta)$ in the remainder of this section.

Let C be a cone in $\mathbf{R}^n$, and let C^* be the dual cone. We denote the integral points in C^* by $C^* \cap \mathbf{Z}^n$.

Example 6.12.4. Let $n = 4$, and let

$$C = \mathbf{R}_{\geq 0}(1,0,0,0) + \mathbf{R}(1,1,0,0) + \mathbf{R}(0,0,1,1)$$
$$+\mathbf{R}(1,0,1,0) + \mathbf{R}(0,1,0,1).$$

The dual cone is $C^* = \mathbf{R}_{\geq 0}(1,-1,-1,1)$. We note that the vector $(1,-1,-1,1)$ is orthogonal to the generating vectors of C, except for the first one. The set of integral points in the cone $C^* \cap \mathbf{Z}^n$ is $\mathbf{Z}_{\geq 0}(1,-1,-1,1)$.

We denote by $\mathbf{C}[C^* \cap \mathbf{Z}^n]$ the set of the formal series for which the support is $C^* \cap \mathbf{Z}^n$. Any element of this set can be written as

$$\sum_{k \in C^* \cap \mathbf{Z}^n} c_k x^k, \quad c_k \in \mathbf{C},$$

where $x^k = \prod_{i=1}^n x_i^{k_i}$.

Example 6.12.5. In the case of the previous example, this series can be expressed as $\sum_{m=0}^{\infty} c_m \left(\frac{x_1 x_4}{x_2 x_3}\right)^m$ in terms of the generator of $C^* \cap \mathbf{Z}^n$.

Let ρ be an n-dimensional real vector. We denote by $x^\rho \mathbf{C}[C^* \cap \mathbf{Z}^n]$ the set of the formal series $\sum_{k \in C^* \cap \mathbf{Z}^n} c_k x^{\rho+k}$ (where $c_k \in \mathbf{C}$) for which the support is $\rho + (C^* \cap \mathbf{Z}^n)$. We call it the *series with support on a dual cone* (shifted by ρ).

Let $w \in C \cap \mathbf{Z}^n$ be a weight vector, and let $f \in x^\rho \mathbf{C}[C^* \cap \mathbf{Z}^n]$ be a series with support on a dual cone. We denote by $\text{start}_w(f)$ the sum of the terms of f for which w degrees $w \cdot (\rho+k)$ are a minimum, and we call this the *starting term* of f. The definition of the starting term can be generalized to the case that ρ is a complex number. For more on this topic, see [29, Sect. 2.5].

Theorem 6.12.6. *Let $w \in C \cap \mathbf{Z}^n$ be a weight vector, and let $\ell \in D$ a differential operator. Suppose that the series $f \in x^\rho \mathbf{C}[C^* \cap \mathbf{Z}^n]$ has support on the dual cone of C and is a solution of ℓ. Then, the staring term of f satisfies the following differential equation*

$$\text{in}_{(-w,w)}(\ell) \bullet \text{start}_w(f) = 0. \tag{6.37}$$

This theorem is fundamental to the application of the theory of Gröbner basis to the analysis of series solutions of differential equations. It is important for characterizing the dominant part of a solution of a system of differential equations. For example, the dominant part of the function $f = x^2 + \sum_{k=3}^{\infty} x^k$ near $x = 0$ is x^2. The graph of f near $x = 0$ is approximated by that of x^2. The starting term when $w = (1)$ is x^2. In general, the dominant part of the series f which has support on the dual cone is $\text{start}_w(f)$, and it is characterized by the $(-w, w)$ initial term of the operator ℓ.

Proof. Replace x_i by $t^{w_i} x_i$ and ∂_i by $t^{-w_i} \partial_i$. Here, t is a new indeterminate that commutes with x_j and ∂_j. The result of replacing ℓ and f independently and then applying the new ℓ to the new f is the same as the result of applying ℓ to f and then doing the replacement for the series $\ell \bullet f$. In particular, we have

$$(\text{in}_{(-w,w)}(\ell)t^p + o(t^p)) \bullet (\text{start}_w(f)t^q + o(t^q)) = 0.$$

This is an identity with respect to t, and we conclude the proof by comparing the lowest degree coefficients of t.

Example 6.12.7. Let $f = x_1 x_2 x_3 \left(1 + \frac{1}{2}\frac{x_1 x_4}{x_2 x_3}\right)$, and let $w = (0, 0, 0, 1)$. For the operator $\ell = x_3 \partial_3 + x_4 \partial_4 - 1$, we have $\text{in}_{(-w,w)}(\ell) = \ell$. We have $\text{start}_w(f) = x_1 x_2 x_3$ and $\ell \bullet f = 0$, and in this case, it is easy to confirm the identity of the theorem. Take $\ell = \partial_1 \partial_4 - \partial_2 \partial_3$. Then $\text{in}_{(-w,w)}(\ell) = \partial_1 \partial_4$ and $\ell \bullet f = 0$. We can see that $\partial_1 \partial_4 \bullet x_1 x_2 x_3 = 0$.

It follows from the theorem that we may find generators of the left ideal $J = \text{in}_{(-w,w)}(H_A(\beta))$ in D to determine the starting terms of series solutions with support on the dual cone of the A-hypergeometric system $H_A(\beta)$. Let us find a set of generators. It is easy to see $E_i - \beta_i \in J$, $\text{in}_w(I_A) \subset J$. Conversely, we have the following theorem.

Theorem 6.12.8. *If β is generic, then the operators $E_i - \beta_i$, $(i = 1, \ldots, d)$, $\text{in}_w(I_A)$ generate $\text{in}_{(-w,w)}(H_A(\beta))$.*

We can prove this theorem by using the Buchberger criterion in $D^{(h)}$; see [29, Theorem 3.1.3] (the proof there is only sketched, and we should apply the criterion in $D^{(h)}$ for a rigorous proof).

We introduce the A-hypergeometric series to use the starting term to determine the higher-order terms with respect to w. Let $\rho = (\rho_1, \ldots, \rho_n)$ be a vector in $\mathbf{C}^n$, and let $u = (u_1, \ldots, u_n)$ an integer vector. We decompose u into two vectors, u_+ and u_-, which have disjoint supports and nonnegative components. They satisfy $u = u_+ - u_-$ and $u_+, u_- \in \mathbf{Z}_{\geq 0}^n$, and the support of u_+ and that of u_- are disjoint. For example, we can decompose $(1, -1, -1, 1)$ as $(1, 0, 0, 1) - (0, 1, 1, 0)$.

For $p \in \mathbf{Z}_{\geq 0}^n$, we define a falling factorial as

$$[\rho]_p = \prod_{i : p_i > 0} \rho_i (\rho_i - 1) \cdots (\rho_i - p_i + 1).$$

When all the p_i are 0, we define it to be equal to 1. We have the following identities:

$$[\rho]_{u_-} = \prod_{i:u_i<0} \prod_{j=1}^{-u_i} (\rho_i - j + 1)$$

$$[u + \rho]_{u_+} = \prod_{i:u_i>0} \prod_{j=1}^{u_i} (u_i + \rho_i - j + 1) = \prod_{i:u_i>0} \prod_{j=1}^{u_i} (\rho_i + j).$$

For example, when $\rho = (\rho_1, \rho_2, 0, \rho_4)$ and $u = (-2, 2, 2, -2)$, we have $\frac{[\rho]_{u_-}}{[\rho+u]_{u_+}} = \frac{\rho_1(\rho_1-1)\rho_4(\rho_4-1)}{(\rho_2+2)(\rho_2+1)2!}$. When $\rho \in (\mathbf{C} \setminus \mathbf{Z}_{<0})^n$, we note that $[u + \rho]_{u_+} \neq 0$ holds. Let $L = \mathrm{Ker}(\mathbf{Z}^n \xrightarrow{A} \mathbf{Z}^d) = \{k \in \mathbf{Z}^n \mid Ak = 0\}$.

Theorem 6.12.9. *Suppose that the vector ρ satisfies $\rho \in (\mathbf{C} \setminus \mathbf{Z}_{<0})^n$, and $A\rho = \beta$. Then, the denominators of the formal series*

$$\phi_\rho := \sum_{u \in L} \frac{[\rho]_{u_-}}{[\rho + u]_{u_+}} \cdot x^{\rho+u} \tag{6.38}$$

are not 0, and the formal series satisfies the A-hypergeometric system $H_A(\beta)$.

We note that the series is determined by I_A; we determine the coefficients of $x^{\rho+u}$ step by step from the starting x^ρ so that the series is a solution of $\partial^p - \partial^q \in I_A$.
Tip on how to remember the expression of the series. To this author, this looks like a sailboat: The expression $[\rho + u]_{u_+}$ looks like the centerboard beneath the boat. Although it can be easy to remember important expressions while proving them, mnemonics like this can aid remembering them later.

Proof. The components of ρ are not negative integers, the components of u are integers, and the denominator $[\rho + u]_{u_+}$ is not 0.
   We note $\theta_i \bullet x^{\rho+u} = (\rho_i + u_i)x^{\rho+u}$. From $A\rho = \beta$, $Au = 0$, we have $(E_i - \beta_i) \bullet x^{\rho+u} = 0$, and thus $(E_i - \beta_i) \bullet \phi_\rho = 0$ holds.
   Assume $\partial^p - \partial^q \in I_A$. Here, we have $\mathrm{supp}(p) \cap \mathrm{supp}(q) = \emptyset$ ($\mathrm{supp}(p) = \{i \mid p_i \neq 0\}$). We note the formulas $\partial^p \bullet x^{\rho+u} = [\rho + u]_p x^{\rho+u-p}$, $\partial^q \bullet x^{\rho+u'} = [\rho + u']_q x^{\rho+u'-q}$. When we take u, u' such that $u - p = u' - q$, it is sufficient to show that

$$[\rho + u]_p \frac{[\rho]_{u_-}}{[\rho + u]_{u_+}} = [\rho + u']_q \frac{[\rho]_{u'_-}}{[\rho + u']_{u'_+}}.$$

We need to check this for several cases.

1. When $(u-p)_i = (u'-q)_i \geq 0$, we have $u_i, u'_i \geq 0$ because $p_i, q_i \geq 0$. Therefore, we have $(u_-)_i = 0 = ((u-p)_-)_i = ((u'-q)_-)_i$. Note $(u_+)_i = u_i$, $[\rho_i]_{(u_-)_i} = 1$. We will omit the subscript i in the following. We have $[\rho+u]_p/[\rho+u]_u = [\rho+u-p]_{u-p} = [\rho+u-p]_{(u-p)_+}$. In summary, we have $[\rho+u]_p \frac{[\rho]_{u_-}}{[\rho+u]_{u_+}} = \frac{[\rho]_{u-p_-}}{[\rho+u-p]_{(u-p)_+}}$. We can make an analogous discussion for $u' - q$ and obtain the conclusion.

2. When $(u-p)_i = (u'-q)_i < 0$ and $u_i < 0, u'_i < 0$, we have a support on i only for u_-. We do not consider the denominator and omit the subscript i. We can see that $[\rho + u]_p [\rho]_{u_-} = [\rho]_{(u-p)_-}$. Therefore, we have $[\rho + u]_p \frac{[\rho]_{u_-}}{[\rho+u]_{u_+}} = \frac{[\rho]_{u-p_-}}{[\rho+u-p]_{(u-p)_+}}$. The discussion for $u' - q$ is analogous.

3. The other cases can be checked analogously. For details, see [29, Proposition 3.4.1].

When there is no negative integer component in ρ, the series ϕ_ρ can be expressed in terms of the Γ function. This series was introduced by Gel'fand et al. [5]. Let $\Gamma(u + \rho + 1) = \prod_{i=1}^{n} \Gamma(u_i + \rho_i + 1)$. Here, if there exists i such that $u_i + \rho_i \in \mathbf{Z}_{<0}$, we set $1/\Gamma(u + \rho + 1) = 0$. Under this setting, the identity

$$\frac{1}{\Gamma(\rho + u + 1)} = \frac{[\rho]_{u_-}}{[\rho + u]_{u_+}} \frac{1}{\Gamma(\rho + 1)}$$

holds for $u \in L$ and $\rho \in (\mathbf{C} \setminus \mathbf{Z}_{<0})^n$. Define

$$\Phi_\rho := \sum_{u \in L} \frac{1}{\Gamma(u + \rho + 1)} x^{\rho+u}. \tag{6.39}$$

We have $\Phi_\rho = \frac{1}{\Gamma(\rho+1)} \phi_\rho$ if there is no negative integer component ρ_i. This can be easily proved by using

$$\Gamma(\alpha + m) = \Gamma(\alpha)(\alpha)_m, \quad \Gamma(\alpha - m + 1) = \Gamma(\alpha + 1)(-1)^m/(-\alpha)_m.$$

When some ρ_i is a negative integer, these series are different. For example, if $\rho_i = -1$ and $u_i = 1$, then we have $[u_i + \rho_i]_{u_i} = 0$ and ϕ_ρ cannot be defined, but we have $\Gamma(u_i + \rho_i + 1) = 1$.

The series ϕ_ρ and Φ_ρ are called *A-hypergeometric series*. Let us construct series solutions of $H_A(\beta)$ in terms of the A-hypergeometric series ϕ_ρ or Φ_ρ. We will use the Gröbner bases of toric ideals. Let I_A be the toric ideal associated with the matrix A. For simplicity, we will assume that I_A is homogeneous in the sequel. Let $w \in \mathbf{Z}^n$ be a generic weight vector with respect to I_A. The cone $C[w]$ introduced in Proposition 5.3.7 is called the *Gröbner cone* for the weight w. Since w is generic, the dimension of the closure C of the Gröbner cone for w is n. Let G be the reduced Gröbner basis of I_A with the order $\prec_w$. Since w is generic, the initial form ideal $\mathrm{in}_w(I_A)$ is a monomial ideal in $\mathbf{C}[\partial_1, \ldots, \partial_n]$. From Theorems 6.12.6 and 6.12.8, solutions of the following system of differential equations are starting terms of series solutions for which the support is on the dual cone of C:

$$(E_i - \beta_i) \bullet s = 0, \quad i = 1, \ldots, d,$$

$$\ell \bullet s = 0, \quad \ell \in \mathrm{in}_w(G).$$

For simplicity, we assume that the solution s can be expressed as x^ρ. The vector ρ is a solution of the following system of algebraic equations:

$$A\rho = \beta, \quad \prod_{i=1}^{n} \rho_i(\rho_i - 1)\cdots(\rho_i - e_i + 1) = 0, \quad \partial^e = \prod_{i=1}^{n} \partial_i^{e_i} \in \text{in}_w(G). \quad (6.40)$$

We call the solutions ρ the *fake exponents*. It is known that when there is no degenerate solution of the system of algebraic equations (6.40), any solution s is a linear combination over $\mathbf{C}$ of solutions of the form x^ρ [29, Sect. 2.3].

Example 6.12.10. A continuation of Examples 6.12.2 and 6.12.3. Take a weight vector $w = (0,0,0,1)$. The initial form ideal $\text{in}_w(I_A)$ of $I_A = \langle \partial_1\partial_4 - \partial_2\partial_3 \rangle$ is generated by $\partial_1\partial_4$. In order to obtain the fake exponents, we solve (6.40). We may solve the system of algebraic equations

$$A\rho = \beta, \rho_1\rho_4 = 0$$

for ρ. It has the two solutions

$$(0, \beta_1 - \beta_3, \beta_1 - \beta_2, \beta_2 + \beta_3 - \beta_1) \text{ and } (\beta_1 - \beta_2 - \beta_3, \beta_2, \beta_3, 0).$$

We consider the second fake exponent. We note $L = \mathbf{Z}(1, -1, -1, 1)$. When no component of the fake exponent is a negative integer, the series ϕ_ρ can be written as

$$x_1^{\beta_1-\beta_2-\beta_3} x_2^{\beta_2} x_3^{\beta_3} \sum_{m=0}^{\infty} c_m \left(\frac{x_1 x_4}{x_2 x_3}\right)^m.$$

Here, we set $c_m = \frac{[\beta_2]_m [\beta_3]_m}{[\beta_1-\beta_2-\beta_3+m]_m [m]_m}$. The sum is taken only over $m \geq 0$, because if the fourth component of L is negative, then $[\rho]_{u_-}/[\rho+u]_{u_+}$ is 0. The series solution Φ_ρ, which is expressed in terms of Γ functions, is

$$x_1^{\beta_1-\beta_2-\beta_3} x_2^{\beta_2} x_3^{\beta_3} \sum_{m=0}^{\infty} c'_m \left(\frac{x_1 x_4}{x_2 x_3}\right)^m, \quad (6.41)$$

where we set $c'_m = \frac{1}{\Gamma(\beta_1-\beta_2-\beta_3+m+1)\Gamma(\beta_2-m+1)\Gamma(\beta_3-m+1)\Gamma(m+1)}$. These series can be expressed in terms of the Gauss hypergeometric series.

Theorem 6.12.11 ([5, 29, Theorem 3.4.2]). *Let w be a generic weight vector, and let ρ be a fake exponent standing for w. If $\rho \in (\mathbf{C} \setminus \mathbf{Z}_{<0})^n$ holds, then the series ϕ_ρ is a formal solution of $H_A(\beta)$, and its support is on $\rho + (C^* \cap L)$. In particular, we have $\text{start}_w(\phi_\rho) = x^\rho$.*

Proof. Let C be the closure of the Gröbner cone of I_A that contains w as a point in its interior. We will prove by contradiction that if u satisfies $[\rho]_{u_-} \neq 0$, then u

belongs to C^*. Let σ be a subset of $\{1,\ldots,n\}$. There exists a σ which satisfies the following conditions. (1) If $i \notin \sigma$, then $\rho_i \in \mathbf{N}_0$. (2) The set σ appears in the regular triangulation obtained by the weight w. We denote by $\bar{\sigma}$ the complement of σ. Suppose $u_i < 0$. If $i \in \bar{\sigma}$ and $\rho_i + u_i \geq 0$, then $[\rho_i]_{-u_i} \neq 0$. Therefore, when u lies in the support of ϕ_ρ, the inequality $\rho_i + u_i \geq 0$, $i \notin \sigma$ holds. Let $L' = \{u \in L \mid \rho_i + u_i \geq 0, i \notin \sigma\}$. The set L' includes the support of ϕ_ρ.

We will prove by contradiction that $L' \cdot w \geq 0$. Suppose that there exists an element u of L' such that $w \cdot u < 0$ holds. We define vectors u_σ and a, which have nonnegative components, by $u_{\sigma i} = -\min(u_i, 0)$ $(i \in \sigma)$, $u_{\sigma i} = 0$ $(i \notin \sigma)$, $a_i = \rho_i$ $(i \notin \sigma)$, $a_i = 0$ $(i \in \sigma)$. The operator $\partial^{a+u_\sigma} - \partial^{a+u_\sigma+u}$ is an element of I_A. Here, by our assumption, the underlined term $\overline{\text{is the leading}}$ term. However, since we have $\mathrm{supp}(u_\sigma) \subseteq \sigma$, the monomial ∂^{a+u_σ} does not belong to $\mathrm{in}_w(I_A)$. Here, we use the fact that ρ is a fake exponent. This is a contradiction, and we have $w \cdot u \geq 0$. We can easily show that if $w \cdot u = 0$, then $u = 0$. The last part of the proof is only sketched. In order to prove it rigorously, we need to use the notion of standard pairs and their properties; see [29, Sects. 3.2, 3.4].

Example 6.12.12. We consider

$$
A = \begin{pmatrix} 1\ 1\ 1\ 1\ 1\ 1 \\ 0\ 0\ 0\ 1\ 1\ 1 \\ 0\ 1\ 0\ 0\ 1\ 0 \\ 0\ 0\ 1\ 0\ 0\ 1 \end{pmatrix}
$$

and construct series solutions. The toric ideal I_A is generated by $\{\partial_2\partial_6 - \partial_3\partial_5, \partial_1\partial_6 - \partial_3\partial_4, \partial_1\partial_5 - \partial_2\partial_4\}$. A Gröbner basis of I_A for $w = (8, 2, 0, 30, 20, 14)$ is

$$
G = \{\underline{\partial_2\partial_4} - \partial_1\partial_5, \underline{\partial_3\partial_4} - \partial_1\partial_6, \underline{\partial_3\partial_5} - \partial_2\partial_6\}.
$$

The underlined terms are generators of the initial ideal with respect to w. Therefore, the fake exponents are solutions of

$$
A\rho = \beta, \rho_2\rho_4 = \rho_3\rho_4 = \rho_3\rho_5 = 0.
$$

The system of equations $\rho_2\rho_4 = \rho_3\rho_4 = \rho_3\rho_5 = 0$ are equivalent to $\rho_4 = \rho_5 = 0$ or $\rho_3 = \rho_4 = 0$ or $\rho_2 = \rho_3 = 0$. This can be shown by the primary ideal decomposition. If two ρ_i's are determined, the other ρ_j's are uniquely determined by $A\rho = \beta$. There are three solutions, which are in the following format: $\begin{pmatrix} \rho_1 & \rho_2 & \rho_3 \\ \rho_4 & \rho_5 & \rho_6 \end{pmatrix}$.

They are:

$$
\rho^{(1)} = \begin{pmatrix} \beta_1 - \beta_3 - \beta_4 & \beta_3 & \beta_4 - \beta_2 \\ 0 & 0 & \beta_2 \end{pmatrix},
$$

$$
\rho^{(2)} = \begin{pmatrix} \beta_1 - \beta_3 - \beta_4 & \beta_3 + \beta_4 - \beta_2 & 0 \\ 0 & \beta_2 - \beta_4 & \beta_4 \end{pmatrix},
$$

$$
\rho^{(3)} = \begin{pmatrix} \beta_1 - \beta_2 & 0 & 0 \\ \beta_2 - \beta_3 - \beta_4 & \beta_3 & \beta_4 \end{pmatrix}.
$$

The lattice L has the same format as ρ, and it is generated by the following vectors:

$$b^{(1)} = \begin{pmatrix} -1 & 0 & 1 \\ 1 & 0 & -1 \end{pmatrix}, \quad b^{(2)} = \begin{pmatrix} 0 & -1 & 1 \\ 0 & 1 & -1 \end{pmatrix}.$$

When β is generic, the series $\phi_{\rho^{(i)}}$ is a solution. This is an example of the Appell function F_1.

Thus, we have seen that series solutions can be obtained by computing a Gröbner basis of a toric ideal. In Sect. 5.5, we proved that a Gröbner basis of a toric ideal gives a regular triangulation. Gel'fand et al. [5] constructed series solutions from regular triangulations, but their method is not intuitive. On the other hand, our method is natural in the sense of solving the principal part $\text{in}_{(-w,w)}(I_A)$ and $E_i - \beta_i$, and then extending solutions of the principal part to solutions of the original system. Regular triangulations appear since the principal part contains the initial ideal of the toric ideal.

When the toric ideal I_A is homogeneous and the parameter β is generic, we can show that the series solutions we have constructed span the solution space on a translate of a secondary cone [29, Theorems 2.4.9, 2.5.16, 3.13]. When the toric ideal I_A is homogeneous and β is not generic, we have solutions containing logarithmic functions. A method of constructing $\text{vol}(A) = \text{degree}(I_A)$ many solutions is described in Sect. 3.5 of [29], but it is an open problem when I_A is not homogeneous. Series solutions are interesting mathematical objects, and they also have applications to the numerical evaluation of hypergeometric functions and to the drawing of graphs of hypergeometric functions (which are of particular interest to the author). For example, Example 7.4.14 uses a series solution to solve the problem of finding a local minimum of a function.

As the last topic of this chapter, we provide an algorithm that uses the Gröbner bases of toric ideals to output the terms of the A-hypergeometric series (6.38) in the order defined by the weight w. Let G be a Gröbner basis of I_A with respect to a generic weight vector w.

Theorem 6.12.13. *We can regard the elements of G as generators of $L = \ker(A : \mathbf{Z}^n \to \mathbf{Z}^d)$. In other words, the Gröbner basis is a set of generators of the lattice L over $\mathbf{Z}$ (see the proof, below, for how this can be regarded as generators).*

Proof. Let $u - v$ be an element of L where $u, v \in \mathbf{N}_0^n$. Since $\partial^u - \partial^v \in I_A$, there exists an element $\underline{\partial^{u'}} - \partial^{v'}$ of G such that $\partial^a \partial^{u'} = \partial^u$. Here, the underlined terms are the leading term with the order $\prec_w$. We identify this element of G and $u' - v'$ below. The reduction of $\partial^u - \partial^v$ by $\partial^{u'} - \partial^{v'}$ represents the following rewriting of vectors:

$$(u - v) - ((u' + a) - (v' + a)) = (v' + a) - v.$$

The remainder by the reduction is $\partial^{v'+a} - \partial^v$. Since a is canceled, the last vector is an expression of $(u - v) - (u' - v')$. The procedure of the reduction is written as

$$\partial^u - \partial^v = \sum_{\partial^{u'} - \partial^{v'} \in G} \partial^a (\partial^{u'} - \partial^{v'}).$$

Here, $\partial^{u'} - \partial^{v'}$ appears more than once. The number of appearances is equal to the number of coefficients of $u' - v'$ when $u - v$ is expressed in terms of a Gröbner basis. In particular, $u - v$ can be expressed as a linear combination of elements of G with nonnegative coefficients.

Let us illustrate this proof with an example. When $G = \{(1, 1, -1, -1)\}$, we reduce $(2, 2, -2, -2)$ by G (we denote ∂_i by x_i in the sequel):

$$(x_1^2 x_2^2 - x_3^2 x_4^2), \quad \overbrace{(2,2,0,0)}^{u} - \overbrace{(0,0,2,2)}^{v}$$

$$\to x_1 x_2 x_3 x_4 - x_3^2 x_4^2 \text{ by } x_1 x_2 (x_1 x_2 - x_3 x_4),$$

$$\overbrace{(2,2,-2,-2)}^{u-v} = \overbrace{(1,1,-1,-1)}^{u'-v'} + \overbrace{(1,1,1,1)}^{a+v'} - \overbrace{(0,0,2,2)}^{v}$$

$$a = (1,1,0,0), u' = (1,1,0,0), v' = (0,0,1,1)$$

$$\to 0 \text{ by } x_3 x_4 (x_1 x_2 - x_3 x_4), \quad \overbrace{(1,1,-1,-1)}^{\text{new } u-v} = \overbrace{(1,1,-1,-1)}^{\text{new } u'-v'} + (0,0,0,0)$$

$$\text{new } a = (1,1,0,0), \text{ new } u = (1,1,1,1), \text{ new } v = (0,0,2,2)$$

$$\text{new } u' = (1,1,0,0), \text{ new } v' = (0,0,1,1).$$

Theorem 6.12.14. *We have*

$$C^* \cap L = \sum_{g \in G} \mathbf{Z}_{\geq 0} g.$$

Here, when $g = \partial^u - \partial^v$, $\mathbf{Z}_{\geq 0} g$ means $\mathbf{Z}_{\geq 0}(u - v)$.

Proof. Let $u - v$ be an element of $C^* \cap L$. Here, we assume $u, v \in \mathbf{Z}_{\geq 0}$. Let $w \in C$ be a weight vector in the interior of C, and we assume $w \cdot u > w \cdot v$. We reduce $\partial^u - \partial^v$ with the order $\prec_w$ by G, and finally it is reduced to 0. As in the proof of the previous theorem, we identify the binomials and vectors. With this identification, $u - v$ can be expressed as a linear combination of elements of G with nonnegative coefficients. Then, the left-hand side is included in the right-hand side of the conclusion. The opposite inclusion follows from the definition of C and $g \in L$.

By virtue of this theorem, the points of the support of a hypergeometric series are output in the order determined by w. In general, this method gives redundant results, which must be removed.

Example 6.12.15. This is a continuation of Example 6.12.12. The vectors $b^{(1)}$, $b^{(2)}$, and

$$b^{(3)} = \begin{pmatrix} -1 & 1 & 0 \\ 1 & -1 & 0 \end{pmatrix}$$

expresses $C^* \cap L$ as $\sum_{i=1}^{3} \mathbf{Z}_{\geq 0} b^{(i)}$. This expression has a redundancy. For example, we have $b^{(1)} = b^{(2)} + b^{(3)}$. In order to express the hypergeometric series associated with $\rho^{(1)}$, we need only $b^{(1)}$ and $b^{(2)}$. For the series for $\rho^{(2)}$, we need only $b^{(2)}$ and $b^{(3)}$. For the series for $\rho^{(3)}$, we need only $b^{(1)}$ and $b^{(3)}$.

We now use this theorem to provide an algorithm for constructing the hypergeometric series $\sum_{u \in L} \frac{[\rho]_{u_-}}{[\rho + u]_{u_+}} x^{\rho + u}$ up to a given w order $u \cdot w$.

In a preparation, we first give a method for enumerating all the pairs of nonnegative integers $(m_1, \ldots, m_s)$ satisfying

$$\sum_{i=1}^{s} p_i m_i \leq N$$

for a given integer $p_i \geq 1$ and a nonnegative integer N. Expanding the polynomial $(1 + x_1 t^{p_1} + \cdots + x_s t^{p_s})^N$, we obtain

$$\sum_{m_0 + m_1 + \cdots + m_s = N} \frac{N!}{m_0! m_1! \cdots m_s!} x_1^{m_1} \cdots x_s^{m_s} t^{\sum_{i=1}^{s} m_i p_i}.$$

Since $p_i \geq 1$, what we wish to enumerate are the exponents $(m_1, \ldots, m_s)$ of x, which appear as coefficients of a power of t and for which the order is less than or equal to N. This method can be easily implemented using computer algebra systems. We note that this is simply the enumeration of the feasible points of integer programs. Then, there are existing algorithms that we can use (see Sect. 1.7 for historical notes, and Examples 7.2.4 and 7.2.5). Although the method above is not efficient when N is large, it works well for small N and is easy to implement.

Let $G = \{g_1, \ldots, g_s\}$ be a Gröbner basis of I_A. With our definitions, the binomial g_i can be expressed as $[u_+, u_-]$. When we need to specify i, we denote it by $[u_+^{(i)}, u_-^{(i)}]$. Here, we have $u_+ w > u_- w$. Put $u_+ w - u_- w$ as p_i. Then, elements of $C^* \cap L$ for which w is of an order less than or equal to N, are written as

$$\sum m_i (u_+^{(i)} - u_-^{(i)}).$$

Here, m_i includes all of the solutions of $\sum m_i p_i \leq N$, and if there are redundancies, they need to be removed. For this purpose, we list all the elements $\sum m_i (u_+^{(i)} - u_-^{(i)})$ of $C^* \cap L$ for which the degrees $\sum m_i p_i$ are equal, and then we remove the redundant ones. Thus, we have an algorithm that generates the hypergeometric series.

6.13 Notes

In the previous section, we explained the construction of series solutions for A-hypergeometric systems. We used the initial ideal $\mathrm{in}_w(I_A)$, which determines the starting term of the series solution. In other words, the first approximate solutions are obtained by $\mathrm{in}_w(I_A)$. Therefore, several properties of $\mathrm{in}_w(I_A)$ which were discussed in Chap. 5 control the properties of the hypergeometric series (for details, see [29]). We did not discuss the related elementary topics or history of the Gauss hypergeometric series; interested reader are referred to the chapters on the hypergeometric function in [2, 8, 13, 14], all of which begin with elementary topics and build up to current research topics. Research on A-hypergeometric systems has advanced significantly since [29], which was published in 2000. A survey of recent research on A-hypergeometric systems is found in [32].

In Chap. 4, we studied the enumeration of contingency tables and the MCMC method. Contingency tables appear in hypergeometric series. For example, we considered A in Example 6.12.12. For the integral representation (6.36) of a solution for $\beta = (5, 3, 2, 2)$, we make the change of variables $t_1 f = -s$ (f will be defined below), and we can see that the integral is formally equal to a constant multiple of

$$\int_{C'} f^\beta t_2^{-\beta_2-1} t_3^{-\beta_3-1} t_4^{-\beta_4-1} dt_2 dt_3 dt_4, \quad f = x_1 + x_2 t_3 + x_3 t_4 + x_4 t_2 + x_5 t_2 t_3 + x_6 t_2 t_4$$

because $t_1 = -sf^{-1}$ and $dt_1 = -f^{-1} ds$. Let C' be the direct product of the circles centered at the origin in the complex plane with radius t_i, where $i = 2, 3, 4$. From the residue theorem, this integral is a constant multiple of the coefficient of $t_2^{\beta_2} t_3^{\beta_3} t_4^{\beta_4}$ of the polynomial f^{β_1} of the t variables. We can easily show that the integral satisfies the A-hypergeometric system for β. The coefficient is

$$30(0, 0, 2, 1, 2, 0) + 120(0, 1, 1, 1, 1, 1) + 60(1, 0, 1, 0, 2, 1)$$

$$+ 30(0, 2, 0, 1, 0, 2) + 60(1, 1, 0, 0, 1, 2),$$

where the vector $(k_1, k_2, k_3, k_4, k_5, k_6)$ represents $x^k = \prod x_i^{k_i}$. If we rewrite the vectors in the format $\begin{pmatrix} k_1 & k_2 & k_3 \\ k_4 & k_5 & k_6 \end{pmatrix}$, then they are the 2×3 contingency tables for which the row and column sums are $(2 = 5 - 3, 3; 1, 2, 2)$ (see Example 7.2.4). Let us generalize this example. For given A and β, which have only nonnegative integer components, we consider the distribution $\frac{|k|! x^k}{k_1! \cdots k_n!} / Z(\beta; x)$ with a parameter vector x, where k runs over the integer vectors with nonnegative components satisfying $Ak = \beta$. Here, $Z(\beta; x)$ is the normalizing constant defined by

$$Z(\beta; x) = \sum_{k \in \mathbb{N}_0^n : Ak = \beta} \frac{|k|! x^k}{k_1! \cdots k_n!}. \tag{6.42}$$

The normalizing constant $Z(\beta; x)$ satisfies the A-hypergeometric system and is a special case of the A-hypergeometric series studied in this chapter [29, p. 131]. Thus, the multiple hypergeometric distribution studied in Sect. 4.1.4 assists in the study of the hypergeometric series.

We started with the ring of differential operators, which looks unrelated to the topics in the other chapters. However, we have come to a happy ending with connections between several topics.

About 24 years ago, the author remembers that Prof. I.M. Gel'fand introduced the A-hypergeometric system and said that if we continued to study this topic and related areas, we would be able to write many volumes of research books. He also told us that combinatorial and computational mathematics would become increasingly important in the future. His predictions are being realized, such as by this book.

Unfortunately, there is little feedback from hypergeometric systems to the combinatorics of I_A and algebraic statistics. The holonomic gradient descent method, presented in this chapter, may be a first step in such feedback. The author hopes that there will be successful advances in this direction, leading to a sequel of this book.

References

1. L. Ahlfors, *Complex Analysis* (McGraw-Hill, New York, 1979)
2. K. Aomoto, M. Kita, *The Theory of Hypergeometric Functions* (Springer, Berlin, 2011)
3. J.E. Björk, *Rings of Differential Operators* (North-Holland, New York, 1979)
4. A. Esterov, K. Takeuchi, Confluent A-hypergeometric functions and rapid decay homology cycles. arxiv:1107.0402
5. I.M. Gel'fand, A.V. Zelevinsky, M.M. Kapranov, Hypergeometric functions and toral manifolds. Funct. Anal. Appl. **23**, 94–106 (1989)
6. G. Granger, T. Oaku, N. Takayama, Tangent cone algorithm for homogenized differential operators. J. Symb. Comput. **39**, 417–431 (2005)
7. G.M. Greuel, G. Pfister, *A Singular Introduction to Commutative Algebra* (Springer, Berlin, 2002)
8. Y. Haraoka, *Hypergeometric Functions* (Asakura, Tokyo, 2002) (in Japanese)
9. H. Hashiguchi, Y. Numata, N. Takayama, A. Takemura, Holonomic gradient method for the distribution function of the largest root of a Wishart matrix. J. Multivar. Anal. **117**, 296–312 (2013)
10. T. Hibi, K. Nishiyama, N. Takayama, Pfaffian systems of A-hypergeometric equations. arxiv:1212.6103
11. R. Hotta, *Introduction to Algebra* (Asakura, Tokyo, 1987) (in Japanese)
12. R. Hotta, K. Takeuchi, T. Tanisaki, *D-Modules, Perverse Sheaves, and Representation Theory* (Birkhauser, Boston, 2008)
13. K. Iwasaki, H. Kimura, S. Shimomura, M. Yoshida, *From Gauss to Painlevé* (Vieweg, Braunschweig, 1991)
14. H. Kimura, *Introduction to Hypergeometric Functions* (Saiensu-sha, Tokyo, 2007) (in Japanese)
15. K. Knopp, *Theory of Functions* (Dover, Mineola, 1996)
16. T. Koyama, A. Takemura, Calculation of orthant probabilities by the holonomic gradient method. arxiv:1211.6822

17. T. Koyama, H. Nakayama, K. Nishiyama, N. Takayama, Holonomic gradient descent for the Fisher-Bingham distribution on the n-dimensional sphere. to appear in Computational Statistics (2013)
18. H. Majima, Solutions around irregular singular points. RIMS Kokyuroku **431**, 192–206 (1981)
19. H. Majima, *Asymptotic Analysis for Integrable Connections with Irregular Singular Points*. Lecture Notes in Mathematics, vol. 1075 (Springer, Berlin, 1984)
20. H. Nakayama, K. Nishiyama, An algorithm of computing inhomogeneous differential equations for definite integrals. Mathematical Software — ICMS 2010. Lecture Notes in Computer Science, vol. 6327 (Springer, Berlin, 2010), pp. 221–232
21. H. Nakayama, K. Nishiyama, M. Noro, K. Ohara, T. Sei, N. Takayama, A. Takemura, Holonomic gradient descent and its application to the Fisher-Bingham integral. Adv. Appl. Math. **47**, 639–658 (2011)
22. J. Norcedal, S. Wright, *Numerical Optimization* (Springer, Berlin, 2007)
23. M. Noro, An efficient modular algorithm for computing the global b-function, in *Mathematical Software—Proceedings of ICMS2002* (World Scientific, Singapore, 2002), pp. 147–157
24. M. Noro, N. Takayama, *Risa/Asir Drill Book 2010*. A Free Book. http://www.math.kobe-u.ac.jp/Asir (2010)
25. T. Oaku, Algorithms for b-functions, restrictions, and algebraic local cohomology groups of D-modules. Adv. Appl. Math. **19**, 61–105 (1997)
26. T. Oaku, *D-Modules and Computational Mathematics* (Asakura, Tokyo, 2002) (in Japanese)
27. T. Oaku, N. Takayama, Algorithms for D-modules—restriction, tensor product, localization, and local cohomology groups. J. Pure Appl. Algebra **156**, 267–308 (2001)
28. T. Oaku, N. Takayama, W. Walther, A localization algorithm for D-modules. J. Symb. Comput. **29**, 721–728 (2000)
29. M. Saito, B. Sturmfels, N. Takayama, *Gröbner Deformations of Hypergeometric Differential Equations* (Springer, Berlin, 2000)
30. T. Sei, H. Shibata, A. Takemura, K. Ohara, N. Takayama, Properties and applications of Fisher distribution on the rotation group. J. Multivar. Anal. **116**, 440–455 (2013)
31. N. Takayama, An approach to the zero recognition problem by Buchberger algorithm. J. Symb. Comput. **14**, 265–282 (1992)
32. N. Takayama, A-hypergeometric function, in *Encyclopedia of Special Functions* (tentative title), ed. by T. Koornwinder. Multivariable Special Functions, vol. 5 (Cambridge University Press, to appear)
33. N. Takayama, Introduction to holonomic gradient method (movie). http://www.youtube.com/watch?v=SgyDDLzWTyI (2013)
34. H. Tsai, Algorithms for associated primes, Weyl closure, and local cohomology of D-modules. Local cohomology and its applications. Lecture Notes in Pure and Applied Mathematics, vol. 226 (Dekker, New York, 2002), pp. 169–194
35. M. Yoshida, K. Takano, On a linear system of pfaffian equations with regular singular points. Funkcialaj Ekvacioj **19**, 175–189 (1976)
36. D. Zeilberger, A holonomic systems approach to special function identities. J. Comput. Appl. Math. **32**, 321–368 (1990)

Chapter 7
Examples and Exercises

Hiromasa Nakayama and Kenta Nishiyama

Abstract There are two aspects to the study of Gröbner bases: theory and computation. For problems which are difficult to solve by theoretical approaches, it may be possible to obtain solutions by computation, using either brute force or more elegant methods. On the other hand, for problems for which the computational methods are difficult, it may be possible to obtain solutions by a combination of theoretical insight and calculations. This is one of the attractions of Gröbner bases. Chapters 4–6 emphasized the theoretical aspect. In this chapter, we present problems and answers which utilize various software systems. It is our hope that readers will perform the calculations on these software systems while studying this chapter. Following these problems and their answers, we provide easy exercises which will help the reader to understand how to use these software systems to study or apply Gröbner bases. We will use computer algebra systems, statistical software systems, and some expert systems for polytopes and toric ideals; this covers several areas related to the theory and applications of Gröbner bases.

H. Nakayama (✉)
Department of Mathematics, Graduate School of Science, Kobe University,
1-1 Rokkodai, Nada-ku, Kobe 657-8501, Japan
e-mail: nakayama@math.kobe-u.ac.jp

K. Nishiyama
School of Management and Information, University of Shizuoka, Shizuoka 422-8526, Japan
e-mail: k-nishiyama@u-shizuoka-ken.ac.jp

T. Hibi (ed.), *Gröbner Bases: Statistics and Software Systems*,
DOI 10.1007/978-4-431-54574-3_7, © Springer Japan 2013

7.1   Software

Software

In this chapter, we will use many of the software packages shown in Table 7.1.
Sections 3.2 and 3.6 explained the use of Macaulay2, Singular, and Risa/Asir. Refer
to the examples in this chapter for the use of other software packages, such as
4ti2, Gfan, LattE, polymake, and TOPCOM, which are all command-line interface
software packages. The statistical software package R was introduced in Chap. 4.
To start R, use the command R in the shell; to quit R, use the command q() while
in R.

Listing 7.1 Starting and quitting R

```
$ R
R version 2.4.0 Patched (2006-11-25 r39997)
Copyright (C) 2006 The R Foundation for Statistical Computing
.... omitted
> q();
Save workspace image? [y/n/c]: n
```

Versions

Table 7.1 shows the versions of software which are used in this chapter. We will use
these when examining the behavior of programs and sample codes when solving the
examples.

Displays of Input and Output

In order to save space, we will modify the display of program input and output in
the following way:

- remove diffuse spaces and blank lines,
- wrap lines,
- reformat,
- add comments.

Table 7.1 Software versions

Software	Versions	Software	Versions
4ti2	1.3.2	Polymake	2.9.9
Gfan	0.4	R	2.4
Kan/sm1	3.050615	Risa/Asir	20110330
LattE	1.2	Singular	3-1-2
Macaulay2	1.4	TOPCOM	0.16.2
Maple	14.00	–	–

How to Get the Program Files

The program files used in the following examples and exercises are available at

http://www.math.kobe-u.ac.jp/OpenXM/Math/dojo-en/.

7.2 Markov Bases and Designed Experiments

This section includes examples and exercises for Chap. 4, "Markov bases in designed experiments". The goals of this section are the computation of Markov bases, the estimation of p values, the enumeration of the nonnegative integer solutions for a system of linear inequalities, and the computer selection of statistical models for designed experiments. We use the statistical software package R [18], the computer algebra system Risa/Asir [15], the software package for computing toric ideals 4ti2 [26], and the software package for enumerating lattice points LattE [3].

Software	Command(or Function)	Computation
Asir (toric.rr)	gr_w(Id, VL, W)	Gröbner basis w.r.t. $<_w$
	toric_ideal(A)	Generators for a toric ideal I_A (Elimination methods by GB)
Asir (Asir-Contrib)	poly_toric_ideal(A)	Generators for a toric ideal I_A (Use 4ti2, fast computation)
Asir (ipp_one.rr)	ipp_one(A,B)	Compute a nonnegative integer solution $\mathbf{x}$ for $A\mathbf{x} = B$
Asir (alias-2.rr)		Compute aliasing relations by Gröbner basis
4ti2	markov	Markov basis for a toric ideal I_A
LattE	count	Count the number of lattice points in a polytope
R (metropolis.r)	metropolis	Random sampling by MCMC
R (2x3mcmc.r)	c2x3mcmc	MCMC for an independence model of 2×3 table
R (5x5mcmc.r)	c5x5mcmc	MCMC for an independence model of 5×5 table
R (cov1_mcmc.r)	cov1_mcmc	MCMC for a fractional design
R (cov2_mcmc.r)	cov2_mcmc	MCMC for a fractional design
C Program (enumerate_fiber.c)	enumerate_fiber	Enumerate the nonnegative integer solutions for linear equations

7.2.1   Conditional Tests of Contingency Tables (Sect. 4.1)

Example 7.2.1. We consider the following 2×2 contingency table.

	Smoking	Nonsmoking	Total
Cases	3	1	4
Controls	2	4	6
Total	5	5	10

1. Let the null hypotheses be H_0: "there is no true relation between the diseases and smoking". Calculate the probability $p(\mathbf{x}) = \frac{1}{Z} \frac{1}{x_{11}! x_{12}! x_{21}! x_{22}!}$ for the data above, where Z is the normalizing constant.
2. Evaluate the p value of the following test: $X_{11} \geq c \implies H_0$ is rejected.

Answer. 1. $\mathbf{x}$ can be taken from one of the five cases:

$$\begin{vmatrix} 4 & 0 \\ 1 & 5 \end{vmatrix}, \begin{vmatrix} 3 & 1 \\ 2 & 4 \end{vmatrix}, \begin{vmatrix} 2 & 2 \\ 3 & 3 \end{vmatrix}, \begin{vmatrix} 1 & 3 \\ 4 & 2 \end{vmatrix}, \begin{vmatrix} 0 & 4 \\ 5 & 1 \end{vmatrix}.$$

Their probabilities are, respectively,

$$\frac{1}{42}, \frac{10}{42}, \frac{20}{42}, \frac{10}{42}, \frac{1}{42}.$$

2. The observed value of x_{11} is 3. Therefore, the p value is

$$p \text{ value} = \Pr(X_{11} \geq 3) = \frac{10}{42} + \frac{1}{42} = \frac{11}{42} = 0.261.$$

Since the p value > 0.05, H_0 is not rejected.

Exercise. Perform the same calculations using R (use the command `fisher.test`).

Example 7.2.2. Let $x_1, x_2, x_3 \in \{+1, -1\}$ be discrete random variables. The joint probability function of $X = (x_1, x_2, x_3)$ is

$$p(x_1, x_2, x_3) = \frac{\exp(0.2(x_1 x_2 + x_2 x_3 + x_1 x_3))}{Z},$$

where

$$Z = \sum_{(x_1, x_2, x_3) \in \{+1, -1\}^3} \exp(0.2(x_1 x_2 + x_2 x_3 + x_1 x_3))$$

is the normalizing constant. Generate random samples from the distribution $p(x_1, x_2, x_3)$ by using the Markov chain Monte Carlo method. Moreover, check that the relative frequency for each of the eight points $\{+1, -1\}^3$ approaches $p(x_1, x_2, x_3)$. (Please refer to [10, p.8].)

Answer. The following shows the application of the Markov chain Monte Carlo method to this example.

1. Initialize $\mathbf{x} = (x_1, x_2, x_3)$.
   (For example, set $\mathbf{x} \leftarrow (-1, -1, -1)$.)
2. Randomly select a variable x_i from $\{x_1, x_2, x_3\}$.
   Set $\mathbf{x}' \leftarrow (x_1, x_2, x_3)$. Replace x_i with $-x_i$ in $\mathbf{x}'$.
3. $r \leftarrow \frac{p(\mathbf{x}')}{p(\mathbf{x})}$.
4. Take a random number R from the uniform distribution from 0 to 1.
5. If $r > R$, then $\mathbf{x}_{next} \leftarrow \mathbf{x}'$;
   else $\mathbf{x}_{next} \leftarrow \mathbf{x}$.
6. Get $\mathbf{x}_{next}$ as a sample.
7. $\mathbf{x} \leftarrow \mathbf{x}_{next}$
   Go to step 2.

The file metropolis.r is a sample program implemented using R.

Listing 7.2 R: executing metropolis.r

```
> source("metropolis.r")
metropolis(number of samples, initial value)
e.g. metropolis(10000, c(-1,-1,-1))
> metropolis(10000, c(-1,-1,-1))
[1] -1 -1 -1   <- output sample values
[1] -1 -1 -1
...
[1]  1 -1 -1
[1] 2127  957  923  987  918  934  924 2230 <- frequency
experimental values
[1] 0.2127 0.0957 0.0923 0.0987 0.0918 0.0934 0.0924 0.2230
exact values
[1] 0.21294838 0.09568387 0.09568387 0.09568387 0.09568387 0.09568387
0.09568387 0.21294838
```

Exercise. Let $x_1, x_2, x_3, x_4 \in \{+1, -1\}$ be discrete random variables. The joint probability function of $X = (x_1, x_2, x_3, x_4)$ is

$$p(x_1, x_2, x_3, x_4) = \frac{\exp(0.2(x_1 x_2 + x_1 x_3 + x_2 x_4 + x_3 x_4))}{Z},$$

where Z is the normalizing constant. Generate random samples from the distribution $p(x_1, x_2, x_3, x_4)$ by using the Markov chain Monte Carlo method. (Please refer to the 2×2 square-lattice model of Ising [10, p. 16].)

7.2.2 Markov Basis (Sect. 4.2)

Example 7.2.3. Use the Gröbner basis method to derive a 2×3 contingency table for which the row sums are 6 and 3 and the column sums are 2, 3, and 4. In other words, compute a nonnegative integer solution for $A\mathbf{x} = \mathbf{b}$, where $A = \begin{pmatrix} 1 & 1 & 1 & 0 & 0 & 0 \\ 0 & 0 & 0 & 1 & 1 & 1 \\ 1 & 0 & 0 & 1 & 0 & 0 \\ 0 & 1 & 0 & 0 & 1 & 0 \\ 0 & 0 & 1 & 0 & 0 & 1 \end{pmatrix}$ and

$$\mathbf{b} = \begin{pmatrix} 6 \\ 3 \\ 2 \\ 3 \\ 4 \end{pmatrix}.$$

Answer. Here, we explain the solution by using the Gröbner basis for a toric ideal. Of course, since this problem is small, we can compute the solution without using the Gröbner basis.

To obtain the toric ideal I_A, we compute the intersection

$$I_A = \langle x_1 - t_1 t_3, x_2 - t_1 t_4, x_3 - t_1 t_5, x_4 - t_2 t_3, x_5 - t_2 t_4, x_6 - t_2 t_5 \rangle \cap \mathbb{Q}[\mathbf{x}].$$

Please refer to Lemma 1.5.11 or Corollary 4.2.11. To compute the intersection, we compute the Gröbner basis G of the ideal

$$I = \langle x_1 - t_1 t_3, x_2 - t_1 t_4, x_3 - t_1 t_5, x_4 - t_2 t_3, x_5 - t_2 t_4, x_6 - t_2 t_5 \rangle$$

with respect to the monomial order $<$ satisfying $t_1, \ldots, t_5 > x_1, \ldots, x_6$. The elements $G \cap \mathbb{Q}[x_1, \ldots, x_6]$ generate I_A. The remainder of the monomial $\mathbf{t}^\mathbf{b}$ when divided by the Gröbner basis G corresponds to a nonnegative solution. In this case, the Gröbner basis G is

$$\begin{aligned} G = \langle & -x_2 x_6 + x_3 x_5, -x_1 x_6 + x_3 x_4, -x_1 x_5 + x_2 x_4, -t_4 x_6 + t_5 x_5, -t_4 x_3 + t_5 x_2, \\ & -t_3 x_6 + t_5 x_4, -t_3 x_5 + t_4 x_4, -t_3 x_3 + t_5 x_1, -t_3 x_2 + t_4 x_1, -t_1 x_6 + t_2 x_3, \\ & -t_1 x_5 + t_2 x_2, -t_1 x_4 + t_2 x_1, x_6 - t_2 t_5, x_3 - t_1 t_5, x_5 - t_2 t_4, x_2 - t_1 t_4, \\ & x_4 - t_2 t_3, x_1 - t_1 t_3 \rangle, \end{aligned}$$

and the remainder of $\mathbf{t}^\mathbf{b} = t_1^6 t_2^3 t_3^2 t_4^3 t_5^4$ when divided by G is $-x_1^2 x_2^3 x_3 x_6^3$. The monomial corresponds to the nonnegative solution $(2, 3, 1, 0, 0, 3)$

The above computation, performed using Risa/Asir, is shown below.

Listing 7.3 Risa/Asir: computing the Gröbner basis

```
[1356] Id=[x1-t1*t3,x2-t1*t4,x3-t1*t5,x4-t2*t3,x5-t2*t4,x6-t2*t5];
[1357] VL=[t1,t2,t3,t4,t5,x1,x2,x3,x4,x5,x6];
[1358] G=nd_gr(Id,VL,0,[[0,5],[0,6]]); <- compute the Groebner basis
```

```
[-x2*x6+x3*x5,-x1*x6+x3*x4,-x1*x5+x2*x4,-t4*x6+t5*x5,-t4*x3+t5*x2,
 -t3*x6+t5*x4,-t3*x5+t4*x4,-t3*x3+t5*x1,-t3*x2+t4*x1,-t1*x6+t2*x3,
 -t1*x5+t2*x2,-t1*x4+t2*x1,x6-t2*t5,x3-t1*t5,x5-t2*t4,x2-t1*t4,
 x4-t2*t3,x1-t1*t3]
[1360] p_nf(t1^6*t2^3*t3^2*t4^3*t5^4, G, VL, 0); <- compute the remainder
-x1^2*x2^3*x3*x6^3
```

The function nd_gr computes the Gröbner basis, and the function p_nf computes the remainder. (For details, please refer to Sects. 3.6.6 and 3.6.8.) The argument [[0,5],[0,6]] of nd_gr is a block-type order which indicates the monomial order satisfying $t_1, \ldots, t_5 > x_1, \ldots, x_6$. (For details, please refer to Sect. 3.6.5)

The Risa/Asir program ipp_one.rr uses a Gröbner basis to compute a nonnegative integer solution for a system of linear equations.

Listing 7.4 Risa/Asir: computing a nonnegative integer solution using ipp_one.rr

```
[1367] load("ipp_one.rr");
[1372] A=[[1,1,1,0,0,0],[0,0,0,1,1,1],[1,0,0,1,0,0],[0,1,0,0,1,0],
[0,0,1,0,0,1]];
[1373] B=[6,3,2,3,4];
[1374] ipp_one(A,B);   <- Compute a nonnegative integer solution for Ax = B
[2,3,1,0,0,3]
```

Exercise. Use the Gröbner basis method to derive a 4×4 contingency table for which the row sums are $3, 4, 5$, and 6 and the column sums are $3, 5, 5$, and 5.

Example 7.2.4. We consider 2×3 contingency tables for which the row and column sums are fixed. In other words, we consider a system of linear equations $Ax = \mathbf{b}$, where

$$A = \begin{pmatrix} 1 & 1 & 1 & 0 & 0 & 0 \\ 0 & 0 & 0 & 1 & 1 & 1 \\ 1 & 0 & 0 & 1 & 0 & 0 \\ 0 & 1 & 0 & 0 & 1 & 0 \\ 0 & 0 & 1 & 0 & 0 & 1 \end{pmatrix},$$

and $\mathbf{b}$ is a vector.

1. Compute a Markov basis for the matrix A by computing generators of the toric ideal I_A.
2. Compute a Markov basis for the matrix A by using the software package 4ti2.
3. Compute all of the nonnegative integer solutions $\mathbf{x} \geq \mathbf{0}$ for $Ax = \mathbf{t}$, where $\mathbf{t} = (2, 3; 1, 2, 2)'$. In other words, enumerate $\mathbf{t}$-fiber $\mathscr{F}_\mathbf{t}$.

Answer. 1. We compute the generators of the toric ideal I_A by using a method similar to that of Example 7.2.3. We can obtain generators

$$\{-x_2 x_6 + x_3 x_5, -x_1 x_6 + x_3 x_4, -x_1 x_5 + x_2 x_4\}$$

for I_A. The set of the vectors corresponding to these binomials is a Markov basis for A (Theorem 4.2.8). A Markov basis for A is

$$\{(0,-1,1,0,1,-1),(-1,0,1,1,0,-1),(-1,1,0,1,-1,0)\}.$$

Here is the above computation performed using Risa/Asir program
`toric.rr`.

Listing 7.5 Risa/Asir: computing a toric ideal using `toric.rr`

```
[1367] load("toric.rr");
[1385] toric_ideal([[1,1,1,0,0,0],[0,0,0,1,1,1],[1,0,0,1,0,0],
[0,1,0,0,1,0],[0,0,1,0,0,1]]);
ideal :
[x6-t1*t4,x5-t1*t3,x4-t1*t2,x3-t0*t4,x2-t0*t3,x1-t0*t2,
t0*t1*t2*t3*t4*t5-1]
gb :
[-x2*x6+x3*x5,-x1*x6+x3*x4,-x1*x5+x2*x4,-t3*x6+t4*x5,-t3*x3+t4*x2,
 -t2*x6+t4*x4,-t2*x5+t3*x4,-t2*x3+t4*x1,-t2*x2+t3*x1,t5*x1*x5*x6-t1,
 t5*x1*x2*x6-t0,t4*t5*x1*x5-1,-t4*t5*x1*x2*x6+x3,t3*t4*t5*x1*x4-t2]
[-x2*x6+x3*x5,-x1*x6+x3*x4,-x1*x5+x2*x4] <- generators for the toric ideal
```

The function `toric_ideal(A)` computes generators for the toric ideal I_A.
The algorithm in this function is the elimination method using the Gröbner basis,
which is explained in Example 7.2.3. This algorithm is not very efficient, and it
is not practical for solving large problems. In Example 7.2.7, we will treat a large
problem.

2. Below, we will compute a Markov basis, using the software package 4ti2.
   We make the following input file `cont2x3` corresponding to the matrix A. The
   first row 5 6 gives the numbers of rows and columns of the matrix; the following
   rows are the matrix elements.

Listing 7.6 4ti2: input file `cont2x3`

```
5 6
1 1 1 0 0 0
0 0 0 1 1 1
1 0 0 1 0 0
0 1 0 0 1 0
0 0 1 0 0 1
```

We execute the 4ti2 command `markov` to compute a Markov basis.

Listing 7.7 Executing the 4ti2 command `markov`

```
$ markov cont2x3
...skip
Size:      3, Time: -0.00 / 0.01 secs. Done.
Total Time: 0.02 secs.
```

The 4ti2 outputs a Markov basis in the file `cont2x3.mar`.

Listing 7.8 4ti2: output file `cont2x3.mar`

```
3 6
 0 -1  1  0  1 -1
 1 -1  0 -1  1  0
 1  0 -1 -1  0  1
```

The first row provides the numbers of vectors in the Markov basis and variables,
and the following rows are vectors in the Markov basis. These vectors correspond
to the binomials $x_3x_5 - x_2x_6, x_1x_5 - x_2x_4, x_1x_6 - x_3x_4$ which generate the toric
ideal I_A.

L

Risa/Asir has the function `poly_toric_ideal` which uses 4ti2 to compute a Gröbner basis for a toric ideal.

Listing 7.9 Risa/Asir: computing a toric ideal using `poly_toric_ideal`

```
[1532] poly_toric_ideal([[1,1,1,0,0,0],[0,0,0,1,1,1],[1,0,0,1,0,0],[0,1,
0,0,1,0],[0,0,1,0,0,1]], [x1,x2,x3,x4,x5,x6]);
-------------------------------------------------
4ti2 version 1.3.2, Copyright (C) 2006 4ti2 team.   <- call 4ti2
... skip
4ti2 Total Time:   0.00 secs.
[-x1*x6+x3*x4,-x1*x5+x2*x4,-x2*x6+x3*x5]
```

The computer algebra systems Macaulay2 and Singular also use 4ti2 to compute toric ideals. (Macaulay2 : `FourTiTwo.m2`, Singular : `sing4ti2.lib`)

3. We now compute a nonnegative integer solution $\mathbf{x} \geq \mathbf{0}$ for $A\mathbf{x} = \mathbf{t}$. We use a method similar to that of Example 7.2.3.

Listing 7.10 Risa/Asir: computing a nonnegative integer solution using `ipp_one.rr`

```
[1367] load("ipp_one.rr");
[1372] A=[[1,1,1,0,0,0],[0,0,0,1,1,1],[1,0,0,1,0,0],[0,1,0,0,1,0],
[0,0,1,0,0,1]];
[1375] B=[2,3,1,2,2];
[1376] ipp_one(A,B);
[1,1,0,0,1,2]
```

We obtained the solution $\mathbf{x} = (1, 1, 0, 0, 1, 2)'$. By adding the elements of the Markov basis to $\mathbf{x}$ or subtracting the elements of the Markov basis from $\mathbf{x}$, we obtain the elements in the fiber. For these elements, we perform similar computations: $\mathbf{t}$-fiber is

$$\mathscr{F}_\mathbf{t} = \{(0, 1, 1, 1, 1, 1)', (0, 2, 0, 1, 0, 2)', (1, 1, 0, 0, 1, 2)', (1, 0, 1, 0, 2, 1)',$$
$$(0, 0, 2, 1, 2, 0)'\}.$$

Exercise. Let

$$A = \begin{pmatrix} 1 & 1 & 1 & 0 & 0 & 0 & 0 & 0 & 0 \\ 0 & 0 & 0 & 1 & 1 & 1 & 0 & 0 & 0 \\ 0 & 0 & 0 & 0 & 0 & 0 & 1 & 1 & 1 \\ 1 & 0 & 0 & 1 & 0 & 0 & 1 & 0 & 0 \\ 0 & 1 & 0 & 0 & 1 & 0 & 0 & 1 & 0 \\ 0 & 0 & 1 & 0 & 0 & 1 & 0 & 0 & 1 \end{pmatrix}, \mathbf{t} = \begin{pmatrix} 3 \\ 3 \\ 3 \\ 3 \\ 3 \\ 3 \end{pmatrix}.$$

Enumerate the $\mathbf{t}$-fiber $\mathscr{F}_\mathbf{t}$. In other words, enumerate all the 3×3 contingency tables for which the row and column sums are each equal to 3.

Example 7.2.5. A Markov basis $\mathscr{B}$ for a configuration A and a vector $\mathbf{x}^o$ are given. Write a program implementing the following algorithm for enumerating all the elements in the fiber $\mathscr{F}_{A\mathbf{x}^o}$.

(Algorithm enumerating all the elements in the fiber $\mathscr{F}_{A\mathbf{x}^o}$)

Input: Markov basis $\mathscr{B}$ and vector $\mathbf{x}^o$

Output: All the elements in the fiber $\mathscr{F}_{A\mathbf{x}^o}$

1. Active $\leftarrow \{\mathbf{x}^o\}$, Fiber $\leftarrow \{\mathbf{x}^o\}$
2. While Active $\neq \phi$

    a. $\mathbf{u} \leftarrow$ (an element in Active), Active $\leftarrow$ Active $\setminus\{\mathbf{u}\}$
    b. For each element $\mathbf{v} = \mathbf{v}^+ - \mathbf{v}^-$ in $\mathscr{B}$

        i. If $\mathbf{u} - \mathbf{v}^+ \geq \mathbf{0}$ and $\mathbf{u} - \mathbf{v} \notin$ Fiber, then
           Active $\leftarrow$ Active $\cup\{\mathbf{u} - \mathbf{v}\}$, Fiber $\leftarrow$ Fiber $\cup\{\mathbf{u} - \mathbf{v}\}$
        ii. If $\mathbf{u} - \mathbf{v}^- \geq \mathbf{0}$ and $\mathbf{u} + \mathbf{v} \notin$ Fiber, then
           Active $\leftarrow$ Active $\cup\{\mathbf{u} + \mathbf{v}\}$, Fiber $\leftarrow$ Fiber $\cup\{\mathbf{u} + \mathbf{v}\}$

3. Output Fiber.

This algorithm is not efficient. For a more efficient algorithm, please see [22, Algorithm 5.7].

Answer. The C program `enumerate_fiber.c` implements the above algorithm and enumerates all the elements in the fiber. This program needs the input files `start.txt` and `markov.txt`. We write an element in the fiber to the file `start.txt` and a Markov basis to `markov.txt`. The format of the input file for this program is similar to that used with 4ti2.

    The following example calculates the fiber of a 2×3 contingency table. Let

$$A = \begin{pmatrix} 1 & 1 & 1 & 0 & 0 & 0 \\ 0 & 0 & 0 & 1 & 1 & 1 \\ 1 & 0 & 0 & 1 & 0 & 0 \\ 0 & 1 & 0 & 0 & 1 & 0 \\ 0 & 0 & 1 & 0 & 0 & 1 \end{pmatrix}, x^o = \begin{pmatrix} 0 \\ 1 \\ 1 \\ 1 \\ 1 \\ 1 \end{pmatrix}, \mathscr{B} = \left\{ \begin{pmatrix} 0 \\ -1 \\ 1 \\ 0 \\ 1 \\ -1 \end{pmatrix}, \begin{pmatrix} 1 \\ -1 \\ 0 \\ -1 \\ 1 \\ 0 \end{pmatrix}, \begin{pmatrix} 1 \\ 0 \\ -1 \\ -1 \\ 0 \\ 1 \end{pmatrix} \right\}.$$

We enumerate all the elements in the fiber. This is a solution using C program `enumerate_fiber` to solve Example 7.2.4 (3). First, we make the input files `start.txt` and `markov.txt`.

Listing 7.11 C program `enumerate_fiber`: Input file `start.txt`

```
1 6
0 1 1 1 1 1    <- an element in the fiber
```

Listing 7.12 C program `enumerate_fiber`: Input file `markov.txt`

```
3 6                    <- the numbers of vectors in the Markov basis and variables
  0 -1  1  0  1 -1  <- vectors in the Markov basis
  1 -1  0 -1  1  0
  1  0 -1 -1  0  1
```

Below, we show how to execute the program `enumerate_fiber`.

Listing 7.13 Executing the C program `enumerate_fiber`

```
$ enumerate_fiber start.txt markov.txt
n_move : 1, msize : 6
start_v :     <- start element in the fiber
  0  1  1  1  1  1
n_move : 3, msize : 6
```

```
move : 3 6    <- moves
  0 -1   1   0   1 -1
  1 -1   0 -1   1   0
  1   0 -1 -1   0   1
  0 0 2 1 2 0
  .... skip
count: 5
fiber :
  0 1 1 1 1 1    <- enumerates elements in the fiber
  0 0 2 1 2 0
  1 0 1 0 2 1
  1 1 0 0 1 2
  0 2 0 1 0 2
```

Exercise. 1. Implement the above algorithm using another programming language, such as R or Risa/Asir.
2. Implement a more efficient algorithm for enumerating all the elements in the fiber. (Please refer to [22, Algorithm 5.7].)

Example 7.2.6. We consider a 3×4 contingency table whose $(1, 1), (2, 2), (3, 3)$, $(3, 4)$ elements are structural zeros,

$$\begin{vmatrix} [0] & * & * & * \\ * & [0] & * & * \\ * & * & [0] & [0] \end{vmatrix}.$$

We assume that

$$\begin{cases} p_{ij} = p_i p_j & (i, j) \neq (1, 1), (2, 2), (3, 3), (3, 4) \\ p_{ij} = 0 & (i, j) = (1, 1), (2, 2), (3, 3), (3, 4) \end{cases}.$$

The sufficient statistics are the row and column sums.

1. What is a configuration A for this contingency table?
2. Compute the toric ideal I_A for the configuration A, by using the elimination method. (Please refer to Corollary 4.2.11 or Lemma 1.5.11.)

Answer. 1. Let x_{ij} be the (i, j)-th element in the contingency table. The rows and columns in the configuration A correspond to

$$(x_{1\cdot}, x_{2\cdot}, x_{3\cdot}, x_{\cdot 1}, x_{\cdot 2}, x_{\cdot 3}, x_{\cdot 4}) \text{ and } (x_{12}, x_{13}, x_{14}, x_{21}, x_{23}, x_{24}, x_{31}, x_{32}).$$

The configuration A is thus

$$A = \begin{pmatrix} 1 & 1 & 1 & 0 & 0 & 0 & 0 & 0 \\ 0 & 0 & 0 & 1 & 1 & 1 & 0 & 0 \\ 0 & 0 & 0 & 0 & 0 & 0 & 1 & 1 \\ 0 & 0 & 0 & 1 & 0 & 0 & 1 & 0 \\ 1 & 0 & 0 & 0 & 0 & 0 & 0 & 1 \\ 0 & 1 & 0 & 0 & 1 & 0 & 0 & 0 \\ 0 & 0 & 1 & 0 & 0 & 1 & 0 & 0 \end{pmatrix}.$$

2. From the configuration A, we make the ideal

$$I_A^* = \langle u_1 - v_1 v_5, u_2 - v_1 v_6, u_3 - v_1 v_7, u_4 - v_2 v_4, u_5 - v_2 v_6,$$

$$u_6 - v_2 v_7, u_7 - v_3 v_4, u_8 - v_3 v_5 \rangle.$$

In order to obtain the intersection $I_A^* \cap \mathbb{Q}[u_1, \ldots, u_8]$, we compute a Gröbner basis of I_A^* with respect to a monomial order satisfying $v_1, \ldots, v_7 > u_1, \ldots, u_8$.

Listing 7.14 Risa/Asir: computing the Gröbner basis

```
[1659] Polys = [u1-v5*v1,u2-v6*v1,u3-v7*v1,u4-v4*v2,u5-v6*v2,u6-v7*v2,
u7-v4*v3,u8-v5*v3];
[1660] VL=[v1,v2,v3,v4,v5,v6,v7,u1,u2,u3,u4,u5,u6,u7,u8];
[1661] GB=nd_gr(Polys, VL, 0, [[0,7],[0,8]]);
[-u6*u2+u5*u3,-u7*u6*u1+u8*u4*u3,-u7*u5*u1+u8*u4*u2,v7*u5-v6*u6,
v7*u2-v6*u3,v7*u1-v5*u3,v6*u1-v5*u2,-v7*u8*u4+v5*u7*u6,
-v6*u8*u4+v5*u7*u5,v5*u7-v4*u8,v7*u4-v4*u6,v6*u4-v4*u5,v3*u4-v2*u7,
v3*u6*u1-v2*u8*u3,v3*u5*u1-v2*u8*u2,v3*u1-v1*u8,v2*u3-v1*u6,
v2*u2-v1*u5,v7*v3*u4-u7*u6,v7*v3*u1-u8*u3,u6-v7*v2,u3-v7*v1,
v6*v3*u4-u7*u5,v6*v3*u1-u8*u2,u5-v6*v2,u2-v6*v1,u8-v5*v3,u1-v5*v1,
u7-v4*v3,u4-v4*v2]
```

The argument $[[0,7],[0,8]]$ of the function $\mathtt{nd\_gr}$ indicates the monomial order satisfying $v_1, \ldots, v_7 > u_1, \ldots, u_8$. For details, please see Sect. 3.6.5. The polynomials whose variables are u in the above output GB are

$$\{u_3 u_5 - u_2 u_6, u_3 u_4 u_8 - u_1 u_6 u_7, u_2 u_4 u_8 - u_1 u_5 u_7\}.$$

These polynomials generate the toric ideal I_A and correspond to the tables

$$\left\{ \begin{bmatrix} [0] & 0 & -1 & 1 \\ 0 & [0] & 1 & -1 \\ 0 & 0 & [0] & [0] \end{bmatrix}, \begin{bmatrix} [0] & -1 & 0 & 1 \\ 1 & [0] & 0 & -1 \\ -1 & 1 & [0] & [0] \end{bmatrix}, \begin{bmatrix} [0] & -1 & 1 & 0 \\ 1 & [0] & -1 & 0 \\ -1 & 1 & [0] & [0] \end{bmatrix} \right\}.$$

This set is a Markov basis for A.

Exercise. 1. For the same contingency table, enumerate all the tables for which the row sums are $3, 3$, and 2 and the column sums are $2, 2, 2$, and 2.
2. We consider a 3×4 contingency table for which the $(1, 1), (1, 3), (2, 2), (2, 4), (3, 1), (3, 3)$ elements are structural zeros,

$$\begin{bmatrix} [0] & * & [0] & * \\ * & [0] & * & [0] \\ [0] & * & [0] & * \end{bmatrix}.$$

For this table, compute the configuration A and a Markov basis for A.

Example 7.2.7. We consider a $3 \times 3 \times 3$ contingency table with the fixed two-dimensional marginal totals $\{x_{ij\cdot}\}, \{x_{i\cdot k}\}, \{x_{\cdot jk}\}$.

1. Find a configuration A for this contingency table.
2. For the configuration A, compute the toric ideal I_A by using the elimination method. (Please refer to Corollary 4.2.11 or Lemma 1.5.11.)
3. Compute a Markov basis for the configuration A by using 4ti2.
4. Enumerate all the contingency tables with $x_{ij.} = 2, x_{i \cdot k} = 2, x_{\cdot jk} = 2$. In other words, enumerate all the elements in the fiber $\mathscr{F}_{\mathbf{t}}$, where $\mathbf{t} = (2, 2, \ldots, 2)'$.

Answer. 1. Let x_{ijk} be the (i, j, k)-th element in the contingency table. The columns and rows in the configuration A correspond to

$$(x_{111}, x_{112}, x_{113}, x_{121}, x_{122}, x_{123}, x_{131}, x_{132}, x_{133}, x_{211}, x_{212}, x_{213}, x_{221}, x_{222}, x_{223},$$

$$x_{231}, x_{232}, x_{233}, x_{311}, x_{312}, x_{313}, x_{321}, x_{322}, x_{323}, x_{331}, x_{332}, x_{333})$$

and

$$(x_{11.}, x_{12.}, x_{13.}, x_{21.}, x_{22.}, x_{23.}, x_{31.}, x_{32.}, x_{33.}, x_{1 \cdot 1}, x_{1 \cdot 2}, x_{1 \cdot 3}, x_{2 \cdot 1}, x_{2 \cdot 2}, x_{2 \cdot 3},$$

$$x_{3 \cdot 1}, x_{3 \cdot 2}, x_{3 \cdot 3}, x_{\cdot 11}, x_{\cdot 12}, x_{\cdot 13}, x_{\cdot 21}, x_{\cdot 22}, x_{\cdot 23}, x_{\cdot 31}, x_{\cdot 32}, x_{\cdot 33}).$$

The configuration A is

Listing 7.15 4ti2: input file `3x3x3cont`

```
27 27
1 1 1 0 0 0 0 0 0 0 0 0 0 0 0 0 0 0 0 0 0 0 0 0 0 0 0
0 0 0 1 1 1 0 0 0 0 0 0 0 0 0 0 0 0 0 0 0 0 0 0 0 0 0
0 0 0 0 0 0 1 1 1 0 0 0 0 0 0 0 0 0 0 0 0 0 0 0 0 0 0
0 0 0 0 0 0 0 0 0 1 1 1 0 0 0 0 0 0 0 0 0 0 0 0 0 0 0
0 0 0 0 0 0 0 0 0 0 0 0 1 1 1 0 0 0 0 0 0 0 0 0 0 0 0
0 0 0 0 0 0 0 0 0 0 0 0 0 0 0 1 1 1 0 0 0 0 0 0 0 0 0
0 0 0 0 0 0 0 0 0 0 0 0 0 0 0 0 0 0 1 1 1 0 0 0 0 0 0
0 0 0 0 0 0 0 0 0 0 0 0 0 0 0 0 0 0 0 0 0 1 1 1 0 0 0
0 0 0 0 0 0 0 0 0 0 0 0 0 0 0 0 0 0 0 0 0 0 0 0 1 1 1
1 0 0 1 0 0 1 0 0 0 0 0 0 0 0 0 0 0 0 0 0 0 0 0 0 0 0
0 1 0 0 1 0 0 1 0 0 0 0 0 0 0 0 0 0 0 0 0 0 0 0 0 0 0
0 0 1 0 0 1 0 0 1 0 0 0 0 0 0 0 0 0 0 0 0 0 0 0 0 0 0
0 0 0 0 0 0 0 0 0 1 0 0 1 0 0 1 0 0 0 0 0 0 0 0 0 0 0
0 0 0 0 0 0 0 0 0 0 1 0 0 1 0 0 1 0 0 0 0 0 0 0 0 0 0
0 0 0 0 0 0 0 0 0 0 0 1 0 0 1 0 0 1 0 0 0 0 0 0 0 0 0
0 0 0 0 0 0 0 0 0 0 0 0 0 0 0 0 0 0 1 0 0 1 0 0 1 0 0
0 0 0 0 0 0 0 0 0 0 0 0 0 0 0 0 0 0 0 1 0 0 1 0 0 1 0
0 0 0 0 0 0 0 0 0 0 0 0 0 0 0 0 0 0 0 0 1 0 0 1 0 0 1
1 0 0 0 0 0 0 0 0 1 0 0 0 0 0 0 0 0 1 0 0 0 0 0 0 0 0
0 1 0 0 0 0 0 0 0 0 1 0 0 0 0 0 0 0 0 1 0 0 0 0 0 0 0
0 0 1 0 0 0 0 0 0 0 0 1 0 0 0 0 0 0 0 0 1 0 0 0 0 0 0
0 0 0 1 0 0 0 0 0 0 0 0 1 0 0 0 0 0 0 0 0 1 0 0 0 0 0
0 0 0 0 1 0 0 0 0 0 0 0 0 1 0 0 0 0 0 0 0 0 1 0 0 0 0
0 0 0 0 0 1 0 0 0 0 0 0 0 0 1 0 0 0 0 0 0 0 0 1 0 0 0
0 0 0 0 0 0 1 0 0 0 0 0 0 0 0 1 0 0 0 0 0 0 0 0 1 0 0
0 0 0 0 0 0 0 1 0 0 0 0 0 0 0 0 1 0 0 0 0 0 0 0 0 1 0
0 0 0 0 0 0 0 0 1 0 0 0 0 0 0 0 0 1 0 0 0 0 0 0 0 0 1
```

2. From the configuration A, we make the ideal

$$I_A^* = \langle u_1 - v_{19}v_{10}v_1, u_2 - v_{20}v_{11}v_1, u_3 - v_{21}v_{12}v_1, u_4 - v_{22}v_{10}v_2, u_5 - v_{23}v_{11}v_2,$$

$$u_6 - v_{24}v_{12}v_2, u_7 - v_{25}v_{10}v_3, u_8 - v_{26}v_{11}v_3, u_9 - v_{27}v_{12}v_3, u_{10} - v_{19}v_{13}v_4,$$

$$u_{11} - v_{20}v_{14}v_4, u_{12} - v_{21}v_{15}v_4, u_{13} - v_{22}v_{13}v_5, u_{14} - v_{23}v_{14}v_5, u_{15} - v_{24}v_{15}v_5,$$

$$u_{16} - v_{25}v_{13}v_6, u_{17} - v_{26}v_{14}v_6, u_{18} - v_{27}v_{15}v_6, u_{19} - v_{19}v_{16}v_7, u_{20} - v_{20}v_{17}v_7,$$

$$u_{21} - v_{21}v_{18}v_7, u_{22} - v_{22}v_{16}v_8, u_{23} - v_{23}v_{17}v_8, u_{24} - v_{24}v_{18}v_8, u_{25} - v_{25}v_{16}v_9,$$

$$u_{26} - v_{26}v_{17}v_9, u_{27} - v_{27}v_{18}v_9 \rangle.$$

To obtain the toric ideal I_A, we eliminate the variables v from the ideal. However, because the ideal has many variables, this computation may be very slow. This algorithm is very inefficient for large problems.

3. By using the 4ti2 command `markov`, it only takes a few seconds to obtain a Markov basis with 81 elements. For more information about algorithms using 4ti2, please see [8].

Listing 7.16 Executing the 4ti2 command `markov`

```
$ markov cont3x3x3
... skip
  Size:       81, Time:  0.00 /  0.00 secs. Done.
4ti2 Total Time:  0.00 secs.
```

Listing 7.17 4ti2: output file `3x3x3cont.mar`

```
81 27
-1  0  1  1 -1  0  0  1 -1  0  0  0  0  0  0  0  0  0  1  0 -1 -1  1  0
 0 -1  1
-1  0  1  1 -1  0  0  1 -1  1  0 -1 -1  1  0  0 -1  1  0  0  0  0  0  0
 0  0  0
....
 1  0 -1  0  0  0 -1  0  1  0  0  0  0  0  0  0  0  0  0 -1  0  1  0  0  0
 1  0 -1
```

4. An element in the fiber $\mathscr{F}_t$ is

$$
\begin{array}{|ccc|ccc|ccc|}
0 & 0 & 2 & 2 & 0 & 0 & 0 & 2 & 0 \\
0 & 2 & 0 & 0 & 0 & 2 & 2 & 0 & 0 \\
2 & 0 & 0 & 0 & 2 & 0 & 0 & 0 & 2 \\
\end{array}.
$$

By using this element and the Markov basis, we can enumerate all the elements in the fiber; see Example 7.2.5. We can then obtain the 132 elements in the fiber.

Exercise. We consider the $3 \times 3 \times K$ contingency table ($K = 4, 5, 6, 7, 8$) with the fixed two-dimensional marginal totals $\{x_{ij\cdot}\}, \{x_{i\cdot k}\}$, and $\{x_{\cdot jk}\}$. What are the configurations fore these tables? Use 4ti2 to compute the toric ideals for the configurations.

Example 7.2.8. We consider the **t**-fiber

$$
\mathscr{F}_t = \{(x_{ij}) \mid x_{ij} \in \mathbb{N}, x_{1\cdot} = 5, x_{2\cdot} = 15, x_{\cdot 1} = 5, x_{\cdot 2} = 5, x_{\cdot 3} = 10\}.
$$

Elements in the fiber are the 2×3 contingency tables with marginal frequencies $\mathbf{t} = (5, 15; 5, 5, 10)'$. Generate samples in the fiber from the distribution $\frac{1}{Z} \frac{1}{x_{11}! x_{12}! x_{13}! x_{21}! x_{22}! x_{23}!}$, by using the Markov chain Monte Carlo method. Here Z is the normalizing constant defined by

$$
Z = \sum_{x \in \mathscr{F}_t} \frac{1}{x_{11}! x_{12}! x_{13}! x_{21}! x_{22}! x_{23}!}.
$$

Evaluate the expected value of $2x_{11} + x_{12}$.

Answer. A Markov basis for a 2×3 contingency table is

$$\mathscr{B} = \left\{ \begin{bmatrix} 1 & -1 & 0 \\ -1 & 1 & 0 \end{bmatrix}, \begin{bmatrix} 1 & 0 & -1 \\ -1 & 0 & 1 \end{bmatrix}, \begin{bmatrix} 0 & 1 & -1 \\ 0 & -1 & 1 \end{bmatrix} \right\}.$$

The following algorithm generates samples in the fiber $\mathscr{F}_t$ from the distribution $f(\mathbf{x}) = \frac{1}{Z} \prod_{i,j} \frac{1}{x_{ij}!}$ by using the Markov chain Monte Carlo method. (Please see Algorithm 4.2.4.)

1. Choose an element $\mathbf{x}$ in the fiber $\mathscr{F}_t$.
   $e \leftarrow 0$.
2. Randomly choose an element $\mathbf{z} \in \mathscr{B} \cup (-\mathscr{B})$.
3. If all the components of $\mathbf{x} + \mathbf{z}$ are nonnegative, then $r \leftarrow \frac{f(\mathbf{x}+\mathbf{z})}{f(\mathbf{x})}$,
   else $r \leftarrow 0$.
4. Let R be a random number selected from the uniform distribution between 0 and 1.
5. If $r > R$, then $\mathbf{x}_{\text{next}} \leftarrow \mathbf{x} + \mathbf{z}$,
   else $\mathbf{x}_{\text{next}} \leftarrow \mathbf{x}$.
6. Output a sample $\mathbf{x}_{\text{next}}$.
   $e \leftarrow e + (2x_{11} + x_{12} \text{ of } \mathbf{x}_{\text{next}})$.
7. $\mathbf{x} \leftarrow \mathbf{x}_{\text{next}}$.
   Go to step 2.
8. Output the expected value $e /$ (number of samples) .

The R program $2\text{x}3\text{mcmc}.\text{r}$ implements this algorithm. Here, we implement Algorithm 4.2.4, but Algorithm 4.2.5 is more efficient. We run the program $2\text{x}3\text{mcmc}.\text{r}$ as follows. The exact expected value is $15/4 = 3.75$.

Listing 7.18 R: the result of $2\text{x}3\text{mcmc}.\text{r}$

```
> source("2x3mcmc.r")
c2x3mcmc(n, burnin, initial data)
e.g. c2x3mcmc(10000, 10000, c(0,5,0,5,0,10))
> c2x3mcmc(10000, 10000, c(0,5,0,5,0,10))
[1] 2 1 2 3 4 8   <- sample
[1] 1 2 2 4 3 8
....
[1] 0 0 5 5 5 5
[1] 0 0 5 5 5 5
average
[1] 3.7485
```

Exercise. Implement Algorithm 4.2.5, and solve the above problem.

Example 7.2.9. We consider the 5×5 contingency table in Example 4.1.12.

Geometry/Probability	5	4	3	2	1	Total
5	2	1	1	0	0	4
4	8	3	3	0	0	14
3	0	2	1	1	1	5
2	0	0	0	1	1	2
1	0	0	0	0	1	1
Total	10	6	5	2	3	26

Let the null hypothesis be H_0: the geometry and probability scores are independent.

1. Enumerate all the elements in the **t**-fiber $\mathscr{F}_t$, where $\mathbf{t} = (4, 14, 5, 2, 1, 10, 6, 5, 2, 3)'$.
2. Calculate the exact p value by using the previous results.
3. Estimate the p value by using the Markov chain Monte Carlo method.

Answer. 1. We can obtain a Markov basis for this model by using Theorem 4.2.6. We can enumerate all the elements in the **t**-fiber $\mathscr{F}_t$ by using the given data

$$\mathbf{x}^o = (2, 1, 1, 0, 0, 8, 3, 3, 0, 0, 0, 2, 1, 1, 1, 0, 0, 0, 1, 1, 0, 0, 0, 0, 1)'$$

and the Markov basis. As Example 7.2.5, we use the program `enumerate_fiber`, and we put the data $\mathbf{x}^o$ in the input file `start.txt`.

Listing 7.19 C program enumerate_fiber: Input file `start.txt`

```
1 25
2 1 1 0 0 8 3 3 0 0 0 2 1 1 1 0 0 0 1 1 0 0 0 0 1
```

We write a Markov basis in the input file `move.txt`.

Listing 7.20 C program enumerate_fiber: Input file `move.txt`

```
100 25
 0  0  0  0  0  0  0  0  0  0  0  0  0  0  0  0  0  0  0  0  1 -1  0  0  0
-1  1
 0  0  0  0  0  0  0  0  0  0  0  0  0  0  0  0  0  0  1 -1  0  0  0 -1
 1  0
 0  0  0  0  0  0  0  0  0  0  0  0  0  0  0  0  0  0  1  0 -1  0  0 -1
 0  1
 .....            <- Markov basis for independence model of 5x5 table
```

We execute the program `enumerate_fiber` as follows.

Listing 7.21 Executing the C program enumerate_fiber

```
$ enumerate_fiber start.txt move.txt > fiber.txt
n_move : 1, msize : 25
start_v :
 2  1  1  0  0  8  3  3  0  0  0  2  1  1  1  0  0  0  1  1  0  0  0  0
 1
n_move : 100, msize : 25
move : 100 25
 0  0  0  0  0  0  0  0  0  0  0  0  0  0  0  0  0  0  0  0  1 -1  0  0  0 -1
```

```
   1
..... skip
count : 229174
fiber :
```

The program outputs the file `fiber.txt`, which includes all the elements in the fiber. The format of the output file is similar to that of 4ti2.

Listing 7.22 C program `enumerate_fiber`: output file `fiber.txt`

```
229174 25
2 1 1 0 0 8 3 3 0 0 0 2 1 1 1 0 0 0 1 1 0 0 0 0 1
2 1 1 0 0 8 3 3 0 0 0 2 1 2 0 0 0 0 0 2 0 0 0 0 1
... skip
```

We now explain how to use the software program LattE. LattE can count the number of integer solutions for a system of linear inequalities. We make the following input file.

Listing 7.23 LattE: input file `5x5cont_latte`

```
10 26
4 -1 -1 -1 -1 -1  0  0  0  0  0  0  0  0  0  0  0  0  0  0  0  0  0  0  0
   0  0
14  0  0  0  0  0 -1 -1 -1 -1 -1  0  0  0  0  0  0  0  0  0  0  0  0  0  0
   0  0
5  0  0  0  0  0  0  0  0  0  0 -1 -1 -1 -1 -1  0  0  0  0  0  0  0  0  0
   0  0
2  0  0  0  0  0  0  0  0  0  0  0  0  0  0  0 -1 -1 -1 -1 -1  0  0  0
   0  0
1  0  0  0  0  0  0  0  0  0  0  0  0  0  0  0  0  0  0  0  0 -1 -1 -1
  -1 -1
10 -1  0  0  0  0 -1  0  0  0  0 -1  0  0  0  0 -1  0  0  0  0 -1  0  0
   0  0
6  0 -1  0  0  0  0 -1  0  0  0  0 -1  0  0  0  0 -1  0  0  0  0 -1  0
   0  0
5  0  0 -1  0  0  0  0 -1  0  0  0  0 -1  0  0  0  0 -1  0  0  0  0 -1
   0  0
2  0  0  0 -1  0  0  0  0 -1  0  0  0  0 -1  0  0  0  0 -1  0  0  0  0
  -1  0
3  0  0  0  0 -1  0  0  0  0 -1  0  0  0  0 -1  0  0  0  0 -1  0  0  0
   0 -1
linearity 10 1 2 3 4 5 6 7 8 9 10
nonnegative 25 1 2 3 4 5 6 7 8 9 10 11 12 13 14 15 16 17 18 19 20 21 22
   23 24 25
```

The first row `10 26` gives the size of the matrix. Each row in the matrix indicates an inequality or an equality. The first column is the constant term, and the subsequent columns are the coefficients of the variables. For example, the first row

```
4 -1 -1 -1 -1 -1  0  0  0  0  0  0  0  0  0  0  0  0  0  0  0  0  0  0  0  0  0
```

represents the equality

$$4 - x_1 - x_2 - x_3 - x_4 - x_5 = 0.$$

The option `linearity` indicates which rows are equalities. In this case, all the rows are equalities. The option `nonnegative` represents which variables are nonnegative. In this case, all the variables are nonnegative. We use the LattE command `count` to count the elements in the fiber.

Listing 7.24 Executing the LattE command `count`

```
$ count 5x5cont_input
This is LattE v1.0 beta.    (September 17, 2002)
Revised version.           (Aug        1, 2003)
....

***** Total number of lattice points: 229174 ****

Computation done.
Time: 102.518 sec
```

LattE outputs the number of the elements in the fiber, which is 229174. This result is equal to the result of `enumerate_fiber`.

2. We will compute

$$\sum_{\mathbf{x}\in\mathscr{F}_t,\ \chi^2(\mathbf{x})\geq 25.3376} \mathbf{x}h(\mathbf{x}) = 0.0609007,$$

where $h(\mathbf{x}) = \frac{(4!14!5!2!1!)(10!6!5!2!3!)}{26!} \prod_{i=1}^{5} \prod_{j=1}^{5} \frac{1}{x_{ij}!}$.

3. We will execute the Markov chain Monte Carlo method as Example 7.2.8. The algorithm is as follows.

1. $\mathbf{x} \leftarrow$ (given data).
   $c \leftarrow 0$.
   $\chi^2 \leftarrow \chi^2(\mathbf{x})$.
2. Randomly choose an element $\mathbf{z} \in \mathscr{B} \cup (-\mathscr{B})$.
   (Here, $\mathscr{B}$ is a Markov basis for the 5×5 table.)
3. If all components of $\mathbf{x} + \mathbf{z}$ are nonnegative, then $r \leftarrow \frac{f(\mathbf{x}+\mathbf{z})}{f(\mathbf{x})}$,
   else $r \leftarrow 0$.
   (Here, $f(\mathbf{x}) = \frac{1}{Z} \frac{1}{\prod_{i,j} x_{ij}!}$, Z is the normalizing constant.)
4. Let R be a random number from the uniform distribution between 0 and 1.
5. If $r > R$, then $\mathbf{x}_{next} \leftarrow \mathbf{x} + \mathbf{z}$,
   else $\mathbf{x}_{next} \leftarrow \mathbf{x}$.
6. Output a sample $\mathbf{x}_{next}$.
   If $\chi^2(\mathbf{x}_{next}) \geq \chi^2$, then $c \leftarrow c + 1$.
7. $\mathbf{x} \leftarrow \mathbf{x}_{next}$.
   Go to step 2.

When the algorithm is completed, $\frac{c}{\text{(number of samples)}}$ is the estimate of the p value. The R program `5x5mcmc.r` implements this algorithm.

Listing 7.25 R: result of `5x5mcmc.r`

```
> source("5x5mcmc.r")
c5x5mcmc(n, burnin, initial data)
e.g. c5x5mcmc(10000, 10000, c(2,1,1,0,0,8,3,3,0,0,0,2,1,1,1,0,0,0,1,1,0,
0,0,0,1))
> c5x5mcmc(10000, 10000, c(2,1,1,0,0,8,3,3,0,0,0,2,1,1,1,0,0,0,1,1,0,0,
0,0,1))
X-squared
25.33762
.... skip
 [1] 2 0 1 0 1 7 2 4 1 0 1 3 0 0 1 0 0 0 1 1 0 1 0 0 0
```

```
.... skip
 [1] 1 1 2 0 0 7 1 3 1 2 1 3 0 1 0 1 1 0 0 0 0 0 0 1
approximate p value
[1] 0.0639
```

Since the p value > 0.05, H_0 is not rejected.

Exercise. We consider the 5×5 contingency table Let null hypothesis be H_0: the

Geometry/Probability	5	4	3	2	1	Total
5	2	0	2	0	2	6
4	0	2	0	2	0	4
3	0	0	2	0	2	4
2	0	0	0	2	0	2
1	0	0	0	0	2	2
Total	2	2	4	4	6	18

scores of Geometry and Probability are independent of each other. Compute the exact p value and estimate the p value.

7.2.3 Design of Experiments and Markov Basis (Sect. 4.3)

Example 7.2.10. We consider the data in Table 4.5, which is a 2^{7-3} fractional factorial design chosen by the aliasing relation $ABDE = ACDF = BCDG = I$.

Run	A	B	C	D	E	F	G	Defects
1	1	1	1	1	1	1	1	69
2	1	1	1	-1	-1	-1	-1	31
3	1	1	-1	1	1	-1	-1	55
4	1	1	-1	-1	-1	1	1	149
5	1	-1	1	1	-1	1	-1	46
6	1	-1	1	-1	1	-1	1	43
7	1	-1	-1	1	-1	-1	1	118
8	1	-1	-1	-1	1	1	-1	30
9	-1	1	1	1	-1	-1	1	43
10	-1	1	1	-1	1	1	-1	45
11	-1	1	-1	1	-1	1	-1	71
12	-1	1	-1	-1	1	-1	1	380
13	-1	-1	1	1	1	-1	-1	37
14	-1	-1	1	-1	-1	1	1	36
15	-1	-1	-1	1	1	1	1	212
16	-1	-1	-1	-1	-1	-1	-1	52

We evaluate the fit of the hierarchical model $AC/BD/E/F/G$ with the following procedure.

1. Find a covariate matrix M for the model.

2. Evaluate the fitted value $(m_1, \ldots, m_{16})$ for the model.
3. For the given data of defects $\mathbf{y} = (y_1, \ldots, y_{16})$, evaluate the log-likelihood ratio:

$$G(\mathbf{y}) = 2 \sum_{i=1}^{16} y_i \log \frac{y_i}{m_i}.$$

Moreover, compare $G(\mathbf{y})$ with the upper 5% point of the χ^2 distribution with 6 degrees of freedom.

4. Find a Markov basis for the matrix M'.
5. Estimate the p value by using the Markov chain Monte Carlo method.

Answer. 1. In the hierarchical model $AC/BD/E/F/G$, the main effects are A, B, C, D, E, F, G and the two-factor interaction effects are AC, BD. Therefore, the covariate matrix is

$$M = (1, d_A, d_B, d_C, d_D, d_E, d_F, d_G, d_{AC}, d_{BD}).$$

The following is the input file format for 4ti2. We note that this matrix represents the transpose matrix M'.

Listing 7.26 4ti2: input file `covariate_mat1`

```
10 16
1  1  1  1  1  1  1  1  1  1  1  1  1  1  1  1
1  1  1  1  1  1  1  1 -1 -1 -1 -1 -1 -1 -1 -1
1  1  1  1 -1 -1 -1 -1  1  1  1  1 -1 -1 -1 -1
1  1 -1 -1  1  1 -1 -1  1  1 -1 -1  1  1 -1 -1
1 -1  1 -1  1 -1  1 -1  1 -1  1 -1  1 -1  1 -1
1 -1  1 -1 -1  1 -1  1 -1  1 -1  1  1 -1  1 -1
1 -1 -1  1  1 -1 -1  1 -1  1  1 -1 -1  1  1 -1
1 -1 -1  1 -1  1  1 -1 -1  1 -1  1 -1  1  1 -1
1  1 -1 -1  1  1 -1 -1 -1 -1  1  1 -1 -1  1  1
1 -1  1 -1 -1  1  1 -1  1 -1 -1  1 -1  1
```

2. We will use R to compute the fitted value. First, we make the following input file:

Listing 7.27 R: input file `2_7-3.dat`

```
A,  B,  C,  D,  E,  F,  G,   x
1,  1,  1,  1,  1,  1,  1,   69
1,  1,  1, -1, -1, -1, -1,   31
1,  1, -1,  1,  1, -1, -1,   55
1,  1, -1, -1, -1,  1,  1,  149
1, -1,  1,  1, -1,  1, -1,   46
1, -1,  1, -1,  1, -1,  1,   43
1, -1, -1,  1, -1, -1,  1,  118
1, -1, -1, -1,  1,  1, -1,   30
-1,  1,  1,  1, -1, -1,  1,   43
-1,  1,  1, -1,  1,  1, -1,   45
-1,  1, -1,  1, -1,  1, -1,   71
-1,  1, -1, -1,  1, -1,  1,  380
-1, -1,  1,  1,  1, -1, -1,   37
-1, -1,  1, -1, -1,  1,  1,   36
-1, -1, -1,  1,  1,  1,  1,  212
-1, -1, -1, -1, -1, -1, -1,   52
```

and execute the following commands in R (please refer to Example 4.3.5)

Listing 7.28 R: computing the fitted value

```
> dat<-read.table(file="2_7-3.dat", header=T, sep=",")
> dat.glm<-glm(x~A+B+C+D+E+F+G+A*C+B*D, dat, family=poisson)
> fitted(dat.glm)
        1         2         3         4         5         6
 64.52677  47.25345  53.14603 151.07960  30.42595  46.79383
        7         8         9        10        11        12
115.24100  32.53337  49.42430  46.13193  70.90290 360.53502
       13        14        15        16
 35.18867  30.25510 232.14438  51.41770
```

3. $G(\mathbf{y}) = 19.09271$ and $\chi^2_{0.05}(6) = 12.59159$.
4. Executing the command `markov covariate_mat1`, we obtain a Markov basis consisting of 23 generators.
5. We will perform the Markov chain Monte Carlo method, as in Example 7.2.8. To calculate the p value, in each step, if $G(\mathbf{x}) \geq 19.09271$ for the sample $\mathbf{x}$, we increase the counter c. At the end of the sampling, $\frac{c}{\text{(number of samples)}}$ is the estimated p-value. The following is a summary of the algorithm.

1. $\mathbf{x} \leftarrow$ (given data).
   Initialize counter $c \leftarrow 0$.
   Set the log-likelihood ratio $G \leftarrow G(\mathbf{x})$.
2. Randomly choose an element $\mathbf{z}$ from the set $\mathscr{B} \cup (-\mathscr{B})$.
   (Here, $\mathscr{B}$ is a Markov basis of the matrix M'.)
3. If all components of $\mathbf{x} + \mathbf{z}$ are nonnegative, then $r \leftarrow \frac{f(\mathbf{x}+\mathbf{z})}{f(\mathbf{x})}$;
   else $r \leftarrow 0$.
   (Here, $f(\mathbf{x}) = \frac{1}{Z} \frac{1}{\prod_{i,j} x_{ij}!}$ and Z is the normalizing constant.)
4. Take a random number R from the uniform distribution from 0 to 1.
5. If $r > R$, then $\mathbf{x}_{\text{next}} \leftarrow \mathbf{x} + \mathbf{z}$;
   else $\mathbf{x}_{\text{next}} \leftarrow \mathbf{x}$.
6. If $G(\mathbf{x}_{\text{next}}) \geq G$, then $c \leftarrow c + 1$.
7. $\mathbf{x} \leftarrow \mathbf{x}_{\text{next}}$.
   Go to step 2.

The file `cov1_mcmc.r` is an example of the implementation of this algorithm using R.

Listing 7.29 R: result of `cov1_mcmc.r`

```
> source("cov1_mcmc.r")
cov1_mcmc(n, burnin)
e.g. cov1_mcmc(10000, 10000)
> cov1_mcmc(10000, 10000)
likelihood ratio stat :
        1
19.09271
 [1] 69  32  54 149  45  43 118  31  44  45  71 379  37  35 213  52
 ...          <- output samples
 [1] 64  39  51 154  34  52 117  30  44  45  79 367  41  31 221  48
n            <- number of samples
[1] 10000
count        <- counter c
[1] 31
estimate of p-value
[1] 0.0031
```

Exercise. Evaluate the fit of the hierarchical model $AB/AC/BD/E/F/G$ for the same data as in Example 7.2.10.

Example 7.2.11. Enumerate all the aliasing relations for the 2^{7-3}-fractional factorial design in Example 4.3.7 with the following procedure.

Each of the factors A, B, C, D, E, F, and G have two levels $\{+1, -1\}$, and they satisfy the relation $ABDE = ACDF = BCDG = 1$ (4.52). We consider the ideal

$$I = \langle A^2 - 1, B^2 - 1, C^2 - 1, D^2 - 1, E^2 - 1, F^2 - 1, G^2 - 1,$$

$$ABDE - 1, ACDF - 1, BCDG - 1 \rangle.$$

In this design, the two-factor interaction effects AB and FG are confounded; this means that $AB - FG \in I$. In order to check this condition, we can check that the normal form (remainder) is equal to 0 when $AB - FG$ is divided by a Gröbner basis G of I with respect to any term order $<$ (ideal membership problem).

Answer. Let us use Risa/Asir to check the aliasing relation for the two two-factor interactions AB, FG and AC, BD, using the Gröbner basis method.

Listing 7.30 Risa/Asir: check the aliasing relation

```
[1361] Id=[a^2-1,b^2-1,c^2-1,d^2-1,e^2-1,f^2-1,g^2-1,a*b*d*e-1,
a*c*d*f-1,b*c*d*g-1];
[1362] VL=[a,b,c,d,e,f,g];
[1364] GB=nd_gr(Id,VL,0,0);          <- Groebner basis GB of Id
[a^2-1,-b*a+g*f,b^2-1,-c*a+g*e,-c*b+g*d,c^2-1,-d*a+f*c,d*b-g*c,-g*b+d*c,
d^2-1,-e*a+g*c,-e*b+f*c,-g*a+e*c,e*d-g*f,e^2-1,-f*a+g*b,-g*a+f*b,f*d-g*e,
g*d-f*e,f^2-1,g^2-1]
[1365] p_nf(a*b-f*g, GB, VL, 0);     <- normal form of a*b-f*g by GB
0                                     <- AB and FG are confounded
[1367] p_nf(a*c-b*d, GB, VL, 0);     <- normal form of a*c-b*d by GB
g*c-g*e                               <- AC and BD are not confounded
```

We can enumerate the aliasing relations by classifying the set of monomials

$$\{A^{i_1} B^{i_2} C^{i_3} D^{i_4} E^{i_5} F^{i_6} G^{i_7} \mid i_1, \ldots, i_7 \in \{0, 1\}\}$$

by using the normal form with respect to the Gröbner basis G of I. The file `alias-2.rr` is a program for Risa/Asir which uses this method to enumerate the relations.

Listing 7.31 Risa/Asir: enumeration of the aliasing relations using `alias-2.rr`

```
[1351] load("alias-2.rr");
[a^2-1,b^2-1,c^2-1,d^2-1,e^2-1,f^2-1,g^2-1,e*d*b*a-1,f*d*c*a-1,
g*d*c*b-1]
....
[[g*f*c,g*e*b,g*d*a,f*e*a,f*d*b,e*d*c,c*b*a,g*f*e*d*c*b*a],
 [f*c,e*b,d*a,g*f*e*a,g*f*d*b,g*e*d*c,g*c*b*a,f*e*d*c*b*a],
 [g*a,f*b,e*c,g*f*d*c,g*e*d*b,f*e*d*a,d*c*b*a,g*f*e*c*b*a],
 [g*b,f*a,d*c,g*f*e*c,g*e*d*a,f*e*d*b,e*c*b*a,g*f*d*c*b*a],
 [g*c,e*a,d*b,g*f*e*b,g*f*d*a,f*e*d*c,f*c*b*a,g*e*d*c*b*a],
 [g*d,f*e,c*b,g*f*c*a,g*e*b*a,f*d*b*a,e*d*c*a,g*f*e*d*c*b],
 [g*e,f*d,c*a,g*f*c*b,g*d*b*a,f*e*b*a,e*d*c*b,g*f*e*d*c*a],
 [g*f,e*d,b*a,g*e*c*b,g*d*c*a,f*e*c*a,f*d*c*b,g*f*e*d*b*a],
 [a,g*f*b,g*e*c,f*d*c,e*d*b,g*f*e*d*a,g*d*c*b*a,f*e*c*b*a],
```

```
[b,g*f*a,g*d*c,f*e*c,e*d*a,g*f*e*d*b,g*e*c*b*a,f*d*c*b*a],
[c,g*e*a,g*d*b,f*e*b,f*d*a,g*f*e*d*c,g*f*c*b*a,e*d*c*b*a],
[d,g*f*e,g*c*b,f*c*a,e*b*a,g*f*d*b*a,g*e*d*c*a,f*e*d*c*b],
[e,g*f*d,g*c*a,f*c*b,d*b*a,g*f*e*b*a,g*e*d*c*b,f*e*d*c*a],
[f,g*e*d,g*b*a,e*c*b,d*c*a,g*f*e*c*a,g*f*d*c*b,f*e*d*b*a],
[g,f*e*d,f*b*a,e*c*a,d*c*b,g*f*e*c*b,g*f*d*c*a,g*e*d*b*a],
[1,g*f*e*d,g*f*b*a,g*e*c*a,g*d*c*b,f*e*c*b,f*d*c*a,e*d*b*a]]
```

Each element of the output list represents an aliasing relation. For example, the ninth element

```
[a,g*f*b,g*e*c,f*d*c,e*d*b,g*f*e*d*a,g*d*c*b*a,f*e*c*b*a]
```

represents the aliasing relation (4.54):

$$A = BFG = CEG = CDF = BDE = ADEFG = ABCDG = ABCEF.$$

Exercise. Enumerate all the aliasing relations for the 2^{4-1} fractional factorial design of four factors with two levels $\{-1, 1\}$ chosen by the relation $ACD = 1$.

7.3 Convex Polytopes and Gröbner Bases

This section includes examples and exercises for Chap. 5, "Convex polytopes and Gröbner bases". The goals of this section are to use a computer to compute various objects of a polytope, the Gröbner fan of a toric ideal, various bases of a configuration matrix, and the triangulations of a polytope. The software used in this section includes the computer algebra system Risa/Asir [15], the program for studying polytopes polymake [6], the program for computing triangulations

Software	Command (or function)	Computation
polymake	polymake FACETS	Facets of a convex polytope
	polymake VERTICES	Vertices of a convex polytope
	polymake VERTICES_IN_FACETS	Vertices in facets
Asir(toric.rr)	gr_w(Id, VL, W)	Gröbner basis w.r.t. $<_w$
	toric_ideal(A)	Generators for a toric ideal I_A
		(Elimination methods by GB)
Asir(Asir-Contrib)	poly_toric_ideal(A)	Generators for a toric ideal I_A
		(Use 4ti2, fast computation)
Gfan	gfan	All reduced Gröbner bases
	groebner_cone	Max dim. cones of the Gröbner fan
4ti2	groebner	Gröbner basis of I_A
	circuits	All circuits of I_A
	graver	Graver basis of I_A
TOPCOM	points2alltriangs	All triangulations of A
	points2triangs -regular	All regular triangulations of A

TOPCOM [19], the program for computing various bases for a toric ideal 4ti2 [26], and the program for computing the Gröbner fan Gfan [11].

7.3.1 Convex Polytopes (Sect. 5.1)

Example 7.3.1. Let the set of points $X = \{[1, 0], [0, 1]\}$, and let the polyhedral convex cone $P = \mathbb{Q}_{\geq 0} X$.

1. Compute the faces $F_i := \text{FACE}_{\mathbf{w}_i}(P)$ with respect to the following weight vectors $\mathbf{w}_i \in \mathbb{Q}^2$:

$$\mathbf{w}_1 = [1, 1], \mathbf{w}_2 = [-1, -1], \mathbf{w}_3 = [-1, 0], \mathbf{w}_4 = [0, -1], \mathbf{w}_5 = [0, 0].$$

2. For each face F_i, compute the normal cone $\mathcal{N}_P(F_i)$.

Answer. By definition, we obtain the following result (Fig. 7.1).

1. • $F_1 = \text{FACE}_{\mathbf{w}_1}(P) = \emptyset$
   • $F_2 = \text{FACE}_{\mathbf{w}_2}(P) = \{[0, 0]\}$
   • $F_3 = \text{FACE}_{\mathbf{w}_3}(P) = [0, 1] \times \mathbb{Q}_{\geq 0}$
   • $F_4 = \text{FACE}_{\mathbf{w}_4}(P) = [1, 0] \times \mathbb{Q}_{\geq 0}$
   • $F_5 = \text{FACE}_{\mathbf{w}_5}(P) = P$
2. • $\mathcal{N}_P(F_1) = \{[w_1, w_2] \mid w_1 > 0 \text{ or } w_2 > 0\}$
   • $\mathcal{N}_P(F_2) = \{[w_1, w_2] \mid w_1 < 0, w_2 < 0\}$
   • $\mathcal{N}_P(F_3) = \{[w_1, 0] \mid w_1 < 0\}$
   • $\mathcal{N}_P(F_4) = \{[0, w_2] \mid w_2 < 0\}$
   • $\mathcal{N}_P(F_5) = \{[0, 0]\}$

Exercise. Let the convex polytope be $P = \text{CONV}(\{[0, 0], [0, 1], [1, 0], [1, 1]\})$, and let the weight vectors be

$$\mathbf{w}_1 = [0, 0], \mathbf{w}_2 = [-1, 0], \mathbf{w}_3 = [1, 0], \mathbf{w}_4 = [0, -1], \mathbf{w}_5 = [0, 1], \mathbf{w}_6 = [-1, -1],$$

$$\mathbf{w}_7 = [-1, 1], \mathbf{w}_8 = [1, -1], \mathbf{w}_9 = [1, 1].$$

Compute the faces $F_i := \text{FACE}_{\mathbf{w}_i}(P)$ and its normal cones $\mathcal{N}_P(F_i)$.

Example 7.3.2. Compute the vertices of the convex polytope

$$P = \{[x_{11}, x_{12}, x_{21}, x_{22}] \in \mathbb{Q}^4 \mid x_{11}, x_{12}, x_{21}, x_{22} \geq 0,$$

$$x_{11} + x_{12} = 1, x_{21} + x_{22} = 1, x_{11} + x_{21} = 1, x_{12} + x_{22} = 1\}$$

by using the software program polymake. This convex polytope is called the **Birkhoff polytope** $P(2)$ and is related to two-dimensional contingency tables. Please refer to [25, Example 0.11].

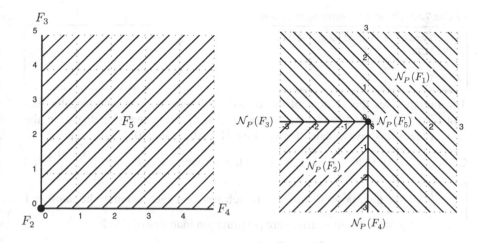

Fig. 7.1 Faces and normal cones of P

Answer. In order to use the software program polymake, we prepare the following input file `birkhoff2.p`, which defines the convex polytope P.

Listing 7.32 Polymake: input file `birkhoff2.p`

```
INEQUALITIES
 0   1   0   0   0
 0   0   1   0   0
 0   0   0   1   0
 0   0   0   0   1
 1  -1  -1   0   0
-1   1   1   0   0
 1   0   0  -1  -1
-1   0   0   1   1
 1  -1   0  -1   0
-1   1   0   1   0
 1   0  -1   0  -1
-1   0   1   0   1
```

In the first row, we write INEQUALITIES, which indicates that we will use linear inequalities to define a convex polytope. The subsequent rows indicate the linear inequalities defining the convex polytope. The first column is the constant term, and the following columns are the coefficients of the variables in the linear inequality. In this case, we set the variables to be x_{11}, x_{12}, x_{21}, and x_{22}. For example, the sixth row 1 -1 -1 0 0 indicates the linear inequality $1 - x_{11} - x_{12} \geq 0$. The eighth row 1 0 0 -1 -1 and the ninth row -1 0 0 1 1 indicate the linear inequalities $1 - x_{21} - x_{22} \geq 0$ and $1 - x_{21} - x_{22} \leq 0$, respectively, and they combine to give the linear equation $1 - x_{21} - x_{22} = 0$.

In order to obtain the vertices of the convex polytope P, we execute the following command.

Listing 7.33 Polymake: computing the vertices

```
$ polymake birkhoff2.p VERTICES
VERTICES
1 1 0 0 1
1 0 1 1 0
```

The output gives the vertices of the convex polytope P. For example, the second row 1 1 0 0 1 gives the vertex $[1, 0, 0, 1]$. Here, we skip the 1 in the first column, and we obtain the vertices $[1, 0, 0, 1]$ and $[0, 1, 1, 0]$, and $P = $ CONV$(\{[1, 0, 0, 1], [0, 1, 1, 0]\})$. We write the vertices as 2×2 tables, $\begin{bmatrix} 1 & 0 \\ 0 & 1 \end{bmatrix}$ and $\begin{bmatrix} 0 & 1 \\ 1 & 0 \end{bmatrix}$. Note that, in these tables, for each row and each column, the number 1 appears only once. These matrices are permutation matrices of size 2.

Exercise. We consider the Birkhoff polytope

$$P(d) = \left\{ [x_{ij}]_{1 \leq i, j \leq d} \in \mathbb{Q}^{d^2} \mid x_{ij} \geq 0, \sum_{k=1}^{d} x_{ik} = 1, \sum_{k=1}^{d} x_{kj} = 1 \text{ for } 1 \leq i, j \leq d \right\}.$$

Compute the vertices of the Birkhoff polytopes $P(3)$ and $P(4)$. What are the vertices of the Birkhoff polytopes $P(d)$? For more about the Birkhoff polytopes, please refer to [21, Theorem 8.6].

Example 7.3.3. We consider the convex polytope

$$P = \text{CONV}(\{[1, 2, 3], [1, 3, 2], [2, 1, 3], [2, 3, 1], [3, 1, 2], [3, 2, 1]\}).$$

What are the linear inequalities defining the convex polytope P? We will compute the linear inequalities by using the software program polymake. The convex polytope is called the **permutahedron** Π_2. For more about permutahedrons, please refer to [25, Example 0.10].

Answer. This example is the reverse transformation of the previous example. In order to use the software program polymake, we prepare the following input file permutation2.p which defines the convex polytope P.

Listing 7.34 Polymake: input file permutation2.p

```
POINTS
1 1 2 3
1 1 3 2
1 2 1 3
1 2 3 1
1 3 1 2
1 3 2 1
```

In the first row, we write POINTS to indicate that we will input the points. The points are written in the following rows, as follows: we write 1 in the first column and the coordinates of points in the following columns. For example, the second row 1 1 2 3 represents the point $[1, 2, 3]$.

In order to obtain the linear inequalities defining the convex polytope, we execute the following command.

Listing 7.35 Polymake: computing the facets

```
$ polymake permutation2.p FACETS
FACETS
-1  0  1  0
-3  1  1  0
-1  1  0  0
 3  0 -1  0
 5 -1 -1  0
 3 -1  0  0
```

This output gives the linear inequalities defining the convex polytope. For example, the third row `-3 1 1 0` indicates the linear inequality $-3 + x_1 + x_2 \geq 0$. After computation, the following result is in the file `permutation2.p`.

Listing 7.36 Polymake: a part of the file `permutation2.p`

```
<property name="AFFINE_HULL">
    <m>
       <v>-6 1 1 1</v>
    </m>
</property>
```

This `-6 1 1 1` means that the linear equation $-6 + x_1 + x_2 + x_3 = 0$ exists in the linear inequalities defining the convex polytope. So we can obtain

$$P = \{[x_1, x_2, x_3] \mid -1 + x_2 \geq 0, -3 + x_1 + x_2 \geq 0, -1 + x_1 \geq 0,$$

$$3 - x_2 \geq 0, 5 - x_1 - x_2 \geq 0, 3 - x_1 \geq 0, x_1 + x_2 + x_3 = 6\}.$$

We can draw the convex polytope shown in Fig. 7.2.

Exercise. We consider the permutahedron

$$\Pi_{d-1} = \text{CONV}(\{[\sigma(1), \ldots, \sigma(d)] \mid \sigma \in S_d\}).$$

Here S_d is the symmetric group of degree d. Compute the linear inequalities defining the permutahedrons Π_3 and Π_4.

Example 7.3.4. By using the software program polymake, determine whether the convex polytopes

$$P_1 = \text{CONV}(\{[1, 0, 0], [0, 1, 0], [0, 0, 1], [-1, -1, -1]\}),$$

$$P_2 = \text{CONV}(\{[-1, 0, 0], [0, 1, 0], [0, 0, 1], [1, -1, 1]\})$$

include the origin.

Answer. In order to obtain the linear inequalities defining the convex polytope P_1, we prepare the following file.

Fig. 7.2 Permutahedron Π_2
on the hyperplane
$x_1 + x_2 + x_3 = 6$

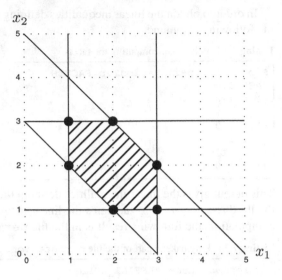

Listing 7.37 Polymake: input file `ip_p1.p`

```
POINTS
1  1  0  0
1  0  1  0
1  0  0  1
1 -1 -1 -1
```

We execute the following command to compute the linear inequalities for the convex polytope.

Listing 7.38 Polymake: computing the facets

```
$ polymake ip_p1.p FACETS
FACETS
1 -1 -1 -1
1 -1 -1  3
1 -1  3 -1
1  3 -1 -1
```

From this output, the convex polytope P_1 is defined by

$$P_1 = \{[x_1, x_2, x_3] \mid 1 - x_1 - x_2 - x_3 \geq 0, 1 - x_1 - x_2 + 3x_3 \geq 0,$$
$$1 - x_1 + 3x_2 - x_3 \geq 0, 1 + 3x_1 - x_2 - x_3 \geq 0\}.$$

Substituting the origin $[x_1, x_2, x_3] = [0, 0, 0]$ into the above, every one of these linear inequalities becomes $1 > 0$. Thus, the origin is an internal point of this convex polytope.

We compute the convex polytope P_2 using the same method as the one we used for P_1.

Listing 7.39 Polymake: input file `ip_p2.p`

```
POINTS
1 -1  0  0
1  0  1  0
1  0  0  1
1  1 -1  1
```

Listing 7.40 Polymake: computing the facets

```
$ polymake ip_p2.p FACETS
FACETS
 1 -1 -1 -1
-1 -1  1  3
 1  1  1 -1
 1  1 -1 -1
```

The third row `-1 -1 1 3` represents the linear inequality $-1 - x_1 + x_2 + 3x_3 \geq 0$. The origin $[x_1, x_2, x_3] = [0, 0, 0]$ does not satisfy this inequality, so the origin is not in the convex polytope P_2.

Exercise. Let the convex polytope be

$$P = \mathrm{CONV}(\{[5, 0, 0], [-5, 0, 0], [0, 5, 0], [0, -5, 0], [0, 0, 5], [0, 0, -5]\}).$$

Determine whether the points $p = [1, 1, 1]$ and $q = [2, 2, 2]$ are internal points of P.

Example 7.3.5. Let the convex polytopes be

$$P_1 = \mathrm{CONV}(\{[1, 0], [0, 1], [-1, -1]\}),$$
$$P_2 = \mathrm{CONV}(\{[1, 0, 0], [0, 1, 0], [0, 0, 1], [1, 1, 1]\}),$$
$$P_3 = \mathrm{CONV}(\{[1, -1, -1], [-1, -1, 0], [0, 1, -1], [0, 0, 1]\}),$$
$$P_4 = \mathrm{CONV}(\{[1, 0, 0], [0, 1, 0], [0, 0, 1], [-1, 0, 0], [0, -1, 0], [0, 0, -1]\}).$$

Compute the **dual polytopes**. Here, for a convex polytope $P = \mathrm{CONV}(\{\mathbf{u}_1, \ldots, \mathbf{u}_m\})$, the dual polytope of P is defined by

$$P^* = \{\mathbf{v} \mid \mathbf{v} \cdot \mathbf{u} \geq -1 \text{ for all } \mathbf{u} \in P\}.$$

Answer. For a convex polytope $P = \mathrm{CONV}(\{\mathbf{u}_1, \ldots, \mathbf{u}_m\})$, we explain how to compute the dual polytope P^*. For any point $\mathbf{u} \in P$, there exist nonnegative rational numbers $t_i \in \mathbb{Q}_{\geq 0}$ such that $\sum_{i=1}^m t_i = 1$ and $\mathbf{u} = \sum_{i=1}^m t_i \mathbf{u}_i$. For some point $\mathbf{v}$,

$$\mathbf{v} \cdot \mathbf{u} \geq -1 \text{ for all } \mathbf{u} \in P \iff \mathbf{v} \cdot \mathbf{u}_i \geq -1 \text{ for } i = 1, \ldots, m.$$

Hence, these linear inequalities $\mathbf{v} \cdot \mathbf{u}_i \geq -1$ ($i = 1, \ldots, m$) define the dual polytope P^*.

In the case of the convex polytope P_1, the above linear inequalities are

$$1 + x_1 \geq 0, 1 + x_2 \geq 0, 1 - x_1 - x_2 \geq 0.$$

In order to compute the vertices of the convex polytope P_1, we prepare the following input file.

Listing 7.41 Polymake: input file `dual_p1.p`

```
INEQUALITIES
1   1   0
1   0   1
1  -1  -1
```

We then execute the following command.

Listing 7.42 Polymake: computing the vertices

```
$ polymake dual_p1.p VERTICES
VERTICES
1  2 -1
1 -1 -1
1 -1  2
```

Hence, we obtain the dual polytope

$$P_1^* = \mathrm{CONV}(\{[2, -1], [-1, -1], [-1, 2]\}).$$

We compute the dual polytopes of P_2, P_3, and P_4 as we did for P_1, and we obtain

$$P_2^* = \mathrm{CONV}(\{[0, 0, 1], [0, 1, 0], [1, 0, 0], [1, -1, -1], [-1, 1, -1], [-1, -1, 1]\}),$$
$$P_3^* = \mathrm{CONV}(\{[3, -2, -1], [-4, -2, -1], [-1/2, 3/2, -1], [2/3, 1/3, 4/3]\}),$$
$$P_4^* = \mathrm{CONV}(\{[1, -1, -1], [1, -1, 1], [1, 1, 1], [1, 1, -1], [-1, 1, 1],$$
$$[-1, 1, -1], [-1, -1, 1], [-1, -1, -1]\}).$$

Exercise. Compute the dual polytopes of the convex polytopes

$$P_1 = \mathrm{CONV}(\{[1, 0, 0, 0], [0, 1, 0, 0], [0, 0, 1, 0], [-1, -1, -1, -1]\}),$$
$$P_2 = \mathrm{CONV}(\{[1, 0, 0, 0], [0, 1, 0, 0], [0, 0, 1, 0], [-1, -1, -1, 3], [0, 0, 0, -1]\}).$$

Example 7.3.6. Let the convex polytopes be

$$P_1 = \mathrm{CONV}(\{[1, 0, 0], [1, 1, 0], [0, 1, 0]\}), P_2 = \mathrm{CONV}(\{[0, 1, 0], [0, 1, 1], [0, 0, 1]\}).$$

Compute the vertices of the Minkowski sum $P = P_1 + P_2$ by using the software program polymake. Compute the faces $\mathrm{FACE}_{[-1,0,0]}(P)$ and $\mathrm{FACE}_{[-1,-1,-1]}(P)$.

Answer. First, we compute the sums of the vertices of P_1 and P_2 and prepare the following input file.

Listing 7.43 Polymake: input file `mink_sum.p`

```
POINTS
1 1 1 0
1 1 2 0
1 0 2 0
1 1 1 1
1 1 2 1
1 0 2 1
1 1 0 1
1 1 1 1
1 0 1 1
```

In order to obtain the vertices of the convex polytope $P = P_1 + P_2$, we execute the following command.

Listing 7.44 Polymake: computing the vertices

```
$ polymake mink_sum.p VERTICES
VERTICES
1 1 2 1
1 1 2 0
1 0 2 1
1 1 1 0
1 0 2 0
1 1 0 1
1 0 1 1
```

We obtain the vertices $[1, 2, 1], [1, 2, 0], [0, 2, 1], [1, 1, 0], [0, 2, 0], [1, 0, 1], [0, 1, 1]$.

In order to obtain the face $\text{FACE}_{[-1,0,0]}(P)$, we take the vertices of the convex polytope P for which the inner product with $[-1, 0, 0]$ is maximized. Please refer to Proposition 5.1.9. Hence, we obtain the face

$$\text{FACE}_{[-1,0,0]}(P) = \text{CONV}(\{[0, 2, 1], [0, 2, 0], [0, 1, 1]\}).$$

In a similar way, we obtain the face

$$\text{FACE}_{[-1,-1,-1]}(P) = \text{CONV}(\{[1, 1, 0], [0, 2, 0], [1, 0, 1], [0, 1, 1]\}).$$

Exercise. 1. For the polynomials $f = x + y + z + 1$ and $g = xy + yz + zx + 1$, compute the Minkowski sum $\text{New}(f) + \text{New}(g)$. Check that the Minkowski sum is equal to the Newton polytope $\text{New}(fg)$. This is an example of Proposition 5.3.2.
2. Prove Proposition 5.1.3. In the above example, we computed the Minkowski sum by using this proposition.

Example 7.3.7. Let a polynomial be $f = x^4 + y^4 + z^4 + xyz + xy + yz + zx + 1$. Which terms do not appear in the initial form $\text{in}_w(f)$ for any nonzero weight vector $\mathbf{w}$?

Answer. From Proposition 5.3.3, we have

$$\text{FACE}_w(\text{New}(f)) = \text{New}(\text{in}_w(f)).$$

Hence, what we want are the terms in f corresponding to the internal points of New(f). We use polymake to obtain the vertices of New(f)

$$[0, 0, 0], [0, 0, 4], [0, 4, 0], [4, 0, 0].$$

These points correspond to the terms in f

$$1, z^4, y^4, x^4$$

and appear in initial forms of f. We next determine if the points $[1, 1, 1], [1, 1, 0]$, $[0, 1, 1], [1, 0, 1]$, corresponding to the terms xyz, xy, yz, zx, are internal points of New(f). As in Example 7.3.4, we obtain linear inequalities defining New(f):

$$4 - x_1 - x_2 - x_3 \geq 0, x_1 \geq 0, x_2 \geq 0, x_3 \geq 0.$$

We next check to see if any of the points, after being substituted into the inequalities, results in a left-hand side which is greater than 0. Since $[1, 1, 1]$ is such a point, the term xyz does not appear in any initial form of f. The other three points are not vertices but lie on the facets. The terms corresponding to these three points appear in the initial forms of f.

Exercise. Let a polynomial be $f = x^4 + y^4 + z^4 + xyz + xy + yz + zx + 1$. Which terms do not appear in the initial form $\text{in}_w(f^2)$ for any nonzero weight vector $\mathbf{w}$?

7.3.2   Initial Ideals (Sect. 5.2)

Example 7.3.8. Let the ideal be

$$I = \langle x^2 + y - 1, x + y^2 - 1 \rangle,$$

and let the weight vectors be

$$\mathbf{w}_1 = [3, 1], \mathbf{w}_2 = [1, 1], \mathbf{w}_3 = [1, 3], \mathbf{w}_4 = [2, 1], \mathbf{w}_5 = [1, 2], \mathbf{w}_6 = [0, 0].$$

Compute the initial form ideal $\text{in}_w(I)$.

Answer. We compute the Gröbner bases $\mathcal{G}_i$ of I with respect to the monomial orders $<_{\mathbf{w}_i}$, where the tie-breaker $<$ is the reverse lexicographic order with $x > y > z$.

$$\mathcal{G}_1 = \{x + y^2 - 1, -y^4 + 2y^2 - y\},$$
$$\mathcal{G}_2 = \{x + y^2 - 1, x^2 + y - 1\},$$
$$\mathcal{G}_3 = \{x^2 + y - 1, -x^4 + 2x^2 - x\},$$

$$\mathscr{G}_4 = \{x^2 + y - 1, x + y^2 - 1\},$$

$$\mathscr{G}_5 = \{x^2 + y - 1, x + y^2 - 1\},$$

$$\mathscr{G}_6 = \{x^2 + y - 1, x + y^2 - 1\}.$$

By Corollary 5.2.5, the sets $\mathrm{in_w}(\mathscr{G}_i) = \{\mathrm{in_w}(g) \mid g \in \mathscr{G}_i\}$ generate the initial form ideal $\mathrm{in_w}(I)$. Hence, we have

$$\mathrm{in_{w_1}}(I) = \langle x, -y^4 \rangle,$$

$$\mathrm{in_{w_2}}(I) = \langle y^2, x^2 \rangle,$$

$$\mathrm{in_{w_3}}(I) = \langle y, -x^4 \rangle,$$

$$\mathrm{in_{w_4}}(I) = \langle x^2, x + y^2 \rangle,$$

$$\mathrm{in_{w_5}}(I) = \langle x^2 + y, y^2 \rangle,$$

$$\mathrm{in_{w_6}}(I) = \langle x^2 + y - 1, x + y^2 - 1 \rangle.$$

We now explain how to compute a Gröbner basis with respect to a monomial order by using Risa/Asir. We will use the program `toric.rr`. In this program, the function `gr_w(Id, VL, W)` returns a Gröbner basis of an ideal `Id` with respect to a monomial order $<_w$. Here, the argument `VL` is a list of variables. The function `in_w(P, VL, W)` returns the initial form of a polynomial `P` with respect to a weight vector `W`.

Listing 7.45 Risa/Asir: computing initial form ideals

```
[1358]  load("toric.rr");              Read the program toric.rr
[1376]  Id=[x+y^2-1, x^2+y-1];
[1377]  VL=[x,y];
[1378]  G1=gr_w(Id,VL,[3,1]);          Compute GB of Id w.r.t. <_[3,1]
[x+y^2-1,-y^4+2*y^2-y]
[1379]  IN1 = map(in_w, G1, VL, [3,1]); Initial forms of G1 w.r.t. [3,1]
[x,-y^4]
[1380]  G2=gr_w(Id,VL,[1,1]);          Compute GB of Id w.r.t. <_[1,1]
[x+y^2-1,x^2+y-1]
[1381]  IN2 = map(in_w, G2, VL, [1,1]); Initial forms of G2 w.r.t. [1,1]
[y^2,x^2]
[1382]  G3=gr_w(Id,VL,[1,3]);          Compute GB of Id w.r.t. <_[1,3]
[x^2+y-1,-x^4+2*x^2-x]
[1383]  IN3 = map(in_w, G3, VL, [1,3]); Initial forms of G3 w.r.t. [1,3]
[y,-x^4]
[1384]  G4=gr_w(Id,VL,[2,1]);          Compute GB of Id w.r.t. <_[2,1]
[x+y^2-1,x^2+y-1]
[1385]  IN4 = map(in_w, G4, VL, [2,1]); Initial forms of G4 w.r.t. [2,1]
[x+y^2,x^2]
[1387]  G5=gr_w(Id,VL,[1,2]);          Compute GB of Id w.r.t. <_[1,2]
[x^2+y-1,x+y^2-1]
[1388]  IN5 = map(in_w, G5, VL, [1,2]); Initial forms of G5 w.r.t. [1,2]
[x^2+y,y^2]
[1390]  G6=gr_w(Id,VL,[0,0]);          Compute GB of Id w.r.t. <_[0,0]
[x+y^2-1,x^2+y-1]
[1391]  IN6 = map(in_w, G6, VL, [0,0]); Initial forms of G6 w.r.t. [0,0]
[x+y^2-1,x^2+y-1]
```

Exercise. Let the ideal be

$$I = \langle x + y + z, xy + yz + zx, xyz - 1 \rangle$$

and let the weight vectors be

$$\mathbf{w}_1 = [3, 2, 1], \mathbf{w}_2 = [3, 1, 2], \mathbf{w}_3 = [2, 1, 3], \mathbf{w}_4 = [1, 2, 3], \mathbf{w}_5 = [2, 3, 1], \mathbf{w}_6 = [1, 3, 2].$$

Compute the initial form ideal $\text{in}_{\mathbf{w}_i}(I)$.

Example 7.3.9. We consider the ideal $I = \langle x^2 + y - 1, x + y^2 - 1 \rangle$ in Example 7.3.8. Compute the universal Gröbner basis of I by using the software program Gfan.

Answer. Gfan is a software program which computes the universal Gröbner basis. We prepare the following input file for Gfan.

Listing 7.46 Gfan: input file gfan_input.txt

```
Q[x,y]
{x^2+y-1, x+y^2-1}
```

Q[x,y] is notation for the polynomial ring in variables x, y, and {x^2+y-1, x+y^2-1} are the generators of an ideal. We execute the command gfan, as follows.

Listing 7.47 Gfan: computing the universal Gröbner basis

```
$ gfan < gfan_input.txt
Q[x,y]
LP algorithm being used: "cddgmp".
{
{
y^4+y-2*y^2,
x-1+y^2}
,
{
y^2-1+x,
x^2-1+y}
,
{
y-1+x^2,
x^4+x-2*x^2}
}
```

These three sets are the reduced Gröbner bases of I. For each polynomial in the set, the first term is the initial term. For example, the first set {y^4+y-2*y^2,x-1+y^2} is the reduced Gröbner basis for which the initial terms are y^4, x. The Gröbner bases $\mathscr{G}_1, \mathscr{G}_2, \mathscr{G}_3$ in Example 7.3.8 correspond to these reduced Gröbner bases.

Exercise. For the ideal $I = \langle x + y + z, xy + yz + zx, xyz - 1 \rangle$, compute the universal Gröbner basis.

7.3.3   Gröbner Fans and State Polytopes (Sect. 5.3)

Example 7.3.10. We consider the ideal $I = \langle x^2 + y^2 - 4, xy - 1 \rangle$.

1. Let $<_{\text{purelex}}$ be the pure lexicographic order with $x > y$. Compute the Gröbner cone $C[<_{\text{purelex}}] = \{\mathbf{w}' \in \mathbb{Q}^2_{\geq 0} \mid \text{in}_{\mathbf{w}'}(I) = \text{in}_{<_{\text{purelex}}}(I)\}$.
2. Let the weight vector be $\mathbf{w} = [3, 1]$. Compute the Gröbner cone $C[\mathbf{w}] = \{\mathbf{w}' \in \mathbb{Q}^2_{\geq 0} \mid \text{in}_{\mathbf{w}'}(I) = \text{in}_{[3,1]}(I)\}$.
3. Draw the Gröbner fan GF(I) in $\mathbb{Q}^2_{\geq 0}$.
4. For each cone in the Gröbner fan GF(I), compute the Gröbner basis corresponding to the cone and the initial form ideal $\text{in}_{\mathbf{w}}(I)$.

Answer. 1. The reduced Gröbner basis of the ideal I with respect to the pure lexicographic order is

$$\{g_1 = \underline{y^4} - 4y^2 + 1, g_2 = \underline{x} + y^3 - 4y\}.$$

Here, the underlined terms are the initial terms. From the conditions $\text{in}_{\mathbf{w}}(g_1) = \text{in}_{<_{\text{purelex}}}(g_1) = y^4$ and $\text{in}_{\mathbf{w}}(g_2) = \text{in}_{<_{\text{purelex}}}(g_2) = x$, we obtain the following linear inequalities with respect to the weight vector $\mathbf{w} = [w_1, w_2]$:

$$[0, 2] \cdot \mathbf{w} > 0, \quad [0, 4] \cdot \mathbf{w} > 0,$$

$$[1, -3] \cdot \mathbf{w} > 0, \; [1, -1] \cdot \mathbf{w} > 0,$$

$$[1, 0] \cdot \mathbf{w} > 0, \quad [0, 1] \cdot \mathbf{w} > 0.$$

For example, the linear equation $[1, -3] \cdot \mathbf{w} > 0$ is obtained from $\text{in}_{\mathbf{w}}(g_2) = x$. More concretely, $\text{in}_{\mathbf{w}}(g_2) = x$ is equivalent to the inequality for the weights of x and y^3, which states that $[1, 0] \cdot \mathbf{w}$ is greater than $[0, 3] \cdot \mathbf{w}$.

We solve these linear inequalities and obtain the cone $C[<_{\text{purelex}}] = \{[w_1, w_2] \mid w_2 > 0, w_2 < 1/3w_1\}$. In order to use polymake to solve these linear inequalities, we prepare the following input file.

Listing 7.48 Polymake: input file `cone1.p`

```
INEQUALITIES
0 0 2
0 0 4
0 1 -3
0 1 -1
0 1 0
0 0 1
```

We then execute the following command.

Listing 7.49 Polymake: computing the facets

```
$ polymake cone1.p FACETS
FACETS
0 0 2
0 1 -3
1 0 0
```

The output indicates the linear inequalities $2w_2 > 0$, $w_1 - 3w_2 > 0$, and $1 > 0$, leading to solution $w_2 > 0$ and $w_2 < 1/3w_1$.

2. The reduced Gröbner basis of the ideal I with respect to the monomial order $<_{[3,1]}$ is

$$\{g_1 = \underline{y^4} - 4y^2 + 1, g_2 = \underline{x + y^3} - 4y\}.$$

Here, underlined terms are the initial forms with respect to the weight vector $[3, 1]$. By Proposition 5.3.5 and the conditions $\text{in}_w(g_1) = \text{in}_{[3,1]}(g_1) = y^4$ and $\text{in}_w(g_2) = \text{in}_{[3,1]}(g_2) = x + y^3$, we obtain the following linear inequalities with respect to $\mathbf{w} = [w_1, w_2]$:

$$[1, -3] \cdot \mathbf{w} = 0, \ [1, -1] \cdot \mathbf{w} > 0,$$

$$[0, 2] \cdot \mathbf{w} > 0, \quad [0, 4] \cdot \mathbf{w} > 0,$$

$$[1, 0] \cdot \mathbf{w} > 0, \quad [0, 1] \cdot \mathbf{w} > 0.$$

For example, the linear inequality $[1, -3] \cdot \mathbf{w} = 0$ is obtained from the condition $\text{in}_w(g_2) = x + y^3$. More concretely, the condition $\text{in}_w(g_2) = x + y^3$ implies that in the polynomial g_2 the weight of x, which is $[1, 0] \cdot \mathbf{w}$, is equal to the weight of y^3, which is $[0, 3] \cdot \mathbf{w}$. We solve the linear inequalities and obtain the cone $C[[3, 1]] = \{[w_1, w_2] \mid w_2 > 0, w_2 = 1/3w_1\}$. In order to use polymake to solve these linear inequalities, we prepare the following input file.

Listing 7.50 Polymake: input file cone2.p

```
INEQUALITIES
0 1 -3
0 -1 3
0 1 -1
0 0 2
0 0 4
0 1 0
0 0 1
```

We then execute the following command.

Listing 7.51 Polymake: computing the facets and the affine hulls

```
$ polymake cone2.p FACETS AFFINE_HULL
FACETS
0 1 -1
1 0 0

AFFINE_HULL
0 1 -3
```

This output represents the linear inequalities $w_1 - w_2 > 0$ and $w_1 - 3w_2 = 0$, and thus we obtain the solution $w_2 > 0$, $w_2 = 1/3w_1$.

3. We consider a point $\mathbf{w}$ not in the closure $\overline{C}[<_{\text{purelex}}]$. For example, we take a point $\mathbf{w} = [3, 2]$. As in the previous example, we compute the Gröbner cone

$$C[\mathbf{w}] = \{[w_1, w_2] \mid w_2 < w_1, w_2 > 1/3w_1\}.$$

Fig. 7.3 Gröbner fan of I

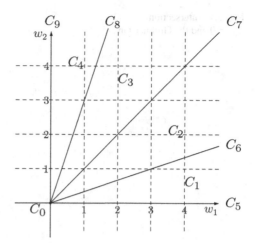

The cones $\overline{C[\mathbf{w}]}$ and $\overline{C[<_{\text{purelex}}]}$ have the common border $\{[w_1, w_2] \mid w_1 \geq 0, w_2 \geq 0, w_2 = 1/3w_1\}$. If there is no common border, we take another point $\mathbf{w}$ which is nearer the closure $\overline{C[<_{\text{purelex}}]}$. We continue in a similar way until the closures of the obtained cones fill $\mathbb{Q}_{\geq 0}^2$. In this case, using the symmetry of x and y, we can easily obtain other cones. The Gröbner fan GF(I) is shown in Fig. 7.3 and consists of 10 cones.

4. We take a point $\mathbf{w}_i$ in a Gröbner cone C_i. Let $<_{\mathbf{w}_i}$ be the monomial order whose tie-breaker $<$ is the pure lexicographic order with $x > y$. We compute the reduced Gröbner basis of I with respect to $<_{\mathbf{w}_i}$.

- $C_1 : \{\underline{x} + y^3 - 4y, \underline{y^4} - 4y^2 + 1\}$
- $C_2 : \{\underline{x^2} + y^2 - 4, \underline{xy} - 1, y^3 + x - 4y\}$
- $C_3 : \{\underline{y^2} + x^2 - 4, \underline{xy} - 1, \underline{x^3} - 4x + y\}$
- $C_4 : \{\underline{x^4} - 4x^2 + 1, \underline{y} + x^3 - 4x\}$
- $C_5 : \{\underline{x} + y^3 - 4y, \underline{y^4} - 4y^2 + 1\}$
- $C_6 : \{x + y^3 - 4y, y^4 - 4y^2 + 1\}$
- $C_7 : \{x^2 + y^2 - 4, xy - 1, y^3 + x - 4y\}$
- $C_8 : \{y^2 + x^2 - 4, xy - 1, x^3 + y - 4x\}$
- $C_9 : \{\underline{x^4} - 4x^2 + 1, \underline{y} + x^3 - 4x\}$
- $C_0 : \{\underline{x} + y^3 - 4y, \underline{y^4} - 4y^2 + 1\}$

Here, the underlined polynomials are the initial forms.

Exercise. Let the ideal be $I = \langle x^5 + y^3, xy - 1 \rangle$. Draw the Gröbner fan GF(I) in $\mathbb{Q}_{\geq 0}^2$.

Example 7.3.11. We consider the ideal $J = \langle x + y + z, y + 2z \rangle$.

1. Compute the Gröbner fan of the ideal J. For a relatively interior point $\mathbf{w}$ in maximal dimensional cones of the fan, compute the initial form ideal $\text{in}_{\mathbf{w}}(J)$.

Fig. 7.4 Intersection of
$w_3 = 0$ and the Gröbner fan

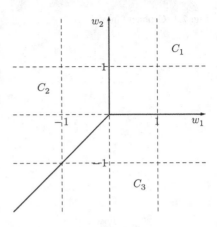

2. We consider the intersection of the hyperplane $z = 0$ and the Gröbner fan. This
   intersection is a fan. Draw this fan and a convex polytope for which this is the
   normal fan.
3. Find a convex polytope for which the normal fan is the Gröbner fan of J; that is,
   find the state polytope of J.

Answer. 1. As in Example 7.3.10, we compute the Gröbner cones.

$$C_1 = \{[w_1, w_2, w_3] \mid w_1 > w_3, w_2 > w_3\},$$

$$C_2 = \{[w_1, w_2, w_3] \mid w_1 < w_2, w_1 < w_3\},$$

$$C_3 = \{[w_1, w_2, w_3] \mid w_1 > w_2, w_2 < w_3\}.$$

The Gröbner fan of J consists of these closures. The reduced Gröbner bases
corresponding to the Gröbner cones are as follows.

- $C_1 : \{\underline{x} - z, \underline{y} + 2z\}$
- $C_2 : \{\underline{y} + 2\underline{x}, \underline{z} - x\}$
- $C_3 : \{\underline{2x} + y, \underline{2z} + y\}$

The underlined terms are the initial forms.

2. The intersection of the hyperplane $w_3 = 0$ and the Gröbner fan are shown in
   Fig. 7.4. The convex polytope which we want is the shaded polytope in Fig. 7.5.
3. Since the ideal J is generated by linear homogeneous polynomials, we have
   $D = 1$, where D is the maximum degree of the elements in the universal Gröbner
   basis. For each Gröbner cone C_i, we take a relatively interior point $\mathbf{w}_i$ in the cone
   C_i and compute $\mathbf{s}_{(J,1,<_{\mathbf{w}_i})}$. By definition, $\mathbf{s}_{(J,1,<_{\mathbf{w}_i})}$ $(i = 1, 2, 3)$ are the points
   $[1, 1, 0]$, $[0, 1, 1]$, and $[1, 0, 1]$. Hence, we have the state polytope

Fig. 7.5 Polytope whose
normal fan is the fan in
Fig. 7.4

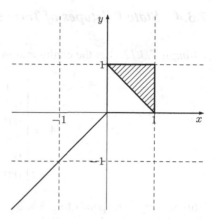

Fig. 7.6 State polytope of J

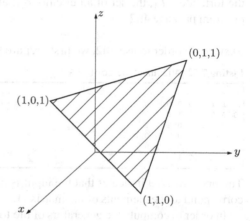

$$\text{State}(J) = \text{State}_1(J) = \text{CONV}(\{[1, 1, 0], [0, 1, 1], [1, 0, 1]\}).$$

The state polytope is shown in Fig. 7.6.

Of course, the projection of the state polytope to xy-space is the polytope in (2.). For more about the properties of state polytopes of ideals generated by linear homogeneous polynomials, please refer to [22, Proposition 2.11].

Exercise. Let the ideal be $J = \langle x + y + z + w, z + 2w \rangle$.

1. Compute the Gröbner fan of the ideal J. For a relatively interior point $\mathbf{w}$ in maximal dimensional cones of the fan, compute the initial form ideal $\text{in}_\mathbf{w}(J)$.
2. Compute the state polytope of the ideal J.

7.3.4 State Polytopes of Toric Ideals (Sect. 5.4)

Example 7.3.12. Let the configuration matrix be

$$A = \begin{bmatrix} 1 & 1 & 1 & 0 & 0 & 0 & 0 & 0 & 0 \\ 0 & 0 & 0 & 1 & 1 & 1 & 0 & 0 & 0 \\ 0 & 0 & 0 & 0 & 0 & 0 & 1 & 1 & 1 \\ 1 & 0 & 0 & 1 & 0 & 0 & 1 & 0 & 0 \\ 0 & 1 & 0 & 0 & 1 & 0 & 0 & 1 & 0 \\ 0 & 0 & 1 & 0 & 0 & 1 & 0 & 0 & 1 \end{bmatrix}.$$

This matrix corresponds to a 3×3 contingency table. Compute the generators of the toric ideal I_A, the set of all circuits $\mathscr{C}_A$, and the Graver basis Gr_A by using the program package 4ti2.

Answer. In order to use 4ti2, we first prepare the following input file.

Listing 7.52 4ti2: input file cont3x3

```
6 9
1 1 1 0 0 0 0 0 0
0 0 0 1 1 1 0 0 0
0 0 0 0 0 0 1 1 1
1 0 0 1 0 0 1 0 0
0 1 0 0 1 0 0 1 0
0 0 1 0 0 1 0 0 1
```

The first row 6 9 indicates that the input is a 6×9 matrix, and the following rows correspond to the elements of the matrix A.

In order to compute the generators of the toric ideal I_A, we execute the command groebner in 4ti2.

Listing 7.53 4ti2: executing the command groebner

```
$ groebner cont3x3
```

This program outputs the file cont3x3.gro.

Listing 7.54 4ti2: output file cont3x3.gro

```
9 9
-1   0   1   0   0   0   1   0 -1
-1   0   1   1   0 -1   0   0   0
-1   1   0   0   0   0   1 -1   0
-1   1   0   1 -1   0   0   0   0
 0 -1   1   0   0   0   0   1 -1
 0 -1   1   0   1 -1   0   0   0
 0   0   0 -1   0   1   1   0 -1
 0   0   0 -1   1   0   1 -1   0
 0   0   0   0 -1   1   0   1 -1
```

In this output file, each row vector corresponds to a binomial. For example, the second row -1 0 1 0 0 0 1 0 -1 corresponds to the binomial $x_3 x_7 - x_1 x_9$. Here, we set that the variable corresponding to the ith column is x_i. The set of these nine binomials is a reduced Gröbner basis of I_A. In fact, without such computations, we can obtain generators of the toric ideals corresponding to $I \times J$ contingency tables. Please refer to [22, Proposition 5.4].

In order to compute the set of all circuits $\mathscr{C}_A$, we execute the command circuits in 4ti2.

Listing 7.55 4ti2: executing the command circuits

```
$ circuits cont3x3
```

This program outputs the file cont3x3.cir.

Listing 7.56 4ti2: output file cont3x3.cir

```
15 9
 0  0  0  0  1 -1  0 -1  1
 0  0  0  1 -1  0 -1  1  0
 0  0  0  1  0 -1 -1  0  1
 0  1 -1 -1  0  1  1 -1  0
 0  1 -1  0 -1  1  0  0  0
 0  1 -1  0  0  0  0 -1  1
 0  1 -1  1 -1  0 -1  0  1
 1 -1  0 -1  0  1  0  1 -1
 1 -1  0 -1  1  0  0  0  0
 1 -1  0  0  0  0 -1  1  0
 1 -1  0  0  1 -1 -1  0  1
 1  0 -1 -1  0  1  0  0  0
 1  0 -1 -1  1  0  0 -1  1
 1  0 -1  0 -1  1 -1  1  0
 1  0 -1  0  0  0 -1  0  1
```

In this output file, each row vector corresponds to a binomial. The set of these 15 binomials is the set of all circuits.

In order to compute the Graver basis of I_A, we execute the command graver in 4ti2.

Listing 7.57 4ti2: executing the command graver

```
$ graver cont3x3
```

This program outputs the file cont3x3.gra. In this case, this output equals the previous output; that is, the set of all circuits equals the Graver basis.

Exercise. Let the configuration matrix be

$$A = \begin{bmatrix} 1 & 1 & 1 & 1 \\ 0 & 1 & 2 & 3 \end{bmatrix}.$$

Compute generators of the toric ideal I_A, the set of all circuits $\mathscr{C}_A$ and the Graver basis Gr_A.

Example 7.3.13. Let the configuration matrix be

$$A = \begin{bmatrix} 1 & 1 & 1 & 1 \\ 0 & 1 & 2 & 3 \end{bmatrix}.$$

Compute an upper bound for the degrees of the circuits: $\frac{1}{2}(d + 1)D(A)$. Generate candidates for the circuits and find the set of all circuits.

Answer. In this case, the configuration matrix A has $d = 2, D(A) = 3$. By Theorem 5.4.4, an upper bound for degrees of circuits is $\frac{1}{2} \cdot (2 + 1) \cdot 3 = \frac{9}{2}$, and, for a circuit f, the set VAR(f) consists of at most three elements. Generators of I_A are binomials corresponding to the vectors $\mathbf{a}_1 = [1, -2, 1, 0]$ and $\mathbf{a}_2 = [2, -3, 0, 1]$. Hence, we generate the vectors

$$k_1 \mathbf{a}_1 + k_2 \mathbf{a}_2 \ (k_i \in \mathbb{Z})$$

such that the number of nonzero components is at most three, and the sum of the positive components and the absolute value of the sum of the negative components are both less than or equal to $\frac{9}{2}$. These vectors are candidates for circuits; from them, we pick the vector g which has the minimal VAR(g), and obtain the set of all circuits $\mathscr{C}_A$.

Exercise. Let the configuration matrix be

$$A = \begin{bmatrix} 1 & 1 & 1 & 1 \\ 0 & 1 & 2 & k \end{bmatrix} \ (k \in \mathbb{N}, k \geq 3).$$

Compute the generators of the toric ideal I_A and an upper bound for the degree of circuits $\frac{1}{2}(d + 1)D(A)$. For $k = 4, 5, 6$, generate candidates for the circuits and compute the set of all circuits $\mathscr{C}_A$. For a general k, what is $\mathscr{C}_A$?

Example 7.3.14. Let the configuration matrix be

$$A = \begin{bmatrix} 1 & 1 & 1 & 1 \\ 0 & 1 & 2 & 10 \end{bmatrix}.$$

Compute the Graver basis of the toric ideal I_A by using the Lawrence lifting $\Lambda(A)$ and Theorem 5.4.11.

Answer. In order to compute a Gröbner basis, we use the command `groebner` in 4ti2. We prepare the following input file, which corresponds to the Lawrence lifting $\Lambda(A)$.

Listing 7.58 4ti2: input file `lawrence1`

```
6 8
1 1 1 1   0 0 0 0
0 1 2 10  0 0 0 0
1 0 0 0   1 0 0 0
0 1 0 0   0 1 0 0
0 0 1 0   0 0 1 0
0 0 0 1   0 0 0 1
```

By using the command `groebner`, we obtain a reduced Gröbner basis of the toric ideal $I_{\Lambda(A)}$.

Listing 7.59 4ti2: output file `lawrence1.gro`

```
11 8
-9 10  0 -1  9 -10  0  1
-8  8  1 -1  8  -8 -1  1
-7  6  2 -1  7  -6 -2  1
```

```
-6   4   3  -1   6  -4  -3   1
-5   2   4  -1   5  -2  -4   1
-4   0   5  -1   4   0  -5   1
-3  -2   6  -1   3   2  -6   1
-2  -4   7  -1   2   4  -7   1
-1  -6   8  -1   1   6  -8   1
-1   2  -1   0   1  -2   1   0
 0  -8   9  -1   0   8  -9   1
```

By Theorem 5.4.11, this output is the Graver basis $\mathrm{Gr}_{\Lambda(A)}$. The additional variables $\mathbf{y}$ for the Lawrence lifting are the 5th, 6th, 7th, and 8th columns. If 1 is substituted for y_i, these parts vanish.

Listing 7.60 Graver basis Gr_A

```
-9  10   0  -1
-8   8   1  -1
-7   6   2  -1
-6   4   3  -1
-5   2   4  -1
-4   0   5  -1
-3  -2   6  -1
-2  -4   7  -1
-1  -6   8  -1
-1   2  -1   0
 0  -8   9  -1
```

Hence, from these vectors, we obtain the Graver basis

$$\mathrm{Gr}_A = \{x_2^{10} - x_1^9 x_4,\ x_2^8 x_3 - x_1^8 x_4,\ x_2^6 x_3^2 - x_1^7 x_4,\ x_2^4 x_3^3 - x_1^6 x_4,\ x_2^2 x_3^4 - x_1^5 x_4,\ x_3^5 - x_1^4 x_4,$$

$$x_3^6 - x_1^3 x_2^2 x_4,\ x_3^7 - x_1^2 x_2^4 x_4,\ x_3^8 - x_1 x_2^6 x_4,\ x_2^2 - x_1 x_3,\ x_3^9 - x_2^8 x_4\}.$$

Exercise. For the configuration matrix in Example 7.3.12, compute the Graver basis by using the Lawrence lifting.

Example 7.3.15. Let the configuration matrix be

$$A = \begin{bmatrix} 1 & 1 & 1 & 1 \\ 0 & 1 & 2 & 3 \end{bmatrix}.$$

By using the Graver basis Gr_A, compute the candidates for the Gröbner degrees.

Answer. From the definition of a Gröbner degree and $\mathscr{U}_A \subset \mathrm{Gr}_A$, for a binomial $\mathbf{x}^{\mathbf{u}} - \mathbf{x}^{\mathbf{v}} \in \mathrm{Gr}_A$, the vector $A\mathbf{u} = A\mathbf{v}$ is a candidate for a Gröbner degree. The Graver basis of the toric ideal I_A is

$$\mathrm{Gr}_A = \{x_1^2 x_4 - x_2^3,\ x_1 x_4 - x_2 x_3,\ x_2 x_4 - x_3^2,\ x_1 x_4^2 - x_3^3\}.$$

For a binomial $\mathbf{x}^{\mathbf{u}} - \mathbf{x}^{\mathbf{v}} \in \mathrm{Gr}_A$, we compute the vector $A\mathbf{u}$ and obtain the following candidates for a Gröbner degree:

$$\begin{bmatrix} 3 \\ 3 \end{bmatrix},\ \begin{bmatrix} 2 \\ 2 \end{bmatrix},\ \begin{bmatrix} 2 \\ 3 \end{bmatrix},\ \begin{bmatrix} 2 \\ 4 \end{bmatrix},\ \begin{bmatrix} 3 \\ 6 \end{bmatrix}.$$

In fact, in this case, the Graver basis Gr_A coincides with the universal Gröbner basis $\mathscr{U}_A$. The above vectors are Gröbner degrees.

Example 7.3.16. Let the configuration matrix be

$$A = \begin{bmatrix} 1 & 1 & 1 & 1 \\ 0 & 1 & 2 & 3 \end{bmatrix}$$

and let the vectors be

$$\mathbf{b}_1 = \begin{bmatrix} 3 \\ 3 \end{bmatrix}, \mathbf{b}_2 = \begin{bmatrix} 2 \\ 2 \end{bmatrix}, \mathbf{b}_3 = \begin{bmatrix} 2 \\ 3 \end{bmatrix}, \mathbf{b}_4 = \begin{bmatrix} 2 \\ 4 \end{bmatrix}, \mathbf{b}_5 = \begin{bmatrix} 3 \\ 6 \end{bmatrix}.$$

Compute the nonnegative integer solutions $\mathbf{u} \in \mathbb{Z}_{\geq 0}^4$ for the linear equations $A\mathbf{u} = \mathbf{b}_i$ for $i = 1, \ldots, 5$.

Answer. By a method similar to that used in Example 7.2.5, we obtain the following solutions:

- For $i = 1$, $[2, 0, 0, 1], [1, 1, 1, 0], [0, 3, 0, 0]$;
- For $i = 2$, $[1, 0, 1, 0], [0, 2, 0, 0]$;
- For $i = 3$, $[1, 0, 0, 1], [0, 1, 1, 0]$;
- For $i = 4$, $[0, 1, 0, 1], [0, 0, 2, 0]$;
- For $i = 5$, $[1, 0, 0, 2], [0, 1, 1, 1], [0, 0, 3, 0]$.

Example 7.3.17. Let the configuration matrix be

$$A = \begin{bmatrix} 1 & 1 & 1 & 1 \\ 0 & 1 & 2 & 3 \end{bmatrix}.$$

By using the answer to Example 7.3.16, compute the state polytope of the toric ideal I_A.

Answer. Using a method similar to that used in Example 7.3.6, compute the Minkowski sum of Gröbner fibers Fiber($\mathbf{b}_i$). The Minkowski sum is the convex polytope

$$\mathrm{CONV}(\{[5, 1, 1, 5], [5, 0, 3, 4], [4, 0, 6, 2], [4, 3, 0, 5], [2, 2, 8, 0], [2, 6, 0, 4],$$
$$[0, 8, 2, 2], [0, 6, 6, 0]\}).$$

By Theorem 5.4.12, this polytope is the state polytope of the toric ideal I_A. We can use the software program polymake to obtain the linear inequalities defining the state polytope

$$0 \leq x_1 \leq 5, \ 0 \leq x_2, \ 4 \leq x_1 + x_2 \leq 8, \ 6 \leq 2x_1 + x_2 \leq 11, \ \frac{3}{2}x_1 + x_2 \leq 9,$$

$$x_3 = 18 - 3x_1 - 2x_2, \ x_4 = -6 + 2x_1 + x_2.$$

This state polytope is shown in Fig. 7.7.

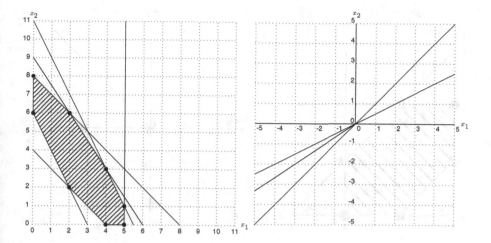

Fig. 7.7 State polytope in the plane $x_3 = 18 - 3x_1 - 2x_2$, $x_4 = -6 + 2x_1 + x_2$ and its normal fan

Example 7.3.18. Let the configuration matrix be

$$A = \begin{bmatrix} 1 & 1 & 1 & 0 & 0 & 0 \\ 0 & 0 & 0 & 1 & 1 & 1 \\ 1 & 0 & 0 & 1 & 0 & 0 \\ 0 & 1 & 0 & 0 & 1 & 0 \\ 0 & 0 & 1 & 0 & 0 & 1 \end{bmatrix}.$$

This matrix corresponds to a 2×3 contingency table. Compute the Graver basis Gr_A, the Gröbner degrees $\mathbf{b}_i$, the Gröbner fibers Fiber($\mathbf{b}_i$), and the state polytope.

Answer. We can compute these by using a method similar to that used in the previous example. We show only the results. The Graver basis of the toric ideal I_A is

$$Gr_A = \{x_{11}x_{22} - x_{12}x_{21}, \ x_{11}x_{23} - x_{13}x_{21}, \ x_{12}x_{23} - x_{13}x_{22}\}.$$

Here, we set the variables of the polynomial ring to be $x_{11}, x_{12}, x_{13}, x_{21}, x_{22}$, and x_{23}. The Gröbner degrees are

$$\mathbf{b}_1 = [1, 1, 1, 1, 0]^T, \mathbf{b}_2 = [1, 1, 1, 0, 1]^T, \mathbf{b}_3 = [1, 1, 0, 1, 1]^T.$$

We then compute the nonnegative integer solutions for the linear equations $A\mathbf{u} = \mathbf{b}_i$; that is, the Gröbner fibers Fiber($\mathbf{b}_i$):

$$\text{Fiber}(\mathbf{b}_1) = \{[1, 0, 0, 0, 1, 0], [0, 1, 0, 1, 0, 0]\},$$

$$\text{Fiber}(\mathbf{b}_2) = \{[1, 0, 0, 0, 0, 1], [0, 0, 1, 1, 0, 0]\},$$

$$\text{Fiber}(\mathbf{b}_3) = \{[0, 1, 0, 0, 0, 1], [0, 0, 1, 0, 1, 0]\}.$$

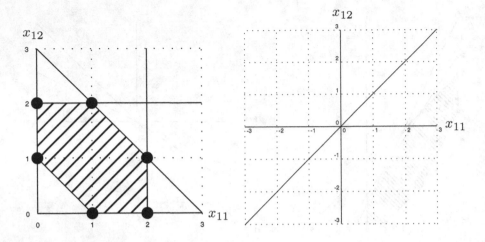

Fig. 7.8 State polytope in the plane $x_{11} + x_{12} + x_{13} = 3, x_{11} + x_{21} = 2, x_{12} + x_{22} = 2, x_{13} + x_{23} = 2$ and its normal fan

Next, we compute the Minkowski sum of the Gröbner fibers:

$$\sum_{i=1}^{3} \text{Fiber}(\mathbf{b}_i) = \text{CONV}(\{[2, 1, 0, 0, 1, 2], [2, 0, 1, 0, 2, 1], [1, 0, 2, 1, 2, 0], [1, 2, 0, 1, 0, 2],$$

$$[0, 2, 1, 2, 0, 1], [0, 1, 2, 2, 1, 0]\}).$$

This polytope is the state polytope of I_A. The linear inequalities defining the state polytope are

$$x_{11} + x_{12} + x_{13} = 3, \ x_{11} + x_{21} = 2, \ x_{12} + x_{22} = 2, \ x_{13} + x_{23} = 2,$$

$$0 \le x_{11} \le 2, \ 0 \le x_{12} \le 2, \ 1 \le x_{11} + x_{12} \le 3.$$

The state polytope is shown in Fig. 7.8. This polytope is isomorphic to the permutahedron Π_2.

Exercise. Choose a configuration matrix A corresponding to a $2 \times n$ contingency table. For $n = 3, 4, 5$, compute the state polytope. For a general n, what is the state polytope?

7.3.5   Triangulations of Convex Polytopes and Gröbner Bases (Sect. 5.5)

Example 7.3.19. Let the configuration matrix be

$$A = \begin{bmatrix} 1 & 1 & 1 \\ 0 & 1 & 2 \end{bmatrix}.$$

1. Compute the toric ideal I_A.
2. For the weight vector $\mathbf{w} = [2, 1, 2]$, compute the initial form ideal $\mathrm{in}_\mathbf{w}(I_A)$.
3. For the weight vector $\mathbf{w} = [2, 1, 2]$, find the regular triangulation $\Delta_\mathbf{w}$.
4. For the weight vector $\mathbf{w} = [1, 2, 1]$, find the regular triangulation $\Delta_\mathbf{w}$.

Answer. 1. The toric ideal is $I_A = \langle x_1 x_3 - x_2^2 \rangle$.
2. The initial form ideal is $\mathrm{in}_\mathbf{w}(I_A) = \langle x_1 x_3 \rangle$.
3. The radical of the initial form ideal I_A is $\sqrt{\mathrm{in}_\mathbf{w}(I_A)} = \langle x_1 x_3 \rangle = \langle x_1 \rangle \cap \langle x_3 \rangle$. By Corollary 5.5.7, the maximal simplices in the regular triangulation $\Delta_\mathbf{w}$ are 23 and 12, where, for example, the symbol 23 means the simplex $\underline{\mathrm{CONV}}(\{\mathbf{a}_2, \mathbf{a}_3\})$.
4. The initial form ideal is $\mathrm{in}_\mathbf{w}(I_A) = \langle x_2^2 \rangle$, and the radical is $\sqrt{\mathrm{in}_\mathbf{w}(I_A)} = \langle x_2 \rangle$. By Corollary 5.5.7, the maximal simplex in the regular triangulation $\Delta_\mathbf{w}$ is 13.

Exercise. Let the configuration matrix be

$$A = \begin{bmatrix} 1 & 1 & 1 & 1 \\ 0 & 1 & 0 & 1 \\ 0 & 0 & 1 & 1 \end{bmatrix}.$$

1. Compute the toric ideal I_A.
2. For the weight vector $\mathbf{w} = [1, 0, 0, 2]$, compute the initial form ideal $\mathrm{in}_\mathbf{w}(I_A)$.
3. Find the regular triangulation $\Delta_\mathbf{w}$.

Example 7.3.20. Let the configuration matrix be

$$A = \begin{bmatrix} 1 & 1 & 1 & 0 & 0 & 0 \\ 0 & 0 & 0 & 1 & 1 & 1 \\ 1 & 0 & 0 & 1 & 0 & 0 \\ 0 & 1 & 0 & 0 & 1 & 0 \\ 0 & 0 & 1 & 0 & 0 & 1 \end{bmatrix} = [\mathbf{a}_1, \mathbf{a}_2, \mathbf{a}_3, \mathbf{a}_4, \mathbf{a}_5, \mathbf{a}_6]$$

corresponding to a 2×3 contingency table.

1. Find a triangulation of A by using a Gröbner basis of the toric ideal I_A.
2. Compute the Gröbner fan of the toric ideal I_A.
3. Generate all the regular triangulations of A.

Answer. 1. The toric ideal I_A is

$$I_A = \langle -x_2x_6 + x_3x_5, -x_1x_6 + x_3x_4, -x_1x_5 + x_2x_4 \rangle.$$

For the use of a computer to determine toric ideals, please refer to Example 7.3.12 (4ti2) or Example 7.2.4 (Asir). Here, we take the weight vector $\mathbf{w} = [2, 1, 0, 0, 0, 0]$ and compute the initial form ideal

$$\text{in}_{\mathbf{w}}(I_A) = \langle x_2x_6, -x_1x_6, -x_1x_5 \rangle.$$

For how to compute the initial form ideal, please refer to Example 7.3.8. The prime decomposition of the radical ideal of the initial form ideal is

$$\sqrt{\text{in}_{\mathbf{w}}(I_A)} = \langle x_1, x_2 \rangle \cap \langle x_1, x_6 \rangle \cap \langle x_5, x_6 \rangle.$$

We can use the function `primedec` in Risa/Asir to compute the prime decompositions of ideals. By Corollary 5.5.7, the maximal simplices of the regular triangulation $\Delta_{\mathbf{w}}$ are 3456, 2345, and 1234. Here, the symbol 3456 means the simplex $\text{CONV}(\{\mathbf{a}_3, \mathbf{a}_4, \mathbf{a}_5, \mathbf{a}_6\})$.

2. We use the software program Gfan to compute the Gröbner fan. We first prepare the input file corresponding to generators of the ideal I_A.

Listing 7.61 Gfan: input file `gfan_cont2x3.txt`

```
Q[x1,x2,x3,x4,x5,x6]
{-x2*x6+x3*x5,-x1*x6+x3*x4,-x1*x5+x2*x4}
```

We then execute the Gfan command `gfan`, using `gfan_cont2x3.txt` as input; the output contains all the reduced Gröbner bases of I_A in the file `gfan_cont2x3.out`.

Listing 7.62 Gfan: executing the command `gfan`

```
$ gfan < gfan_cont2x3.txt > gfan_cont2x3.out
```

Listing 7.63 Gfan: output file `gfan_cont2x3.out`

```
Q[x1,x2,x3,x4,x5,x6]
{{
x2*x6-x3*x5,
x1*x6-x3*x4,
x1*x5-x2*x4}
,
{
x3*x5-x2*x6,
x1*x6-x3*x4,
x1*x5-x2*x4}
,
{
x3*x5-x2*x6,
x3*x4-x1*x6,
x1*x5-x2*x4}
```

```
,
{
x3*x5-x2*x6,
x3*x4-x1*x6,
x2*x4-x1*x5}
,
{
x2*x6-x3*x5,
x2*x4-x1*x5,
x1*x6-x3*x4}
,
{
x3*x4-x1*x6,
x2*x6-x3*x5,
x2*x4-x1*x5}
}
```

We obtain six reduced Gröbner bases for I_A. In order to compute the Gröbner cones from these reduced Gröbner bases, we use the Gfan command gfan_groebnercone. For example, we take the reduced Gröbner basis $\{x_3x_4 - x_1x_6, x_2x_6 - x_3x_5, x_2x_4 - x_1x_5\}$ in the output file gfan_cont2x3.out, and compute its Gröbner cone. We prepare the following input file cont2x3_6.gb to use the Gfan command gfan_groebner.

Listing 7.64 Gfan: input file cont2x3_6.gb

```
Q[x1,x2,x3,x4,x5,x6]
{x3*x4-x1*x6, x2*x6-x3*x5, x2*x4-x1*x5}
```

We then execute the Gfan command gfan_groebnercone.

Listing 7.65 Gfan: computing the Gröbner cone

```
$ gfan_groebnercone < cont2x3_6.gb
LP algorithm being used: "cddgmp".
_application PolyhedralCone
_version 2.2
_type PolyhedralCone

AMBIENT_DIM
6

DIM
6

IMPLIED_EQUATIONS

LINEALITY_DIM
4

LINEALITY_SPACE
1 0 0 0 -1 -1
0 1 0 0 1 0
0 0 1 0 0 1
0 0 0 1 1 1

FACETS
-1 0 1 1 0 -1
0 1 -1 0 -1 1

RELATIVE_INTERIOR_POINT
-1 1 0 0 0 0
```

The property RELATIVE_INTERIOR_POINT indicates a relatively interior point of the Gröbner cone. In this case, we take the point $\mathbf{w} = [-1, 1, 0, 0, 0, 0]$. The two row vectors with the property FACETS are the normal vectors of the hyperplanes defining the facets of the Gröbner cone. In other words, the linear inequalities defining the Gröbner cone are

$$-w_1 + w_3 + w_4 - w_6 > 0,$$

$$w_2 - w_3 - w_5 + w_6 > 0.$$

In a similar way, for each reduced Gröbner basis $\mathcal{G}_i$, we obtain a relative interior point $\mathbf{w}_i$ of the Gröbner cone corresponding to $\mathcal{G}_i$:

$$\mathcal{G}_1 = \{x_2x_6 - x_3x_5, x_1x_6 - x_3x_4, x_1x_5 - x_2x_4\}, \quad \mathbf{w}_1 = [2, 1, 0, 0, 0, 0];$$
$$\mathcal{G}_2 = \{x_3x_5 - x_2x_6, x_1x_6 - x_3x_4, x_1x_5 - x_2x_4\}, \quad \mathbf{w}_2 = [1, -1, 0, 0, 0, 0];$$
$$\mathcal{G}_3 = \{x_3x_5 - x_2x_6, x_3x_4 - x_1x_6, x_1x_5 - x_2x_4\}, \quad \mathbf{w}_3 = [-1, -2, 0, 0, 0, 0];$$
$$\mathcal{G}_4 = \{x_3x_5 - x_2x_6, x_3x_4 - x_1x_6, x_2x_4 - x_1x_5\}, \quad \mathbf{w}_4 = [-2, -1, 0, 0, 0, 0];$$
$$\mathcal{G}_5 = \{x_3x_4 - x_1x_6, x_2x_6 - x_3x_5, x_2x_4 - x_1x_5\}, \quad \mathbf{w}_5 = [-1, 1, 0, 0, 0, 0];$$
$$\mathcal{G}_6 = \{x_2x_6 - x_3x_5, x_2x_4 - x_1x_5, x_1x_6 - x_3x_4\}, \quad \mathbf{w}_6 = [1, 2, 0, 0, 0, 0].$$

Here, for each binomial, the first term is the initial form with respect to the weight vector $\mathbf{w}_i$.

3. From the previous result (2.), we can obtain all the initial form ideals $\mathrm{in}_\mathbf{w}(I_A)$ which are generated by the monomials.

$$\mathrm{in}_{\mathbf{w}_1}(I_A) = \langle x_2x_6, x_1x_6, x_1x_5 \rangle, \quad \mathbf{w}_1 = [2, 1, 0, 0, 0, 0];$$
$$\mathrm{in}_{\mathbf{w}_2}(I_A) = \langle x_3x_5, x_1x_6, x_1x_5 \rangle, \quad \mathbf{w}_2 = [1, -1, 0, 0, 0, 0];$$
$$\mathrm{in}_{\mathbf{w}_3}(I_A) = \langle x_3x_5, x_3x_4, x_1x_5 \rangle, \quad \mathbf{w}_3 = [-1, -2, 0, 0, 0, 0];$$
$$\mathrm{in}_{\mathbf{w}_4}(I_A) = \langle x_3x_5, x_3x_4, x_2x_4 \rangle, \quad \mathbf{w}_4 = [-2, -1, 0, 0, 0, 0];$$
$$\mathrm{in}_{\mathbf{w}_5}(I_A) = \langle x_3x_4, x_2x_6, x_2x_4 \rangle, \quad \mathbf{w}_5 = [-1, 1, 0, 0, 0, 0];$$
$$\mathrm{in}_{\mathbf{w}_6}(I_A) = \langle x_2x_6, x_2x_4, x_1x_6 \rangle, \quad \mathbf{w}_6 = [1, 2, 0, 0, 0, 0].$$

For each initial form ideal $\mathrm{in}_{\mathbf{w}_i}(I_A)$ $(i = 1, \ldots, 6)$, we compute the regular triangulation of A in the same way as in the previous example (1.). For each weight vector $\mathbf{w}_i$ $(i = 1, \ldots, 6)$, the maximal simplices of the regular triangulation $\Delta_{\mathbf{w}_i}$ are as follows:

$$3456, 2345, 1234, \quad \mathbf{w}_1 = [2, 1, 0, 0, 0, 0];$$
$$2456, 2346, 1234, \quad \mathbf{w}_2 = [1, -1, 0, 0, 0, 0];$$
$$2456, 1246, 1236, \quad \mathbf{w}_3 = [-1, -2, 0, 0, 0, 0];$$
$$1456, 1256, 1236, \quad \mathbf{w}_4 = [-2, -1, 0, 0, 0, 0];$$
$$1456, 1356, 1235, \quad \mathbf{w}_5 = [-1, 1, 0, 0, 0, 0];$$
$$3456, 1345, 1235, \quad \mathbf{w}_6 = [1, 2, 0, 0, 0, 0].$$

Fig. 7.9 All regular
triangulations of A

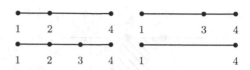

Example 7.3.21. Let the configuration matrix be

$$A = \begin{bmatrix} 1 & 1 & 1 & 1 \\ 0 & 1 & 2 & 3 \end{bmatrix}.$$

Find all the regular triangulations and sketch the triangulations.

Answer. In the same way as in Example 7.3.20, we can obtain the Gröbner bases $\mathscr{G}_i$ $(i = 1, \ldots, 8)$ and the relatively interior points $\mathbf{w}_i$ of the Gröbner cones corresponding to $\mathscr{G}_i$. The results are as follows:

$$\mathscr{G}_1 = \langle x_2 x_4 - x_3^2, x_1 x_4 - x_2 x_3, x_1 x_3 - x_2^2 \rangle, \qquad \mathbf{w}_1 = [3, 1, 0, 0];$$
$$\mathscr{G}_2 = \langle x_3^2 - x_2 x_4, x_1 x_4 - x_2 x_3, x_1 x_3 - x_2^2 \rangle, \qquad \mathbf{w}_2 = [0, -1, 0, 0];$$
$$\mathscr{G}_3 = \langle x_3^2 - x_2 x_4, x_2 x_3 - x_1 x_4, x_1 x_3 - x_2^2, x_1^2 x_4 - x_2^3 \rangle, \qquad \mathbf{w}_3 = [-4, -3, 0, 0];$$
$$\mathscr{G}_4 = \langle x_3^2 - x_2 x_4, x_2 x_3 - x_1 x_4, x_2^3 - x_1^2 x_4, x_1 x_3 - x_2^2 \rangle, \qquad \mathbf{w}_4 = [-5, -3, 0, 0];$$
$$\mathscr{G}_5 = \langle x_3^2 - x_2 x_4, x_2 x_3 - x_1 x_4, x_2^2 - x_1 x_3 \rangle, \qquad \mathbf{w}_5 = [-3, -1, 0, 0];$$
$$\mathscr{G}_6 = \langle x_2 x_4 - x_3^2, x_2^2 - x_1 x_3, x_1 x_4 - x_2 x_3 \rangle, \qquad \mathbf{w}_6 = [3, 2, 0, 0];$$
$$\mathscr{G}_7 = \langle x_2 x_4 - x_3^2, x_2 x_3 - x_1 x_4, x_2^2 - x_1 x_3, x_1 x_4^2 - x_3^3 \rangle, \qquad \mathbf{w}_7 = [1, 2, 0, 0];$$
$$\mathscr{G}_8 = \langle x_3^3 - x_1 x_4^2, x_2 x_4 - x_3^2, x_2 x_3 - x_1 x_4, x_2^2 - x_1 x_3 \rangle, \qquad \mathbf{w}_8 = [-1, 1, 0, 0].$$

Here, for each binomial, the first term is the initial form with respect to the weight vector.

For the weight vector $\mathbf{w}_i$ $(i = 1, \ldots, 8)$, the maximal simplices of the regular triangulation $\Delta_{\mathbf{w}_i}$ are as follows:

$$\Delta_{\mathbf{w}_1} : 34, 23, 12; \quad \Delta_{\mathbf{w}_2} : 24, 12; \quad \Delta_{\mathbf{w}_3} : 24, 12; \quad \Delta_{\mathbf{w}_4} : 14;$$
$$\Delta_{\mathbf{w}_5} : 14; \quad \Delta_{\mathbf{w}_6} : 34, 13; \quad \Delta_{\mathbf{w}_7} : 34, 13; \quad \Delta_{\mathbf{w}_8} : 14.$$

All the regular triangulations of A are shown in Fig. 7.9.

Exercise. For the configuration matrix A in Example 7.3.12, find all the regular triangulations.

Example 7.3.22. Let the configuration matrix be

$$A = \begin{bmatrix} 1 & 1 & 1 & 1 \\ 0 & 1 & 2 & 3 \end{bmatrix}.$$

From the results of Example 7.3.21, compute the secondary polytope $\Sigma(A)$.

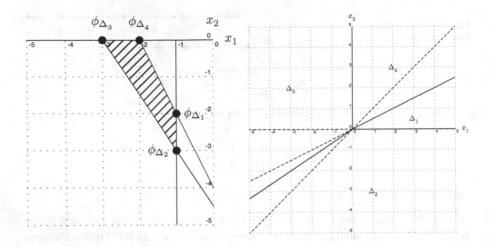

Fig. 7.10 Secondary polytope $\Sigma(A)$ in the plane $x_3 = -9 - 3x_1 - 2x_2$, $x_4 = 3 + 2x_1 + x_2$, its normal fan, and the Gröbner fan of the toric ideal I_A

Answer. From the result of Example 7.3.21, the regular triangulations of A are

$$\Delta_1 : 34, 23, 12; \Delta_2 : 24, 12; \Delta_3 : 14; \Delta_4 : 34, 13.$$

These are the maximal simplices of the regular triangulations. We next compute the normalized volume of each simplex:

$$\text{VOL}(12) = 1; \text{VOL}(13) = 2; \text{VOL}(14) = 3;$$

$$\text{VOL}(23) = 1; \text{VOL}(24) = 2; \text{VOL}(34) = 1.$$

Using the definition of the GKZ vector, we compute the points ϕ_{Δ_i} ($i = 1, \ldots, 4$):

$$\phi_{\Delta_1} = [1, 2, 2, 1]; \phi_{\Delta_2} = [1, 3, 0, 2]; \phi_{\Delta_3} = [3, 0, 0, 3]; \phi_{\Delta_4} = [2, 0, 3, 1].$$

Hence, we obtain the secondary polytope

$$\Sigma(A) = \text{CONV}(\{[-1, -2, -2, -1], [-1, -3, 0, -2], [-3, 0, 0, -3], [-2, 0, -3, -1]\}).$$

The linear inequalities defining the secondary polytope are

$$x_3 = -9 - 3x_1 - 2x_2, \ x_4 = 3 + 2x_1 + x_2,$$

$$x_1 \le -1, \ x_2 \le 0, \ x_2 \ge -\frac{3}{2}x_1 - \frac{9}{2}, \ x_2 \le -2x_1 - 4.$$

The secondary polytope is shown in Fig. 7.10. For more about how to compute these linear inequalities, please refer to Example 7.3.3.

Example 7.3.23. Consider the configuration matrix A in Example 7.3.20, and from the results of this example, compute the secondary polytope $\Sigma(A)$.

Answer. We can compute this secondary polytope $\Sigma(A)$ in the same way as that in the previous example. We only show the results. The regular triangulations of A are

$$\Delta_1 : 3456, 2345, 1234; \Delta_2 : 2456, 2346, 1234; \Delta_3 : 2456, 1246, 1236;$$

$$\Delta_4 : 1456, 1256, 1236; \Delta_5 : 1456, 1356, 1235; \Delta_6 : 3456, 1345, 1235.$$

These are the maximal simplices of regular triangulations. Since A is unimodular, the normalized volume of each simplex of A is 1. We compute the points ϕ_{Δ_i}:

$$\phi_{\Delta_1} = [1, 2, 3, 3, 2, 1]; \phi_{\Delta_2} = [1, 3, 2, 3, 1, 2]; \phi_{\Delta_3} = [2, 3, 1, 2, 1, 3];$$

$$\phi_{\Delta_4} = [3, 2, 1, 1, 2, 3]; \phi_{\Delta_5} = [3, 1, 2, 1, 3, 2]; \phi_{\Delta_6} = [2, 1, 3, 2, 3, 1].$$

The secondary polytope is

$$\Sigma(A) = \mathrm{CONV}(\{ - [1, 2, 3, 3, 2, 1], -[1, 3, 2, 3, 1, 2], -[2, 3, 1, 2, 1, 3],$$
$$- [3, 2, 1, 1, 2, 3], -[3, 1, 2, 1, 3, 2], -[2, 1, 3, 2, 3, 1]\}).$$

The linear inequalities defining the secondary polytope are

$$x_3 = -6 - x_1 - x_2, \ x_4 = -4 - x_1, \ x_5 = -4 - x_2, \ x_6 = 2 + x_1 + x_2,$$
$$- 3 \le x_1 \le -1, \ -3 \le x_2 \le -1, \ -5 \le x_1 + x_2 \le -3.$$

The secondary polytope $\Sigma(A)$ is isomorphic to the permutahedron Π_2, which is the state polytope of the toric ideal I_A. The secondary polytope $\Sigma(A)$ is shown in Fig. 7.11, and the regular triangulations corresponding to the vertices are shown in Fig. 7.12.

Exercise. 1. Let the configuration matrix be $A = \begin{bmatrix} 1 & 1 & 1 & 1 & 1 \\ 0 & 1 & 2 & 3 & 4 \end{bmatrix}$. Compute the secondary polytope $\Sigma(A)$.

2. Let the configuration matrix A correspond to a 3×3 contingency table. Compute the secondary polytope $\Sigma(A)$.

Example 7.3.24. Let the configuration matrix be

$$A = \begin{bmatrix} 4 & 0 & 0 & 2 & 1 & 1 \\ 0 & 4 & 0 & 1 & 2 & 1 \\ 0 & 0 & 4 & 1 & 1 & 2 \end{bmatrix} = [\mathbf{a}_0, \ldots, \mathbf{a}_5].$$

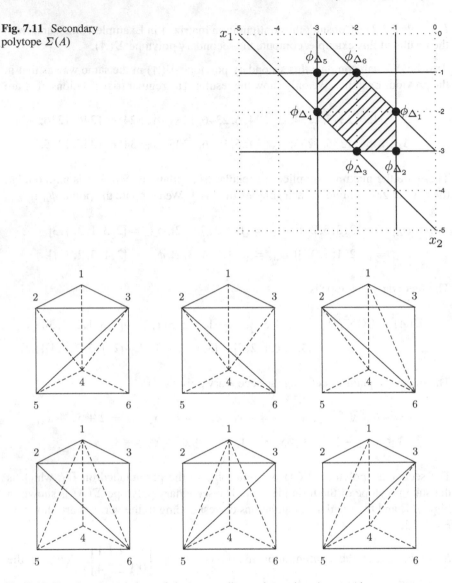

Fig. 7.11 Secondary polytope $\Sigma(A)$

Fig. 7.12 Regular triangulations of A corresponding to the vertices $\phi_{\Delta_1}, \ldots, \phi_{\Delta_6}$

1. Compute the regular triangulation $\Delta_{\mathbf{w}}$ with respect to the weight vector $\mathbf{w} = [3, 2, 1, 0, 0, 0]$, by the geometric method, using the software program polymake.
2. Use the software program TOPCOM to find all of the triangulations.

Answer. These six points $\mathbf{a}_0, \mathbf{a}_1, \ldots, \mathbf{a}_5$ lie on the plane $x + y + z = 4$ in the xyz-space, as shown in Fig. 7.13 and Fig. 7.14. Note that we index the points by $0, 1, \ldots, 5$ in order to match them to the results of the software programs polymake and TOPCOM.

Fig. 7.13 Points
$a_0, a_1, \ldots, a_5$ in the plane
$x + y + z = 4$

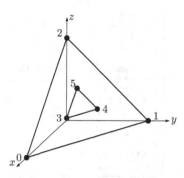

Fig. 7.14 Points
$a_0, a_1, \ldots, a_5$

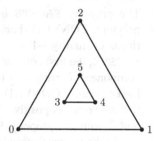

1. We lift the configuration matrix A into the next dimension by using the the height vector $\mathbf{w} = [3, 2, 1, 0, 0, 0]$; that is, we increase the dimensionality A by one. We consider resulting configuration matrix

$$\widehat{A} = \begin{bmatrix} 4 & 0 & 0 & 2 & 1 & 1 \\ 0 & 4 & 0 & 1 & 2 & 1 \\ 0 & 0 & 4 & 1 & 1 & 2 \\ 3 & 2 & 1 & 0 & 0 & 0 \end{bmatrix}.$$

The regular triangulation $\Delta_{\mathbf{w}}$ is now the projection of the set of all the lower faces of $\mathrm{CONV}(\widehat{A})$. We use the software program polymake to compute this. We prepare the following input file corresponding to the matrix $\widehat{A}$.

Listing 7.66 Polymake: input file `reg_tri_1.p`

```
POINTS
1 4 0 0 3
1 0 4 0 2
1 0 0 4 1
1 2 1 1 0
1 1 2 1 0
1 1 1 2 0
```

We compute the vertices and facets of $\mathrm{CONV}(\widehat{A})$.

Listing 7.67 Polymake: computing the vertices and the facets

```
$ polymake reg_tri_1.p VERTICES
VERTICES
1 4 0 0 3
1 0 4 0 2
```

```
1 0 0 4 1
1 2 1 1 0
1 1 2 1 0
1 1 1 2 0

$ polymake reg_tri_1.p FACETS
FACETS
-1 0 1 0 1
0 0 0 0 1
6 -2 -2 0 1
-4 6 -1 0 4
-1 1 0 0 1
4 2 1 0 -4
8 -11/4 -5/2 0 1
-2 -1 4 0 2
```

The property FACETS indicates the linear inequalities defining the convex polytope $\text{CONV}(\widehat{A})$. For example, the 0th row -1 0 1 0 1 represents the linear inequality $-1 + x_2 + x_4 \geq 0$. Hence, for each row with the property FACETS, the first component corresponds to a constant term, and the remaining components corresponds to the normal vector of the facet. In order to take the lower facets of $\text{CONV}(\widehat{A})$, we take the facets for which the last component of the normal vector is positive. Thus, the facets corresponding to rows $0, 1, 2, 3, 4, 6$, and 7 with the property FACETS are the lower facets of $\text{CONV}(\widehat{A})$. We can obtain the vertices on each facet as follows.

Listing 7.68 Polymake: computing the vertices of the facets

```
$ polymake reg_tri_1.p VERTICES_IN_FACETS
VERTICES_IN_FACETS
{2 3 5}
{3 4 5}
{1 3 4}
{1 2 4}
{2 4 5}
{0 1 2}
{0 1 3}
{0 2 3}
```

For example, the second row $\{1, 3, 4\}$ indicates that the vertices corresponding to rows $1, 3$, and 4 with the property VERTICES lie on the facet corresponding to the second row with the property FACETS. In other words, the vertices $[0, 4, 0, 2], [2, 1, 1, 0]$, and $[1, 2, 1, 0]$ lie on the facets with the normal vector $[-2, -2, 0, 1]$. Since the lower facets correspond to rows $0, 1, 2, 3, 4, 6$, and 7 with FACETS, we take rows $0, 1, 2, 3, 4, 6$, and 7 with VERTICES_IN_FACETS. Thus, the simplices

$$235, 345, 134, 124, 245, 013, 023$$

are the maximal simplices of the regular triangulation $\Delta_{\mathbf{w}}$; the regular triangulation is shown in Fig. 7.15. Here, we define $ijk = \text{CONV}(\{\mathbf{a}_i, \mathbf{a}_j, \mathbf{a}_k\})$.

2. By using the software program TOPCOM, we can obtain all the triangulations of a configuration matrix. We prepare the following input file reg_tri_1.top.

Fig. 7.15 Regular
triangulation with respect to
$\mathbf{w} = [3, 2, 1, 0, 0, 0]$

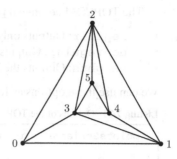

Listing 7.69 TOPCOM: input file `reg_tri_1.top`

```
[[4,0,0],
 [0,4,0],
 [0,0,4],
 [2,1,1],
 [1,2,1],
 [1,1,2]]
```

First, we use the TOPCOM command `points2nalltriangs` to count
the number of triangulations of A. Next, we generate all the triangulations
connected to the regular triangulations, by using the TOPCOM command
`points2triangs`.

Listing 7.70 TOPCOM: generate triangulations

```
$ points2nalltriangs < reg_tri_1.top
Evaluating Commandline Options ...
... done.
18

$ points2triangs < reg_tri_1.top
Evaluating Commandline Options ...
... done.
T[1]:=[6,3:{{0,1,2}}];
T[2]:=[6,3:{{1,2,4},{0,2,4},{0,1,4}}];
T[3]:=[6,3:{{1,2,5},{0,2,5},{0,1,5}}];
T[4]:=[6,3:{{1,2,3},{0,2,3},{0,1,3}}];
T[5]:=[6,3:{{1,2,4},{0,2,3},{0,1,3},{2,3,4},{1,3,4}}];
T[6]:=[6,3:{{1,2,5},{0,2,3},{0,1,3},{2,3,5},{1,3,5}}];
T[7]:=[6,3:{{1,2,4},{0,1,4},{0,2,3},{2,3,4},{0,3,4}}];
T[8]:=[6,3:{{1,2,4},{0,1,4},{0,2,5},{2,4,5},{0,4,5}}];
T[9]:=[6,3:{{1,2,5},{0,2,5},{0,1,3},{1,3,5},{0,3,5}}];
T[10]:=[6,3:{{0,1,4},{1,2,5},{0,2,5},{0,4,5},{1,4,5}}];
T[11]:=[6,3:{{1,2,4},{0,1,4},{0,2,5},{0,3,4},{2,4,5},{0,3,5},{3,4,5}}];
T[12]:=[6,3:{{1,2,5},{0,2,5},{0,1,3},{1,3,5},{0,3,5},{1,4,5},{3,4,5}}];
T[13]:=[6,3:{{1,2,4},{0,2,3},{0,1,3},{1,3,4},{2,3,5},{2,4,5},{3,4,5}}];
T[14]:=[6,3:{{1,2,5},{0,2,3},{0,1,3},{1,3,4},{2,3,5},{1,4,5},{3,4,5}}];
T[15]:=[6,3:{{1,2,4},{0,1,4},{0,2,3},{2,3,5},{0,3,4},{2,4,5},{3,4,5}}];
T[16]:=[6,3:{{0,1,4},{1,2,5},{0,2,5},{0,3,4},{0,3,5},{1,4,5},{3,4,5}}];
T[17]:=[6,3:{{0,1,4},{1,2,5},{0,2,3},{2,3,5},{0,3,4},{1,4,5},{3,4,5}}];
T[18]:=[6,3:{{1,2,4},{0,2,5},{0,1,3},{1,3,4},{2,4,5},{0,3,5},{3,4,5}}];
```

From this output, we obtain all the triangulations and see that there are a total of
18 of them. For example, {0, 1, 3} represents the simplex whose vertices are the
points in rows 0, 1, and 3 of the input file `reg_tri_1.top`. The triangulation
`T[13]` in this output is the triangulation in (1.).

The TOPCOM command `points2triangs` has the following options:

- `--regular`: Outputs only regular triangulations.
- `--nonregular`: Outputs only nonregular triangulations.
- `--heights`: Outputs the height vectors of regular triangulations.

We can use these options as follows.

Listing 7.71 Options of the TOPCOM command `points2triangs`

```
$ points2triangs --regular < reg_tri_1.top

$ points2triangs --heights < reg_tri_1.top

$ points2triangs --nonregular < reg_tri_1.top
```

In order to compute the triangulations of a convex polytope with symmetries, we add the information about the symmetries to the input file. By doing this, the software works efficiently, and the output is simplified. We prepare the following input file.

Listing 7.72 TOPCOM: input file `reg_tri_1_sym.top`

```
[[4,0,0],
 [0,4,0],
 [0,0,4],
 [2,1,1],
 [1,2,1],
 [1,1,2]]

[[1,2,0,4,5,3],
 [1,0,2,4,3,5]]
```

The second matrix contains information about the symmetries of the configuration. In this case, the configuration has a 120° rotational symmetry and left–right symmetry. [1,2,0,4,5,3] means that the points indexed by $0, 1, 2, 3, 4, 5$ are transformed to the points indexed by $1, 2, 0, 4, 5, 3$; that is, a 120° rotation. [1,0,2,4,3,5] means that the points indexed by $0, 1, 2, 3, 4, 5$ are transformed to the points indexed by $1, 0, 2, 4, 3, 5$; that is, a reflection. With this input file, we execute the TOPCOM command `points2triangs`.

Listing 7.73 TOPCOM: generate triangulations

```
$ points2triangs < reg_tri_1_sym.top
Evaluating Commandline Options ...
... done.
T[1]:=[6,3:{{0,1,2}}];
T[2]:=[6,3:{{1,2,5},{0,2,5},{0,1,5}}];
T[3]:=[6,3:{{1,2,5},{0,2,5},{0,1,4},{1,4,5},{0,4,5}}];
T[4]:=[6,3:{{1,2,5},{0,2,5},{0,1,4},{1,4,5},{0,3,4},{0,3,5},{3,4,5}}];
T[5]:=[6,3:{{1,2,5},{0,1,4},{0,2,3},{1,4,5},{0,3,4},{2,3,5},{3,4,5}}];
```

Exercise. Consider the configuration matrix A in Example 7.3.12 and set the weight vector to $\mathbf{w} = [1, 0, 0, 0, 1, 0, 0, 3, 1]$. Geometrically compute the regular triangulation $\Delta_{\mathbf{w}}$. Use the software program TOPCOM to find all the triangulations.

7.3.6   Ring-Theoretic Properties and Triangulations (Sect. 5.6)

Example 7.3.25. Let the configuration matrix be

$$A = \begin{bmatrix} 1 & 1 & 1 & 1 & 1 & 1 \\ 0 & 1 & 0 & 1 & 0 & 1 \\ 0 & 0 & 1 & 1 & 2 & 2 \end{bmatrix} = [\mathbf{a}_1, \ldots, \mathbf{a}_6].$$

Let $<_{\text{lex}}$ be a lexicographic order with $x_1 > x_2 > x_3 > x_4 > x_5 > x_6$. Compute a lexicographic triangulation $\Delta_{\text{lex}}(A)$ by using Proposition 5.6.2.

Answer. We denote a set $\{\mathbf{a}_{i_1}, \ldots, \mathbf{a}_{i_m}\}$ by $i_1 \ldots i_m$, and denote all the faces of the convex polytope $\text{CONV}(\{\mathbf{a}_{i_1}, \ldots, \mathbf{a}_{i_m}\})$ by $\overline{i_1 \ldots i_m}$.

Since $\mathbf{a}_1 \notin \text{CONV}(23456)$, by Proposition 5.6.2, we have

$$\Delta_{\text{lex}}(123456) = \Delta_{\text{lex}}(23456) \cup \Delta_{23456}^1.$$

Here, we set

$$\Delta_{23456}^1 = \{\text{CONV}(1j_1 \ldots j_m) \mid j_1 \ldots j_m \subset 23456, \text{CONV}(j_1 \ldots j_m) \in \Delta_{\text{lex}}(23456),$$

$$\text{CONV}(j_1 \ldots j_m) \text{ is visible from } \mathbf{a}_1\}.$$

We compute $\Delta_{\text{lex}}(23456)$. Since $\mathbf{a}_2 \notin \text{CONV}(3456)$, by Proposition 5.6.2, we have

$$\Delta_{\text{lex}}(23456) = \Delta_{\text{lex}}(3456) \cup \Delta_{3456}^2.$$

Here, we set

$$\Delta_{3456}^2 = \{\text{CONV}(2j_1 \ldots j_m) \mid j_1 \ldots j_m \subset 3456, \text{CONV}(j_1 \ldots j_m) \in \Delta_{\text{lex}}(3456),$$

$$\text{CONV}(j_1 \ldots j_m) \text{ is visible from } \mathbf{a}_2\}.$$

We compute $\Delta_{\text{lex}}(3456)$. Since $\mathbf{a}_3 \notin \text{CONV}(456)$, by Proposition 5.6.2, we have

$$\Delta_{\text{lex}}(3456) = \Delta_{\text{lex}}(456) \cup \Delta_{456}^3.$$

Here, we set

$$\Delta_{456}^3 = \{\text{CONV}(3j_1 \ldots j_m) \mid j_1 \ldots j_m \subset 456, \text{CONV}(j_1 \ldots j_m) \in \Delta_{\text{lex}}(456) = \overline{456},$$

$$\text{CONV}(j_1 \ldots j_m) \text{ is visible from } \mathbf{a}_3\}.$$

By Proposition 5.6.1, it holds that $\Delta_{\text{lex}}(456) = \overline{456}$. Since the set of visible elements in $\overline{456}$ from $\mathbf{a}_3$ is $\overline{45}$, we have $\Delta_{456}^3 = \overline{345}$. Hence, it holds that

$$\Delta_{\text{lex}}(3456) = \overline{456} \cup \overline{345}.$$

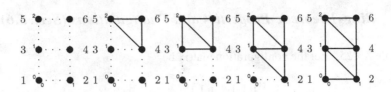

Fig. 7.16 Computational processes of a lexicographic triangulation of A

Since we have $\Delta_{\text{lex}}(3456)$, in the same way, we have $\Delta^2_{3456} = \overline{234}$ and

$$\Delta_{\text{lex}}(23456) = \overline{456} \cup \overline{345} \cup \overline{234}.$$

Since we have $\Delta_{\text{lex}}(23456)$, in the same way, we have $\Delta^1_{23456} = \overline{123}$ and

$$\Delta_{\text{lex}}(123456) = \overline{456} \cup \overline{345} \cup \overline{234} \cup \overline{123}.$$

These processes are shown in Fig. 7.16.

Exercise. 1. Let the configuration matrix be $A = \begin{bmatrix} 1 & 1 & 1 & 1 \\ 0 & 1 & 2 & 3 \end{bmatrix}$. Let $<_{\text{lex}}$ be a lexicographic order with $x_1 > x_2 > x_3 > x_4$. Compute a lexicographic triangulation $\Delta_{\text{lex}}(A)$. Consider the $2 \times (k+1)$ configuration matrix $A_k = \begin{bmatrix} 1 & 1 & 1 & \cdots & 1 \\ 0 & 1 & 2 & \cdots & k \end{bmatrix}$. Find a lexicographic triangulation $\Delta_{\text{lex}}(A_k)$.

2. Let the configuration matrix be $A = \begin{bmatrix} 1 & 1 & 1 & 1 & 1 & 1 & 1 & 1 \\ 0 & 1 & 0 & 1 & 0 & 1 & 0 & 1 \\ 0 & 0 & 1 & 1 & 0 & 0 & 1 & 1 \\ 0 & 0 & 0 & 0 & 1 & 1 & 1 & 1 \end{bmatrix}$. Let $<_{\text{lex}}$ be a lexico-

graphic order with $x_1 > x_2 > x_3 > x_4 > x_5 > x_6 > x_7 > x_8$. Compute a lexicographic triangulation $\Delta_{\text{lex}}(A)$.

Example 7.3.26. Let the configuration matrix be

$$A = \begin{bmatrix} 1 & 1 & 1 & 1 & 1 & 1 & 1 & 1 \\ 0 & 1 & 0 & 1 & 0 & 1 & 0 & 1 \\ 0 & 0 & 1 & 1 & 0 & 0 & 1 & 1 \\ 0 & 0 & 0 & 0 & 1 & 1 & 1 & 1 \end{bmatrix} = [\mathbf{a}_1, \ldots, \mathbf{a}_8].$$

Let $<_{\text{rev}}$ be a reverse lexicographic order with $x_1 > x_2 > x_3 > x_4 > x_5 > x_6 > x_7 > x_8$. Compute a reverse lexicographic triangulation $\Delta_{\text{rev}}(A)$ by using Proposition 5.6.5.

Answer. We denote $\text{CONV}(\{\mathbf{a}_{i_1}, \ldots, \mathbf{a}_{i_m}\})$ by $i_1 \ldots i_m$. Since the facets of $\text{CONV}(A)$ which do not include the point $\mathbf{a}_8$ are 1256, 1234, and 1375, we have

$$\text{FCT}_8^{12345678} = \{\{\mathbf{a}_1, \mathbf{a}_2, \mathbf{a}_5, \mathbf{a}_6\}, \{\mathbf{a}_1, \mathbf{a}_2, \mathbf{a}_3, \mathbf{a}_4\}, \{\mathbf{a}_1, \mathbf{a}_3, \mathbf{a}_5, \mathbf{a}_7\}\}.$$

Fig. 7.17 Computational processes of a reverse lexicographic triangulation of A (3-cube)

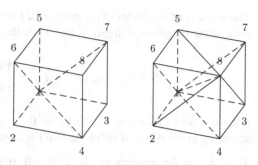

We next compute reverse lexicographic triangulations $\Delta_{\text{rev}}(1256)$, $\Delta_{\text{rev}}(1234)$ and $\Delta_{\text{rev}}(1375)$. For $\Delta_{\text{rev}}(1234)$, since the facets of 1234 which do not include the point $\mathbf{a}_4$ are 12 and 13, we have

$$\text{FCT}_4^{1234} = \{\{\mathbf{a}_1, \mathbf{a}_2\}, \{\mathbf{a}_1, \mathbf{a}_3\}\}.$$

By Proposition 5.6.5, the maximal simplices of $\Delta_{\text{rev}}(1234)$ are 124 and 134. In the same way, the maximal simplices of $\Delta_{\text{rev}}(1256)$ are 126 and 156, and those of $\Delta_{\text{rev}}(1357)$ are 137 and 157. Hence, the maximal simplices of $\Delta_{\text{rev}}(A)$ are

$$\bigcup_{A' \in \text{FCT}_8^{12345678}} \{\text{CONV}(\{\mathbf{a}_8\} \cup B) \mid \text{CONV}(B) \text{ is a maximal simplex of } \Delta_{\text{rev}}(A')\}$$

$$= \{1248, 1348, 1268, 1568, 1378, 1578\}.$$

These are shown in Fig. 7.17.

Exercise. In the previous example, we computed the reverse lexicographic triangulation of a three-dimensional unit cube. We now consider a four-dimensional unit cube; that is, the configuration matrix

$$A = \begin{bmatrix} 1 & 1 & 1 & 1 & 1 & 1 & 1 & 1 & 1 & 1 & 1 & 1 & 1 & 1 & 1 & 1 \\ 0 & 1 & 0 & 1 & 0 & 1 & 0 & 1 & 0 & 1 & 0 & 1 & 0 & 1 & 0 & 1 \\ 0 & 0 & 1 & 1 & 0 & 0 & 1 & 1 & 0 & 0 & 1 & 1 & 0 & 0 & 1 & 1 \\ 0 & 0 & 0 & 0 & 1 & 1 & 1 & 1 & 0 & 0 & 0 & 0 & 1 & 1 & 1 & 1 \\ 0 & 0 & 0 & 0 & 0 & 0 & 0 & 0 & 1 & 1 & 1 & 1 & 1 & 1 & 1 & 1 \end{bmatrix}.$$

Let $<_{\text{rev}}$ be a lexicographic order with $x_1 > x_2 > \cdots > x_{16}$. Compute the reverse lexicographic triangulation $\Delta_{\text{rev}}(A)$.

Example 7.3.27. Consider the Birkhoff polytope

$$P(d) = \left\{ [x_{ij}]_{1 \le i, j \le d} \in \mathbb{Q}^{d^2} \mid x_{ij} \ge 0, \sum_{k=1}^{d} x_{ik} = 1, \sum_{k=1}^{d} x_{kj} = 1 \text{ for } 1 \le i, j \le d \right\}.$$

The vertices of the Birkhoff polytope $P(d)$ are the set $\{P_\sigma \mid \sigma \in S_d\}$, where S_d is the symmetric group of degree d and P_σ is the permutation matrix

$$P_\sigma = [p_{ij}]_{1 \le i, j \le d}, \quad p_{ij} = \begin{cases} 1 & (j = \sigma(i)) \\ 0 & (\text{otherwise}) \end{cases}.$$

(Please refer to Example 7.3.2.) Prove that the configuration matrix $A(d)$ corresponding to the Birkhoff polytope $P(d)$ is compressed, by using Theorem 5.6.6.

Answer. All the vertices of the Birkhoff polytope $P(d)$ are vertices of a d^2-dimensional unit cube. The intersection of a d^2-dimensional unit cube and the affine subspace

$$\left\{ [x_{lm}] \in \mathbb{Q}^{d^2} \mid \sum_{k=1}^{d} x_{ik} = 1, \sum_{k=1}^{d} x_{kj} = 1 \text{ for } 1 \le i, j \le d \right\}$$

is the Birkhoff polytope $P(d)$. By Theorem 5.6.6 (ii), the configuration matrix $A(d)$ is compressed.

Example 7.3.28. Let the configuration matrices be

$$A_1 = \begin{bmatrix} 1 & 1 & 1 \\ 0 & 1 & 2 \end{bmatrix}; A_2 = \begin{bmatrix} 1 & 1 & 1 & 1 \\ 0 & 1 & 0 & 1 \\ 0 & 0 & 1 & 1 \end{bmatrix}; A_3 = \begin{bmatrix} 1 & 1 & 1 & 1 & 1 & 1 & 1 & 1 \\ 0 & 1 & 0 & 1 & 0 & 1 & 0 & 1 \\ 0 & 0 & 1 & 1 & 0 & 0 & 1 & 1 \\ 0 & 0 & 0 & 0 & 1 & 1 & 1 & 1 \end{bmatrix}; A_4 = \begin{bmatrix} 1 & 1 & 1 & 0 & 0 & 0 \\ 0 & 0 & 0 & 1 & 1 & 1 \\ 1 & 0 & 0 & 1 & 0 & 0 \\ 0 & 1 & 0 & 0 & 1 & 0 \\ 0 & 0 & 1 & 0 & 0 & 1 \end{bmatrix}.$$

Use Theorem 5.6.3 to determine which are unimodular configuration matrices.

Answer. We compute all circuits of A_1, A_2, A_3, and A_4. For any circuit f of A_2 and A_4, the monomials appearing in f are squarefree. By Theorem 5.6.3, A_2 and A_4 are unimodular configuration matrices, but A_1 and A_3 are not.

Exercise. Prove that the configuration matrix A corresponding to a $m \times n$ contingency table is unimodular.

Example 7.3.29. Let the configuration matrices be

$$A_1 = \begin{bmatrix} 1 & 1 & 1 \\ 0 & 1 & 2 \end{bmatrix}; A_2 = \begin{bmatrix} 1 & 1 & 1 & 0 & 0 & 0 \\ 0 & 1 & 2 & 0 & 0 & 0 \\ 1 & 0 & 0 & 1 & 0 & 0 \\ 0 & 1 & 0 & 0 & 1 & 0 \\ 0 & 0 & 1 & 0 & 0 & 1 \end{bmatrix}; A_3 = \begin{bmatrix} 1 & 1 & 1 \\ 0 & 1 & 3 \end{bmatrix}.$$

Determine which of the toric rings $K[A_1], K[A_2]$, and $K[A_3]$ are normal.

Answer. The toric ideal of A_1 is $I_{A_1} = \langle x_1x_3 - x_2^2 \rangle$. Let $<$ be a monomial order with $x_1 > x_3$. The initial ideal is $\mathrm{in}_<(I_{A_1}) = \langle x_1x_3 \rangle$. By Corollary5.6.8, the toric ring $K[A_1]$ is normal.

The toric rings $K[A_2]$ and $K[A_3]$ are not normal. Since A_2 is the Lawrence lifting $\Lambda(A_1)$ and A_1 is not unimodular, by Theorem 5.6.10, the toric ring $K[\Lambda(A_1)] = K[A_2]$ is not normal.

The toric ideal of A_3 is $I_{A_3} = \langle x_2^3 - x_1^2x_3 \rangle$. The generator $x_2^3 - x_1^2x_3$ has no squarefree monomials. By Proposition 5.6.9, the normal ring $K[A_3]$ is not normal. We also see that the toric ring $K[A_3]$ is not normal, since the vector $-\mathbf{a}_1 + 2\mathbf{a}_2 =$

$$\tfrac{1}{3}\mathbf{a}_1 + \tfrac{2}{3}\mathbf{a}_3 = \begin{bmatrix} 1 \\ 2 \end{bmatrix} \in \mathbb{Z}A_3 \cap \mathbb{Q}_{\geq 0}A_3 \text{ is not in } \mathbb{Z}_{\geq 0}A_3.$$

Exercise. Let the configuration matrices be

$$A_4 = \begin{bmatrix} 1 & 1 & 1 & 1 \\ 0 & 1 & 2 & 3 \end{bmatrix}; A_5 = \begin{bmatrix} 1 & 1 & 1 & 1 \\ 0 & 1 & 3 & 4 \end{bmatrix}; A_6 = \begin{bmatrix} 1 & 1 & 1 & 1 \\ 0 & 1 & 2 & 4 \end{bmatrix}.$$

Determine which of the toric rings $K[A_4]$, $K[A_5]$, and $K[A_6]$ are normal.

7.3.7 Examples of Configuration Matrices (Sect. 5.7)

Example 7.3.30. What is the configuration matrix corresponding to the graph G in Example 5.7.7?

Answer. By definition, we obtain the configuration matrix

$$A_G = \begin{bmatrix} 1 & 0 & 0 & 0 & 1 & 1 & 0 & 0 & 0 & 0 & 0 & 0 & 0 & 0 & 1 \\ 1 & 1 & 0 & 0 & 0 & 0 & 1 & 1 & 0 & 0 & 0 & 0 & 0 & 0 & 0 \\ 0 & 1 & 1 & 0 & 0 & 0 & 0 & 0 & 1 & 1 & 0 & 0 & 0 & 0 & 0 \\ 0 & 0 & 1 & 1 & 0 & 0 & 0 & 0 & 0 & 0 & 1 & 1 & 0 & 0 & 0 \\ 0 & 0 & 0 & 1 & 1 & 0 & 0 & 0 & 0 & 0 & 0 & 0 & 1 & 1 & 0 \\ 0 & 0 & 0 & 0 & 0 & 1 & 1 & 0 & 0 & 0 & 0 & 0 & 0 & 0 & 0 \\ 0 & 0 & 0 & 0 & 0 & 0 & 0 & 1 & 1 & 0 & 0 & 0 & 0 & 0 & 0 \\ 0 & 0 & 0 & 0 & 0 & 0 & 0 & 0 & 0 & 1 & 1 & 0 & 0 & 0 & 0 \\ 0 & 0 & 0 & 0 & 0 & 0 & 0 & 0 & 0 & 0 & 0 & 1 & 1 & 0 & 0 \\ 0 & 0 & 0 & 0 & 0 & 0 & 0 & 0 & 0 & 0 & 0 & 0 & 0 & 1 & 1 \end{bmatrix}.$$

Example 7.3.31. What is the graph G corresponding to the configuration matrix

Fig. 7.18 Graph
corresponding to a 3 × 3
contingency table

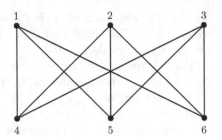

$$A = \begin{bmatrix} 1 & 1 & 1 & 0 & 0 & 0 & 0 & 0 & 0 \\ 0 & 0 & 0 & 1 & 1 & 1 & 0 & 0 & 0 \\ 0 & 0 & 0 & 0 & 0 & 0 & 1 & 1 & 1 \\ 1 & 0 & 0 & 1 & 0 & 0 & 1 & 0 & 0 \\ 0 & 1 & 0 & 0 & 1 & 0 & 0 & 1 & 0 \\ 0 & 0 & 1 & 0 & 0 & 1 & 0 & 0 & 1 \end{bmatrix},$$

which corresponds to a 3 × 3 contingency table?

Answer. The graph G is the complete bipartite graph $K_{3,3}$, as shown in Fig. 7.18. In general, the graph corresponding to the configuration matrix of an $m \times n$ contingency table is a complete bipartite graph $K_{m,n}$.

Example 7.3.32. Using Proposition 5.7.3, compute the circuits of the following matrices:

$$A_1 = \begin{bmatrix} 1 & 1 & 1 & 0 & 0 \\ 1 & 0 & 0 & 1 & 0 \\ 0 & 1 & 0 & 1 & 1 \\ 0 & 0 & 1 & 0 & 1 \end{bmatrix}; A_2 = \begin{bmatrix} 1 & 0 & 1 & 1 & 0 & 1 \\ 1 & 1 & 0 & 0 & 0 & 0 \\ 0 & 1 & 1 & 0 & 0 & 0 \\ 0 & 0 & 0 & 1 & 1 & 0 \\ 0 & 0 & 0 & 0 & 1 & 1 \end{bmatrix}; A_3 = \begin{bmatrix} 1 & 0 & 1 & 1 & 0 & 0 & 0 \\ 1 & 1 & 0 & 0 & 0 & 0 & 0 \\ 0 & 1 & 1 & 0 & 0 & 0 & 0 \\ 0 & 0 & 0 & 1 & 1 & 0 & 1 \\ 0 & 0 & 0 & 0 & 1 & 1 & 0 \\ 0 & 0 & 0 & 0 & 0 & 1 & 1 \end{bmatrix}.$$

Answer. The graphs corresponding to these configuration matrices are shown in Figs. 7.19–7.21. For these graphs, we need to find even closed walks which satisfy the conditions of Proposition 5.7.3. For the graph corresponding to A_1, there exists an even walk which satisfies condition (i), and its circuit is $x_1 x_5 - x_3 x_4$. For the graph corresponding to A_2, there exists an even walk which satisfies condition (ii), and its circuit is $x_1 x_3 x_5 - x_2 x_4 x_6$. For the graph corresponding to A_3, there exists an even walk which satisfies condition (iii), and its circuit is $x_1 x_3 x_5 x_7 - x_2 x_4^2 x_6$.

Example 7.3.33. For the graph in Example 5.7.1, use Proposition 5.7.3 to find the circuits.

Answer. In the graph, there are five even cycles of length 4, five even cycles of length 6, and five even closed walks which consist of two odd cycles having only a single point in common. Hence, we obtain the following 15 circuits.

Fig. 7.19 Graph
corresponding to A_1

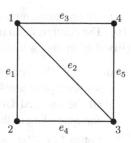

Fig. 7.20 Graph
corresponding to A_2

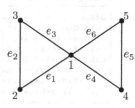

Fig. 7.21 Graph
corresponding to A_3

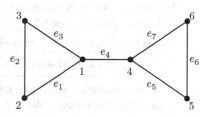

- The even cycles of length 4 correspond to

  $[1, 0, -1, 0, 0, -1, 1, 0, 0, 0]$, $[1, 0, 0, -1, 0, 0, 0, 0, 1, -1]$, $[0, 1, 0, -1, 0, 0, -1, 1, 0, 0]$,
  $[0, 1, 0, 0, -1, -1, 0, 0, 0, 1]$, $[0, 0, 1, 0, -1, 0, 0, -1, 1, 0]$.

- The even cycles of length 6 correspond to

  $[1, 0, 0, 0, -1, -1, 1, -1, 1, 0]$, $[0, 0, 0, 1, -1, -1, 1, -1, 0, 1]$, $[0, 0, 1, -1, 0, 1, -1, 0, 1, -1]$,
  $[0, 1, -1, 0, 0, -1, 0, 1, -1, 1]$, $[1, -1, 0, 0, 0, 0, 1, -1, 1, -1]$.

- The even closed walks which consist of two odd cycles having only a single point
  in common correspond to

  $[0, 1, 1, -1, -1, 0, -1, 0, 1, 0]$, $[1, 0, -1, -1, 1, 0, 0, 1, 0, -1]$, $[1, 1, 0, -1, -1, -1, 0, 0, 1, 0]$,
  $[1, -1, -1, 0, 1, 0, 1, 0, 0, -1]$, $[1, 1, -1, -1, 0, -1, 0, 1, 0, 0]$.

Exercise. For the graph in Example 5.7.7, use Proposition 5.7.3 to find the circuits.
For the graph in Example 7.3.31, find the circuits.

Example 7.3.34. For the graph G in Example 5.7.7, prove the following properties
by using the software programs 4ti2, Gfan, and TOPCOM.

(i) For any monomial order $<$, $\sqrt{in_<(I_{A_G})} \neq in_<(I_{A_G})$.

(ii) The configuration matrix A_G has unimodular triangulations.

(iii) Any triangulation of A_G having the fewest maximal simplices is nonregular.

Answer. (i) We compute the generators of the toric ideal I_{A_G} by using the software program 4ti2. In order to use the software program Gfan to compute all the reduced Gröbner bases of I_{A_G}, we prepare the following input file corresponding to the generators of I_{A_G}.

Listing 7.74 Gfan: input file eg577_gfan.txt

```
Q[a,b,c,d,e,f,g,h,i,j,k,l,m,n,o]
{o*m*d-n*l*e,n*f*a-o*g*e,n*c*a-o*d*b,m*k*c-l*j*d,m*c*a-l*e*b,k*i*b-j*h*c,
-j*d*a+k*e*b,-i*g*a+h*f*b,-i*d*a+h*e*c,f*d*b-g*e*c,-n*l*j*a+o*m*k*b,
-n*l*i*a+o*m*h*c,n*l*f*b-o*m*g*c,-n*k*i*a+o*j*h*d,-n*k*f*b+o*j*g*d,
n*k*e*c-o*j*d^2,-n*i*a^2+o*h*e*b,-n*h*f*c+o*i*g*d,m*k*i*a-l*j*h*e,
-m*k*f*b+l*j*g*e,m*h*f*c-l*i*g*e,-l*i*d*b+m*h*c^2,-m*f*d*a+l*g*e^2,
-j*h*f*d+k*i*g*e,j*g*c*a-k*f*b^2,-n*l*j*h*f+o*m*k*i*g,
-n*k^2*i*e*b+o*j^2*h*d^2,l*i^2*g*d*a-m*h^2*f*c^2}
```

We execute the Gfan command gfan with the input file eg577_gfan.txt, and the output file is eg577_gfan.result.

Listing 7.75 Gfan: executing the command gfan

```
$ gfan < eg577_gfan.txt > eg577_gfan.result
```

This computation is lengthy (it takes about half an hour), and we obtain 15,090 reduced Gröbner bases. We can check that, for each reduced Gröbner basis, there exists an element whose initial term is not squarefree.

(ii) From the property (i), none of the regular triangulations of A_G are unimodular. Hence, all the unimodular triangulations of A_G are nonregular. By using the software program TOPCOM, we can generate all the nonregular triangulations which are connected to regular triangulations. We prepare the following input file for TOPCOM.

Listing 7.76 TOPCOM: input file graph_ag.top

```
[
[1,1,0,0,0,0,0,0,0,0],
[0,1,1,0,0,0,0,0,0,0],
[0,0,1,1,0,0,0,0,0,0],
[0,0,0,1,1,0,0,0,0,0],
[1,0,0,0,1,0,0,0,0,0],
[1,0,0,0,0,1,0,0,0,0],
[0,1,0,0,0,1,0,0,0,0],
[0,1,0,0,0,0,1,0,0,0],
[0,0,1,0,0,0,1,0,0,0],
[0,0,1,0,0,0,0,1,0,0],
[0,0,0,1,0,0,0,1,0,0],
[0,0,0,1,0,0,0,0,1,0],
[0,0,0,0,1,0,0,0,1,0],
[0,0,0,0,1,0,0,0,0,1],
[1,0,0,0,0,0,0,0,0,1]
]

[
[1,2,3,4,0,7,8,9,10,11,12,13,14,5,6],
[4,3,2,1,0,14,13,12,11,10,9,8,7,6,5]
]
```

The first matrix corresponds to the configuration matrix A_G. The second matrix contains information about the symmetries of the configuration. The first row [1,2,3,4,0,7,8,9,10,11,12,13,14,5,6] indicates a 72° rotational symmetry, and the second row [4,3,2,1,0,14,13,12,11,10,9, 8,7,6,5] indicates left–right symmetry. We execute the TOPCOM command points2triangs--nonregular as follows.

Listing 7.77 TOPCOM: generate triangulations

```
$ points2triangs --nonregular < graph_ag.top > graph_ag.nonreg
```

This computation takes about 10 minutes, and the output is contained in the file graph_ag.nonreg, which is 1.3 M bytes.

Listing 7.78 TOPCOM: output file graph_ag.nonreg

```
T[1]:=[15,10:{{0,1,2,4,5,7,8,9,11,13},...
...
T[1054]:=[15,10:{{0,2,3,4,5,7,10,11,13,14},...
```

We thus obtain 1054 nonregular triangulations. We need to find the unimodular triangulations in these nonregular triangulations. We begin by finding the triangulations whose simplices have the same volume. The triangulation satisfying this condition is T[16], which is unimodular.

Listing 7.79 Unimodular triangulation of A_G

```
T[16]:=[15,10:
{{0,1,2,3,4,5,7,9,11,13}   ,{0,1,3,4,5,6,7,9,11,13},
{0,1,2,4,5,7,8,9,11,13}   ,{0,3,4,5,6,7,9,10,11,13},
{2,3,5,6,7,8,9,10,11,13}   ,{0,1,4,5,6,7,9,11,12,13},
{0,1,3,4,6,7,9,11,12,13}   ,{0,1,4,5,7,8,9,11,12,13},
{3,4,5,6,7,9,10,11,12,13}  ,{0,4,5,6,7,9,10,11,12,13},
{0,3,4,6,7,9,10,11,12,13}  ,{0,1,3,6,7,9,10,11,12,13},
{3,4,5,7,8,9,10,11,12,13}  ,{3,5,6,7,8,9,10,11,12,13},
{0,1,2,5,7,8,9,11,13,14}   ,{0,1,2,4,5,8,9,11,13,14},
{1,2,5,6,7,8,9,11,13,14}   ,{0,2,3,5,6,7,10,11,13,14},
{2,5,6,7,8,9,10,11,13,14}  ,{0,1,5,6,7,9,11,12,13,14},
{1,2,4,5,8,9,11,12,13,14}  ,{0,4,5,7,8,9,11,12,13,14},
{0,1,5,7,8,9,11,12,13,14}  ,{0,1,4,5,8,9,11,12,13,14},
{1,5,6,7,8,9,11,12,13,14}  ,{0,5,6,7,9,10,11,12,13,14},
{0,4,5,7,9,10,11,12,13,14} ,{4,5,7,8,9,10,11,12,13,14},
{0,2,3,4,5,7,9,11,13,14}   ,{2,3,4,5,7,8,9,11,13,14},
{0,3,4,5,7,9,10,11,13,14}  ,{0,2,3,4,5,7,9,10,11,14},
{2,3,5,7,8,9,10,11,13,14}  ,{3,4,5,7,8,9,10,11,13,14},
{2,3,4,5,7,8,9,10,11,14}   ,{0,2,4,5,7,8,9,10,11,14},
{0,1,2,3,5,7,9,10,11,13}   ,{0,1,2,5,7,9,10,11,13,14},
{1,2,3,5,6,7,9,10,11,13}   ,{0,1,2,3,5,6,7,10,11,13},
{1,3,4,5,6,7,8,9,12,13}    ,{0,1,5,6,7,9,10,11,13,14},
{1,2,5,6,7,9,10,11,13,14}  ,{0,1,2,5,6,7,10,11,13,14},
{1,2,3,4,5,7,9,11,12,13}   ,{1,2,3,5,6,7,9,11,13,13},
{2,3,4,5,7,8,9,11,12,13}   ,{1,2,3,4,5,7,8,9,12,13},
{1,2,5,6,7,8,9,11,12,13}   ,{2,3,5,6,7,8,9,11,12,13},
{1,2,3,5,6,7,8,9,12,13}    ,{0,1,3,5,6,7,9,10,11,13},
{0,2,4,5,7,8,9,11,13,14}   ,{1,2,4,5,7,8,9,11,12,13},
{0,2,3,5,7,9,10,11,13,14}  ,{1,3,4,5,6,7,9,11,12,13}}];
```

(iii) From the output file graph_ag.top, in order to find triangulations with the fewest maximal simplices, we execute the TOPCOM command points2triangs, as follows.

Listing 7.80 TOPCOM: generate the triangulations with i maximal simplices ($i = 1, \ldots, 45$)

```
$ for i in `seq 1 1 45`;
> do
> points2triangs --cardinality $i < graph_ag.top;
> done
Evaluating Commandline Options ...
--cardinality  : restrict to triangulations with 1 simplex
... done.
Evaluating Commandline Options ...
--cardinality  : restrict to triangulations with 2 simplices
... done.
...
Evaluating Commandline Options ...
--cardinality  : restrict to triangulations with 45 simplices
... done.
```

The option -cardinality i means to output only triangulations with i maximal simplices. In the above computation, we check $i = 1, \ldots, 45$ and find no triangulations with i maximal simplices in the output graph_ag.top.

Listing 7.81 TOPCOM: triangulation with the fewest maximal simplices

```
$ points2triangs --cardinality 46 < graph_ag.top
Evaluating Commandline Options ...
--cardinality  : restrict to triangulations with 46 simplices
... done.
T[1]:=[15,10:{{0,1,2,3,4,5,7,9,11,13}, ...
```

From this output, we find only one triangulation with the fewest maximal simplices, and the total number of maximal simplices is 46. We now use TOPCOM to check that this triangulation is nonregular, and we prepare the following input file graph_ag.check.

Listing 7.82 TOPCOM: input file graph_ag.check

```
[                                      <-+
[1,1,0,0,0,0,0,0,0,0],                   |
[0,1,1,0,0,0,0,0,0,0],                   |
...                                      |
]                                        | these parts are the same as
[                                        | the input file "graph_ag.top"
[1,2,3,4,0,7,8,9,10,11,12,13,14,5,6],    |
[4,3,2,1,0,14,13,12,11,10,9,8,7,6,5]     |
]                                      <-+
{{0,1,2,3,4,5,7,9,11,13}, ...          <-- triangulation T[1] in
                                           the above output
```

We execute the TOPCOM command checkregularity as follows.

Listing 7.83 TOPCOM: check the regularity of a triangulation

```
$ checkregularity < graph_ag.check
Evaluating Commandline Options ...
... done.
```

```
{{0,1,2,3,4,5,7,9,11,13},...
is non-regular.
Checked 1 triangulations, 1 nonregular so far.
```

From this output, this triangulation is nonregular, and so we have proved property (iii).

Exercise. Let G be the graph in Example 5.7.7. Prove $\sqrt{\mathrm{in}_<(I_{A_G})} \neq \mathrm{in}_<(I_{A_G})$ for any monomial order $<$, without using a computer. (Please refer to [9, p.102].)

7.4 Gröbner Basis of Rings of Differential Operators and Applications

This section includes examples and exercises for Chap. 6. We consider that it is important to use computer software to examine these problems; relevant software programs include Macaulay2 [7], Singular [4], and Risa/Asir [15], all contained in MathLibre. The computer algebra system Maple [12] is also used in some examples. The results from these programs will assist in the understanding of Chap. 6. Although readers might prefer that only one software program is used, we note that each computer system has its own areas of specialization. We should thus use the appropriate system for each operation. Once we are familiar with one computer algebra system, it is easy to use other systems. However, it is inefficient to do so without a *dictionary* which shows how a function is realized on other systems. We will show how to perform functions with various software programs so that this chapter can be used as a dictionary for software.

Some of the programming exercises below are treated on each of several systems (e.g., Examples and Exercises 7.4.3, 7.4.5, and 7.4.6). These are advanced exercises for readers interested in the inner structure of software; these can be skipped by readers not interested. However, we believe it is important to understand the mechanisms of computation, and so we hope that readers will follow these.

The following is a table of the commands for each of the software programs discussed in this chapter.

Commands	Macaulay2	Singular	Risa/Asir
Packages	Dmodules	dmod.lib	nk_restriction.rr
Gröbner basis	gb, gbw	groebner, GBWeight	nd_weyl_gr
Initial ideal	inw	initialIdealW	initilal_w
Holonomic rank	holonomicRank	— (kbase)	sm1.rank
Holonomicity	isHolonomic	isHolonomic	—
Hilbert polynomial	hilbertPolynomial	hilbPoly	sm1.hilbert
Fourier transform	Fourier	fourier	fourier_trans
b-Function	bFunction	bfctIdeal	generic_bfct
Integration ideal	DintegrationIdeal	integrationIdeal	integration_ideal
Annihilating ideal	AnnFs	Sannfs	ann

7.4.1 Gröbner Basis for the Ring of Differential Operators with Rational Function Coefficients R (Sect. 6.1)

Example 7.4.1. Expand $\partial_x^2 x \partial_x$ both by hand and by using the computer systems Macaulay2, Singular, and Risa/Asir.

Answer. We expand it using the noncommutative relation $\partial_x x = x \partial_x + 1$.

Listing 7.84 By hand: multiplication of differential operators

$$\partial_x^2 x \partial_x = \partial_x (x \partial_x + 1) \partial_x = (x \partial_x + 1) \partial_x^2 + \partial_x^2 = x \partial_x^3 + 2 \partial_x^2.$$

Kan/sm1 [23] outputs the following:

Listing 7.85 Kan/sm1

```
sm1>[(x) ring_of_differential_operators 0] define_ring ;
sm1>(Dx^2*x*Dx).  ;
sm1>dehomogenize :  :
x*Dx^3+2*Dx^2
```

We do not discuss Kan/sm1 in detail in this book, because it is no longer being maintained. It is, however, the earliest system for manipulating differential operators, and other systems use it as a reference. Since Kan/sm1 has many commands which are not found in other systems, it will continue to be used.

In Chap. 3, we learned how to use Macaulay2 and Singular to declare a polynomial ring. In this chapter, we will primarily use the ring of differential operators. We begin by declaring the ring of differential operators.

Listing 7.86 Macaulay2: declaration of the ring of differential operators and the product of differential operators

```
i1 : R = QQ[x,dx,WeylAlgebra => {x=>dx}];
i2 : dx^2*x*dx
            3      2
o2 = x*dx  + 2dx
o2 : R

i3 : loadPackage "Dmodules";
i4 : R1=QQ[x];
i5 : D1=makeWA R1;
i6 : dx^2*x*dx
            3      2
o6 = x*dx  + 2dx
o6 : D1
```

In Macaulay2, it is similar to the case of a polynomial ring; moreover, `WeylAlgebra => {x=>dx}` is required after the variables to specify that `dx` is the derivative symbol with respect to the variable `x`. The operator symbol "`*`" is assigned to both noncommutative and ordinal multiplication.

We can also define the ring of differential operators from the polynomial ring by using the command `makeWeylAlgebra` (`makeWA` in short) in the `Dmodules` package. Then, adding "d" at the beginning of a variable creates a differential operator. Although the latter method is useful, the former method is often required for setting the advanced options.

The command describe returns variables and options for the base ring.

Listing 7.87 Macaulay2: obtaining information about a ring

```
i7 : describe D
o7 = QQ[x, dx, Degrees => {2:1}, Heft => {1}, MonomialOrder =>
-------------------------------------------------------------
{MonomialSize => 32}, DegreeRank => 1, WeylAlgebra => {x => dx}]
{GRevLex => {2:1}  }
{Position => Up    }
```

Various types of information are displayed; for example, GRevLex means that the graded lexicographic order is selected by default as the term order.

Listing 7.88 Singular: declaration of the ring of differential operators and the product of differential operators

```
> LIB "dmod.lib";
> ring r=0,(x,dx),dp;
> def D=Weyl();
> setring D;
> poly f = dx^2*x*dx;
> f;
x*dx^3+2*dx^2
```

With Singular, it is easy to use the command Weyl, which is defined in the nctools.lib library. However, in the examples in this chapter, we always load dmod.lib library, because it includes nctools.lib and other useful libraries for D-modules.

We begin by declaring the polynomial ring with variables and differential operators. Then we define the ring of differential operators by using the command Weyl.

The command Weyl, either with no arguments or with 0 as the argument, generates the Weyl algebra in which the second half of the variables corresponds to the differential operators for the first half of variables. Weyl with nonzero arguments generates a Weyl algebra with a corresponding variable and operator, in order from the front. In other words, under the polynomial ring with the variables $(x_1, \ldots, x_n, x_{n+1}, \ldots, x_{2n})$, Weyl() or Weyl(0) generates the ring of differential operators and treats x_{i+n} as the differential operators for the first n variables x_i ($1 \leq i \leq n$). Weyl(1) treats x_{i+1} as the differential operators for the variables x_i (i : odd).

Since some commands assume the former format on the variable, we always use Weyl() in this chapter.

We note that Weyl() only generates the Weyl algebra and does not set the ring. To set the ring, use the command setring. Then, the operator symbol "*" is assigned to noncommutative multiplication as well as to ordinal multiplication.

In Singular, to obtain information about a ring, simply input the variable name of the ring.

Listing 7.89 Singular: obtaining information about a ring

```
> D;
//    characteristic : 0
//    number of vars : 2
//         block   1 : ordering dp
//                   : names    x dx
//         block   2 : ordering C
//    noncommutative relations:
//       dxx=x*dx+1
> basering;
(omitted)
> nameof(basering);
D
```

If you forgot the variable name of the ring, the information can be obtained by using the command basering, which is an alias of the name of the base ring. You can also obtain the variable name of the base ring by using the command nameof(basering);.

Listing 7.90 Risa/Asir: product of a differential operator

```
[1371]  P=dx^2;
dx^2
[1372]  Q=x*dx;
x*dx
[1373]  V=[x,dx];
[x,dx]
[1374]  DP=dp_ptod(P,V);
(1)*<<0,2>>
[1375]  DQ=dp_ptod(Q,V);
(1)*<<1,1>>
[1376]  DPQ=dp_weyl_mul(DP,DQ);
(1)*<<1,3>>+(2)*<<0,2>>
[1377]  PQ=dp_dtop(DPQ,V);
x*dx^3+2*dx^2
```

Risa/Asir does not require setting the base ring, but has dedicated commands for differential operators. To compute the product of differential operators, use the command dp_weyl_mul. The first argument acts as an operator on the second argument from the left. Since dp_weyl_mul only accepts distributed representations (see Sect. 3.6.4), we must convert this to a distributed representation. Moreover, the variables in the last half of the exponential part of the distributed representation are the differential operators for the variables in the first half. That is, the command dp_weyl_mul for the distributed representation with the list of variables $V = [v_1, \ldots, v_n, v_{n+1}, \ldots, v_{2n}]$ treats the variable v_{i+n} as the differential operator for the variable v_i.

Exercise. Expand the following expressions:

1. $\partial_x^4 x^3 \partial_x$,
2. $(x^2 \partial_x^2 - 2\partial_x)(x^3 \partial_x - x)$.

Example 7.4.2. Let $F_m = \{\sum_{|k| \le m} a_k(x) \partial^k \mid a_k(x) \in \mathbf{C}(x_1, \dots, x_n)\}$ be a subset of R^n. Prove that F_m is a finite-dimensional vector space over $\mathbf{C}(x_1, \dots, x_n)$, and find the dimension of F_m.

Answer. It is easy to check that F_m is a vector space. Since we can take $B = \{\partial^k \mid |k| \le m\}$ as a basis, F_m is finite-dimensional. Let us count the number of elements in the set B. The number of ways to sample up to m elements from a set of n elements, allowing for duplicates, is equal to the number of ways to sample exactly m elements from a set of $n + 1$ elements, allowing for duplicates. Thus, $\binom{(n+1)+m-1}{m} = \binom{n+m}{m}$, and the dimension of F_m is $\binom{n+m}{m} = \frac{(m+n)!}{m!n!}$.

Exercise. Find the dimension of the set

$$F_m \setminus \left\{ \sum_{m_1, m_2 \ge 0} a_{m_1, m_2}(x) \partial_1^{m_1 + 1} \partial_2^{m_2 + 1} \mid a_{m_1, m_2}(x) \in \mathbf{C}(x_1, \dots, x_n) \right\}$$

as a vector space over $\mathbf{C}(x_1, \dots, x_n)$.

Example 7.4.3. Implement a multiplying function for univariate differential operators, using Leibniz's rule with Risa/Asir or Maple.

Answer.

Listing 7.91 Maple: multiplication by Leibniz's rule (dm1.ml)

```
dmult:=proc(F,G)
    local N1,A,K;
    N1:=degree(numer(F),dx);
    A:=0;
    for K from 0 by 1 to N1 do
      A:=A+(1/factorial(K))*diff2(F,dx,K)*diff2(G,x,K);
    od:
    RETURN(simplify(A));
end:

diff2:=proc(F,G,K)
    local i,A;
    A:=F;
    for i from 0 by 1 to K-1 do
      A:=diff(A,G);
        od:
    RETURN(A);
end:
dmult(dx^2,x*dx);
```

The function dmult is the multiplying function for the differential operators. The symbols dx and x represent the differential operator ∂_x and the variable x, respectively. Although the symbol dx is displayed to the left of x, we consider mathematically that dx is on the right of x.

Listing 7.92 Risa/Asir: multiplication by Leibniz's rule (dm1.rr)

```
/* Multiplication F by G */
def d_mult(F, G)
{
    N = deg(nm(F), dx); /* the degree of F w.r.t. dx */
    A = 0;
    for (K = 0; K <= N; K++) {
        /* Leibniz's rule */
        A = A + (1/fac(K)) * diff2(F, dx, K) * diff2(G, x, K);
        A = red(A);       /* reduce A */
    }
    return A;
}

/* K-th derivative of F w.r.t. G */
def diff2(F, G, K)
{
    for (I = 0; I < K; I++)
        F = diff(F, G);
    return F;
}

d_mult(dx^2, x*dx);      /* --> x*dx^3+2*dx^2 */
```

The symbols dx and x represent the differential operator ∂_x and the variable x, respectively. dmult is the function for the multiplication of differential operators.

Implementing a multiplying function for n-variate differential operators, and implementing the above with languages which do not have a multiplying function for polynomials (e.g., the programming language C) are advanced exercises.

Example 7.4.4. Expand $\theta_x(\theta_x - 1) \cdots (\theta_x - k)$.

Answer.

- In the case of $k = 1$, we have

$$\theta_x(\theta_x - 1) = x\partial_x(x\partial_x - 1) = x(x\partial_x + 1)\partial_x - x\partial_x = x^2\partial_x^2.$$

- In the case of $k = 2$, we have

$$\theta_x(\theta_x - 1)(\theta_x - 2) = x^2\partial_x^2(x\partial_x - 2) = x^2\partial_x^2 x\partial_x - 2x^2\partial_x$$
$$= x^2\partial_x(x\partial_x + 1)\partial_x - 2x^2\partial_x^2 = x^2\{(x\partial_x + 1)\partial_x + \partial_x\}\partial_x - 2x^2\partial_x^2$$
$$= x^3\partial_x^3.$$

- In the case of k, we will show by induction that $\theta_x(\theta_x - 1) \cdots (\theta_x - k) = x^{k+1}\partial_x^{k+1}$. Assume that it is true in the case of $k - 1$: we have

$$\theta_x(\theta_x - 1) \cdots (\theta_x - k) = x^k\partial_x^k(x\partial_x - k)$$

$$= x^{k+1}\partial_x^{k+1} + kx^k\partial_x^k - kx^k\partial_x^k$$
$$= x^{k+1}\partial_x^{k+1}.$$

We applied Leibniz's rule to the expansion of $x^k\partial_x^k(x\partial_x - k)$.

Exercise. Prove $\partial_x^k x^k = (\theta_x + 1)(\theta_x + 2)\cdots(\theta_x + k)$.

Example 7.4.5. Implement the normal form algorithm (Algorithm 6.1.4) in R_1.

Answer. We introduce a sample program in Risa/Asir. For simplicity, x and dx are fixed as variables. We use the three subfunctions: in, in_, and c_in, to extract an initial term, an initial monomial, and an initial coefficient, respectively.

Listing 7.93 Risa/Asir: computing the initial term, initial monomial, and initial coefficient (in nf_r1.rr)

```
/* initial term (with coefficient) */
def in(F)
{
    NM = nm(F);          /* numerator of F */
    DN = dn(F);          /* denominator of F */
    Deg = deg(NM, dx);
    return coef(NM, Deg, dx)/DN * dx^Deg;
}

/* initial monomial (without coefficient) */
def in_(F)
{
    Deg = deg(nm(F), dx);
    return dx^Deg;
}

/* initial coefficient */
def c_in(F)
{
    NM = nm(F);          /* numerator of F */
    DN = dn(F);          /* denominator of F */
    Deg = deg(NM, dx);
    return coef(NM, Deg, dx)/DN;
}
```

The following commands output the initial term $(x + \frac{1}{x})\xi_x^2$, the initial monomial ξ_x^2, and the initial coefficient $(x + \frac{1}{x})$ of $f = (x + \frac{1}{x})\partial_x^2 + x\partial_x + 1$, in $R_1 = \mathbf{Q}(x)\langle\partial_x\rangle$.

Listing 7.94 Risa/Asir: computing the initial term, initial monomial, and initial coefficient, using nf_r1.rr

```
[1230] load("nf_r1.rr");
[1246] F = (x+1/x)*dx^2+x*dx+1;
((x^2+1)*dx^2+x^2*dx+x)/(x)
[1247] in(F);
((x^2+1)*dx^2)/(x)
[1248] in_(F);
```

```
dx^2
[1250]  c_in(F);
(x^2+1)/(x)
```

According to Algorithms 6.1.4 and 6.1.5, sample programs for the weak normal form algorithm (w_normal_form) and the normal form algorithm (normal_form) are as follows.

Listing 7.95 Risa/Asir: computing the weak normal form and the normal form (in nf_r1.rr)

```
/* normal form */
def normal_form(F, G)
{
    M = length(G);
    Q = newvect(M);            /* vector for holding quotient */
    R = 0;                     /* variable for holding normal form */
    while (F != 0) {
        L = w_normal_form(F, G); /* compute weak normal form of F by G */
        RR = L[0];             /* weak normal form */
        QQ = L[1];             /* quotient */
        F = RR - in(RR);
        R = R + in(RR);
        Q = Q + QQ;
        R = red(R);               /* reduce the rational function R */
        Q = map(red, Q);          /* reduce each component of vector Q */
    }
    return [R, Q];
}

/* weak normal form */
def w_normal_form(F, G)
{
    M = length(G);
    Q = newvect(M);            /* vector for holding quotient */
    R = F;                     /* variable for holding weak normal form */
    while ((Index = reducible(R, G)) != -1) {
        S = G[Index];
        D = tdiv(in_(R), in_(S)); /* quotient of division in_(R) by in_(S) */
        X = c_in(R) / c_in(S);
        T = d_mult(X*D, S);
        Q[Index] = Q[Index] + X*D;
        R = R - T;
        Q[Index] = red(Q[Index]); /* reduce the rational function Q[Index] */
        R = red(R);               /* reduce the rational function R */
    }
    return [R, Q];
}

/* Is R divisible by initial term of G ? */
def reducible(R, G)
{
    M = length(G);
    InR = in_(R);
    for (I = 0; I < M; I++) {
        InG = in_(G[I]);
        T = tdiv(InR, InG); /* divisibility test: InG | InR ?*/
        if (T != 0)          /* the case of InG | InR */
            return I;
    }
    return -1;
}
```

Listing 7.96 Risa/Asir: computing the normal form, using `nf_r1.rr`

```
[1206] load("nf_r1.rr");
[1226] F = dx^3-x$
[1227] G1 = dx^2-1$
[1228] G2 = x*dx-1$
[1229] normal_form(F, [G1, G2]);
[(-x^2+1)/(x),[ dx (1)/(x) ]]
```

The command `normal_form` returns the pair `[(normal form), (list of quotients)]`. From the output, we see that the normal form of $F = \partial_x^3 - x$ by $G_1 = \partial_x^2 - 1$ and $G_2 = x\partial_x - 1$ is $\frac{-x^2+1}{x}$. Moreover, we obtain the relation

$$F = \partial_x G_1 + \frac{1}{x} G_2 + \frac{-x^2 + 1}{x}$$

from the list of quotients.

Exercise. Implement the algorithm of Example 7.4.5 on Macaulay2, Singular, Maple, and other systems.

Example 7.4.6. Implement a normal form algorithm over R_2 and Buchberger's algorithm in R_2 (refer to [14, Chap. 16]).

Answer. We provide a sample program on Risa/Asir. For simplicity, x, y, dx, and dy are fixed as variables. We begin by implementing a multiplying function `d_mult2` in $R_2 = \mathbf{Q}(x, y)\langle \partial_x, \partial_y \rangle$, which uses Leibniz's rule.

Listing 7.97 Risa/Asir: multiplication in R_2 (in `nf_r2.rr`)

```
def d_mult2(F, G)
{
    NX = deg(nm(F), dx);
    NY = deg(nm(F), dy);
    A = 0;
    for (KX = 0; KX <= NX; KX++) {
        for (KY = 0; KY <= NY; KY++) {
            /* Leibniz's rule */
            A = A + 1/(fac(KX) * fac(KY))
                * diff3(F, dx, KX, dy, KY)
                * diff3(G, x, KX, y, KY);
            A = red(A);
        }
    }
    return A;
}

/* (M,N)-th derivative of F w.r.t. variables (X, Y) */
def diff3(F, X, M, Y, N)
{
    for (I = 0; I < M; I++)
        F = diff(F, X); /* derivative of F w.r.t. X */
    for (I = 0; I < N; I++)
```

```
        F = diff(F, Y); /* derivative of F w.r.t. Y */
    return F;
}
```

The following code shows that $\frac{y}{x}\partial_x^2$ multiplied by $x\partial_x\partial_y + x$ from the left is $y\partial_x^3\partial_y + \partial_x^3 - \frac{y}{x}\partial_x^2\partial_y + \frac{xy-1}{x}\partial_x^2$.

Listing 7.98 Risa/Asir: multiplication in R_2 by nf_r2.rr

```
[1251] load("nf_r2.rr");
[1275] d_mult2(x*dx*dy + x, (y/x)*dx^2);
((y*x*dy+x)*dx^3+(-y*dy+y*x-1)*dx^2)/(x)
```

Next, we implement three subfunctions, in2, in2_, and c_in2, for extracting an initial term, an initial monomial, and an initial coefficient, respectively. Then, we note that, unlike the univariate case, it is necessary to specify a term order. Our program follows the specifications of Risa/Asir: Order = 0 indicates the reverse lexicographic order (the graded reverse lexicographic order), Order = 1 indicates the lexicographic order (the graded lexicographic order), and Order = 2 indicates the pure lexicographic order (the lexicographic order). Here, the terminology in parentheses follows that of Chap. 3. A variable order is given by a list VL. See Sect. 3.6.5 for details.

Listing 7.99 Risa/Asir: initial term, initial monomial, and initial coefficient in R_2 (in nf_r2.rr)

```
/* initial term (with coefficient) */
def in2(F, VL, Order)
{
    OldOrder = dp_ord();      /* save the original order to OldOrder */
    dp_ord(Order);            /* set a new order Order                */
    NM = nm(F);               /* numerator of F                       */
    DNM = dp_ptod(NM, VL);    /* distributed representation of NM     */
    DIN = dp_hm(DNM);         /* initial term of polynomial DNM       */
    IN = dp_dtop(DIN, VL);    /* recursive representation of DIN      */
    IN = IN / dn(F);          /* divide by denominator of F           */
    IN = red(IN);             /* reduce rational function IN          */
    dp_ord(OldOrder);         /* load the original order              */
    return IN;
}

/* initial monomial (without coefficient) */
def in2_(F, VL, Order)
{
    OldOrder = dp_ord();      /* save the original order to OldOrder */
    dp_ord(Order);            /* set a new order Order                */
    NM = nm(F);               /* numerator of F                       */
    DNM = dp_ptod(NM, VL);    /* distributed representation of NM     */
    DIN = dp_ht(DNM);         /* initial term of polynomial DNM       */
    IN = dp_dtop(DIN, VL);    /* recursive representation of DIN      */
    dp_ord(OldOrder);         /* load the original order              */
    return IN;
}

/* initial coefficient */
def c_in2(F, VL, Order)
{
    OldOrder = dp_ord();      /* save the original order to OldOrder */
    dp_ord(Order);            /* set a new order Order                */
    NM = nm(F);               /* numerator of F                       */
    DNM = dp_ptod(NM, VL);    /* distributed representation of NM     */
    LC = dp_hc(DNM);          /* initial coefficient of polynomial DNM */
```

```
    LC = LC / dn(F);          /* divide by denominator of F        */
    LC = red(LC);             /* reduce rational function LC        */
    dp_ord(OldOrder);         /* load the original order            */
    return LC;
}
```

The following sample code outputs the initial term $in_\prec(F) = (\frac{1}{x} + y)\xi_x^3$, the initial monomial ξ_x^3, and the initial coefficient $(\frac{1}{x} + y)$, of $F = \frac{1}{x}\partial_x^2\partial_y^2 + (\frac{1}{x} + y)\partial_x^3 + \partial_x + \partial_y + 1$ with respect to the (pure) lexicographic order with $\partial_x \succ \partial_y$.

Listing 7.100 Risa/Asir: computation of the initial term, initial monomial, and initial coefficient, by using nf_r2.rr

```
[1278] load("nf_r2.rr");
[1302] F = 1/x*dx^2*dy^2+(1/x+y)*dx^3+dx+dy+1;
((y*x^2+x)*dx^3+x*dy^2*dx^2+x^2*dx+x^2*dy+x^2)/(x^2)
[1303] in2(F, [dx,dy], 2);
((y*x+1)*dx^3)/(x)
[1304] in2_(F, [dx,dy], 2);
dx^3
[1305] c_in2(F, [dx,dy], 2);
(y*x+1)/(x)
```

According to Algorithms 6.1.4 and 6.1.5, sample programs for the weak normal form algorithm (w_normal_form2) and the normal form algorithm (normal_form2) are as follows.

Listing 7.101 Risa/Asir: computing the weak normal form and the normal form in R_2 (in nf_r2.rr)

```
/* normal form */
def normal_form2(F, G, VL, Order)
{
    M = length(G);
    Q = newvect(M);              /* vector for holding quotient */
    R = 0;                       /* variable for holding normal form */
    while (F != 0) {
                                 /* compute weak normal form of F by G */
        L = w_normal_form2(F, G, VL, Order);
        RR = L[0];               /* weak normal form */
        QQ = L[1];               /* quotient */
        F = RR - in2(RR, VL, Order);
        R = R + in2(RR, VL, Order);
        Q = Q + QQ;
        R = red(R);              /* reduce the rational function R */
        Q = map(red, Q);         /* reduce each component of vector Q */
    }
    return [R, Q];
}

/* weak normal form */
def w_normal_form2(F, G, VL, Order)
{
    M = length(G);
    Q = newvect(M);                  /* vector for holding quotient */
    R = F;                           /* variable for holding weak normal form */
    while ((Index = reducible2(R, G, VL, Order)) != -1) {
        S = G[Index];
        D = tdiv(in2_(R, VL, Order), in2_(S, VL, Order));
        X = c_in2(R, VL, Order) / c_in2(S, VL, Order);
```

```
        T = d_mult2(X*D, S);
        Q[Index] = Q[Index] + X*D;
        R = R - T;
        Q[Index] = red(Q[Index]);    /* reduce the rational function Q[Index] */
        R = red(R);                   /* reduce the rational function R */
    }
    return [R, Q];
}

/* Is R divisible by initial term of G ? */
def reducible2(R, G, VL, Order)
{
    M = length(G);
    InR = in2_(R, VL, Order);
    for (I = 0; I < M; I++) {
        InG = in2_(G[I], VL, Order);
        T = tdiv(InR, InG);          /* divisibility test: InG | InR ?*/
        if (T != 0)                   /* the case of InG | InR */
            return I;
    }
    return -1;
}
```

We see from the output that the normal form of $F = \partial_x \partial_y^3$ by $G_1 = \partial_x \partial_y + 1$ and $G_2 = 2y\partial_y^2 - \partial_x + 3\partial_y + 2x$ with respect to the graded reverse lexicographic order with $\partial_x \succ \partial_y$ is $-\frac{1}{2y}\partial_x + \frac{3}{2y}\partial_y + \frac{1}{y}x$. Moreover, we obtain the relation

$$F = \partial_y^2 G_1 + \left(-\frac{1}{2y}\right) G_2 + \left(-\frac{1}{2y}\partial_x + \frac{3}{2y}\partial_y + \frac{1}{y}x\right)$$

from the list of quotients.

Listing 7.102 Risa/Asir: computing the normal form by using `nf_r2.rr`

```
[1306] load("nf_r2.rr");
[1330] F = dx*dy^3;
dy^3*dx
[1331] G = [dx*dy + 1, 2*y*dy^2-dx+3*dy+2*x];
[dy*dx+1,-dx+2*y*dy^2+3*dy+2*x]
[1332] normal_form2(F, G, [dx,dy], 0);
[(-1/2*dx+3/2*dy+x)/(y),[ dy^2 (-1/2)/(y) ]]
```

Finally, we implement the function `sp2`, which computes S-pairs and Buchberger's algorithm in R_2 (see Algorithm 3.1.1).

Listing 7.103 Risa/Asir: Buchberger's algorithm in R_2 (with `nf_r2.rr`)

```
/* S-pair sp(F,G) */
def sp2(F, G, VL, Order)
{
    InF = in2_(F, VL, Order);
    InG = in2_(G, VL, Order);
    LCF = c_in2(F, VL, Order);
    LCG = c_in2(G, VL, Order);
    LCM = lcm(InF, InG);
    MF = tdiv(LCM, InF);         /* quotient of division LCM by InF */
    MG = tdiv(LCM, InG);         /* quotient of division LCM by InG */
    Sp = d_mult2(MF, F) - red(LCF/LCG) * d_mult2(MG, G);
```

```
        return red(Sp);
}

/* Buchberger's algorithm in $R_2$ */
def buchberger2(L, VL, Order)
{
    N = length(L);
    Jlist = [];                    /* list for holding index pairs */
    for (I = 0; I < N; I++)
        for (J = I + 1; J < N; J++)
            Jlist = append(Jlist, [[I, J]]);   /* generate index pairs */
    while (Jlist != []) {
        Index = car(Jlist);        /* the first component of Jlist */
        Jlist = cdr(Jlist);        /* remove the first component from Jlist */
        Sp = sp2(L[Index[0]], L[Index[1]], VL, Order);
        Result = normal_form2(Sp, L, VL, Order);
        R = Result[0];             /* normal form of Sp by L */
        if (R != 0) {              /* normal form of Sp by L is non zero */
            L = append(L, [R]);    /* add R to L as the last component */
            N = length(L);
            /* generate index pair for the new element R */
            for (I = 0; I < N - 1; I++)
                Jlist = append(Jlist, [[I, N - 1]]);
        }
    }
    return L;
}
```

The following sample code computes the S-pair of $f_1 = \partial_x^2 + y^2$ and $f_2 = \partial_y^2 + x^2$, and shows that $G = \{\partial_x^2 + y^2, \partial_y^2 + x^2, -4x\partial_x + 4y\partial_y\}$ is a Gröbner basis of $\langle f_1, f_2 \rangle \subset R_2$ with respect to the graded reverse lexicographic order with $\partial_x \succ \partial_y$.

Listing 7.104 Risa/Asir: computation of a Gröbner basis using `nf_r2.rr`

```
[1333] load("nf_r2.rr");
[1357] F1 = dx^2 + y^2$
[1358] F2 = dy^2 + x^2$
[1359] sp2(F1, F2, [dx,dy], 0);
-x^2*dx^2-4*x*dx+y^2*dy^2+4*y*dy
[1360] buchberger2([F1, F2], [dx,dy], 0);
[dx^2+y^2,dy^2+x^2,-4*x*dx+4*y*dy]
```

Exercise. Implement the algorithm of Example 7.4.6 on Macaulay2, Singular, Maple, and other systems.

Example 7.4.7. Prove Buchberger's criterion in R by referring to the proof on a polynomial ring.

Answer. This is analogous to the proofs of Lemma 1.3.2 and Theorem 1.3.3, by replacing the polynomial ring with the ring of differential operators. However, we note that multiplication is noncommutative on the ring of differential operators.

Exercise. Explain the reason why Lemma 1.3.1 does not hold on the ring of differential operators with rational function coefficients R (see Example 6.1.9).

7.4.2 Zero-Dimensional Ideals in R and Pfaffian Equations (Sect. 6.2)

Example 7.4.8. Compute a Gröbner basis in R for the system of differential equations in Example 6.5.4:

$$L_1 \bullet f = L_2 \bullet f = 0, \quad L_1 = y\partial_x - x\partial_y, L_2 = \partial_x\partial_y + 4xy,$$

and derive a Pfaffian system both by hand calculations and by using computer software.

Answer. (hand calculation) We perform Buchberger's algorithm with respect to the graded reverse lexicographic order with $\partial_x \succ \partial_y$:

$$\begin{aligned}
\text{sp}(L_1, L_2) &= \partial_y L_1 - yL_2 \\
&= \partial_y(y\partial_x - x\partial_y) - y(\partial_x\partial_y + 4xy) \\
&= (y\partial_y + 1)\partial_x - x\partial_y^2 - y\partial_x\partial_y - 4xy^2 \\
&= -x\partial_y^2 + \partial_x - 4xy^2 \\
&\longrightarrow -x\partial_y^2 + \frac{x}{y}\partial_y - 4xy^2 =: L_3 \quad \text{by } L_1;
\end{aligned}$$

$$\begin{aligned}
\text{sp}(L_1, L_3) &= x\partial_y^2 L_1 + y\partial_x L_3 \\
&= x\partial_y^2(y\partial_x - x\partial_y) + y\partial_x\left(-x\partial_y^2 + \frac{x}{y}\partial_y - 4xy^2\right) \\
&= x(y\partial_y^2 + 2\partial_y)\partial_x - x^2\partial_y^3 \\
&\qquad - y(x\partial_x + 1)\partial_y^2 + (x\partial_x + 1)\partial_y - 4y(x\partial_x + 1)y^2 \\
&= 2x\partial_x\partial_y - x^2\partial_y^3 - y\partial_y^2 + x\partial_x\partial_y + \partial_y - 4xy^3\partial_x - 4y^3 \\
&\longrightarrow 2x(-4xy) + x\partial_y\left(-\frac{x}{y}\partial_y + 4xy^2\right) + y\left(-\frac{1}{y}\partial_y + 4y^2\right) \\
&\qquad + x(-4xy) + \partial_y - 4xy^3\partial_y - 4y^3 \quad \text{by } L_1, L_2, L_3 \\
&= -4xy^3\partial_x - \frac{x^2}{y}\partial_y^2 + \left(\frac{x^2}{y^2} + 4x^2y^2\right)\partial_y - 4x^2y \\
&\longrightarrow -4xy^2(x\partial_y) + \frac{x}{y}\left(-\frac{x}{y}\partial_y + 4xy^2\right) \\
&\qquad + \left(\frac{x^2}{y^2} + 4x^2y^2\right)\partial_y - 4x^2y \quad \text{by } L_1, L_3 \\
&= 0;
\end{aligned}$$

$$\mathrm{sp}(L_2, L_3) = x\partial_y L_2 + \partial_x L_3$$

$$= x\partial_y(\partial_x\partial_y + 4xy) + \partial_x\left(-x\partial_y^2 + \frac{x}{y}\partial_y - 4xy^2\right)$$

$$= x\partial_x\partial_y + 4x^2(y\partial_y + 1) - (x\partial_x + 1)\partial_y^2 + \frac{1}{y}(x\partial_x + 1)\partial_y - 4y^2(x\partial_x + 1)$$

$$= \frac{x}{y}\partial_x\partial_y - 4xy^2\partial_x - \partial_y^2 + \left(4x^2y + \frac{1}{y}\right)\partial_y + 4x^2 - 4y^2$$

$$\longrightarrow \frac{x}{y}(-4xy) - 4xy(x\partial_y) - \partial_y^2 + \left(4x^2y + \frac{1}{y}\right)\partial_y + 4x^2 - 4y^2 \quad \text{by } L_1, L_2$$

$$= -4x^2 - 4x^2y\partial_y - \partial_y^2 + \left(4x^2y + \frac{1}{y}\right)\partial_y + 4x^2 - 4y^2$$

$$\longrightarrow -4x^2 - 4x^2y\partial_y - \left(\frac{1}{y}\partial_y - 4y^2\right) + \left(4x^2y + \frac{1}{y}\right)\partial_y + 4x^2 - 4y^2 \quad \text{by } L_3$$

$$= 0.$$

Therefore, $\{L_1, L_2, L_3\}$ is a Gröbner basis of the system.

Since the set of standard monomials is $\{1, \partial_y\}$, we calculate the normal forms of ∂_x, $\partial_x\partial_y$, ∂_y, and ∂_y^2 by $\{L_1, L_2, L_3\}$:

$$
\begin{aligned}
\partial_x &\longrightarrow^* \frac{x}{y}\partial_y && \text{by } G, \\
\partial_x\partial_y &\longrightarrow^* -4xy && \text{by } G, \\
\partial_y &\longrightarrow^* \partial_y && \text{by } G, \\
\partial_y^2 &\longrightarrow^* -4y^2 + \frac{1}{y}\partial_y && \text{by } G.
\end{aligned}
$$

Setting $F = \begin{pmatrix} f \\ \frac{\partial f}{\partial y} \end{pmatrix}$, we obtain the Pfaffian system:

$$\frac{\partial F}{\partial x} = \begin{pmatrix} 0 & \frac{x}{y} \\ -4xy & 0 \end{pmatrix} F, \qquad \frac{\partial F}{\partial y} = \begin{pmatrix} 0 & 1 \\ -4y^2 & \frac{1}{y} \end{pmatrix} F.$$

(by computer) We compute a Gröbner basis of the ideal $I = \langle L_1, L_2 \rangle$ generated by the operators $L_1 = y\partial_x - x\partial_y$ and $L_2 = \partial_x\partial_y + 4xy$ with respect to the graded reverse lexicographic order with $\partial_x \succ \partial_y$ by using the program nf_r2.rr on Risa/Asir.

Listing 7.105 Risa/Asir: Gröbner basis computation in R_2

```
[1407]  load("nf_r2.rr");
[1431]  L1=y*dx-x*dy;
[1432]  L2=dx*dy+4*x*y;
[1434]  GB=buchberger2([L1,L2],[dx,dy],0);
[y*dx-x*dy,dy*dx+4*y*x,(-y*x*dy^2+x*dy-4*y^3*x)/(y)]
```

We see from the output that the Gröbner basis is $G = \{y\partial_x - x\partial_y, \partial_x\partial_y + 4xy, -x\partial_y^2 + \frac{x}{y}\partial_y - 4xy^2\}$.

The initial ideal of I is $\text{in}_{\prec}(I) = \langle \xi_x, \xi_x \xi_y, \xi_y^2 \rangle \subset \mathbf{C}(x, y)[\xi_x, \xi_y]$. We find that the set of standard monomials of I is $\{1, \partial_y\}$. We then compute normal forms of ∂_x, $\partial_x \partial_y$, ∂_y, $\partial_y \partial_y$ by G.

Listing 7.106 Risa/Asir: computation of the normal form in R_2

```
[1435] normal_form2(dx,GB,VL=[dx,dy],0);
[(x*dy)/(y),[ (1)/(y) 0 0 ]]
[1436] normal_form2(dx*dy,GB,VL=[dx,dy],0);
[-4*y*x,[ (y*dy-1)/(y^2) 0 (-1)/(y) ]]
[1437] normal_form2(dy,GB,VL=[dx,dy],0);
[dy,[ 0 0 0 ]]
[1438] normal_form2(dy*dy,GB,VL=[dx,dy],0);
[(dy-4*y^3)/(y),[ 0 0 (-1)/(x) ]]
```

The output represents

$$
\begin{array}{lll}
\partial_x & \longrightarrow^* & \frac{x}{y} \partial_y & \text{by } G, \\
\partial_x \partial_y & \longrightarrow^* & -4xy & \text{by } G, \\
\partial_y & \longrightarrow^* & \partial_y & \text{by } G, \\
\partial_y \partial_y & \longrightarrow^* & -4y^2 + \frac{1}{y} \partial_y & \text{by } G.
\end{array}
$$

Setting $F = \begin{pmatrix} f \\ \frac{\partial f}{\partial y} \end{pmatrix}$, we obtain the Pfaffian system:

$$
\frac{\partial F}{\partial x} = \begin{pmatrix} 0 & \frac{x}{y} \\ -4xy & 0 \end{pmatrix} F, \quad \frac{\partial F}{\partial y} = \begin{pmatrix} 0 & 1 \\ -4y^2 & \frac{1}{y} \end{pmatrix} F.
$$

Exercise. Compute a Gröbner basis of $I = \langle L_1 := x\partial_x - (2x + y)\partial_y, L_2 := x\partial_x \partial_y - \partial_y + 2x^3 + x^2 y \rangle$ in R, and derive a Pfaffian system for the system of differential equations: $L_1 \bullet f = L_2 \bullet f = 0$.

Example 7.4.9. Check the integrability condition for the Pfaffian system in Example 7.4.8. That is, show that the Pfaffian system satisfies the condition in Theorem 6.2.3.

Answer. Set

$$
P_1 = \begin{pmatrix} 0 & \frac{x}{y} \\ -4xy & 0 \end{pmatrix}, P_2 = \begin{pmatrix} 0 & 1 \\ -4y^2 & \frac{1}{y} \end{pmatrix}.
$$

We will prove that

$$
\frac{\partial P_1}{\partial y} + P_1 P_2 = \frac{\partial P_2}{\partial x} + P_2 P_1.
$$

Because the derivatives are

$$\frac{\partial P_1}{\partial y} = \begin{pmatrix} 0 & -\frac{x}{y^2} \\ -4x & 0 \end{pmatrix}, \frac{\partial P_2}{\partial x} = \begin{pmatrix} 0 & 0 \\ 0 & 0 \end{pmatrix},$$

and products are

$$P_1 P_2 = \begin{pmatrix} -4xy & \frac{x}{y^2} \\ 0 & -4xy \end{pmatrix}, P_2 P_1 = \begin{pmatrix} -4xy & 0 \\ -4x & -4xy \end{pmatrix},$$

we find that $\frac{\partial P_1}{\partial y} + P_1 P_2$ and $\frac{\partial P_2}{\partial x} + P_2 P_1$ both coincide with

$$\begin{pmatrix} -4xy & 0 \\ -4x & -4xy \end{pmatrix}.$$

Exercise. Check the integrability condition for the Pfaffian system in Exercise 7.4.8.

Example 7.4.10. Compute a Gröbner basis for the system of differential equations for Appell's F_1 in R, and derive a Pfaffian system.

Answer. (using the program nf_r2.rr) The system of differential equations for the Appell's function

$$F_1(\alpha, \beta, \beta', \gamma; x, y) = \sum_{m,n=0}^{\infty} \frac{(\alpha)_{m+n} (\beta)_m (\beta')_n}{(\gamma)_{m+n} (1)_m (1)_n} x^m y^n$$

consists of two operators:

$$(x(1-x)\partial_x^2 + y(1-x)\partial_x\partial_y + (\gamma - (\alpha + \beta + 1)x)\partial_x - \beta y\partial_y - \alpha\beta) \bullet F_1 = 0,$$

$$(y(1-y)\partial_y^2 + x(1-y)\partial_x\partial_y + (\gamma - (\alpha + \beta' + 1)y)\partial_y - \beta'x\partial_x - \alpha\beta') \bullet F_1 = 0.$$

(Note: The ideal generated by the above two operators is not 0-dimensional in R when the parameters $\alpha, \beta, \beta', \gamma$ are special values. If we add the operator

$$(x - y)\partial_x\partial_y - \beta'\partial_x + \beta\partial_y,$$

it is 0-dimensional for any $\alpha, \beta, \beta', \gamma$. However, regarding it as a Pfaffian system in R with $\mathbf{Q}(\alpha, \beta, \beta', \gamma)(x, y)$ coefficients, we consider the ideal without the third operator.)

Let P_1 and P_2 be the operators given above, respectively. We compute a Gröbner basis for the ideal $I = \langle P_1, P_2 \rangle$ with respect to the graded reverse lexicographic order with $\partial_x \succ \partial_y$ in R. In the following code samples, a, b1, b2, c represent the parameters $\alpha, \beta, \beta', \gamma$, respectively.

Listing 7.107 Risa/Asir: computing the normal form in R_2

```
[1440] Id=[((-x^2+x)*dx^2+((-y*x+y)*dy+(-a-b1-1)*x+c)*dx-b1*y*dy-b1*a,
((-y+1)*x*dy-b2*x)*dx+(-y^2+y)*dy^2+((-a-b2-1)*y+c)*dy-b2*a];
[1441] GB=buchberger2(Id,[dx,dy],0);
[(-x^2+x)*dx^2+((-y*x+y)*dy+(-a-b1-1)*x+c)*dx-b1*y*dy-b1*a,
...(omitted)...]
[1442] map(in2_, GB, [dx,dy], 0);
[dx^2,dy*dx,dy^2]
```

We thus obtain the initial ideal $\mathrm{in}_{\prec}(I) = \langle \xi_x^2, \xi_x\xi_y, \xi_y^2 \rangle$ and the set of standard monomials $\{1, \partial_x, \partial_y\}$. We next compute the normal forms of ∂_x, $\partial_x\partial_x$, $\partial_x\partial_y$, ∂_y, $\partial_y\partial_x$, and $\partial_y\partial_y$ by G.

Listing 7.108 Risa/Asir: computing the normal form in R_2

```
[1444] normal_form2(dx,GB,VL=[dx,dy],0);
[dx,[ 0 0 0 ]]
[1445] normal_form2(dx*dx,GB,VL=[dx,dy],0);
[((((-a-b1-1)*x^2+((a+b1-b2+1)*y+c)*x+(-c+b2)*y)*dx
+(b1*y^2-b1*y)*dy-b1*a*x+b1*a*y)/(x^3+(-y-1)*x^2+y*x),
[ (-1)/(x^2-x) (y)/((y-1)*x^2) (-y^2+y)/((a-c+1)*x^2+(-a+c-1)*y*x)]]
[1446] normal_form2(dx*dy,GB,VL=[dx,dy],0);
[(b2*dx-b1*dy)/(x-y),
[ 0 (-1)/((y-1)*x) (y-1)/((a-c+1)*x+(-a+c-1)*y) ]]
[1447] normal_form2(dy,GB,VL=[dx,dy],0);
[dy,[ 0 0 0 ]]
[1448] normal_form2(dy*dx,GB,VL=[dx,dy],0);
[(b2*dx-b1*dy)/(x-y),[ 0 (-1)/((y-1)*x) (y-1)/((a-c+1)*x+(-a+c-1)*y)]]
[1449] normal_form2(dy*dy,GB,VL=[dx,dy],0);
[((-b2*x^2+b2*x)*dx+(((-a+b1-b2-1)*y+c-b1)*x+(a+b2+1)*y^2-c*y)*dy
-b2*a*x+b2*a*y)/((y^2-y)*x-y^3+y^2),
[ 0 0 ((-y+1)*x)/((a-c+1)*y*x+(-a+c-1)*y^2) ]]
```

The output represents

$$
\begin{aligned}
\partial_x \;\; &\longrightarrow^* \;\; \partial_x & \text{by } G, \\[4pt]
\partial_x\partial_x \;\; &\longrightarrow^* \;\; \frac{-\alpha\beta}{x(x-1)} + \frac{\beta y(y-1)}{x(x-1)(x-y)}\partial_y \\[4pt]
&\qquad + \frac{(-\alpha-\beta-1)x^2+((\alpha+\beta-\beta'+1)y+\gamma)x+(-\gamma+\beta')y}{x(x-1)(x-y)}\partial_x & \text{by } G, \\[4pt]
\partial_x\partial_y \;\; &\longrightarrow^* \;\; \frac{\beta'}{x-y}\partial_x + \frac{-\beta}{x-y}\partial_y & \text{by } G, \\[4pt]
\partial_y \;\; &\longrightarrow^* \;\; \partial_y & \text{by } G, \\[4pt]
\partial_y\partial_x \;\; &\longrightarrow^* \;\; \frac{\beta'}{x-y}\partial_x + \frac{-\beta}{x-y}\partial_y & \text{by } G, \\[4pt]
\partial_y\partial_y \;\; &\longrightarrow^* \;\; \frac{-\alpha\beta'}{y(y-1)} + \frac{-\beta'x(x-1)}{y(y-1)(x-y)}\partial_x \\[4pt]
&\qquad + \frac{((-\alpha+\beta-\beta'-1)y+\gamma-\beta)x+(\alpha+\beta'+1)y^2-\gamma y}{y(y-1)(x-y)}\partial_y & \text{by } G.
\end{aligned}
$$

Setting $F = \begin{pmatrix} f \\ \frac{\partial f}{\partial x} \\ \frac{\partial f}{\partial y} \end{pmatrix}$, where $f = F_1$, we obtain the Pfaffian system:

$$
\frac{\partial F}{\partial x} = \begin{pmatrix} 0 & 1 & 0 \\[6pt] \dfrac{-\alpha\beta}{x(x-1)} & \dfrac{(-\alpha-\beta-1)x^2+((\alpha+\beta-\beta'+1)y+\gamma)x+(-\gamma+\beta')y}{x(x-1)(x-y)} & \dfrac{\beta y(y-1)}{x(x-1)(x-y)} \\[6pt] 0 & \dfrac{\beta'}{x-y} & \dfrac{-\beta}{x-y} \end{pmatrix} F,
$$

$$\frac{\partial F}{\partial y} = \begin{pmatrix} 0 & 0 & 1 \\ 0 & \frac{\beta'}{x-y} & \frac{-\beta}{x-y} \\ \frac{-\alpha\beta'}{y(y-1)} & \frac{-\beta'x(x-1)}{y(y-1)(x-y)} & \frac{((-\alpha+\beta-\beta'-1)y+\gamma-\beta)x+(\alpha+\beta'+1)y^2-\gamma y}{y(y-1)(x-y)} \end{pmatrix} F.$$

(by using the program `yang.rr`) `yang.rr` is a package of Risa/Asir which is used to compute Gröbner bases in the rings generated by Euler operators ($\theta_x = x\partial_x$), difference operators, or q-difference operators. We will show how to use it by using Appell's function F_1 as an example.

First, to load the package, use the command `load("yang.rr");`.[1] We note that, with this package, the computations will be performed over the ring $R = Q\langle\theta_x, \theta_y\rangle$ (the symbols in the display, dx and dy, represent the Euler operators θ_x and θ_y). Moreover we can take the rational function field, for example, $\mathbf{Q}(\alpha, \beta, \beta', \gamma)$, as the coefficient field. Second, to declare the ring, use the command `yang.define_ring`. To multiply one operator by another, use the command `yang.mul`. Rewriting the system of differential equations for Appell's function F_1 by using the Euler operators, we have

$$(x(\theta_x + \theta_y + \alpha)(\theta_x + \beta) - (\theta_x + \theta_y + \gamma - 1)\theta_x) \bullet F_1 = 0,$$
$$(y(\theta_x + \theta_y + \alpha)(\theta_y + \beta') - (\theta_x + \theta_y + \gamma - 1)\theta_y) \bullet F_1 = 0.$$

Let P_1 and P_2 be the operators given above, respectively. Third, we compute a Gröbner basis G for the ideal $I = \langle P_1, P_2 \rangle$, by using the command `yang.gr`. With the command `yang.stdmon`, we get the set of standard monomials $S = \{1, \theta_x, \theta_y\}$. Finally, using the command `yang.pf`, we obtain the Pfaffian system for Appell's F_1. (We can derive the Pfaffian system from the normal form, which can be computed by the command `yang.nf`.) Setting $F = \begin{pmatrix} f \\ \theta_x \bullet f \\ \theta_y \bullet f \end{pmatrix}$, we obtain

the Pfaffian system:

$$\frac{\partial F}{\partial x} = \begin{pmatrix} 0 & \frac{1}{x} & 0 \\ \frac{-\alpha\beta}{x-1} & \frac{\beta'-\gamma+1}{x} + \frac{\gamma-1-\alpha-\beta}{x-1} & \frac{-\beta'}{x-y} \frac{-\beta}{x-1} + \frac{\beta}{x-y} \\ 0 & \frac{-\beta'}{x} + \frac{\beta'}{x-y} & \frac{-\beta}{x-y} \end{pmatrix} F,$$

$$\frac{\partial F}{\partial y} = \begin{pmatrix} 0 & 0 & \frac{1}{y} \\ 0 & \frac{\beta'}{x-y} & \frac{-\beta}{y} + \frac{-\beta}{x-y} \\ \frac{-\alpha\beta'}{y-1} & \frac{-\beta'}{y-1} + \frac{-\beta'}{x-y} & \frac{\beta-\gamma+1}{y} + \frac{\gamma-1-\alpha-\beta'}{y-1} + \frac{\beta}{x-y} \end{pmatrix} F.$$

[1]Error messages may be displayed if Asir Contrib is not loaded since the package `yang.rr` uses some functions defined in Asir Contrib. If this happens, use the command `import("names.rr");`.

Here, for simplicity, each component of a Pfaffian matrix has been decomposed into partial fractions.

Listing 7.109 Risa/Asir: `yang.rr`

```
[1371] load("yang.rr");
[1838] yang.define_ring([x,y],0);
{[euler,[x,y]],[x,y],[0,0],[0,0],[dx,dy]}
[1839] P1 = x*yang.mul(dx+dy+a,dx+b1) - yang.mul(dx+dy+c-1,dx);
(x-1)*dx^2+((x-1)*dy+(a+b1)*x-c+1)*dx+b1*x*dy+b1*a*x
[1840] P2 = y*yang.mul(dx+dy+a,dy+b2) - yang.mul(dx+dy+c-1,dy);
((y-1)*dy+b2*y)*dx+(y-1)*dy^2+((a+b2)*y-c+1)*dy+b2*a*y
[1841] GB = yang.gr([P1,P2]);
[((b2*y*x-b2*y)*dx+((y-1)*x-y^2+y)*dy^2+(((a-b1+b2)*y-c+b1+1)*x+(-a-b2)*y^2+
(c-1)*y)*dy+b2*a*y*x-b2*a*y^2)/((y-1)*x-y^2+y),
(((x-y)*dy-b2*y)*dx+b1*x*dy)/(x-y),
((x^2+(-y-1)*x+y)*dx^2+((a+b1)*x^2+((-a-b1+b2)*y-c+1)*x+(c-b2-1)*y)*dx+(-b1
*y+b1)*x*dy+b1*a*x^2-b1*a*y*x)/(x^2+(-y-1)*x+y)]
[1842] Std = yang.stdmon(GB);
[dx,dy,1]
[1845] Std=[1,dx,dy];
[1,dx,dy]
[1846] yang.pf(Std, GB);
[ [ 0 (1)/(x) 0 ]
[ (-b1*a)/(x-1) ((-a-b1)*x^2+((a+b1-b2)*y+c-1)*x+(-c+b2+1)*y)/(x^3+(-y-1)*x^
2+y*x) (b1*y-b1)/(x^2+(-y-1)*x+y) ]
[ 0 (b2*y)/(x^2-y*x) (-b1)/(x-y) ]
[ 0 0 (1)/(y) ]
[ 0 (b2)/(x-y) (-b1*x)/(y*x-y^2) ]
[ (-b2*a)/(y-1) (-b2*x+b2)/((y-1)*x-y^2+y) (((-a+b1-b2)*y+c-b1-1)*x+(a+b2)*
y^2+(-c+1)*y)/((y^2-y)*x-y^3+y^2) ] ]
```

Exercise. 1. Check the integrability condition for the Pfaffian system of Appell's F_1 in Example 7.4.10.

2. Derive the system of differential equations annihilating Appell's series F_1.

3. Derive Pfaffian systems for the hypergeometric functions in Horn's list (see [5, 24]).

The answer to Exercise 3 is available at the website of Prof. Ohara, who developed `yang.rr`:
http://air.s.kanazawa-u.ac.jp/~ohara/HG-Pfaffian/index.html.

7.4.3 Solutions of Pfaffian Equations (Sect. 6.3)

Example 7.4.11. Calculating by hand, find a series solution at the origin for the differential equation: $(3\partial_x^2 + 6\partial_x + (3 - x)) \bullet f = \exp(-x + 1)$. Next, use Maple to compute the solution.

Answer. Let $f = \sum_{k=s}^{\infty} a_k x^k$ be a power series. To determine the initial term of f, apply the differential operator:

$$(3\partial_x^2 + 6\partial_x + (3 - x)) \bullet f$$

$$= 3\sum_{k=s} k(k-1)a_k x^{k-2} + 6\sum_{k=s} ka_k x^{k-1} + 3\sum_{k=s} a_k x^k - \sum_{k=s} a_k x^{k+1}$$

$$= 3s(s-1)a_s x^{s-2} + \{3s(s+1)a_{s+1} + 6sa_s\}x^{s-1}$$

$$+ \{3(s+1)(s+2)a_{s+2} + 6(s+1)a_{s+1} + 3a_s\}x^s$$

$$+ \left\{3\sum_{k=s}(k+2)(k+3)a_{k+3} + 6\sum_{k=s}(k+2)a_{k+2} + 3\sum_{k=s} a_{k+1} - \sum_{k=s} a_k\right\} x^{k+1}.$$

The Taylor series for $\exp(-x + 1)$ at the origin is

$$\exp(-x + 1) = \sum_{k=0}^{\infty} \frac{(-1)^k e}{k!} x^k, \quad \text{where } e = \exp(1).$$

By comparing the exponent parts, we can determine the exponent s of the initial term for the following three cases.

(1) When $s = 2$, the initial term is $3s(s - 1)a_s x^{s-2}$, which implies that $a_0 = 0$, $a_1 = 0$, and $6a_2 = e$.
(2) When $s = 1$, the initial term is $\{3s(s + 1)a_{s+1} + 6sa_s\}x^{s-1}$, which implies that $a_0 = 0$ and $6a_2 + 6a_1 = e$.
(3) When $s = 0$, the initial term is $\{3(s + 1)(s + 2)a_{s+2} + 6(s + 1)a_{s+1} + 3a_s\}x^s$, which implies that $6a_2 + 6a_1 + 3a_0 = e$.

Cases (1) and (2) are special cases of (3) under the condition that the constants a_0 and a_1 are restricted to 0. Therefore, the coefficient a_k of the series solution is given by the following recurrence relation:

$$3(k + 2)(k + 3)a_{k+3} = \frac{(-1)^{k+1}}{(k + 1)!}e - 6(k + 2)a_{k+2} - 3a_{k+1} + a_k \quad (k \geq 0),$$

where a_0 and a_1 are arbitrary constants and $a_2 = \frac{1}{6}e - a_1 - \frac{1}{2}a_0$.

To use Maple to compute a series solution of an inhomogeneous differential equation, use the command dsolve with the option series.

Listing 7.110 Maple: computation of a series solution

```
> ode:=3*diff(f(x),x,x)+6*diff(f(x),x)+(3-x)*f(x);
                / 2      \
               |d        |         /d       \
       ode := 3 |--- f(x) | + 6  |-- f(x) | + (3 - x) f(x)
               | 2       |         \dx      /
               \dx       /

> dsolve({ode=exp(-x+1)},f(x),series);
```

```
f(x) = f(0) + D(f)(0) x +
                                    2
        (-D(f)(0) - 1/2 f(0) + 1/6 exp(1)) x  +
                                         3
        (1/2 D(f)(0) + 7/18 f(0) - 1/6 exp(1)) x  +

        /                  11             \  4
        |-5/36 D(f)(0) -  -- f(0) + 1/12 exp(1)| x  +
        \                  72             /
                                               5        6
        (1/72 D(f)(0) + 1/30 f(0) - 1/40 exp(1)) x  + O(x )
```

In order to get higher-order terms, set `Order:=10;` and use `dsolve` again.

Maple has a powerful package, `DEtools`, for solving differential equations. Although it does not support inhomogeneous differential equations, we recommended using it for homogeneous equations. For the homogeneous case of this example, we obtain the solution by using the command `formal_sol`.

Listing 7.111 Maple: DEtools

```
> with(DEtools):
> ode:=3*diff(f(x),x,x)+6*diff(f(x),x)+(3-x)*f(x);
> formal_sol(ode,f(x),x=0,order=6);
         2      3        4        5       6
[x  - x  + 1/2 x  - 5/36 x  + 1/72 x  + O(x ),

                   2        3        4        5         6
     - 1 + x - 1/2 x  + 1/9 x  + 1/72 x  - 7/360 x  + O(x )]
```

We now check that the series solution derived from the recurrence relation coincides with the output from Maple. The following small program on Risa/Asir uses the recurrence relation to return the coefficients of a series solution up to N-th (> 2). Here, the symbol e is an indeterminate, not the base of the natural logarithm.

Listing 7.112 Risa/Asir: `series_rec.rr`

```
def series_rec(N)
{
    A = newvect(++N);
    A0 = a0; A1 = a1;
    A[0]=A0; A[1]=A1; A[2]=1/6*e-A1-1/2*A0;
    for ( I = 0; I < N-3; I++ ) {
        A[I+3] = (-1)^(I+1)/fac(I+1)*e
                        -6*(I+2)*A[I+2]-3*A[I+1]+A[I];
        A[I+3] /= 3*(I+3)*(I+2);
    }
    return A;
}
end$
```

Listing 7.113 Risa/Asir: computation of coefficients of the series solution

```
[1355] load("series_rec.rr")$
[1356] series_rec(5);
[ a0 a1 1/6*e-1/2*a0-a1 -1/6*e+7/18*a0+1/2*a1
  1/12*e-11/72*a0-5/36*a1 -1/40*e+1/30*a0+1/72*a1 ]
```

We see that at least the first five coefficients coincide with the output from Maple.

Exercise. Find a series solution at the origin for the differential equation: $(3\partial_x^3 + 9\partial_x^2 + (9 - x)\partial_x - x + 1) \bullet f = 4\exp(x - 7)$.

Example 7.4.12. Let U_i be a vector-valued function which is holomorphic at the origin, and let P_i be a matrix which satisfies the integrability condition (6.10). We consider the inhomogeneous Pfaffian system:

$$\frac{\partial F}{\partial x_i} = P_i F + U_i, \quad i = 1, \dots, n,$$

where the inhomogeneous part U_i is annihilated by the operator $\frac{\partial}{\partial x_j} - Q_j^i$ ($j = 1, \dots, n$). Applying this to (6.14), we obtain

$$\left(\frac{\partial^2}{\partial x_j \partial x_i} - Q_j^i \frac{\partial}{\partial x_j} \right) F = \left(\frac{\partial}{\partial x_j} - Q_j^i \right) P_i F.$$

Now, find a homogeneous Pfaffian system for the new vector-valued function $\hat{F} = \left(F^T, \frac{\partial F^T}{\partial x_1}, \dots, \frac{\partial F^T}{\partial x_n} \right)^T$. (Note: This situation commonly appears when we consider a system of differential operators annihilating a definite integral whose integral domain is not a cycle.)

Answer. The relation

$$\frac{\partial^2 F}{\partial x_j \partial x_i} = Q_j^i \frac{\partial F}{\partial x_j} + \frac{\partial (P_i F)}{\partial x_j} - Q_j^i P_i F$$

$$= Q_j^i \frac{\partial F}{\partial x_j} + P_i \frac{\partial F}{\partial x_j} + \frac{\partial P_i}{\partial x_j} F - Q_j^i P_i F$$

$$= \left(P_i + Q_j^i \right) \frac{\partial F}{\partial x_j} + \left(\frac{\partial P_i}{\partial x_j} - Q_j^i P_i \right) F$$

is derived from above expression. Therefore, the desired Pfaffian system: $\frac{\partial \hat{F}}{\partial x_i} = \hat{Q}_i \hat{F}$ ($1 \le i \le n$) is given by

$$
\begin{pmatrix}
\frac{\partial F}{\partial x_j} \\
\frac{\partial^2 F}{\partial x_j \partial x_1} \\
\vdots \\
\frac{\partial^2 F}{\partial x_j \partial x_i} \\
\vdots \\
\frac{\partial^2 F}{\partial x_j \partial x_n}
\end{pmatrix}
=
\begin{pmatrix}
O & O \cdots O & E & O \cdots O \\
\frac{\partial P_1}{\partial x_j} - Q_j^1 P_1 & O \cdots O & P_1 + Q_j^1 & O \cdots O \\
\vdots & \vdots \ddots \vdots & \vdots & \vdots \ddots \vdots \\
\frac{\partial P_i}{\partial x_j} - Q_j^i P_i & O \cdots O & P_i + Q_j^i & O \cdots O \\
\vdots & \vdots \ddots \vdots & \vdots & \vdots \ddots \vdots \\
\frac{\partial P_n}{\partial x_j} - Q_j^n P_n & O \cdots O & P_n + Q_j^n & O \cdots O
\end{pmatrix}
\begin{pmatrix}
F \\
\frac{\partial F}{\partial x_1} \\
\vdots \\
\frac{\partial F}{\partial x_j} \\
\vdots \\
\frac{\partial F}{\partial x_n}
\end{pmatrix}.
$$

The Pfaffian matrix consists of square matrices whose sizes are the same as that of P_i. Except for those in row 1 and $(j + 1)$, the component matrices are the zero matrix. Here, the symbol E is the identity matrix.

Example 7.4.13. By using Example 7.4.12, homogenize the inhomogeneous Pfaffian system in Example 7.4.11:

$$
\frac{dF}{dx} = \begin{pmatrix} 0 & 1 \\ (-3 + x)/3 & -2 \end{pmatrix} F + \begin{pmatrix} 0 \\ \exp(-x + 1)/3 \end{pmatrix}, \text{ where } F = \begin{pmatrix} f \\ \frac{df}{dx} \end{pmatrix}.
$$

Answer. The inhomogeneous part $U = \begin{pmatrix} 0 & \exp(-x + 1)/3 \end{pmatrix}^T$ satisfies the differential equation

$$
\frac{dU}{dx} = \begin{pmatrix} 0 & 0 \\ 0 & -1 \end{pmatrix} U.
$$

Set

$$
P = \begin{pmatrix} 0 & 1 \\ (-3 + x)/3 & -2 \end{pmatrix}, Q = \begin{pmatrix} 0 & 0 \\ 0 & -1 \end{pmatrix},
$$

and $\hat{F} = \begin{pmatrix} f & \frac{df}{dx} & \frac{df}{dx} & \frac{d^2 f}{dx^2} \end{pmatrix}^T$. From the result in Example 7.4.12, we obtain the homogeneous Pfaffian system

$$
\frac{d\hat{F}}{dx} = \begin{pmatrix} O & E_2 \\ \frac{dP}{dx} - QP & P + Q \end{pmatrix} \hat{F} = \begin{pmatrix} 0 & 0 & 1 & 0 \\ 0 & 0 & 0 & 1 \\ 0 & 0 & 0 & 1 \\ (-2 + x)/3 & -2 & (-3 + x)/3 & -3 \end{pmatrix} \hat{F}.
$$

Since the second and third components of $\hat{F}$ agree, the second and third columns of the Pfaffian matrix also agree. After reducing the redundant components, we obtain the following simpler Pfaffian system:

$$\begin{pmatrix} \frac{df}{dx} \\ \frac{d^2 f}{dx^2} \\ \frac{d^3 f}{dx^3} \end{pmatrix} = \begin{pmatrix} 0 & 1 & 0 \\ 0 & 0 & 1 \\ (-2+x)/3 & (-9+x)/3 & -3 \end{pmatrix} \begin{pmatrix} f \\ \frac{df}{dx} \\ \frac{d^2 f}{dx^2} \end{pmatrix}.$$

The function $\exp(-x+1)$ is annihilated by $\partial_x + 1$. We obtain the homogeneous differential equation $\{3\partial_x^3 + 9\partial_x^2 + (9-x)\partial_x + (2-x)\} \bullet f = 0$ by applying $\partial_x + 1$ from the left to the inhomogeneous differential equation. The set of standard monomials is $\{1, \partial_x, \partial_x^2\}$ and $3\partial_x^3$ is reduced as follows:

$$3\partial_x^3 \longrightarrow^* -9\partial_x^2 + (-9+x)\partial_x + (-2+x).$$

Thus, the same Pfaffian system is obtained.

Exercise. Derive an inhomogeneous Pfaffian system for the inhomogeneous differential equation which appears in Exercise 7.4.11:

$$(3\partial_x^3 + 9\partial_x^2 + (9-x)\partial_x - x + 1) \bullet f = 4\exp(x-7).$$

Moreover, derive a homogenized Pfaffian system by using Example 7.4.12.

Example 7.4.14. Find a holomorphic solution of the system of differential equations:

$$\frac{\partial F}{\partial x} = A_1 F, \frac{\partial F}{\partial y} = A_2 F, \quad A_1 = \begin{pmatrix} 0 & x/y \\ -4xy & 0 \end{pmatrix}, A_2 = \begin{pmatrix} 0 & 1 \\ -4y^2 & 1/y \end{pmatrix}.$$

Is the minimum value of the first component of the solution at the origin? We note that the equations appear in Example 6.5.4 and have the solution $F = (-\cos(x^2 + y^2), 2y\sin(x^2 + y^2))^T$.

Answer. It is difficult to calculate a series solution because the matrices A_1 and A_2 both have a y in the denominator of a component. Multiplying both sides by y, we consider $yF_y = (yA_2)F$. Thus, $yA_2 = \begin{pmatrix} 0 & 0 \\ 0 & 1 \end{pmatrix} + O(y)$ implies the new equation $F = TG$, where $T = \begin{pmatrix} 1 & 0 \\ 0 & y \end{pmatrix}$ is the transformation matrix. Then, the system of differential equations for G is

$$\frac{\partial G}{\partial x} = B_1 G, \frac{\partial G}{\partial y} = B_2 G, \quad B_1 = \begin{pmatrix} 0 & x \\ -4x & 0 \end{pmatrix}, B_2 = \begin{pmatrix} 0 & y \\ -4y & 0 \end{pmatrix}.$$

If we choose the transformation matrix T so that the pole of B_2 at $y = 0$ vanishes, then, fortunately, the pole of B_1 also vanishes. We obtain the series solution

$$G = \sum_{m,n=0}^{\infty} c_{mn} R^{m+n} G_0 x^{2m} y^{2n}, \quad R = \begin{pmatrix} 0 & 1 \\ -4 & 0 \end{pmatrix}, c_{mn} = \frac{1}{2^{m+n} m! n!}$$

by a straightforward calculation. Here, G_0 is an arbitrary row vector of length 2.

Let us consider the existence of a minimum of the function F at the origin. Since the first component of F is equal to the first component of TG, we obtain $(1,0) \sum c_{mn} R^{m+n} G_0 x^{2m} y^{2n}$ by multiplying the first row $(1,0)$ of T to G from the left. We have

$$(1,0)G_0 + (0,1)G_0 c_{10} x^2 + (0,1)G_0 c_{01} y^2 + O(4).$$

Here, the symbol $O(4)$ represents a fourth-order growth rate with respect to x and y.

When the terms with degree 2 are not zero, F has a minimum value if and only if $G_0 = (d,c)^T, c > 0$. When the terms with degree 2 are zero, F has a minimum value if $G_0 = (c,0)^T, c \neq 0$. In this case, the solution is expressed as

$$(1,0)G_0 + (-4,0)G_0 c_{11} x^2 y^2 + (-4,0)G_0 c_{20} x^4 + (-4,0)G_0 c_{02} y^4 + O(6).$$

Therefore, F has a minimum when $c < 0$.

In fact, the system of differential equations $L_1 \bullet f = L_2 \bullet f = 0$, $L_1 = y\partial_x - x\partial_y$, $L_2 = \partial_x \partial_y + 4xy$ has a general solution $f = A\cos(x^2+y^2) + B\sin(x^2+y^2)$. The vector-valued function $F = (f, f_y)^T$ satisfies the Pfaffian system. We consider the Taylor expansion of F at the origin. The function f has a minimum at $(x,y) = (0,0)$ for any real numbers A and $B > 0$. Moreover, it holds when $B = 0$ and $A < 0$. We reach the same conclusion with the Pfaffian system.

To draw a graph of the function, we use the program gnuplot. For easier viewing, we draw the graph as a univariate function with respect to r, by using the transformation $r^2 = x^2 + y^2$. The following command displays the graph of $-2\cos(r^2) + \sin(r^2)$. Since the independent variable on gnuplot is x, we replace r by x.

Listing 7.114 gnuplot: graphs

```
plot -2*cos(x**2)+sin(x**2);
```

To output graphs in postscript format for LATEX, use the following commands. Note that we must use the command `set terminal windows` instead of the last command `set terminal x11` in the Windows environment.

Listing 7.115 gnuplot: graphs (EPS format)

```
set terminal postscript
set output "test.eps"
plot -2*cos(x**2)+sin(x**2);
set terminal x11
```

When y is fixed, we can use the series solution obtained in this example to find the conditions for a univariate function with respect to x to have a minimum at the origin.

7.4.4 Holonomic Functions (Sect. 6.4)

Example 7.4.15. Let f be a rational function. Is the exponential function $\exp(f)$ a holonomic function? If it is holonomic, find a 0-dimensional ideal of R satisfying $I \bullet \exp(f) = 0$.

Answer. The derivative of $\exp(f)$ is $\frac{\partial \exp(f)}{\partial x_i} = \frac{\partial f}{\partial x_i} \exp(f)$. Because $f_i := \frac{\partial f}{\partial x_i}$ is a rational function, the ideal $I = \langle \partial_i - f_i \mid 1 \le i \le n \rangle \subset R$ annihilates $\exp(f)$, and it is 0-dimensional on R. Therefore, $\exp(f)$ is holonomic.

Exercise. Let f be a rational function. Is the exponential function $\log(f)$ a holonomic function? If it is holonomic, find a 0-dimensional ideal of R satisfying $I \bullet \log(f) = 0$.

Example 7.4.16. Let f and g be holonomic functions, and let I and J be ideals annihilating f and g, respectively. Find an algorithm for computing the ideals in R annihilating the sum $f + g$ and one annihilating the product fg.

Answer. For simplicity, let f and g be univariate holonomic functions with respect to x. We present an algorithm for the product fg. Assume that f satisfies

$$(\partial^l + p_{l-1}\partial^{l-1} + \cdots + p_0) \bullet f = 0 \quad (p_i \in \mathbf{Q}(x)),$$

and g satisfies

$$(\partial^m + q_{m-1}\partial^{m-1} + \cdots + q_0) \bullet g = 0 \quad (q_i \in \mathbf{Q}(x)).$$

From Leibniz's rule, the k-th derivative of fg is

$$\partial^k \bullet (fg) = \sum_{i=0}^{k} \binom{k}{i} (\partial^i \bullet f)(\partial^{k-i} \bullet g).$$

$\partial^i \bullet f$ can be expressed as a linear combination of $\partial^{l-1} \bullet f, \ldots, \partial \bullet f, f$ over $\mathbf{Q}(x)$, and $\partial^{k-i} \bullet g$ can be expressed as a linear combination of $\partial^{m-1} \bullet g, \ldots, \partial \bullet g, g$ over $\mathbf{Q}(x)$. Thus, we can rewrite the k-th derivative of fg as

$$\partial^k \bullet (fg) = \sum_{0 \le i \le l-1, 0 \le j \le m-1} d_{ij} (\partial^i \bullet f)(\partial^j \bullet g) \quad (d_{ij} \in \mathbf{Q}(x)).$$

From the expression on the right-hand side, $\partial^k \bullet (fg)$ can be regarded as an element $(d_{ij})_{0 \le i \le l-1, 0 \le j \le m-1}$ of the vector space $(\mathbf{Q}(x))^{lm}$. When k is sufficiently large, $fg, \partial \bullet (fg), \ldots, \partial^k \bullet (fg)$ is linearly dependent over $\mathbf{Q}(x)$. Then we have the linear relation

$$r_k \partial^k \bullet (fg) + r_{k-1}\partial^{k-1} \bullet (fg) + \cdots + r_0 fg = 0 \quad (r_i \in \mathbf{Q}(x)).$$

This is the differential equation for the product fg. The ideal $\langle r_k \partial^k + r_{k-1}\partial^{k-1} + \cdots + r_0 \rangle$ has the product fg as a solution.

For multivariate functions, by performing this algorithm for each variable, we obtain the univariate differential equations for all the variables. The ideal generated by them is 0-dimensional and has the product fg as a solution.

We can apply an analogous algorithm for the sum $f + g$.

Exercise. Derive a differential equation annihilating the sum $x + \exp(x)$ and one annihilating the product $x \sin(x)$, by using the algorithm in Example 7.4.16.

7.4.5 Gradient Descent for Holonomic Functions (Sect. 6.5)

Example 7.4.17. Draw a graph of the function in Example 6.5.3:

$$g(x) = \exp(-x + 1) \int_0^\infty \exp(xt - t^3)dt.$$

Answer. It is easy to draw a graph when using a system which has a definite integral calculator. Here, as an example of such a system, we use Maple.

Listing 7.116 Maple: drawing a graph

```
> h := x -> int(exp(x*t-t^3),t=0..infinity):
> g := x -> exp(-x+1)*h(x):
> g(3.4);
                                                    -13
                1.016334715 - 0.4834824802 10      I
> plot(Re(g(x)),x=3..4);
```

To define a function g on Maple, use the arrow symbol ->; it is then easy to obtain the value $g(x)$ for any number x. Although the command $g(3.4)$ is expected to return a real number, it returns a complex number (the symbol I is the imaginary unit). Sometimes Maple evaluates an integral as a complex-valued function. We can extract the real part by using the command Re. To draw a graph, use the command plot; the first argument specifies the functions, and the second argument specifies the domain.

The result of this command is shown in Fig. 7.22. However, the graph shows that $g(3.4)$ is about 0.018. This is different from the value obtained by the previous command. We note that it is necessary to pay close attention to possible errors in the results. Since we cannot solve this by referring to the manual of Maple, we will graph it in a different way (Fig. 7.23).

Listing 7.117 Maple: drawing a graph

```
> a:=3: b:=4: k:=30: L:=[]:
> for n from a to b by 1/k
  do
      L := [op(L),[n,Re(evalf((g(n))))]];
  od:
> plot(L,x=a..b);
```

Fig. 7.22 Incorrect graph of $g(x)$

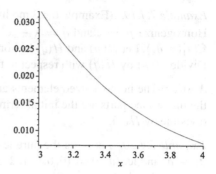

Fig. 7.23 Correct graph of $g(x)$

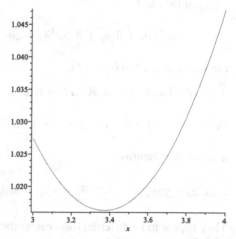

This method will give the correct values. We calculate the points by using a for loop and save the values to a list; we then draw the graph by plotting the points and connecting them with a line. The symbols a and b represent the endpoints of the domain, and 1/k is the interval used for plotting.

7.4.6 Gröbner Bases in the Ring of Differential Operators with Polynomial Coefficients D (Sect. 6.6)

Example 7.4.18. Rewrite $\partial_x^2 x \partial_x$ as a standard form $\sum c_{\alpha\beta\gamma} x^\alpha \partial_x^\beta h^\gamma$ of the homogenized Weyl algebra $D_1^{(h)} = \mathbf{C}[h]\langle x, \partial_x \rangle$.

Answer.

$$\partial_x^2 x \partial_x = \partial_x \partial_x x \partial_x = \partial_x (x \partial_x + h^2) \partial_x = \partial_x x \partial_x^2 + h^2 \partial_x^2$$
$$= (x \partial_x + h^2) \partial_x^2 + h^2 \partial_x^2 = x \partial_x^3 + 2 h^2 \partial_x^2.$$

Example 7.4.19. (Example 6.6.3 modified by homogenization)

Homogenize $p = x$ and $q = x + x^2$ on the homogenized Weyl algebra $D_1^{(h)} = C[h]\langle x, \partial_x \rangle$. Let $H(p)$ and $H(q)$ be homogenized elements of p and q, respectively. Divide $H(p)$ by $H(q)$ with respect to the weighted order $\prec_{(-1,1)}$.

Answer. The homogenized elements are $H(p) = \underline{x}$ and $H(q) = \underline{hx} + x^2$. Here, the underlined parts are the initial terms with respect to $\prec_{(-1,1)}$. Thus, $H(p)$ is not divisible by $H(q)$.

Example 7.4.20. Let $\prec$ be the pure lexicographic order with $x \succ y \succ z \succ \partial_x \succ \partial_y \succ \partial_z$ and let $(u, v) = (0, 0, 0, 1, 2, 3)$. Compute a Gröbner basis with respect to $\prec_{(u,v)}$ of the ideal

$$I = \langle x\partial_x + 3z\partial_z + 3, 3x^2\partial_y + z\partial_x, 3x^2\partial_z + y\partial_x, y\partial_y - z\partial_z \rangle \subset D_3.$$

Find the initial ideal $\text{in}_{(u,v)}(I)$.

Answer. Macaulay2 and Risa/Asir return

$$\langle x^3\partial_x - yz\partial_x + 3x^2, 3y\partial_y + x\partial_x + 3, 3x^2\partial_y + z\partial_x, \underline{3z\partial_z + x\partial_x + 3}, 3x^2\partial_z + y\partial_x \rangle,$$

and Singular returns

$$\langle x^3\partial_x - yz\partial_x + 3x^2, 3y\partial_y + x\partial_x + 3, 3x^2\partial_y + z\partial_x, \underline{z\partial_z - y\partial_y}, 3x^2\partial_z + y\partial_x \rangle.$$

They appear to be different from each other. However, they are both Gröbner bases of I, because both of the initial terms are $z\xi_z$. In this case, the relation

$$3(\underline{z\partial_z - y\partial_y}) = (\underline{3x\partial_z + x\partial_x + 3}) - (3y\partial_y + x\partial_x + 3)$$

implies that they generate the same ideal.

We now show how to perform this computation on the systems Macaulay2, Singular, and Risa/Asir.

Listing 7.118 Macaulay2: computing the Gröbner basis in D

```
i1 : R=QQ[x,y,z,dx,dy,dz,WeylAlgebra=>{x=>dx,y=>dy,z=>dz},
                  MonomialOrder=>{Weights=>{3:0,1,2,3},Lex}];
i2 : L=ideal(x*dx+3*z*dz+3,3*x^2*dy+z*dx,3*x^2*dz+y*dx,y*dy-z*dz);
o2 : Ideal of R
i3 : gens gb L
o3 = | x3dx-yzdx+3x2 3ydy+xdx+3 3x2dy+zdx 3zdz+xdx+3 3x2dz+ydx |

i4 : loadPackage "Dmodules";
i5 : S=QQ[x,y,z,dx,dy,dz,WeylAlgebra=>{x=>dx,y=>dy,z=>dz},
                  MonomialOrder=>Lex];
i4 : M = (map(S,R))(L);
o4 : Ideal of S
```

```
i5 : gbw(M,{0,0,0,1,2,3})
               3      2
o5 = ideal (x dx + 3x   - y*z*dx, x*dx + 3y*dy + 3,
     ------------------------------------------------
           2                               2
        3x dy + z*dx, x*dx + 3z*dz + 3, 3x dz + y*dx)
```

After correctly declaring the ring, we can compute a Gröbner basis in D by using the command gb, as in the case of the polynomial ring. To display the generators of the Gröbner basis, use the command gens. To specify a term order, use the option MonomialOrder, as in Sect. 3.2.3. The term order is applied from left to right. In this case, the ordering is first with respect to the weight vector Weights=>{3:0,1,2,3} and then with respect to Lex. Here, {3:0,1,2,3} is an abbreviated representation of {0,0,0,1,2,3}. We can abbreviate it further as {3:0,1..3}.

Alternatively, we can use the command gbw to compute a Gröbner basis with respect to a weighted order. Since the weight vector is specified as the second argument, Lex is used as a tie breaker in the ring declaration.

Listing 7.119 Singular: computing the Gröbner basis in D

```
> LIB "dmod.lib";
> ring r=0,(x,y,z,dx,dy,dz),(a(0,0,0,1,2,3),1p);
> def D=Weyl();
> setring D;
> ideal L=x*dx+3*z*dz+3,3*x^2*dy+z*dx,3*x^2*dz+y*dx,y*dy-z*dz;
> groebner(L);
_[1]=x^3*dx-y*z*dx+3*x^2
_[2]=3*y*dy+x*dx+3
_[3]=3*x^2*dy+z*dx
_[4]=z*dz-y*dy
_[5]=3*x^2*dz+y*dx

> ring rr=0,(x,y,z,dx,dy,dz),1p;
> def DD=Weyl();
> setring DD;
> ideal L=fetch(D,L);
> intvec u=0,0,0;
> intvec v=1,2,3;
> GBWeight(L,u,v);
_[1]=-y*dy+z*dz
_[2]=x*dx+3*y*dy+3
_[3]=3*x^2*dy+z*dx
_[4]=3*x^2*dz+y*dx
_[5]=x^3*dx+3*x^2-y*z*dx
```

In Singular, a weighted term order is specified with the syntax (a(0,0,0,1, 2,3),1p). This represents $\prec_{(u,v)}$ since the term order is applied from left to right, as with Macaulay2.

We can also use the command GBWeight. This command accepts intvec-type variables so that the second and third arguments correspond to u and v, respectively.

Listing 7.120 Risa/Asir: computing the Gröbner basis in D

```
[1309] L=[x*dx+3*z*dz+3,3*x^2*dy+z*dx,3*x^2*dz+y*dx,y*dy-z*dz]$
[1310] V=[x,y,z,dx,dy,dz]$
[1311] M=newmat(7,6,[[0,0,0,1,2,3],[1],[0,1],[0,0,1],[0,0,0,1],
[0,0,0,0,1],[0,0,0,0,0,1]]);
[ 0 0 0 1 2 3 ]
[ 1 0 0 0 0 0 ]
[ 0 1 0 0 0 0 ]
[ 0 0 1 0 0 0 ]
[ 0 0 0 1 0 0 ]
[ 0 0 0 0 1 0 ]
[ 0 0 0 0 0 1 ]
[1312] G=nd_weyl_gr(L,V,0,M);
[(x^3-z*y)*dx+3*x^2,-x*dx-3*y*dy-3,z*dx+3*x^2*dy,
x*dx+3*z*dz+3,y*dx+3*x^2*dz]
```

With Risa/Asir, we use the dedicated command nd_weyl_gr. The arguments are the same as for the command nd_gr (see Sect. 3.6.6). The weighted term order is specified by an order matrix.

We consider the computation of the initial ideal of I. It is generated by the initial term of each element of the Gröbner basis. Thus we have

$$\text{in}_{(u,v)}(I) = \langle x^3\xi_x - yz\xi_x, 3y\xi_y, 3x^2\xi_y, 3z\xi_z, 3x^2\xi_z \rangle \subset \mathbf{C}[x,y,z,\xi_x,\xi_y,\xi_z].$$

Here, we replace the derivatives $\partial_x, \partial_y, \partial_z$ by the commutative variables ξ_x, ξ_y, ξ_z, respectively. We will show how to compute each system. We note that these three systems do not make the replacement $\partial \mapsto \xi$.

Listing 7.121 Macaulay2: computing the initial ideal

```
i1 : loadPackage "Dmodules";
i2 : R=QQ[x,y,z,dx,dy,dz,WeylAlgebra=>{x=>dx,y=>dy,z=>dz}];
i3 : L=ideal(x*dx+3*z*dz+3,3*x^2*dy+z*dx,3*x^2*dz+y*dx,y*dy-z*dz);
o3 : Ideal of R
i4 : inw(L,{0,0,0,1,2,3})
                 3              2             2
o4 = ideal (x dx - y*z*dx, 3y*dy, 3x dy, 3z*dz, 3x dz)
o4 : Ideal of QQ[x, y, z, dx, dy, dz]
```

To compute the initial ideal with respect to the weighted term order, use the command inw. The second argument is the weight vector, as with the command gbw. We do not set the term order in the ring declaration, because the initial ideal is unique for any tie breaker. In this case, GRevLex is used as a default setting. The line at o4 shows that the initial ideal is regarded as the ideal of the polynomial ring because the weight vector satisfies $u_i + v_i > 0$. However, the variables are not replaced as described above.

Listing 7.122 Singular: computing the initial ideal

```
> LIB "dmod.lib";
> ring r=0, (x,y,z,dx,dy,dz), (a(0,0,0,1,2,3),lp);
> def D=Weyl();
> setring D;
```

```
> ideal L=x*dx+3*z*dz+3,3*x^2*dy+z*dx,3*x^2*dz+y*dx,y*dy-z*dz;
> ideal G=groebner(L);
> intvec w=0,0,0,1,2,3;
> inForm(G,w);
_[1]=x^3*dx-y*z*dx
_[2]=3*y*dy
_[3]=3*x^2*dy
_[4]=z*dz
_[5]=3*x^2*dz
```

The command inForm returns the initial parts of polynomials (or generators of the
ideal) with respect to the weight vector. The second argument accepts the intvec-
type variable corresponding to the weight vector. This command returns the initial
terms of the generators, but it does not return the initial ideal when an ideal is given
as the first argument. This command must be used after we compute a Gröbner basis
with respect to the desired weighted term order. [2]

Listing 7.123 Risa/Asir: computing the initial ideal

```
[1336]  load("nk_restriction.rr");
[1337]  L=[x*dx+3*z*dz+3,3*x^2*dy+z*dx,3*x^2*dz+y*dx,y*dy-z*dz]$
[1338]  V=[x,y,z,dx,dy,dz]$
[1339]  nk_restriction.initial_w(L,V,[0,0,0,1,2,3]);
[(x^3-z*y)*dx,-3*y*dy,3*x^2*dy,3*z*dz,3*x^2*dz]
```

In Risa/Asir, the command nk_restriction.initial_w returns the initial
ideal with respect to the weighted term order. It is defined in the library
nk_restriction.rr. The weight vector is specified as the second and third
arguments.

Exercise. For the ideal in Example 7.4.20, compute a Gröbner basis and an initial
ideal with respect to the weighted term order $\prec_{(u,v)}$, where $(u, v) = (2, 1, 0, 1, 2, 3)$.

7.4.7 Holonomic Systems (Sect. 6.8)

Example 7.4.21. Set $A = \begin{pmatrix} 1 & 1 & 1 & 1 \\ 0 & 1 & 3 & 4 \end{pmatrix}$. Compute the holonomic rank of the
A-hypergeometric system $H_A(\beta)$ for $\beta = (1, 2)^T$ and $\beta = (1, 3)^T$.

Answer.

Listing 7.124 Macaulay2: holonomic rank

```
i1 : loadPackage "Dmodules";
i2 : A=matrix{{1,1,1,1},{0,1,3,4}};
```

---

[2]The command initialIdealW accomplishes this procedure in Singular. However, there may
be a bug in version 3-1-2.

```
               2          4
o2 : Matrix ZZ  <--- ZZ
i3 : b={1,2};
i4 : G=gkz(A,b);
o4 : Ideal of QQ[x , x , x , x , D , D , D , D ]
                 1   2   3   4   1   2   3   4
i5 : holonomicRank G
o5 = 5

i6 : b={1,3};
i7 : G=gkz(A,b);
o7 : Ideal of QQ[x , x , x , x , D , D , D , D ]
                 1   2   3   4   1   2   3   4
i8 : holonomicRank G
o8 = 4
```

In Macaulay2, the command holonomicRank returns the holonomic rank. It is 5 when $\beta = (1, 2)^T$ and 4 when $\beta = (1, 3)^T$.

Listing 7.125 Singular: holonomic rank

```
LIB "dmod.lib";
LIB "toric.lib";
> ring r = (0,x(1..4)),(dx(1..4)),dp;
> intmat A[2][4] = 1,1,1,1, 0,1,3,4;
> ideal IA = toric_ideal(A,"ect");
> intmat ORD[8][8]=
. 0,0,0,0,1,1,1,1,
. 0,0,0,0,0,0,0,-1,
. 0,0,0,0,0,0,-1,0,
. 0,0,0,0,0,-1,0,0,
. 1,1,1,1,0,0,0,0,
. 0,0,0,-1,0,0,0,0,
. 0,0,-1,0,0,0,0,0,
. 0,-1,0,0,0,0,0,0;
> ring r8 = 0, (x(1..4),dx(1..4)),M(ORD);
> def D4 = Weyl();
> setring D4;
> vector T = [x(1)*dx(1),x(2)*dx(2),x(3)*dx(3),x(4)*dx(4)];
> vector B = [1,2];
> matrix AT = A*T-B;
> ideal HA = imap(r,IA)+AT[1,1]+AT[2,1];
> ideal G = std(HA);
> setring r;
> ideal GB = imap(D4,G);
> kbase(GB);
// ** GB is no standard basis
_[1]=dx(4)^2
_[2]=dx(3)*dx(4)
_[3]=dx(4)
_[4]=dx(3)
_[5]=1
```

Singular does not have a command which directly computes the holonomic rank. However, it can be computed by using the command kbase, which returns a base of the vector space. We can compute a Gröbner basis with respect to a block order with $\partial \gg x$ and regard it as a Gröbner basis of R (see Example 7.4.23 for specification of the block order). When the command kbase is executed, a warning is displayed, but it can be disregarded for this purpose.

Executing the command with vector B = [1, 3] ;, the holonomic rank is 4.

Listing 7.126 Risa/Asir: holonomic rank

```
[1371]  A=[[1,1,1,1],[0,1,3,4]]$
[1372]  B=[1,2]$
[1373]  G=sm1.gkz([A,B]);
[[x4*dx4+x3*dx3+x2*dx2+x1*dx1-1,4*x4*dx4+3*x3*dx3+x2*dx2-2,
-dx1*dx4+dx2*dx3,-dx2^2*dx4+dx1*dx3^2,dx1^2*dx3-dx2^3,
-dx2*dx4^2+dx3^3],[x1,x2,x3,x4]]
[1374]  sm1.rank(G);
5

[1375]  B=[1,3]$
[1376]  G=sm1.gkz([A,B]);
[[x4*dx4+x3*dx3+x2*dx2+x1*dx1-1,4*x4*dx4+3*x3*dx3+x2*dx2-3,
-dx1*dx4+dx2*dx3,-dx2^2*dx4+dx1*dx3^2,dx1^2*dx3-dx2^3,
-dx2*dx4^2+dx3^3],[x1,x2,x3,x4]]
[1377]  sm1.rank(G);
4
```

Risa/Asir does not have a command for computing the holonomic rank, but we can do so by using the command sm1, sm1.rank.

For the Gröbner basis computation, in general, Risa/Asir is more efficient than sm1. Thus, for larger problems, it is better to perform computations on Risa/Asir, as we did with Singular. The command in Risa/Asir that corresponds to kbase in Singular is dp_mbase (see Sect. 3.6.7).

Exercise. Set $A = \begin{pmatrix} 1 & 1 & 1 & 1 \\ 0 & 1 & 3 & 4 \end{pmatrix}$. Compute the holonomic rank of the A-hypergeometric system $H_A(\beta)$ for $\beta = (1,0)^T$ and $\beta = (1,4)^T$. (Note: The holonomic rank for a generic parameter β coincides with vol(A). See [20, Chap. 4] for details.)

Example 7.4.22. Provide an algorithm for determining whether D^r/L is holonomic. Find the appropriate programs or commands in Macaulay2, Singular, and Risa/Asir.

Answer. This is somewhat difficult when $r > 1$ (refer to [16]). Macaulay2 and Singular have the command isHolonomic to determine holonomicity.

It is easy, however, when $r = 1$. D/L is holonomic if the Krull dimension of the characteristic variety of the ideal L in the Weyl algebra $D = D_n$ is n. Hence, we compute the $(0, 1)$-initial ideal and examine the degree of the Hilbert polynomial. For example, we can check the holonomicity of the ideal

$$I = \langle x\partial_y, y\partial_x \rangle \subset D_2.$$

Listing 7.127 Macaulay2: check the holonomicity

```
i1 : loadPackage "Dmodules";
i2 : D2=QQ[x,y,dx,dy,WeylAlgebra=>{x=>dx,y=>dy},
                                    MonomialOrder=>GRevLex];
i3 : I=ideal(x*dy,y*dx);
o3 : Ideal of D2
i4 : isHolonomic I
o4 = true

i5 : CI=charIdeal I
o5 = ideal (x*dy, y*dx, x*dx - y*dy)
o5 : Ideal of QQ [x, y, dx, dy]

i6 : IN=inw(I1,{0,0,1,1})
                                      2          2
o6 = ideal (x*dy, y*dx, x*dx - y*dy, y dy, y*dy )
o6 : Ideal of QQ [x, y, dx, dy]
i7 : dim IN
o7 : 2

i8 : hilbertPolynomial(IN)
o8 = 2*P
         1
i9 : hilbertPolynomial(IN,Projective=>false)
o9 = 2i + 2
o9 : QQ[i]
i10 : R4=ring IN;
i11 : R5=QQ[x,y,dx,dy,h];
i12 : IN1=(map(R5,R4))(IN);
o12 : Ideal of R5
i13 : hilbertPolynomial(IN1,Projective=>false)
        2
o13 = i  + 3i + 2
o13 : QQ[i]
```

The command `isHolonomic` returns `true`. Thus, the ideal I is holonomic. To compute the characteristic ideal, use the command `charIdeal` or compute the $(0, 1)$-initial ideal by using the command `inw`. To compute the Krull dimension of the characteristic variety, use the command `dim`. To compute the Hilbert polynomial, use the command `hilbertPolynomial`. This command accepts only homogeneous ideals. In this example, `IN` is homogeneous. The degree of output 1 indicates the dimension of the projective variety. Thus, the dimension of the affine variety is 2. To compute the Hilbert polynomial as the affine variety, we need to compute on the homogenized Weyl algebra with the homogenized variable h.

By default, the output of `hilbertPolynomial` is represented by the symbol $P_k := \binom{i+k}{k}$. The option `Projective=>false` changes the output to a polynomial representation.

Listing 7.128 Singular: check the holonomicity

```
> LIB "dmod.lib";
> ring r=0,(x,y,dx,dy),dp;
> def D2=Weyl();
```

```
> setring D2;
> ideal I=x*dy,y*dx;
> isHolonomic(I);
1
> def A=charVariety(I);
> setring A;
> charVar;
charVar[1]=x*dy
charVar[2]=y*dx
charVar[3]=x*dx-y*dy
charVar[4]=y*dy^2
charVar[5]=y^2*dy
> dim(charVar);
2
> ring B=0,(x,y,dx,dy,h),dp;
> setring B;
> ideal CI=imap(A,charVar);
> hilbPoly(CI);
4,6,2
```

Singular has the command isHolonomic to check holonomicity; 1 indicates it is holonomic, 0 indicates it is not. To compute the characteristic variety, use the command charVariety, move the ring, and read the variable charVar. After computing the Krull dimension by using the command dim, we can check the holonomicity of I. To compute the Hilbert polynomial, use the command hilbPoly. This command accepts only homogeneous ideals, as with Macaulay2. It returns the integer sequence which represents the Hilbert polynomial $\frac{1}{n!}\sum_{i=0}^{n}c_i t^i$. In this example, the Hilbert polynomial is $(4+6t+2t^2)/2! = 2+3t+t^2$.

Listing 7.129 Risa/Asir: check the holonomicity

```
[1371]  load("nk_restriction.rr")$
[1579]  I=[x*dy,y*dx]$
[1580]  V=[x,y,dx,dy]$
[1581]  IN=nk_restriction.initial_w(I,V,[0,0,1,1]);
[dy*x,dx*y,-dx*x+dy*y,-dy*y^2,-dy^2*y]
[1582]  sml.hilbert([IN,V]);
h^2+3*h+2

[1587]  GI=nd_gr(IN,V,0,0);
[-y^2*dy,-y*dy^2,-x*dx+y*dy,y*dx,x*dy]
[1588]  dp_ord(0);
0
[1589]  GIN=map(dp_dtop,map(dp_ht,map(dp_ptod,GI,V)),V);
[y^2*dy,y*dy^2,x*dx,y*dx,x*dy]
[1590]  sml.hilbert([GIN,V]);
h^2+3*h+2
```

Risa/Asir does not have a command to check holonomicity. To compute the characteristic variety, use the command initial_w. To compute the Hilbert polynomial, use the command sml.hilbert, which is defined in sml. The degree of the Hilbert polynomial is 2. Thus the ideal I is holonomic. It is possible to make a user-defined command to check the holonomicity by combining the above procedures.

The computation of a Gröbner basis is used in computing the Hilbert polynomial. As mentioned above, in general, Risa/Asir is more efficient than sm1 for the computation of a Gröbner basis. For larger problems, we recommend that you use Risa/Asir to compute the initial ideal before using the command sm1.hilbert.

Exercise. Check the holonomicity of the following ideals:

1. $I_1 = \langle x^2 \partial_y, y^2 \partial_x \rangle \subset D_2$,
2. $I_2 = \langle (x^3 - y^2)\partial_x + 3x^2, (x^3 - y^2)\partial_y - 2y \rangle \subset D_2$,

3. A-hypergeometric ideal $H_A(\beta) \subset D_4$ associated with $A = \begin{pmatrix} 1 & 1 & 0 & 0 \\ 0 & 0 & 1 & 1 \\ 0 & 1 & 0 & 1 \end{pmatrix}$ and

   $\beta = (1, 1, 1)^T$.

7.4.8 Relationship of D and R (Sect. 6.9)

Example 7.4.23. Use Theorem 6.9.3 to compute a Gröbner basis in R of the ideal in Example 7.4.8.

Answer. We take a term order as a block order $\partial_x \succ \partial_y \gg x \succ y$, where $\succ$ represents the pure lexicographic order. We will show how to set the block order on each system. We cannot use the method shown in Sects. 3.2.3 and 3.6.5, because the derivatives must be placed on the right of the variables. The matrix order is available in this case. The matrix corresponding to the block order in this example is

$$\begin{pmatrix} 0 & 0 & 1 & 1 \\ 0 & 0 & 0 & -1 \\ 1 & 1 & 0 & 0 \\ 0 & -1 & 0 & 0 \end{pmatrix}.$$

Listing 7.130 Macaulay2: computing the Gröbner basis in R

```
i1 : loadPackage "Dmodules";
i2 : R=QQ[x,y,dx,dy,WeylAlgebra=>{x=>dx,y=>dy},
     MonomialOrder=>{Weights=>{0,0,1,1},Weights=>{0,0,0,-1},
                     Weights=>{1,1},Weights=>{0,-1}}];
i3 : L=ideal(y*dx-x*dy,dx*dy+4*x*y);
o3 : Ideal of R
i4 : gens gb L
o4 = | ydx-xdy ydy^2-dy+4y3 xdy^2-dx+4xy2 dxdy+4xy
     -------------------------------------------------
       dx^2-dy^2+4x2-4y2 dy^3+4y2dy+12y |
```

In Macaulay2, use the option Weights. The terms are compared with respect to the first weight, the second weight in the case of drawing, and so on. Thus, we can set the matrix order by enumerating the weight corresponding to each row in the matrix.

When the length of the weight is shorter than that of the variables, the weights of the remaining variables are regarded as 0.

Listing 7.131 Singular: computing the Gröbner basis in R

```
> LIB "dmod.lib";
> intmat ORD[4][4]=
. 0,0,1,1,
. 0,0,0,-1,
. 1,1,0,0,
. 0,-1,0,0;
> ring r=0,(x,y,dx,dy),M(ORD);
> def D=Weyl();
> setring D;
> ideal L=y*dx-x*dy,dx*dy+4*x*y;
> groebner(L);
_[1]=y*dx-x*dy
_[2]=y*dy^2-dy+4*y^3
_[3]=x*dy^2-dx+4*x*y^2
_[4]=dx*dy+4*x*y
_[5]=dx^2-dy^2+4*x^2-4*y^2
_[6]=dy^3+4*y^2*dy+12*y
```

In Singular, after setting the `intmat`-type variable corresponding to the matrix order, we declare the ring by using the command `M()`.

Listing 7.132 Risa/Asir: computing the Gröbner basis in R

```
[1437] L1=y*dx-x*dy$
[1438] L2=dx*dy+4*x*y$
[1439] V=[x,y,dx,dy]$
[1440] M=newmat(4,4,[[0,0,1,1],[0,0,0,-1],[1,1],[0,-1]]);
[ 0 0 1 1 ]
[ 0 0 0 -1 ]
[ 1 1 0 0 ]
[ 0 -1 0 0 ]
[1441] G=nd_weyl_gr([L1,L2],V,0,M);
[-dy*x+dx*y,-4*y^3-dy^2*y+dy,(-4*y^2-dy^2)*x+dx,
4*y*x+dy*dx,-4*x^2+4*y^2-dx^2+dy^2,-4*dy*y^2-12*y-dy^3]
```

In Risa/Asir, the command `nd_weyl_gr` accepts the matrix as the term order at the fourth argument (see Sect. 3.6.5).

Exercise. Let I be an ideal in R_2 defined by

$$I = \langle \partial_x \partial_y + 1, 2y\partial_y^2 - \partial_x + 3\partial_y + 2x \rangle.$$

Compute Gröbner bases of I with respect to the pure lexicographic order and the degree reverse lexicographic order, with $\partial_x \succ \partial_y$.

7.4.9 Integration Algorithm (Sect. 6.10)

Example 7.4.24. Set $L = x^2 y \partial_x \partial_y^2 - z \partial_z$. Calculate the Fourier transform $\mathscr{F}(L)$ of L with respect to the variables x, y, z. Calculate the Fourier transform $\mathscr{F}_x(L)$ of L with respect to the variable x.

Answer. By a simple calculation, we obtain

$$\mathscr{F}(L) = (-\partial_x)^2(-\partial_y)xy^2 - (-\partial_z)z$$

$$= -(x\partial_x^2 + 2\partial_x)(2y + y^2\partial_y) + (z\partial_z + 1)$$

$$= -2xy\partial_x^2 - xy^2\partial_x^2\partial_y - 4y\partial_x - 2y^2\partial_x\partial_y + z\partial_z + 1,$$

$$\mathscr{F}_x(L) = (-\partial_x)^2 yx\partial_y^2 - z\partial_z$$

$$= (x\partial_x^2 + 2\partial_x)y\partial_y^2 - z\partial_z$$

$$= xy\partial_x^2\partial_y^2 + 2y\partial_x\partial_y^2 - z\partial_z.$$

We also can compute the Fourier transform by using a computer.

Listing 7.133 Macaulay2: the Fourier transform

```
i1 : loadPackage "Dmodules";
i2 : R=QQ[x,y,z,dx,dy,dz,WeylAlgebra=>{x=>dx,y=>dy,z=>dz}];
i3 : L=x^2*y*dx*dy^2-z*dz;
i4 : Fourier L
            2  2              2         2
o4 = - x*y dx dy - 2x*y*dx  - 2y dx*dy - 4y*dx + z*dz + 1
o4 : R

i6 : R1=QQ[x,y,z,dx,dy,dz,WeylAlgebra=>{x=>dx}];
i15 : M=(map(R1,R))(L)
          2       2
o15 = x y*dx*dy  - z*dz
o15 : R1
i14 : Fourier(M)
              2  2              2
o14 = x*y*dx dy  + 2y*dx*dy  - z*dz
```

The command `Fourier` returns the Fourier transform with respect to the variables specified as derivatives in the ring declaration by the option `WeylAlgebra=>{...}`. Thus, when we apply the transform with respect to certain variables, they must be defined in a ring. This command accepts an ideal as an input parameter. In this case, it returns the Fourier transform for each generator.

Listing 7.134 Singular: the Fourier transform

```
> LIB "dmod.lib";
> ring r=0,(x,y,z,dx,dy,dz),dp;
> def D=Weyl();
> setring D;
> poly L=x^2*y*dx*dy^2-z*dz;
> fourier(L);
```

```
_[1]=-x*y^2*dx^2*dy-2*x*y*dx^2-2*y^2*dx*dy-4*y*dx+z*dz+1

> intvec w=1;
> fourier(L,w);
_[1]=x*y*dx^2*dy^2+2*y*dx*dy^2-z*dz
```

The command `fourier` computes the Fourier transform. When we set as the second argument an `intvec`-type variable whose entries are the indices of the variable list, it computes the Fourier transform restricted to the variable given in the second argument. It accepts ideals, as does Macaulay2.

Listing 7.135 Risa/Asir: the Fourier transform

```
[1371] load("nk_restriction.rr");
[1583] L=x^2*y*dx*dy^2-z*dz$
[1584] nk_restriction.fourier_trans(L,[x,y,z],[dx,dy,dz]);
(-y^2*x*dy-2*y*x)*dx^2+(-2*y^2*dy-4*y)*dx+z*dz+1
[1585] nk_restriction.fourier_trans(L,[x],[dx]);
y*x*dy^2*dx^2+2*y*dy^2*dx-z*dz
```

In Risa/Asir, the command `nk_restriction.fourier_trans`, defined in `nk_restriction.rr`, computes the Fourier transform. It applies the transform with respect to the variables given by the second and third arguments. The third argument must contain the derivatives for the second argument.

Exercise. Set $L = x^2 y \partial_x \partial_y^2 - z\partial_z$. Calculate the Fourier transform $\mathscr{F}(L)$ of L with respect to the variables x, y, z. Calculate the Fourier transform $\mathscr{F}_x(L)$ of L with respect to the variable x.

Example 7.4.25. Compute the b-function of $I = \langle 3x^2\partial_t^4 + \partial_t^2 - \partial_t\partial_x, 3x^3\partial_t^3 + x\partial_t + t\partial_t + 2\rangle \subset D\langle t, x, \partial_t, \partial_x\rangle$ with respect to the weight vector $(-w, w)$, where $w = (1, 0)$.

Answer. The b-function of I is $s^5 - 2s^4 - s^3 + 2s^2 = (s+1)s^2(s-1)(s-2)$.

Listing 7.136 Macaulay2: the b-function with respect to $(-w, w)$

```
i1 : loadPackage "Dmodules";
i2 : R=QQ[t,x,dt,dx,WeylAlgebra=>{t=>dt,x=>dx}];
i3 : I=ideal(3*x^2*dt^4+dt^2-dx*dt,3*x^3*dt^3+x*dt+t*dt+2);
o3 : Ideal of R
i4 : bFunction(I,{1,0})
         5      4    3      2
o4 = s  - 2s  - s  + 2s
o4 : QQ[s]
```

The command `bFunction` returns the b-function. The weight w is specified as the second argument.

Listing 7.137 Singular: the b-function with respect to $(-w, w)$

```
> LIB "dmod.lib";
> ring r=0,(t,x,dt,dx),dp;
> def D=Weyl();
> setring D;
```

```
> ideal I=3*x^2*dt^4+dt^2-dx*dt,3*x^3*dt^3+x*dt+t*dt+2;
> intvec w=1,0;
> bfctIdeal(I,w);
[1]:
   _[1]=2
   _[2]=1
   _[3]=0
   _[4]=-1
[2]:
   1,1,2,1
```

The command $bfctIdeal$ returns the b-function. The weight w is specified as the second argument by an $intvec$-type variable. The first component of the output is a list of the roots of the b-function. The second component is a list of the multiplicities of the roots. In this example, the output represents the b-function $(s - 2)^1(s - 1)^1 s^2(s + 1)^1$.

Listing 7.138 Risa/Asir: the b-function with respect to $(-w, w)$

```
[1371]  I=[3*x^2*dt^4+dt^2-dx*dt,3*x^3*dt^3+x*dt+t*dt+2]$
[1372]  generic_bfct(I,[t,x],[dt,dx],[1,0]);
s^5-2*s^4-s^3+2*s^2
```

In Risa/Asir, the command $generic\_bfct$ computes the b-function. We set the second and third arguments to the variables and derivatives, respectively, and the fourth argument to the weight vector.

Exercise. Set $f = x^3 - y^2 z^2$ and

$$I_f = \left\langle t - f, \partial_x + \frac{\partial f}{\partial x} \partial_t, \partial_y + \frac{\partial f}{\partial y} \partial_t, \partial_z + \frac{\partial f}{\partial z} \partial_t \right\rangle$$

in $D_4 = \mathbf{C}\langle t, x, y, z, \partial_t, \partial_x, \partial_y, \partial_z \rangle$. Compute the b-function of I_f with respect to $(-w, w)$, where $w = (1, 0, 0, 0)$. (Note: The b-function of I_f is called the b-function of f or the Bernstein-Sato polynomial. See [17, 20] for details.)

Example 7.4.26. Find a differential operator annihilating the definite integrals

$$F(x) = \int_{-\infty}^{\infty} \exp(xy^2 - y^4) dy.$$

Answer. We compute the integration ideal of $I := \langle \partial_x - y^2, \partial_y - 2xy + 4y^3 \rangle$ with respect to y because the derivatives of the integrand $f(x, y) := \exp(xy^2 - y^4)$ are $\partial_x f = y^2 f$ and $\partial_y f = (2xy - 4y^3) f$.

From the output of the integration algorithm by the following sample codes, we have the integration ideal $\langle 4\partial_x^2 - 2x\partial_x - 1 \rangle$. There exists a differential operator $P \in D_2$ such that $4\partial_x^2 - 2x\partial_x - 1 - \partial_y P \in I$. Therefore, we have the following equation:

$$(4\partial_x^2 - 2x\partial_x - 1) \bullet F(x) = [P \bullet \exp(xy^2 - y^4)]_{y=-\infty}^{y=\infty} = 0.$$

The last equality holds because the integrand is a rapidly decreasing function as $y \to \infty$ or $y \to -\infty$.

Listing 7.139 Macaulay2: computing the integration ideal

```
i1 : loadPackage "Dmodules";
i2 : R=QQ[x,y,dx,dy,WeylAlgebra=>{x=>dx,y=>dy}];
i3 : I=ideal(dx-y^2,dy-2*x*y+4*y^3);
o3 : Ideal of R
i4 : DintegrationIdeal(I,{0,1})
                          2
o4 = ideal(2x*dx - 4dx  + 1)
o4 : Ideal of QQ[x, dx]
```

The command `DintegrationIdeal` returns the integration ideal. We set the second argument to the weight vector. When computing the integration ideal with respect to y, we must set the weight vector such that the weight of ∂_y is positive and the other weights are 0; for example, $w = (0, 1)$.

Listing 7.140 Singular: computing the integration ideal

```
> LIB "dmod.lib";
> ring r=0,(x,y,dx,dy),dp;
> def D=Weyl();
> setring D;
> ideal I=dx-y^2,dy-2*x*y+4*y^3;
> intvec w=0,1;
> def J=integralIdeal(I,w);
> setring J;
> intIdeal;
intIdeal[1]=2*x*dx-4*dx^2+1
```

The command `integralIdeal` returns the integration ideal. The weight vector is specified by the second argument, as with Macaulay2. After moving the ring by using `setring`, we can refer to the variable as `intIdeal`.

Listing 7.141 Risa/Asir: computing the integration ideal

```
[1371] load("nk_restriction.rr")$
[1583] I=[dx-y^2,dy-2*x*y+4*y^3]$
[1584] nk_restriction.integration_ideal(I,[y,x],[dy,dx],[1,0]);
-- nd_weyl_gr :0sec(0.0005872sec)
-- weyl_minipoly :0sec(0.000603sec)
-- generic_bfct_and_gr :0sec(0.001509sec)
generic bfct : [[1,1],[s,1],[s-1,1]]
S0 : 1
B_{S0} length : 2
-- fctr(BF) + base :0sec(0.0003839sec)
-- integration_ideal_internal :0sec(0.000494sec)
[4*dx^2-2*x*dx-1]
```

The command `nk_restriction.integration_ideal` defined in the library `nk_restriction.rr` returns the integration ideal. We set the variables and derivatives to the second and third arguments, and the weight vector to the fourth argument. We note that after the first 0 appears, the following components of the weight vector must also be 0. Thus, in the variable lists, y and ∂_y are put before x and ∂_x, respectively.

Exercise. Find a differential operator annihilating the definite integral

$$F(x) = \int_{-\infty}^{\infty} \exp(xy^3 - y^4) dy.$$

Example 7.4.27. Find an inhomogeneous differential equation for the definite integral

$$F(x) = \int_0^1 \exp(xy^2 - y^4) dy.$$

Answer. This is a modification of Example 7.4.26, in which the integration domain $[-\infty, \infty]$ has been replaced with $[0, 1]$. From the result in Example 7.4.26, we have

$$(4\partial_x^2 - 2x\partial_x - 1) \bullet F(x) = [P \bullet \exp(xy^2 - y^4)]_{y=0}^{y=1}.$$

However, we cannot know the values at the end points without the differential operator P. Risa/Asir can compute P by using the package nk_restriction.rr. Maple has a package Mgfun [2] which computes P by the method of undetermined coefficients.

Listing 7.142 Risa/Asir: computing the inhomogeneous differential equation

```
[1371] load("nk_restriction.rr")$
[1583] I=[dx-y^2,dy-2*x*y+4*y^3]$
[1598] nk_restriction.integration_ideal(I,[y,x],[dy,dx],[1,0]|
inhomo=1);
-- nd_weyl_gr :0sec(0.0003309sec)
-- weyl_minipoly :0sec(0.000675sec)
-- generic_bfct_and_gr :0sec(0.001267sec)
generic bfct : [[1,1],[s,1],[s-1,1]]
S0 : 1
B_{S0} length : 2
-- fctr(BF) + base :0.004sec(0.0003982sec)
-- integration_ideal_internal :0sec(0.0005929sec)
[[4*dx^2-2*x*dx-1],[[[[dy,-y]],1]]]
```

The command nk_restriction.integration_ideal with the option inhomo=1 returns the inhomogeneous parts of the generators of the integration ideal. The output represents

$$(4\partial_x^2 - 2x\partial_x - 1) - (1/\underline{1}) \cdot \partial_y(\underline{-y}) \in I.$$

Thus, we obtain the inhomogeneous differential equation

$$(4\partial_x^2 - 2x\partial_x - 1) \bullet F(x) = [-y \exp(xy^2 - y^4)]_{y=0}^{y=1} = -\exp(x - 1).$$

The package Mgfun is not included in Maple by default, so we will show how to install it. First, download the program and help files (algolib.mla, algolib.hdb) from the web page [2] or http://algo.inria.fr/libraries/. Next, place these files in the directory which is the library path for Maple. To determine the library path, use the command libname.

Listing 7.143 Maple: determine the library path

```
> libname;
                    "/usr/local/maple14/lib"
```

If you do not have write-access permission, you will need to add a new directory as a library path. We will assume that the package is located at /home/username/algolib in a UNIX environment.

Listing 7.144 Maple: add a new library path

```
> libname:="/home/username/algolib",libname;
  libname := "/home/username/algolib", "/usr/local/maple14/lib"
```

This path will only be valid for the current session. Finally, we configure the system so that it is executed automatically when Maple is started. In the UNIX environment, locate the file named .mapleinit in the home directory.

Listing 7.145 .mapleinit

```
libname:="/home/username/algolib",libname:
MapleInitRead:=true:
```

In the Windows environment, we locate the file named maple.ini in C:/Documents and Settings/username (2000, XP) or C:/Users/username (Vista, 7).

Listing 7.146 Maple: computing the inhomogeneous differential equation

```
> with(Mgfun):
> creative_telescoping(exp(x*y^2-y^4),[x::diff],[y::diff]);
                        / 2     \
             /d    \    |d      |
  [[-_F(x) - 2 x |-- _F(x)| + 4 |--- _F(x)|, -y _f(x, y)]]
             \dx   /    | 2     |
                        \dx     /
```

To load the package, use the command with(Mgfun). The command creative_telescoping returns inhomogeneous differential operators with P. The first argument is the integrand, the second argument is the list of the remaining variables in the equation, and the third argument is the list of the eliminated variables. diff means that the ring includes the derivative for the variable. In the output, _f(x,y) represents the integrand and _F(x) represents the integral of _f(x,y). Therefore, the output shows

$$(4\partial_x^2 - 2x\partial_x - 1) \bullet \_F(x) = [-y \cdot \_f(x,y)].$$

Mgfun can also treat difference operators and q-difference operators by using the commands shift and qshift. See the manual of Mgfun and [1] for details.

Example 7.4.28. Verify Example 6.10.13 by using a computer software package.

Answer.

Listing 7.147 Macaulay2: nonholonomic example

```
i1 : loadPackage "Dmodules";
i2 : D3=QQ[x,y,z,dx,dy,dz,WeylAlgebra => {x=>dx,y=>dy,z=>dz}];
i3 : F=x^3-y^2*z^2;
i4 : I=ideal(F^2*dx+diff(x,F),F^2*dy+diff(y,F),F^2*dz+diff(z,F));
o4 : Ideal of D3
i5 : isHolonomic I
o5 = false
i6 : Dy=QQ[x,y,z,dx,dy,dz,WeylAlgebra => {y=>dy}];
i7 : FyI=(map(D3,Dy))(Fourier((map(Dy,D3))(I)));
o7 : Ideal of D3
i8 : bFunction(FyI,{0,1,0})
o8 = 0
o8 : QQ[s]
i9 : DintegrationIdeal(I,{0,1,0})
(omitted)/Macaulay2/Dmodules/Drestriction.m2:662:11:(1):[7]:
Module not specializable. Restriction cannot be computed.
(displayed error messages)
i10 : Dx=QQ[x,y,z,dx,dy,dz,WeylAlgebra => {x=>dx}];
i11 : FxI=(map(D3,Dx))(Fourier((map(Dx,D3))(I)));
o11 : Ideal of D3
i12 : bFunction(FxI,{1,0,0})
        4    3    2
o12 = s  - 2s  - s  + 2s
o12 : QQ[s]
```

It is trivial that the ideal *I* is 0-dimensional in *R*. However, isHolonomic says that it is not holonomic. We compute the *b*-function of the Fourier transformed ideal. The command bFunction shows that the *b*-function with respect to *y* is 0. Since the *b*-function does not exist, error messages are displayed when we compute the integration ideal with respect to *y*. If the *b*-function exists, the integration algorithm works fine for nonholonomic ideals. In this example, the integration ideal with respect to *x* is computable because the *b*-function with respect to *x* exists.

Listing 7.148 Singular: nonholonomic example

```
> LIB "dmod.lib";
> ring r=0,(x,y,z,dx,dy,dz),dp;
> def D3=Weyl();
> setring D3;
> poly F=x^3-y^2*z^2;
> ideal I=F^2*dx+diff(F,x),F^2*dy+diff(F,y),F^2*dz+diff(F,z);
> isHolonomic(I);
0
> intvec py = 2;
> ideal FyI = fourier(I,py);
> intvec wy=0,1,0;
> bfctIdeal(FyI,wy);
WARNING: given ideal is not holonomic
... setting bound for degree of b-function to 10 and proceeding
```

```
// Intersection is zero
(omitted)
> integralIdeal(I,wy);
   ? Given ideal is not holonomic
(displayed error messages)
> intvec px=1;
> ideal FxI = fourier(I,px);
> intvec wx=1,0,0;
> bfctIdeal(FxI,wx);
WARNING: given ideal is not holonomic
... setting bound for degree of b-function to 10 and proceeding
[1]:
  _[1]=2
  _[2]=1
  _[3]=0
  _[4]=-1
[2]:
  1,1,1,1
> integralIdeal(I,wx);
   ? Given ideal is not holonomic
(displayed error messages)
```

We can find that the ideal I is not holonomic by using the command isHolonomic. The command bfctIdeal displays a warning for nonholonomic ideals. The default is to search the b-function up to degree 10 using Noro's method [13]. In this example, it detects the b-function since the its degree is 3. Error messages are displayed when we compute the integration ideal with respect to x.

integralIdeal first checks the holonomicity and returns an error message if the input ideal is not holonomic. This is because the integration ideal is not computable for nonholonomic ideals if the b-function exists.

Listing 7.149 Risa/Asir: nonholonomic example

```
[1563] load("nk_restriction.rr")$
[1771] F=x^3-y^2*z^2$
[1772] I=[F^2*dx+diff(F,x),F^2*dy+diff(F,y),F^2*dz+diff(F,z)];
[dx*x^6-2*dx*z^2*y^2*x^3+3*x^2+dx*z^4*y^4,
dy*x^6-2*dy*z^2*y^2*x^3+dy*z^4*y^4-2*z^2*y,
dz*x^6-2*dz*z^2*y^2*x^3+dz*z^4*y^4-2*z*y^2]
[1773] V = [x,y,z]$ DV=[dx,dy,dz]$ VDV=append(V,DV)$
[1774] CI=nk_restriction.initial_w(I,VDV,[0,0,0,1,1,1])$
[1775] sml.hilbert([CI],VDV]);
1/6*h^4+15/2*h^3-170/3*h^2+214*h-251

[1779] FyI=map(nk_restriction.fourier_trans,I,[y],[dy])$
[1780] generic_bfct(FyI,V,DV,[0,1,0]);
(interrupted)
[1781] nk_restriction.generic_bfct_and_gr(FyI,[y,x,z],
                              [dy,dx,dz],[1,0,0]|param=[]);
-- nd_weyl_gr :0.004sec(0.004671sec)
weyl_minipoly_by_elim : b-function does not exist
(omitted)
[1782] nk_restriction.integration_ideal(I,[y,x,z],[dy,dx,dz],
                              [1,0,0]);
```

```
-- nd_weyl_gr :0sec(0sec)
(interrupted)
[1783] nk_restriction.integration_ideal(I,[y,x,z],[dy,dx,dz],
                                          [1,0,0]|param=[]);
-- nd_weyl_gr :0sec(0sec)
weyl_minipoly_by_elim : b-function does not exist

[1781] nk_restriction.integration_ideal(I,[x,y,z],[dx,dy,dz],
                                          [1,0,0])$
-- nd_weyl_gr :1.776sec + gc : 0.188sec(1.974sec)
-- weyl_minipoly :0.024sec(0.01972sec)
-- generic_bfct_and_gr :1.84sec + gc : 0.188sec(2.033sec)
generic bfct : [[1,1],[s,1],[s-2,1],[s-1,1],[s+1,1]]
S0 : 2
B_{S0} length : 3
-- fctr(BF) + base :0.116sec + gc : 0.008001sec(0.1251sec)
-- integration_ideal_internal :0.112sec + gc : 0.01sec(0.122sec)
```

In Risa/Asir, we can check the holonomicity by checking the degree of the Hilbert polynomial, as described above. Since the degree is 4, the ideal I is not holonomic. If the b-function does not exist, computation of the integration ideal by nk_restriction.integration_ideal is not stopped because, by default, it uses Noro's method [13]. The command will display an error message if we use the option param=[].

Example 7.4.29. Let $f(t) = t(1-t)$.

1. Compute the annihilating ideal Ann $f^s := \{P \in D[s] \mid P \bullet f^s = 0\}$.
2. Let E_s be a difference operator such that $E_s \bullet F(s) = F(s+1)$. Then, the trivial relation $(E_s - f) \bullet f^s = 0$ holds. We define the Mellin transform by the following maps: $E_s \leftrightarrow p$, $s+1 \leftrightarrow -p\partial_p$. Let J be an ideal generated by the Mellin transformed elements of $E_s - f$, Ann f^s. Prove that J is a holonomic ideal in $\mathbf{C}\langle t, p, \partial_t, \partial_p \rangle$.
3. Compute the integration ideal of J with respect to t. Find a recurrence relation of $B(s) = \int_0^1 f^s(t)dt$.
4. Evaluate $B(10000)$.

Answer.

1. We compute the annihilating ideal Ann $f^s = \langle t^2\partial_t - t\partial_t - 2ts + s \rangle$ with each system.

Listing 7.150 Macaulay2: computing the annihilating ideal

```
i1 : loadPackage "Dmodules";
i2 : R=QQ[t,dt,WeylAlgebra=>{t=>dt}];
i3 : f=t*(1-t);
i4 : AnnFs f
            2
o4 = ideal(t dt - t*dt - 2t*s + s)
o4 : Ideal of QQ[t, dt, s]
```

Use the command AnnFs. We must first declare the ring of differential operators even though f is an object in the polynomial ring.

Listing 7.151 Singular: computing the annihilating ideal

```
> LIB "dmod.lib";
> ring r=0,x,dp;
> poly f=x*(1-x);
> def A=Sannfs(f);
> setring A;
> LD;
LD[1]=x^2*Dx-x*Dx-2*x*s+s
```

Use the command Sannfs The return is held in the variable LD. We note that the variable x is used instead of t because t is reserved by the command.

Listing 7.152 Risa/Asir: computing the annihilating ideal

```
[1371]  F=t*(1-t);
-t^2+t
[1372]  ann(F);
[(-t^2+t)*dt+(2*t-1)*s]
```

Use the command ann.

2. The Mellin transforms of $E_s - t(1-t)$ and $t^2\partial_t - t\partial_t - 2ts + s$ are $p - t(1-t)$ and $t^2\partial_t - t\partial_t - 2t(-p\partial_p - 1) + (-p\partial_p - 1)$, respectively. The transformed ideal is $J = \langle p - t(1-t), t^2\partial_t - t\partial_t - 2t(-p\partial_p - 1) + (-p\partial_p - 1)\rangle$. We can check that it is holonomic by using the same procedure as was used in Example 7.4.22. We will show only the code for Macaulay2.

Listing 7.153 Macaulay2: check the holonomicity

```
i1 : loadPackage "Dmodules";
i2 : R=QQ[t,p,dt,dp,WeylAlgebra=>{t=>dt,p=>dp}];
i3 : J=ideal(p-t*(1-t),t^2*dt-t*dt-2*t*(-p*dp-1)+(-p*dp-1));
o3 : Ideal of R
i4 : isHolonomic J
o4 = true
```

3. Following Example 7.4.26, the integration ideal can be computed by the command DintegrationIdeal.

Listing 7.154 Macaulay2: the integration ideal of J with respect to t (continuation)

```
i5 : DintegrationIdeal(J,{1,0})
                  2
o5 = ideal(- 4p dp + p*dp - 2p)
o5 : Ideal of QQ[p, dp]
```

Applying the inverse Mellin transform to the generator $-4p^2\partial_p + p\partial_p - 2p$, we obtain the difference operator

$$-4E_s(-s-1) + (-s-1) - 2E_s = 4(s+2)E_s - (s+1) - 2E_s$$

$$= 2(2s+3)E_s - (s+1).$$

This annihilates $B(s)$. In other words, the difference equation $(2(2s + 3)E_s - (s + 1)) \bullet B(s) = 0$ holds. Therefore, we obtain the recurrence relation

$$2(2s + 3)B(s + 1) = (s + 1)B(s).$$

4. From the recurrence relation, we have

$$B(s) = \frac{s}{2(2s + 1)}B(s - 1) = \cdots = \frac{s!}{2^s(2s + 1)!!}B(0)$$

for any integer s. Since $B(0) = \int_0^1 dt = 1$, we obtain

$$B(s) = \frac{s!}{2^s(2s + 1)!!} = \frac{(s!)^2}{(2s + 1)!}.$$

The *inverse* of $B(10000)$ is given by the following for-loop code.

Listing 7.155 Risa/Asir: computing the inverse of $B(10000)$ by for-loop

```
[1217] for (B=1, S=1; S<=10000; S++) B*=2*(2*S+1)/S;
[1218] B;
omitted (a integer with 6023 digits)
```

Exercise. For the function $f(t_1, t_2) = t_1 t_2(1 - t_1 - t_2)$, compute a recurrence relation of the integral $B_2(s) = \int_{T_2} f^s(t_1, t_2)dt_1dt_2$, $(T_2 = \{(t_1, t_2) \mid 0 \le t_1 \le 1, 0 \le t_2 \le 1 - t_1\})$ and evaluate $B_2(10000)$ by using the method in Example 7.4.29.

7.4.10 Finding a Local Minimum of a Function Defined by a Definite Integral (Sect. 6.11)

Example 7.4.30. We consider a distribution in proportion to the function

$$p(t, n) = t^{\frac{1}{2}}(1 - t)^{\frac{1}{2} + n}(1 + 2t)^{\frac{1}{2}}, \quad t \in [0, 1].$$

The normalizing constant $c(n)$ can be expressed as

$$c(n) = \int_0^1 p(t, n)dt = \frac{\Gamma(b)\Gamma(c - b)}{\Gamma(c)}{}_2F_1(a, b, c; -2)$$

from the integral representation of the Gauss hypergeometric function ${}_2F_1$, where $b - 1 = \frac{1}{2}, c - b - 1 = \frac{1}{2} + n, -a = \frac{1}{2}$. For sample data generated by the command test3() in the program rneuman.rr, estimate the value of n by using the difference version of the holonomic gradient descent. The command test3() uses the Neumann rejection method.

1. Derive the second-order difference equation for $c(n)$ from the contiguity relation for the Gauss hypergeometric function:

$$c(c + 1)(z - 1)_2F_1(a, b, c; z)$$
$$= (c + 1)\{(2c - a - b + 1)z - c\}_2F_1(a, b, c + 1; z)$$
$$+ (a - c - 1)(c - b + 1)z_2F_1(a, b, c + 2; z).$$

2. Let $\{t_1, t_2, \ldots, t_m\}$ be observation data. We will find the maximum point n of the likelihood function

$$d(n) = \frac{\left(\prod_{i=1}^{m} p(t_i, n)\right)^{1/m}}{c(n)}.$$

Determine the direction for n which increases the value of $d(n)$, based on the given initial values $c(0) = 0.550836$ and $c(1) = 0.257771$.

3. Find the maximum point of $d(n)$ by using the difference version of the holonomic gradient descent.

Answer. 1. Substituting

$$_2F_1\left(-\frac{1}{2}, \frac{3}{2}, n + 3; -2\right) = \frac{\Gamma(n + 3)}{\Gamma(\frac{3}{2})\Gamma(n + \frac{3}{2})}c(n)$$

and the recurrence relation of the gamma function $\Gamma(z + 1) = z\Gamma(z)$ for the contiguity relation, we obtain

$$3(2n + 3)c(n) = 10(n + 3)c(n + 1) - 2(2n + 9)c(n + 2).$$

2. We load the program file `rneuman.rr` on Risa/Asir and generate sample data by using the command `test3()`.

Listing 7.156 Risa/Asir: generating sample data by using `rneuman.rr`

```
[1371] load("rneuman.rr");
88
[1384] T=test3();
99
...(omitted)...
[0.241507,0.377635,0.126418,...(omitted)...,0.0276227,0.178898]
[1385] length(T);
100
```

The command `test3()` generates a random sample of 100 data points.
   We implement a small program to compute the numerators of $d(0)$ and $d(1)$.

Listing 7.157 Risa/Asir: `d_hgd.rr`

```
def nume_dn(T,N)
{
    F = (t^(1/2))*((1-t)^(1/2+N))*((1-t*(-2))^(1/2));
```

```
    M = length(T);
    for ( P = 1; T != []; T = cdr(T) )
        P *= eval(subst(F,t,car(T))^(1/M));
    return P;
}
end$
```

Listing 7.158 Risa/Asir: computing $d(0), d(1)$ (continuation)

```
[1386] load("d_hgd.rr");
[1388] C0=0.550836$
[1389] C1=0.257771$
[1390] nume_dn(T,0)/C0;
0.5845663141894453936
[1391] nume_dn(T,1)/C1;
1.089577168373801683
```

The inequality $d(0) < d(1)$ shows that $d(n)$ increases as n increases from 0.

3. We evaluate the values of $c(n)$ and $d(n)$ by using the recurrence relation until $d(n)$ begins to decrease.

Listing 7.159 Risa/Asir: add to the file d_hgd.rr

```
def d_hgd(T)
{
    N = 0;
    C0 = 0.550836;   C1 = 0.257771;
    D0 = nume_dn(T,N)/C0;
    print([N,C0,D0]);
    D1 = nume_dn(T,++N)/C1;
    print([N,C1,D1]);
    while ( D0 < D1 ) {
        TMP = (-3*(2*N+1)*C0+10*(N+2)*C1)/(2*(2*N+7));
        C0 = C1; C1 = TMP; D0 = D1;
        D1 = nume_dn(T,++N)/C1;
        print([N,C1,D1]);
    }
    return N-1;
}
end$
```

In this program, the recurrence relation replaces $n - 1$ with n. For simplicity, the numerator of $d(n)$ is evaluated each time through the while-loop. However, it can also be easily evaluated from the previous value.

Listing 7.160 Risa/Asir: estimation of n (continuation)

```
[1393] load("d_hgd.rr");
[1396] d_hgd(T);
[0,0.550836,0.5845663141894453936]
[1,0.257771,1.089577168373801683]
[2,0.1542,1.588702896284702378]
[3,0.104611,2.042609748576398825]
[4,0.0766291,2.432237613685558761]
```

```
[5,0.0591081,2.750355506159823718]
[6,0.0473178,2.996726319687802393]
[7,0.0389529,3.175178559400170999]
[8,0.0327728,3.291775218856224361]
[9,0.0280583,3.353651551523593603]
[10,0.0243673,3.368277176674134017]
[11,0.0214149,3.342994658156209513]
10
```

Each line in the data log represents $[n, c(n), d(n)]$. The inequality $d(9) <$ $d(10) > d(11)$ shows that the maximum point of $d(n)$ is $n = 10$. Thus, we estimate $n = 10$ by the likelihood estimation method. (The procedure outputs values around $n = 9, 10, 11$ for the other sample data generated by rneuman.rr.) In fact, we find that the parameter of the sample data distribution is $n = 10$ by referring to the following code in the program rneuman.rr.

Listing 7.161 Risa/Asir: rneuman.rr

```
def whg(X) {
   F=(X^(1/2))*((1-X)^(1/2+10))*((1-X*(-2))^(1/2));
   Y=deval(F)*20/0.137;
   return(Y);
}
```

Exercise. Modify the program d_hgd.rr in Example 7.4.30 to estimate n by using the initial values $c(30) = 0.00517171$ and $c(29) = 0.00543296$.

References

1. F. Chyzak, An extension of Zeilberger's fast algorithm to general holonomic functions. Discrete Math. **217**, 115–134 (2000)
2. F. Chyzak, Mgfun, http://algo.inria.fr/chyzak/mgfun.html
3. J.A. De Loera et al., LattE. http://www.math.ucdavis.edu/~latte/
4. W. Decker, G.-M. Greuel, G. Pfister, H. Schönemann, SINGULAR 3-1-2 — A computer algebra system for polynomial computations. http://www.singular.uni-kl.de/
5. A. Erdélyi, W. Magnus, F. Oberhettinger, F.G. Tricomi, *Higher Transcendental Functions, Vol. 1* (McGraw-Hill, 1953)
6. E. Gawrilow, M. Joswig, polymake: a framework for analyzing convex polytopes, in *Polytopes — Combinatorics and Computation* (2000), pp. 43–74. http://polymake.org/
7. D.R. Grayson, M.E. Stillman, Macaulay2, a software system for research in algebraic geometry. http://www.math.uiuc.edu/Macaulay2/
8. R. Hemmecke, P.N. Malkin, Computing generating sets of lattice ideals and Markov bases of lattices. J. Symb. Comput. **44**, 1463–1476 (2009)
9. T. Hibi, *Gröbner Bases* (Asakura, Tokyo, 2003) (in Japanese)
10. Y. Iba, M. Tanemura, Y. Omori, H. Wago, S. Sato, A. Takahashi, *Computational Statistics II. Frontiers of Statistical Science*, vol. 12 (Iwanami, Tokyo, 2005) (in Japanese)
11. A.N. Jensen, Gfan, a software system for Gröbner fans and tropical varieties. http://www.math.tu-berlin.de/~jensen/software/gfan/gfan.html
12. Maple. http://www.maplesoft.com/products/maple/

13. M. Noro, An efficient modular algorithm for computing the global b-function, mathematical software, in *Proceedings of ICMS2002* (World Scientific, Singapore, 2002), pp. 147–157
14. M. Noro, N. Takayama, Risa/Asir drill. http://www.math.kobe-u.ac.jp/Asir/ (2012, in Japanese)
15. M. Noro et al., Risa/Asir. http://www.math.kobe-u.ac.jp/Asir/asir-ja.html
16. T. Oaku, Computation of the characteristic variety and the singular locus of a system. Jpn. J. Ind. Appl. Math. **11**, 485–497 (1994)
17. T. Oaku, *D-modules and Computational Mathematics* (Asakura, Tokyo, 2002) (in Japanese)
18. R Development Core Team, R: A language and environment for statistical computing. http://www.r-project.org/
19. J. Rambau, TOPCOM: triangulations of point configurations and oriented matroids, in *Mathematical Software—ICMS 2002* (2002), pp. 330–340. http://www.rambau.wm.uni-bayreuth.de/TOPCOM/
20. M. Saito, B. Sturmfels, N. Takayama, *Gröbner Deformations of Hypergeometric Differential Equations* (Springer, Berlin, 2000)
21. A. Schrijver, *Theory of Linear and Integer Programming* (Wiley Interscience, New York, 1986)
22. B. Sturmfels, *Gröbner Bases and Convex Polytopes*. University lecture series 8 (American Mathematical Society, Providence, 1995)
23. N. Takayama, Kan/sm1. http://www.math.kobe-u.ac.jp/KAN/
24. Wolfram MathWorld, Horn Function. http://mathworld.wolfram.com/HornFunction.html
25. G. M. Ziegler, *Lectures on Polytopes* (Springer, Berlin, 1995)
26. 4ti2 team, 4ti2—a software package for algebraic, geometric and combinatorial problems on linear spaces. http://www.4ti2.de/

Index

Printed in the United States
By Bookmasters